Practical Recording Techniques

Fifth Edition

Practical Recording Techniques

The Step-by-Step Approach to Professional Audio Recording

Fifth Edition

Bruce Bartlett
Jenny Bartlett

AMSTERDAM • BOSTON • HEIDELBERG • LONDON • NEW YORK • OXFORD
PARIS • SAN DIEGO • SAN FRANCISCO • SINGAPORE • SYDNEY • TOKYO

Focal Press is an imprint of Elsevier

Focal Press is an imprint of Elsevier
30 Corporate Drive, Suite 400, Burlington, MA 01803, USA
Linacre House, Jordan Hill, Oxford OX2 8DP, UK

Library of Congress Cataloging-in-Publication Data
Bartlett, Bruce.
Practical recording techniques / Bruce Bartlett, Jenny Bartlett. — 5th ed.
p. cm.
Includes index.
ISBN 978-0-240-81144-4 (pbk. : alk. paper) 1. Magnetic recorders and recording—Handbooks, manuals, etc. 2. Sound—Recording and reproducing—Digital techniques—Handbooks, manuals, etc.
3. Sound—Recording and reproducing—Handbooks, manuals, etc.
I. Bartlett, Jenny. II. Title.
TK7881.6.B367 2008
621.389'3–dc22

2008030765

British Library Cataloguing-in-Publication Data
A catalogue record for this book is available from the British Library.

ISBN: 978-0-240-81144-4

For information on all Focal Press publications visit our website at www.books.elsevier.com

11 5 4 3

Printed in the United States of America

To family, friends, and music.

Contents

ONLINE at www.elsevierdirect.com/companions/9780240811444
See the Practical Recording Techniques Demo Web Site section for the passcode to access this site.

Preface

Recording is a highly skilled craft combining art and science. It requires technical knowledge as well as musical understanding and critical listening ability. By learning these skills, you can capture a musical performance and reproduce it with quality sound for the enjoyment and inspiration of others.

Your recordings will become carefully tailored creations of which you can be proud. They will be a legacy that can bring pleasure to many people for years to come.

This book is intended as a hands-on, practical guide for beginning recording engineers, producers, musicians—anyone who wants to make better music recordings by understanding recording equipment and techniques. I hope to prepare the reader for work in a home studio, a small professional studio, or an on-location recording session.

Practical Recording Techniques offers up-to-date information on the latest music recording technology, such as hard-disk and Flash memory recorders, computer recording, loop-based recording, keyboard and digital workstations, MIDI, surround sound, Web audio, and online collaboration. But it also guides the beginner through the basics, showing how to make quality recordings with the new breed of inexpensive home-studio equipment.

The first chapter answers the question, "Why do we record?" Next, the book overviews the recording chain to instill a system concept. The basics of sound, signals and studio acoustics are explained so that you'll know what you're controlling when you adjust the controls on a piece of recording equipment or build a studio. Then advice is given on equipping a home studio for any style of recording, with examples of equipment manufacturers.

Studio setup is covered next, including cables and connections, choosing monitor speakers, and preventing hum.

Each piece of recording equipment is explained in detail, as well as the control-room techniques you'll use during actual sessions. Two chapters are devoted to the technology of digital recording, audio for video and MIDI sequencing. A major chapter on computer recording covers the latest ways of creating and recording music. Two sections on remote recording cover techniques for both popular and classical music.

A special chapter explains how to judge recordings and improve them. The engineer must know not only how to use the equipment, but also how to tell good sound from bad.

The latest developments in recording are surround sound, Web audio, and online collaboration. All these topics are covered in detail in their own chapters.

Finally, four appendices explain the decibel, suggest how to optimize your computer for multitrack recording, explain impedance, and suggest further education.

On the book's Web site at www.elsevierdirect.com/companions/9780240811444 are audio samples that demonstrate various topics explained in the book. Throughout the text, references to specific Web site tracks guide the reader to relevant audio demonstrations.

Based on my work as a professional recording engineer, this book is full of tips and shortcuts for making great-sounding recordings, whether in a professional studio, project studio, on-location, or at home. You'll find many topics not covered in similar texts:

Loop-based recording
Hum prevention tips
The latest monitoring methods
Examples of mic models by type
Microphone selection guide
Tonal effects of microphone placement
Glossary of sound-quality descriptions
The latest types of digital recorders
Up-to-date coverage of computer recording
Optimizing your computer for multitrack recording
Documenting the recording session
Audio-for-video techniques
On-location recording

Troubleshooting bad sound; guidelines for good sound
Audio on the Web
Selling your music online
File sharing for online collaboration
Surround sound, DVD, and Blu-ray

Acknowledgments

Thank you to Nick Batzdorf of *Recording* magazine for giving me permission to draw from my "Take One" series. For my education, thank you to the College of Wooster, Crown International, Shure Brothers Inc., Astatic Corporation, and all the studios I've worked for.

Thank you to Catharine Steers, Beth Howard, Emma Baxter, Becky Golden-Harrow, Dawnmarie Simpson, and Joanne Tracy at Focal Press/Elsevier for their fine work and support.

My deepest thanks to Jenny Bartlett for her many helpful suggestions as a layperson consultant and editor. She made sure the book could be understood by beginners.

A note of appreciation goes to Sting, Uncle Earl, The Beatles, Led Zepplin, the Pat Metheny Group, and Samuel Barber ("Adagio for Strings"), among many others, whose music inspired the chapter "Music: Why You Record."

We appreciate the following manufacturers who provided photos of their products: Edirol, Boss, Korg, Alesis, Mackie, Frontier Design, Crown International, Audio-Technica, M-Audio, TASCAM, Zoom, Eventide, Rane, Lexicon, CamelAudio, Izotope, Line 6, RML Labs, E-MU, PreSonus, Digidesign, Cakewalk, Propellerhead Software, CD Baby, TuneCore, and imeem (Snocap, Inc.).

Finally, to the musicians I've recorded and played with, a special thanks for teaching me indirectly about recording.

CHAPTER 1

Music: Why You Record

As you learn about recording techniques for music, it's wise to remember that music is a wonderful reason for recording.

Music can be exalting, exciting, soothing, sensuous, and fulfilling. It's marvelous that recordings can preserve it. As a recording engineer or recording musician, it's to your advantage to better understand what music is all about.

Music starts as musical ideas or feelings in the mind and heart of its composer. Musical instruments are used to translate these ideas and feelings into sound waves. Somehow, the emotion contained in the music—the message—is coded in the vibrations of air molecules. Those sounds are converted to electricity and stored magnetically or optically. The composer's message manages to survive the trip through the mixing console and recorders; the signal is transferred to disc or computer files. Finally, the original sound waves are reproduced in the listening room, and miraculously the original emotion is reproduced in the listener as well.

Of course, not everyone reacts to a piece of music the same way, so the listener may not perceive the composer's intent. Still, it's amazing that anything as intangible as a thought or feeling can be conveyed by tiny magnetic patterns on a hard disk or by pits on a compact disc.

The point of music lies in what it's doing now, in the present. In other words, the meaning of "Doo wop she bop" is "Doo wop she bop." The meaning of an Am7 chord followed by an Fmaj7 chord is the experience of Am7 followed by Fmaj7.

INCREASING YOUR INVOLVEMENT IN MUSIC

Sometimes, to get involved in music, you must relax enough to lie back and listen. You have to feel unhurried, to be content to sit between your stereo speakers or wear headphones, and listen with undivided attention. Actively analyze or feel what the musicians are playing.

Music affects people much more when they are already feeling the emotion expressed in the song. For example, hearing a fast Irish reel when you're in a party mood, or hearing a piece by Debussy when you're feeling sensuous, is more moving because your feelings resonate with those in the music. When you're falling in love, any music that is meaningful to you is enhanced.

If you identify strongly with a particular song, that tells you something about yourself and your current mood. And the songs that other people identify with tell you something about them. You can understand individuals better by listening to their favorite music.

DIFFERENT WAYS OF LISTENING

There are so many levels on which to listen to music—so many ways to focus attention. Try this. Play one of your favorite records several times while listening for these different aspects:

- Overall mood and rhythm
- Lyrics
- Vocal technique
- Bass line
- Drum fills
- Sound quality
- Technical proficiency of musicians
- Musical arrangement or structure
- Reaction of one musician to another musician's playing
- Surprises versus predictable patterns

By listening to a piece of music from several perspectives, you'll get much more out of it than if you just hear it as background. There's a lot going on in any song that usually goes unnoticed. Sometimes when you play an old familiar record and listen to the lyrics for the first time, the whole meaning of the song changes for you.

Most people react to music on the basic level of mood and rhythmic motivation. But as a recording enthusiast, you hear much more detail because your focus demands sustained critical listening. The same is true of trained musicians focusing on the musical aspects of a performance.

It's all there for anyone to hear, but you must train yourself to hear selectively and to focus attention on a particular level of the multidimensional musical event. For example, instead of just feeling excited while listening to an impressive lead-guitar solo, listen to what the guitarist is actually playing. You may hear some amazing things.

Here's one secret of really involving yourself in recorded music: Imagine yourself playing it! For example, if you're a bass player, listen to the bass line in a particular record, and imagine that you're playing the bass line. You'll hear the part as never before. Or respond to the music visually—see it as you do in the movie Fantasia.

Follow the melody line and see its shape. Hear where it reaches up, strains, and then relaxes. Hear how one note leads into the next. How does the musical expression change from moment to moment?

There are times when you can almost touch music: some music has a prickly texture (many transients, emphasized high frequencies); some music is soft and sinuous (sine-wave synthesizer notes, soaring vocal harmonies); and some music is airy and spacious (lots of reverberation).

WHY RECORD?

Recording is a real service. Without it, people would be exposed to much less music. They would be limited to the occasional live concert or to their own live music, played once and forever gone.

With recordings, you can preserve a performance for thousands of listeners. You can hear an enormous variety of musical expressions whenever you want. Unlike a live concert, a record can be played over and over for analysis. Tapes or discs are also a way to achieve a sort of immortality. The Beatles may be gone, but their music lives on.

Records can even reveal your evolving consciousness as you grow and change. A computer audio file or CD stays the same physically, but you hear it differently over the years as your perception changes. Recordings are a constant against which you can measure change in yourself.

Be proud that you're contributing to the recording art—it is done in the service of music.

CHAPTER 2

The Recording Chain

Welcome to the brave new world of 21st-century recording! The digital technologies of the past few years have given us possibilities undreamed of just 10 years ago. This book will show you an overview of current recording technology, help you choose the equipment that best suits your needs, and guide you in using it to create great recordings. And it will explain the technical jargon in plain English.

Thanks to the shift from analog to digital technology, the excitement and satisfaction of recording are accessible to more people than ever before. It used to take a whole roomful—or truckful—of expensive equipment to produce a good recording. But the new generation of smaller, cheaper gear means you may be able to tuck your studio into a corner of your bedroom or the back seat of your Toyota. As a result, many more people are involved in the process of recording—as musicians recording their own albums, or as engineers offering services to others.

As a recording engineer, you're a key player. Your skills help artists realize their visions in sound. Your miking techniques capture the vibrancy of the performance, whether it's the shimmering overtones of a string quartet or the sonic assault of an electric blues band. Your "post" work in the studio—adding effects, tweaking levels, etc.—will take the raw material of the performance and shape and blend it into a polished musical statement. By mastering the technology and becoming fluent with the audio tools at hand, you will produce exciting recordings that will delight your clients and give you a real sense of pride and achievement.

Be sure to practice what you learn in this book. There's no substitute for hands-on experience. You might offer to record a band's rehearsal for free while you experiment and master the gear. Be patient, let yourself make mistakes, and above all, listen to how the sound changes when you move a mic or tweak a knob.

TYPES OF RECORDING

Let's get started. Currently there are six main ways to record music:

1. Live stereo recording
2. Live mix recording
3. Multitrack recorder and mixer
4. Stand-alone Digital Audio Workstation (DAW; recorder–mixer)
5. Computer DAW
6. MIDI sequencing

Live Stereo Recording

Record with a stereo microphone or two microphones into a recorder. This method is most common when used to record an orchestra, symphonic band, pipe organ, small ensemble, quartet, or soloist. The microphones pick up the overall sound of the instruments and the concert-hall acoustics from several feet away. You might use this minimalist technique to record a folk group, rock group, or acoustic jazz group in a good-sounding room.

Figure 2.1 shows the stages of this method—the links in the recording chain. From left to right,

1. The musical instruments or voices make sound waves.
2. The sound waves travel through the air and bounce or reflect off the walls, ceiling, and floor of the concert hall. These reflections add a pleasing sense of spaciousness.
3. The sound waves from the instruments and the room reach the microphones, which convert the sound into electrical signals.
4. The sound quality is greatly affected by mic technique: microphone choice and placement.
5. The signals from the microphones go to a 2-track recorder. It may be a hard-drive recorder, CD-R burner, DVD-R burner, Flash memory recorder, or computer hard drive. The signal changes to a pattern stored on a medium, such as magnetic patterns on

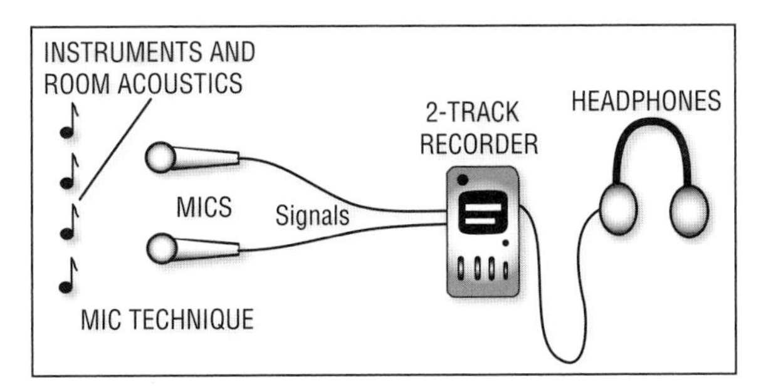

FIGURE 2.1
The recording chain for live stereo recording.

a hard disk. During playback, the patterns on the medium are converted back into a signal.
During recording, signals are stored along a track—a path or channel on the medium containing a recorded signal. A single medium can record one or more tracks. For example, a 2-track hard-disk recording stores two tracks on hard disk, such as the two different audio signals required for stereo recording.

6. To hear the signal you're recording, you need a monitor system: headphones or a stereo power amplifier and loudspeakers. You use the monitors to judge how well your mic technique is working.

The speakers or headphones convert the signal back into sound. This sound resembles that of the original instruments. Also, the acoustics of the listening room affect the sound reaching the listener.

Live Mix Recording

This method is seldom used except for live broadcasts or recordings of PA mixes. Using a mixer, you set up a mix of several microphones and record the mixer's output signal on a 2-track recorder (CD-R, Flash memory recorder, or computer hard drive). Each mic is close to its sound source. Figure 2.2 shows this method.

Multitrack Recorder and Mixer

Record with several mics into a mixer, which is connected to a multitrack hard-disk recorder. You record the signal of each microphone on its own track, then mix these recorded signals after the performance

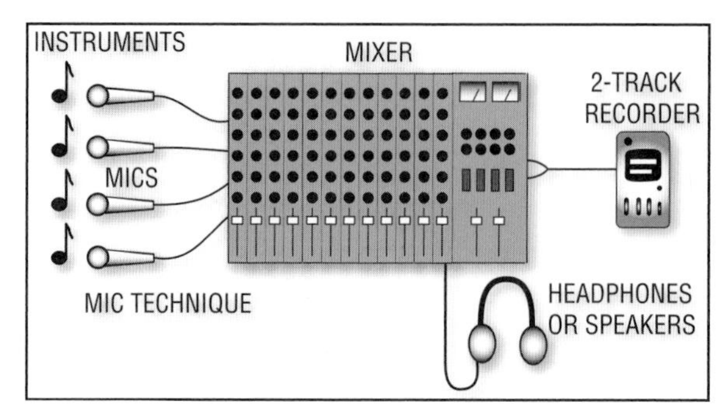

FIGURE 2.2
The recording chain for live-mix recording.

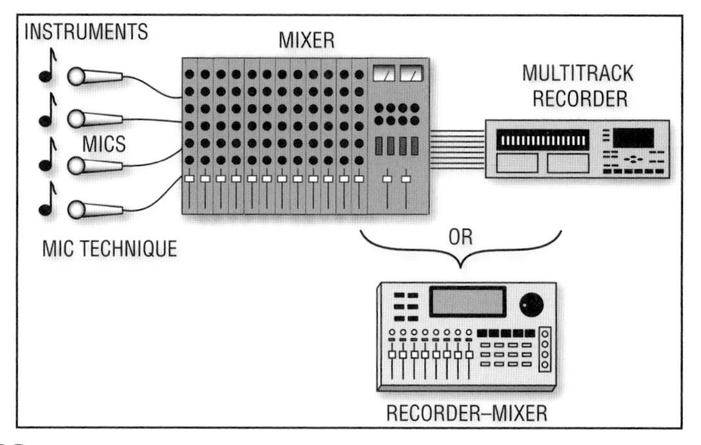

FIGURE 2.3
The recording chain for multitrack recording.

is done. You can also record different groups of instruments on each track. Figure 2.3 shows the stages in this method.

1. Place microphones near the instruments.
2. Plug the mics into a mixing console: a big, sophisticated mixer. During multitrack recording, the mixing console amplifies the weak microphone signals up to the level needed by the recorder. The console also sends each microphone signal to the desired track.
3. Record the amplified mic signals on the multitrack recorder.

You can record more instruments later on unused tracks—a process called overdubbing. Wearing headphones, the performer listens to

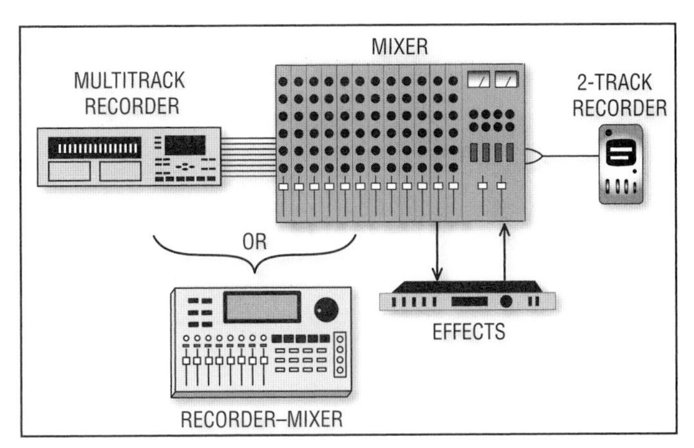

FIGURE 2.4
The recording chain for a multitrack mixdown.

the recorded tracks and plays or sings along with them. You record the performance on an unused track.

After the recording is done, you play all the tracks through the mixing console to mix them with a pleasing balance (Figure 2.4). Here are the steps:

1. Play back the multitrack recording of the song several times, adjusting the track volumes and tone controls until the mix is just the way you want it. You can add effects to enhance the sound quality. Some examples are echo, reverberation, and compression (explained in Chapter 10). Effects are made by signal processors that connect to your mixer, or by software applications (plug-ins) that are part of a recording program.
2. Record or export your final stereo mix onto your computer hard drive or to an external 2-track recorder such as the Alesis Masterlink.

Stand-alone DAW (Recorder–Mixer)

This is a multitrack recorder and a mixer combined in one portable chassis. It's relatively easy to use. The recording medium is a hard drive or a Flash memory card. Other names for a recorder–mixer are "stand-alone Digital Audio Workstation," "digital multitracker," "personal digital studio," or "portable studio." Most recorder–mixers have built-in effects. Figure 2.5 shows an example of a recorder–mixer.

FIGURE 2.5
TASCAM DP-02, an example of a recorder–mixer.

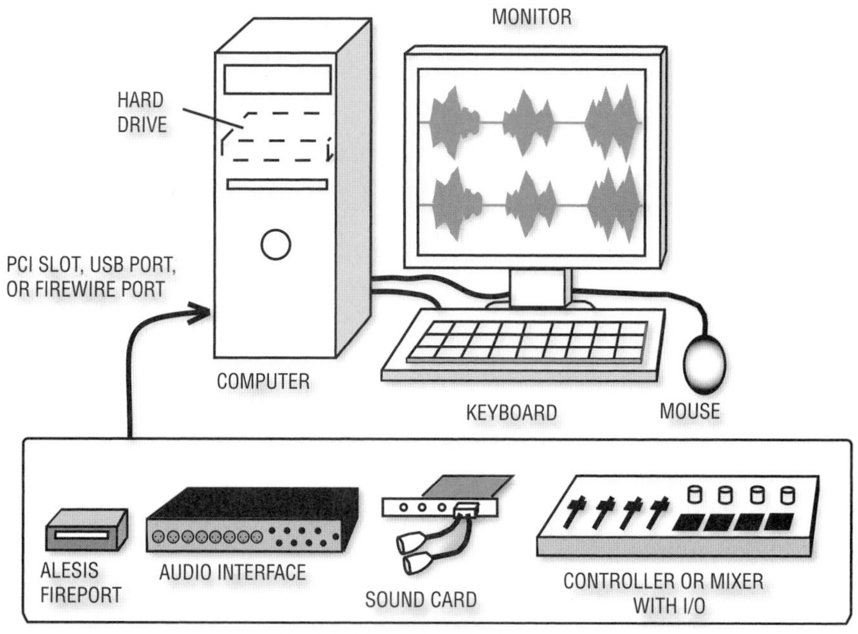

FIGURE 2.6
Computer with a choice of audio interface and recording/editing software.

Computer DAW

This low-cost system includes a computer, recording software, and a sound card or audio interface that gets audio into and out of your computer (Figure 2.6). You record on the computer's hard drive.

Using the recording software, you perform these operations:

1. Record music on the computer's hard drive.
2. Edit the tracks to fix mistakes, delete unwanted material, or to copy/move song sections.
3. Mix the tracks with a mouse or controller by adjusting virtual controls that appear on your computer screen.

You might also assemble a song from samples or from loops. Samples are recordings of single notes of various instruments. Loops are repeating musical patterns.

MIDI Sequencing

With this recording method, a musician performs on a MIDI controller, such as a piano-style keyboard or drum pads. The controller puts out a MIDI signal, a series of numbers that indicates which keys were pressed and when they were pressed. The MIDI signal is recorded into computer memory by a sequencer or sequencer program in a computer. When you play back the MIDI sequence, it plays the tone generators in a synthesizer or sound module. The synthesizer can be hardware or software (as a "soft synth," also called a "virtual instrument"). A MIDI sequence also can play samples: digital recordings of musical notes played by real instruments.

Like a player piano, MIDI sequencing records your performance gestures rather than audio. Figure 2.7 shows the process.

MIDI/digital-audio software lets you record MIDI sequences and digital audio on hard disk. First record a few tracks of MIDI sequences, then add audio tracks: lead vocal, sax solo, or whatever. All these elements stay synchronized.

PROS AND CONS OF EACH METHOD

Live stereo recording is simple, cheap, and fast. But it usually sounds too muddy with rock music, and you must adjust balances by moving musicians. It can work well with classical music, and sometimes with folk or acoustic jazz music.

Live mix recording is fairly simple and quick. However, loud instruments might sound distant in the recording because their sound "leaks" into other instruments' mics. You can't correct mistakes in

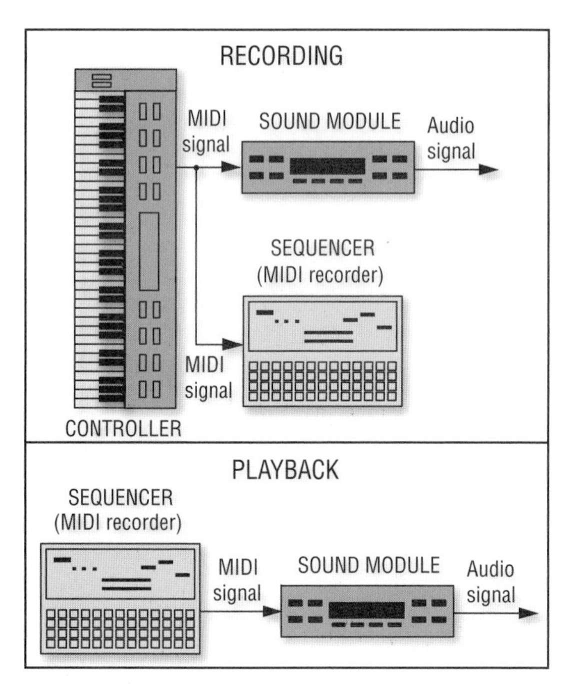

FIGURE 2.7
MIDI sequencing system.

the mix or performance unless you re-record. Also, the live sound of the band can make it hard to hear the monitored sound clearly.

Multitrack recording has many advantages. You can punch-in: fix a musical mistake on a track by recording a new, correct part over the mistake. You can overdub: record one instrument at a time. This reduces leakage and gives a tighter sound. Also, you can postpone mixing decisions until after the performance. Then you can monitor the mix in quiet surroundings. This method is more complex and expensive than live-mix recording.

If you use a separate multitrack recorder and mixer, each component can be used independently. For example, you might do a PA job with just the mixer. Or, if you already have a mixer, all you need to buy is a recorder. You'll need to connect cables between the mixer and recorder, and between the mixer and outboard effects units. It's a reliable, intuitive system for live recording or studio tracking.

A stand-alone DAW (recorder–mixer) is easy to use because it is a single portable chassis that includes most of your studio equipment: recorder, mixer, effects, and often a CD burner. It doesn't require cables except for the mics, instruments, and monitor speakers. High-end units let you edit the music. They also have automated mixing: memory chips in the mixer remember your mixdown settings, and reset the mixer accordingly the next time you play back the recording.

A computer DAW is inexpensive, powerful, and flexible. It lets you do sophisticated editing and automated mixing. Several plug-in (software) effects are included, and you can purchase and install other plug-ins. Recording software can be updated at little cost. As for drawbacks, computers can crash and can be difficult to set up and optimize for audio work.

MIDI sequencing lets you record musical parts by entering notes slowly or one at a time if you wish. After the sequencing is finished, you can edit notes to correct mistakes. You can even change the instrument sounds or the tempo. A huge variety of sounds are available in synthesizers, sound modules, and soft synths. However, you're limited to their sounds unless you use MIDI/digital-audio software, which lets you add miked instruments to the mix.

The accompanying Web site contains samples of several types of recordings. *Play audio clips 1 and 2 at www.elsevierdirect.com/companions/9780240811444*. Clip 2 demonstrates:

- Live stereo recording—orchestra
- Live stereo recording—rock group
- Live mix recording—jazz group
- Multitrack recording—pop group
- MIDI sequencing—synthesizer funk

RECORDING THE MIXES

No matter which recording method you use, eventually you'll mix each song and record the mix on a 2-track recorder, or record the mix as a stereo WAVE file on your hard drive. You can convert the WAVE file to an MP3 or WMA format for uploading to the Web (explained in Chapter 20).

You might want to assemble an album of your recorded mixes. To do this remove noises and count-offs between songs, put the songs in the desired order, and put a few seconds of silence between songs. This is done with a computer and editing software (a DAW, Figure 2.6). The last step is to copy the album to a blank CD. There's your final product, ready to duplicate or replicate.

No matter what type of recording you do, each stage contributes to the sound quality of the finished recording. A bad-sounding CD can be caused by any weak link: low-quality microphones, ineffective mic placement or mixer settings, and so on. You can strive for quality recordings by optimizing each stage. This book will help you reach that goal.

CHAPTER 3

Sound, Signals, and Studio Acoustics

When you make a recording, you deal with at least two kinds of invisible energy: sound waves and electrical signals. For example, a microphone converts sound into a signal. A signal is a varying voltage that carries information. In our case, it's musical information.

This chapter covers some characteristics of sound and audio signals. These facts will help you work with room acoustics, and will help you know what's going on inside your mixer as you adjust the controls. With this knowledge you can make better recordings.

SOUND WAVE CREATION

To produce sound, most musical instruments vibrate against air molecules, which pick up the vibration and pass it along as sound waves. When these vibrations strike your ears, you hear sound.

To illustrate how sound waves are created, imagine a vibrating speaker cone in a guitar amp. When the cone moves out, it pushes the adjacent air molecules closer together. This forms a compression. When the cone moves in, it pulls the molecules farther apart, forming a rarefaction. As shown in Figure 3.1, the compressions have a higher pressure than normal atmospheric pressure; the rarefactions have a lower pressure than normal.

These disturbances pass from one molecule to the next in a spring-like motion—each molecule vibrates back and forth to pass the wave along. The sound waves travel outward from the sound source at 1130 feet per second (344 meters per second), which is the speed of sound in air at room temperature.

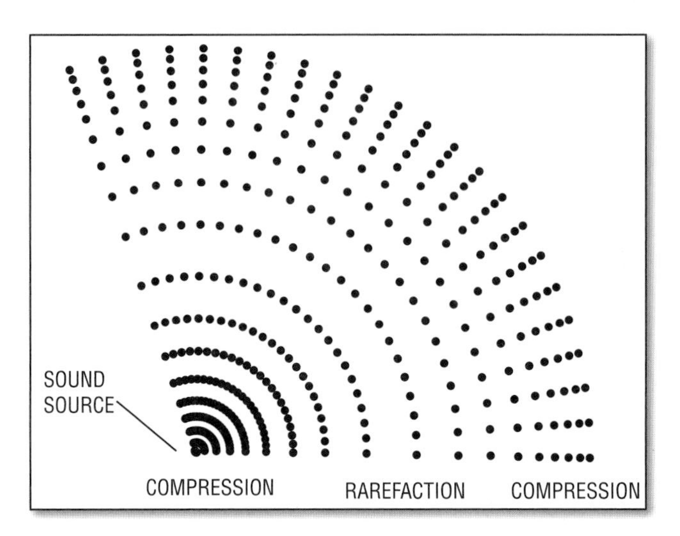

FIGURE 3.1
A sound wave.

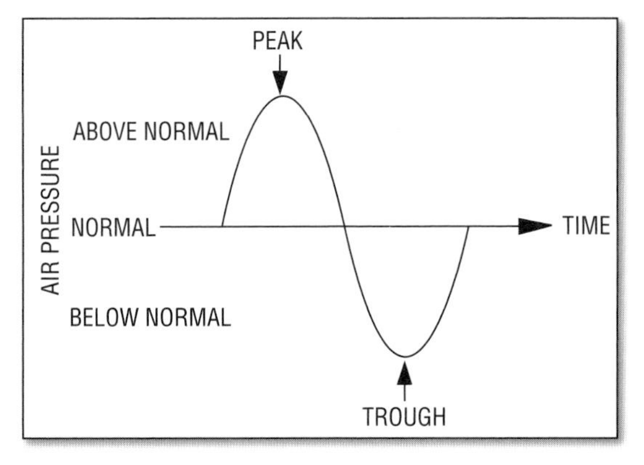

FIGURE 3.2
Sound pressure vs. time of one cycle of a sound wave.

At some receiving point, such as an ear or a microphone, the air pressure varies up and down as the disturbances pass by. Figure 3.2 is a graph showing how sound pressure varies with time, like a wave. The high point of the graph is called a peak; the low point is called a trough. The horizontal center line of the graph is normal atmospheric pressure.

Sound waves tend to spread out as they travel away from the source. The compressions and rarefactions move out as expanding spheres.

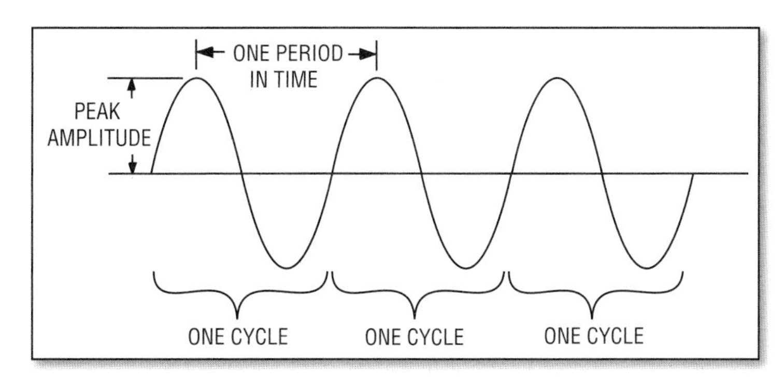

FIGURE 3.3
Three cycles of a wave.

As the spherical waves expand, the sound pressure spreads over a larger area, so the pressure becomes weaker with distance from the source. That means, the farther you're from a sound source, the quieter its sound. Specifically, the sound pressure halves (drops 6 decibels; dB) each time the distance from the source doubles. This phenomenon is called the inverse square law.

CHARACTERISTICS OF SOUND WAVES

Figure 3.3 shows three waves in succession. One complete vibration from normal to high to low pressure and back to the starting point is called one cycle. The time it takes to complete one cycle—from the peak of one wave to the next—is called the period of the wave. One cycle is one period long.

Amplitude

The height of the wave is its amplitude. Loud sounds have high amplitudes (big pressure changes); quiet sounds have low amplitudes (small pressure changes). *Play track 3 on the Web site www.elsevierdirect.com/companions/9780240811444 to hear an example.*

Frequency

The sound source (in this case, the guitar-amp loudspeaker) vibrates back and forth many times a second. The number of cycles completed in one second is called frequency. The faster the speaker vibrates, the higher the frequency of the sound. Frequency is measured in hertz (Hz), which stands for cycles per second. One thousand hertz is called "one kilohertz," abbreviated kHz.

The higher the frequency, the higher the perceived pitch of the sound. Low-frequency tones have a low pitch (like low E on a bass, which is 41 Hz). High-frequency tones have a high pitch (like four octaves above middle C, or 4186 Hz). *Track 4 on the Web site www.elsevierdirect.com/companions/9780240811444 illustrates this.* Doubling the frequency raises the pitch one octave.

Children can hear frequencies from 20 Hz to 20 kHz, and most adults with good hearing can hear up to 15 kHz or higher. Each musical instrument produces a range of frequencies, say, 41 Hz to 9 kHz for a string bass, or 196 Hz to 15 kHz for a violin.

Wavelength

When a sound wave travels through the air, the physical distance from one peak (compression) to the next is called a wavelength (refer to Figure 3.1). Low-pitched sounds have long wavelengths (several feet); high-pitched sounds have short wavelengths (a few inches or less). Wavelength is the speed of sound divided by frequency. So the wavelength of a 1000-Hz wave is 1.13 feet (0.344 meters); 100 Hz is 11.3 feet (3.44 meters), and 10 kHz is 1.35 inches (3.45 centimeters).

Phase and Phase Shift

The phase of any point on the wave is its degree of progression in the cycle—the beginning, the peak, the trough, or anywhere between. Phase is measured in degrees, with 360 degrees being one complete cycle. The beginning of a wave is 0 degrees; the peak is 90 degrees (one-quarter cycle), and the end is 360 degrees. Figure 3.4 shows the phases of various points on the wave.

If there are two identical waves traveling together, but one is delayed with respect to the other, there is a phase shift between the two waves. The more delay, the more phase shift. Phase shift is measured in degrees. Figure 3.5 shows two waves separated by 90 degrees (one-quarter of a cycle) of phase shift. The dashed wave lags the solid wave by 90 degrees.

If you combine two identical sound waves, such as a sound and its reflection off a wall, the peaks of the two waves add together at certain

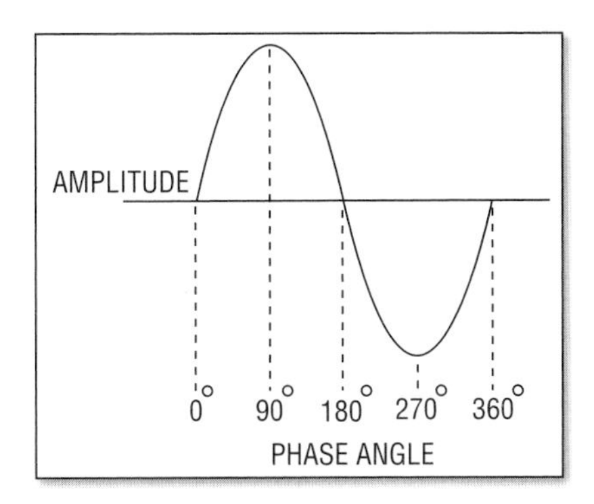

FIGURE 3.4
The phases of various points on a wave.

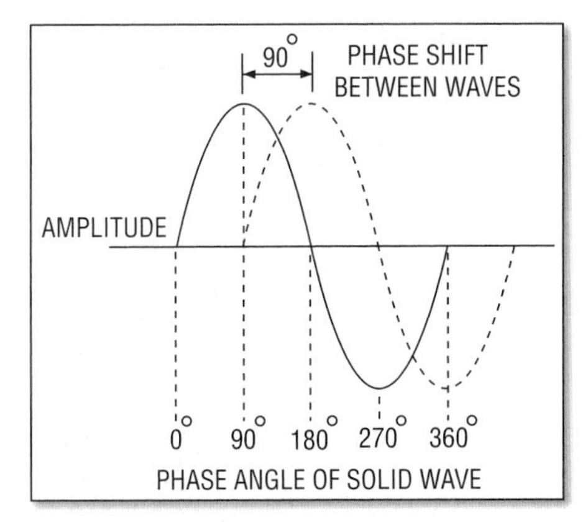

FIGURE 3.5
Two waves that are 90 degrees out-of-phase.

points in the room. This doubles the sound pressure or amplitude, creating areas of louder sound at certain frequencies.

Phase Interference

When there is a 180-degree phase shift between two identical waves, the peak of one wave coincides with the trough of another (Figure 3.6). If these two waves are combined, they cancel out. This phenomenon is called phase cancellation or interference.

Suppose you have a signal with a wide range of frequencies, such as the singing voice. If you delay this signal and combine it with the original undelayed signal, some frequencies will be 180 degrees out-of-phase and will cancel. This makes a hollow, filtered tone quality.

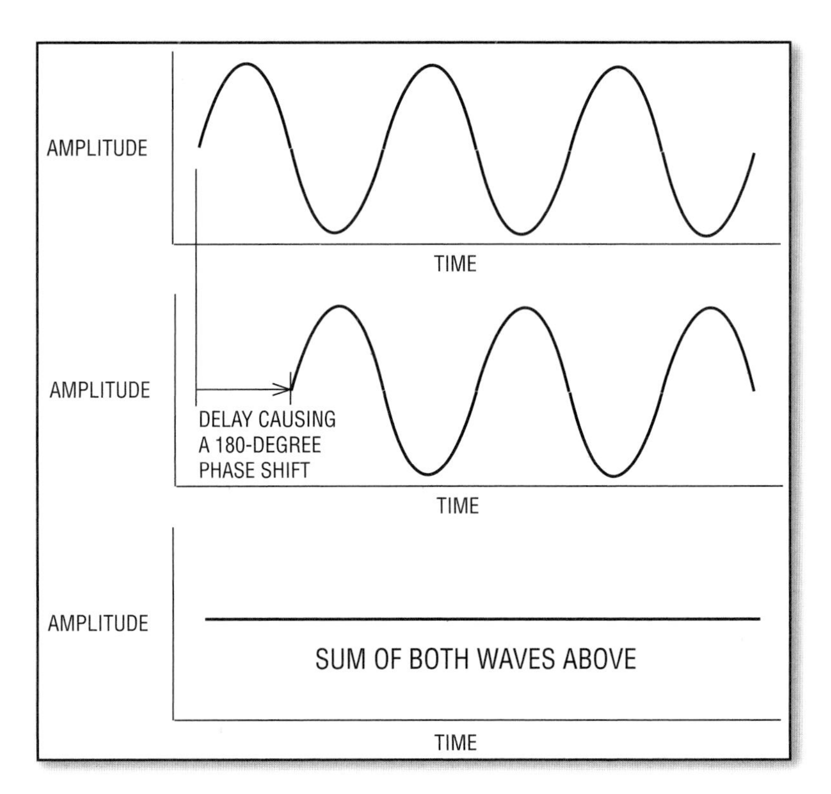

FIGURE 3.6
Phase interference: adding two waves that are out-of-phase cancels the sound at that frequency.

Here's an example of how this can happen. Suppose you're recording a singer/guitarist with one mic near the singer and another mic near the guitar. Both mics pick up the singer. The singer's mic is close to the mouth, and you hear it with no delay in the signal. The guitar mic is farther from the mouth, so its voice signal is delayed. When you mix the two mics, you often hear a colored tone quality caused by phase cancellations between the two mics.

Suppose you're recording a stage play with a mic on a short stand on the floor. The mic picks up the direct sound from the actors, but it also picks up delayed reflections off the floor. Direct and delayed sounds combine at the mic, causing phase cancellations. You hear it as a hollow, filtered sound that changes when the actor walks while talking.

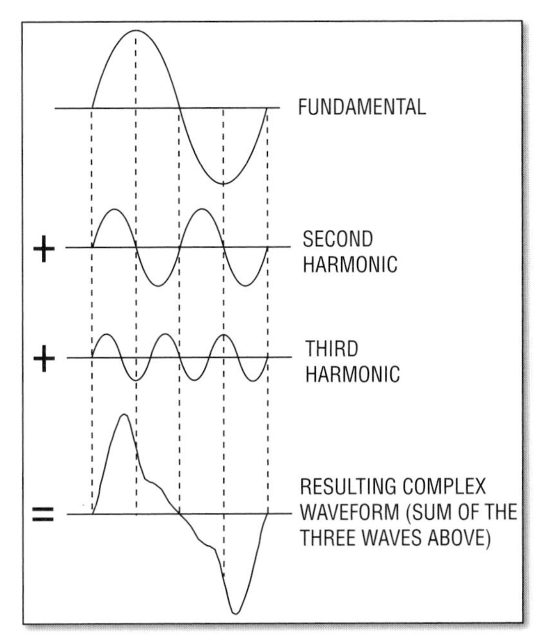

FIGURE 3.7
Adding fundamental and harmonic waveforms to form a complex waveform.

Harmonics

The type of wave shown in Figure 3.2 is called a sine wave. It is a pure tone of a single frequency, like a signal from a tone generator. In contrast, most musical tones have a complex waveform, which has more than one frequency component. All sounds are combinations of sine waves of different frequencies and amplitudes. Figure 3.7 shows sine waves of three frequencies combined to form a complex wave.

The lowest frequency in a complex wave is called the fundamental frequency. It determines the pitch of the sound. Higher frequencies in the complex wave are called overtones or upper partials. If the overtones are multiples of the fundamental frequency, they are called harmonics. For example, if the fundamental frequency is 200 Hz, the second harmonic is 400 Hz, and the third harmonic is 600 Hz.

Harmonics and their amplitudes partly determine the tone quality or timbre of a sound. They help you identify the sound as a drum, piano, organ, voice, etc. *Play audio clip 5 at www.elsevierdirect.com/*

companions/9780240811444. In general, instruments with few or weak harmonics—such as a flute—tend to sound pure and smooth. Instruments with many or strong harmonics—such as a trumpet or distorted guitar—tend to sound bright and edgy.

Explained in Chapter 10, equalization can change the tonal balance of a recorded instrument by boosting or cutting its harmonics and fundamental frequencies. Boosting the fundamentals tends to make the sound warm; cutting the fundamentals makes the sound thin; boosting the harmonics makes the sound bright, defined, or trebly; and cutting the harmonics makes the sound dark or muffled.

Playing an instrument louder usually increases its harmonics. So a piano played loudly sounds brighter than the same piano played softly.

Noise (such as tape hiss) contains a wide band of frequencies and has an irregular, nonrepeating waveform. Some musical sounds with a hissy or noise-like character are a cymbal, snare drum, or the "s" sounds a singer makes (sibilance).

Envelope

Another characteristic that identifies a sound is its envelope. When a note sounds, it doesn't last forever—it rises in volume, lasts a short time, then falls back to silence. This rise and fall in volume of one note is called the note's envelope. The envelope connects the peaks of successive waves that make up a note. Each musical instrument has a different envelope.

Most envelopes have four sections: attack, decay, sustain, and release (see Figure 3.8). During the attack, a note rises from silence to its maximum volume. Then it decays from maximum to some mid-range level. This middle level is the sustain portion. During release, the note falls from its sustain level back to silence.

Percussive sounds, such as drum hits, are so short that they have only a rapid attack and decay. Other sounds, such as organ or violin notes, last longer. They have slower attacks and longer sustains. Guitar plucks and cymbal crashes have quick attacks and slow releases. They hit hard then fade out slowly. *Play audio clip 6 at www.elsevierdirect.com/companions/9780240811444*.

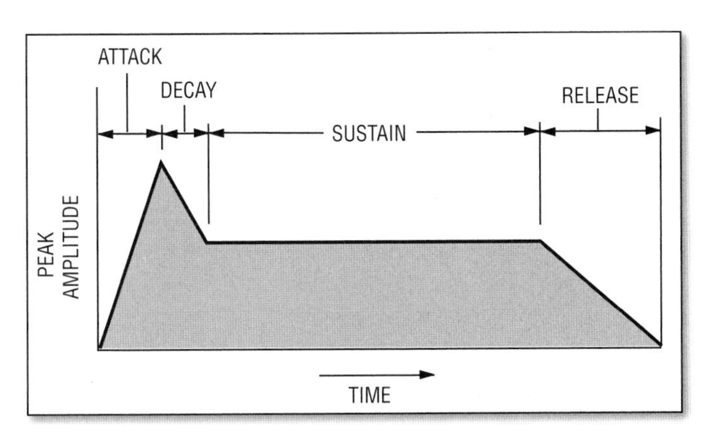

FIGURE 3.8
The four sections of the envelope of a note.

You can shorten a guitar string's decay or ringing by damping the string with the side of your hand. You can press a blanket against a kick drum head to damp the decay and get a tighter sound.

Harmonics usually change during a note's envelope. For example if an instrument has a percussive attack—like a guitar pluck or tom hit—the harmonics are strongest at the attack then become weaker during the decay.

BEHAVIOR OF SOUND IN ROOMS

Because most music is recorded in rooms, you need to understand how room surfaces affect sound.

Echoes

Musical instruments make sound waves that travel outward in all directions. Some of the sound travels directly to your ears (or to a microphone) and is called direct sound. The rest strikes the walls, ceiling, floor, and furnishings of the recording room. At those surfaces, some of the sound is absorbed, some is transmitted through the surface, and the rest is reflected back into the room.

Because sound waves take time to travel (moving at about 1 foot per millisecond), the reflected sound reaches you after the direct sound. The reflection repeats the original sound after a short delay. If the

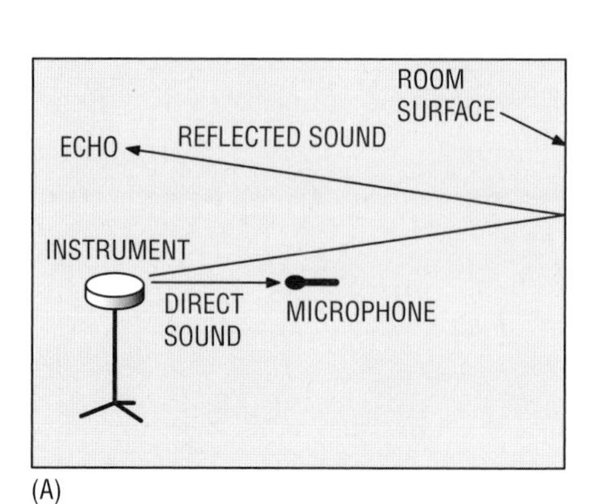

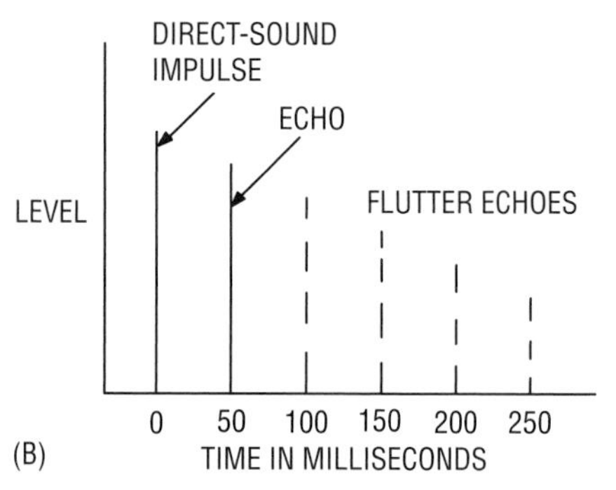

FIGURE 3.9
Echoes: (A) echo formation; (B) intensity vs. time of direct sound and its echoes.

sound is delayed about 50 milliseconds or more, we call it an echo (Figure 3.9). In some concert halls we hear single echoes; in small rooms we often hear a short, rapid succession of echoes called flutter echoes. You can detect them by clapping your hands next to a wall. Flutter echoes happen when sound bounces back and forth between two parallel walls.

Reverberation

Sound reflects many times from all the surfaces in the room. These reflections sustain the sound of each note the musician plays. This persistence of sound in a room after the original sound has stopped is called reverberation (reverb). For example, reverberation is the sound you hear just after you shout in an empty gymnasium. The sound of your shout stays in the room and gradually dies away (decays). *Play audio clip 10 at www.elsevierdirect.com/companions/9780240811444 to hear echoes and reverberation.*

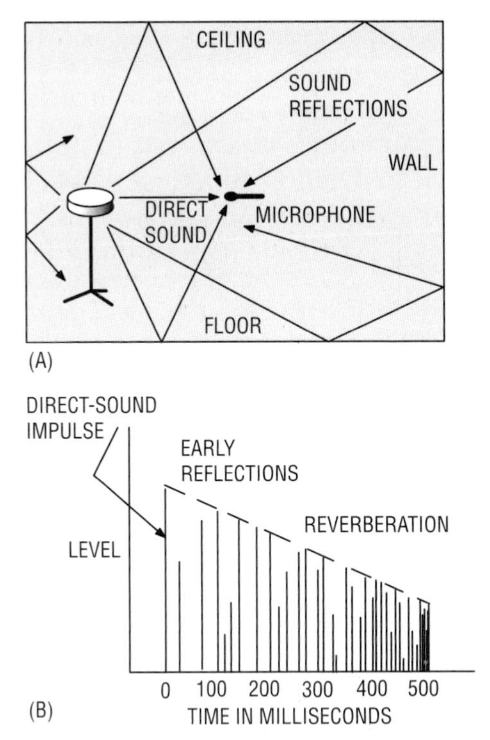

FIGURE 3.10
Reverberation: (A) multiple sound reflections create reverberation; (B) intensity vs. time of direct sound, early reflections, and reverberation.

Reverb is hundreds of echoes that gradually get quieter. The echoes follow each other so rapidly that they merge into a single continuous sound. Eventually, the room surfaces completely absorb the echoes. The timing of the echoes is random, and the echoes increase in number as they decay. Figure 3.10 shows how reverberation develops in a recording room.

Reverberation is a continuous fade-out of sound ("HELLO-O-O-o-o"), while an echo is a discrete repetition of a sound ("HELLO hello hello hello").

Reverberation time (RT60) is the time it takes for reverb to decay 60 dB. Too long a reverb time makes a recording sound distant, muddy, and washed-out. That's why pop-music recordings are usually made in a fairly "dead" or nonreverberant studio, which has an RT60 of about 0.4 second or less. In contrast, classical music is recorded in "live," reverberant concert halls (RT60 about 1 to 3 seconds) because we want to hear reverb with classical music—it's part of the sound.

Reverberation comes to you from every direction because it is a pattern of many sound reflections off the walls, ceiling, and floor. Because we can tell where sounds come from, we can distinguish between the direct sound of an instrument coming to us from a single location and the reverberation coming to us from everywhere else. So we can ignore the reverb and concentrate on the sound source. In fact, we normally aren't aware of reverberation.

But suppose you put a mic next to your ears, record an instrument in that room, and play back the recording. You'll hear a lot more reverb than what you heard live. What's going on? The reverb you recorded and played back isn't all around you. Instead, it's all up front between the speakers. So it's more audible; you can't discriminate against the reverb spatially. To reduce the amount of reverb in your recordings, you need to place mics close to instruments, and maybe add some sound-absorbing materials to the room.

Diffusion

When sound waves strike and reflect off a very bumpy or convex surface, they spread out or diffuse. This diffusion can be used to weaken sound reflections. Sound waves also spread out when they travel through a small opening.

HOW TO TAME ECHOES AND REVERB

Echoes and reverb can make your recordings sound mushy and distant. There are two ways to prevent these problems: with recording techniques and with acoustic treatment.

Controlling Room Problems with Recording Techniques

Sometimes you can make clean recordings in an ordinary room, such as a club, living room, or basement, if you follow these suggestions:

- Mike close. Place each mic 1 to 6 inches from each instrument or voice. Then the mics will hear more of the instruments and less of the room reflections. You might want to use mini mics, which attach directly to instruments.
- Use directional mics—cardioid, supercardioid, or hypercardioid—which reject room acoustics.

- Record bass guitar and synth directly with a guitar cord or a direct box. Since you omit the microphone, you pick up no room acoustics. To get a good sound when recording electric guitar direct, record off the effects boxes or use a guitar-amp simulator.

Also see the tips on reducing leakage in Chapter 7 under the heading, Leakage (Bleed or Spill).

Controlling Room Problems with Acoustic Treatments

When should you apply acoustic treatment to a room or build a studio?

- You clap your hands next to a wall and you hear flutter echoes (a fluttering sound). These are caused by sounds bouncing back and forth between hard parallel walls.
- Your studio is a very live environment, such as a garage or concrete-block basement, so you hear too much room reverberation.
- Your studio is very small.
- You hear outside noises in your recordings.
- Bass-guitar amps and monitor speakers sound boomy.
- You want the freedom to mike several feet away without picking up noise or excess room reverb.
- You hear a lot of leakage in the mic signals. Examples of leakage are drums picked up by the guitar mic or electric guitar picked up by the cymbal mics.

If those conditions apply, upgrade the acoustics of your studio. Here are some suggestions.

Reverb and echoes are caused by sound reflections off room surfaces. So any surface that is highly sound-absorbent helps to reduce those problems.

To absorb high frequencies, use porous materials such as convoluted (bumpy) foam mattresses. They can be highly flammable, so cover them with flame-retardant treatment (see natfire.com or flamestop .com). Nail or glue them to the walls, or mount them on frames. Thick foam works better than thin. Four-inch foam on the wall absorbs frequencies from about 200 to 800 Hz and up, depending on

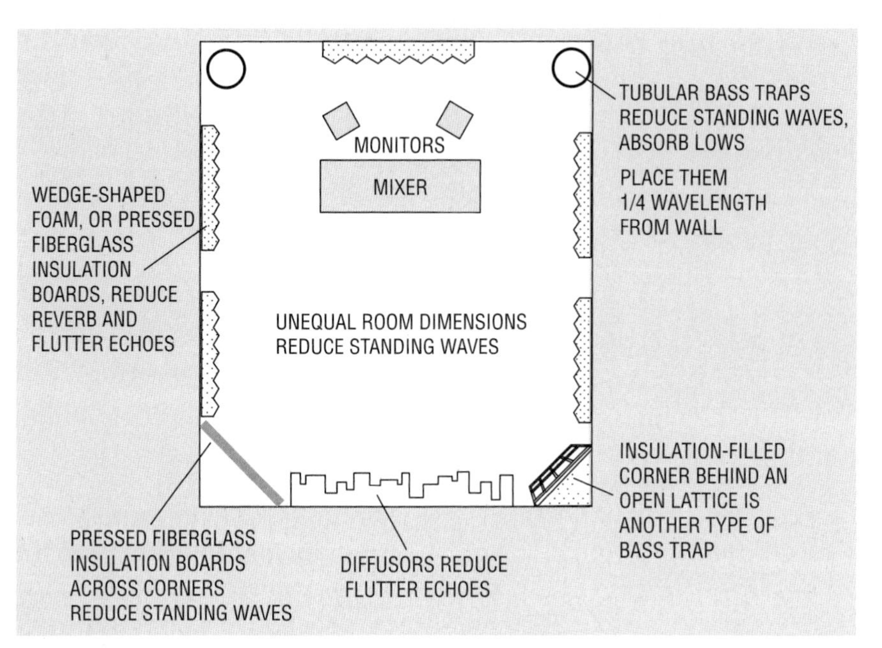

FIGURE 3.11
Acoustic treatments.

the angle at which sound strikes the foam. Leave some space between the foam panels (Figure 3.11); this helps to diffuse or spread out the sound in the room. Don't overdo the foam padding. A stuffed, dead room is uncomfortable to play in. Keep some reflections because they add "air" and liveliness to the sound.

Other high-frequency absorbers are sleeping bags, moving blankets, carpeting, curtains, and dense fiberglass insulation covered with muslin or burlap. If possible, space these materials several inches from the wall. The spacing helps them absorb mid-bass frequencies. A wide-range absorber is 4-inch pressed fiberglass board (Owens-Corning Type 703, 3 lb/ft^3) covered in muslin or burlap.

Start with just a little absorption behind or above the musician you're recording. Add more absorbers a few at a time, until your recordings sound as dead as you wish—typically covering 50–60% of the surfaces. Place absorption where you see mirror images of the monitors from the mix position. Put an absorber on the wall behind the monitors, and also hang an absorbent panel centered halfway between the mix position and monitors, suspended on hooks and wire.

Another absorber is a stand-mounted acoustical panel placed near a microphone. Examples are the ModTrap and the sE Relexion Filter.

To absorb low frequencies, you can make bass traps. Here are four types:

1. Resonant tube trap: Take a 35- to 55-gallon rubber trash-can, stuff it with fiberglass insulation (wear a dust mask and gloves), and cover the open end with muslin or canvas. This tube trap absorbs frequencies near 1130/2H, where H is the height of the trashcan in feet. For example, a 3-foot-high trash-can absorbs 188 Hz. Placement is not critical.
2. Frictional tube trap: Make a 2-foot-diameter canvas bag, 8 feet tall, and fill it with fiberglass insulation. Hang one bag a few feet out from each room corner (Figure 3.11). The distance should be 1130/4F, where F is the frequency in Hz that you want to absorb. For example, to absorb 80 Hz, hang the bag 3.5 feet out from a room corner. (Thanks to David Moulton for these ideas).
3. Lattice: Build a 3-foot-wide flat lattice frame of wood slats, the same height as the ceiling (Figure 3.11). Cover the frame with muslin or burlap. Put the frame diagonally across a room corner and fill the corner with R-30 insulation. Leave the foil on, foil side out toward the room. Put one assembly in each corner. (Thanks to Chips Davis for this idea).
4. Insulation panel: Get 8 pieces of rigid fiberglass insulation, type 705, 2 × 4 feet and 4 inches thick, from an insulation supplier. Cover each piece with muslin or burlap to contain the fibers. Place a piece across each room corner with the 2-foot edge touching the floor. Stack two panels to make them 8 feet high. (Thanks to Ethan Winer for this idea).

There are other ways to absorb bass. Wood paneling works well. It also helps to open closet doors and place couches and books a few inches from the walls. In a basement studio, nail acoustic tile to the ceiling joists with fiberglass insulation in the air space between tiles and ceiling.

You may not need any bass traps if you don't put any bass into the room. For example, don't turn up the bass-guitar amp—just record the bass direct and have the musicians wear headphones to hear the bass.

Controlling Standing Waves

Let's look at another acoustic problem: standing waves. If you play an amplified bass guitar through a speaker in a room, and do a bass run up the scale, you may hear some notes that boom out in the room. The room is resonating at those frequencies. These resonance frequencies, which are strongest below 300 Hz, occur in patterns called "standing waves." They can give a tubby or boomy coloration to musical instruments and monitor speakers.

Room resonances are worst in a cubical room. They are less of a problem if the room's length, width, and height are not multiples of each other. Table 3.1 shows several room dimensions in feet that reduce boomy-sounding standing waves.

For example, if the room width is 9.1 feet, and the ceiling is 8 feet high, the length should be 11.1 feet for best reduction of boominess.

Try to record in a large room because the room resonance frequencies are likely to be below the musical range. Use bass traps to absorb

Table 3.1 Room Dimensions in Feet to Reduce Standing Waves

Height	Width	Length
8	9.1	11.1
8	9.4	11.8
8	10.1	11.3
8	10.2	12.3
8	11.6	16.8
8	11.8	13.6
8	12.8	18.6
8	13.0	21.0
10	11.4	13.9
10	11.7	14.7
10	12.6	14.1
10	12.8	15.4
10	14.5	21.0
10	14.7	17.0
10	16.0	23.3
10	16.2	26.2

room resonances. Contrary to popular opinion, nonparallel walls don't prevent standing waves.

MAKING A QUIETER STUDIO

Ideally the noise level in a studio should be 23–28 dB A-weighted as measured on an SPL meter. The following tips will keep noises out of your recordings:

- Consider having the studio in a basement, because the earth blocks noises from the outside. The furnace or air-conditioning might be a problem, however.
- Turn off appliances, air conditioning, and telephones while recording.
- Pause for ambulances and airplanes to pass.
- Close windows. Consider covering them with thick plywood.
- Close doors and seal with towels.
- Remove small objects that can rattle or buzz.
- Weather-strip doors all around, including underneath.
- Replace hollow doors with solid doors.
- Block openings in the room with thick plywood and caulking.
- Put several layers of plywood and carpet on the floor above the studio, and put insulation in the air space between the studio ceiling and the floor above.
- Place microphones close to instruments and use directional microphones. This won't reduce noise in the studio, but it will reduce noise picked up by the microphones.

When building a new studio, you might want to make the walls of plastered concrete block (because massive walls reduce sound transmission) or make the walls of gypsum board and staggered studs. Nail gypsum board to 2 × 4 staggered studs on 2 × 6 footers as shown in Figure 3.12. Staggering the studs prevents sound transmission through the studs. Fill the air space between walls with insulation.

The ideal home-recording room for pop music is a large, well-sealed room with optimum dimensions. This room is in a quiet neighborhood. It should have some soft surfaces (carpet, acoustic-tile ceiling, drapes, couches), and some hard vibrating surfaces (wood paneling or gypsum board walls on studs).

Your home studio may not need acoustic treatment. Do some trial recordings to find out. But if your room could stand some improvement, the suggestions above should point you in the right direction.

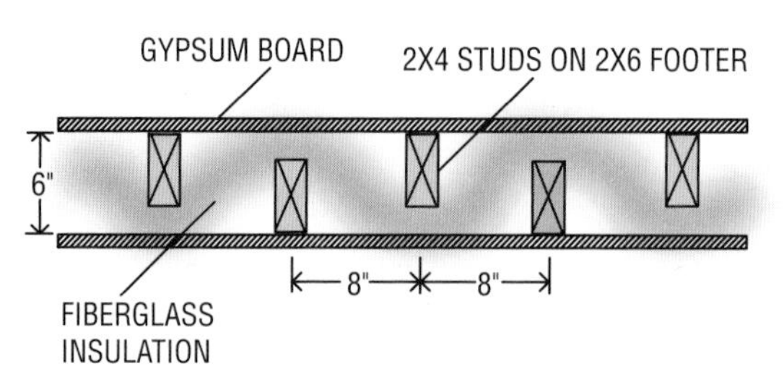

FIGURE 3.12
Staggered-stud construction reduces noise transmission.

For better results and a more professional appearance, consider buying some acoustic treatments from the following companies. Their Web sites are atsacoustics.com, realtraps.com, acousticsciences.com, acousticalsolutions.com, primacoustic.com, auralex.com, acousticsfirst.com, wallmate.net, pinta-acoustic.com, and rpginc.com. Atsacoustics.com has a free online room acoustics analysis.

SIGNAL CHARACTERISTICS OF AUDIO DEVICES

When a microphone converts sound to electricity, this electricity is called the signal. It has the same frequencies and the same amplitude changes as the incoming sound wave.

When this signal passes through an audio device, the device may alter the signal. It might change the level of some frequencies or add unwanted sounds that aren't in the original signal. Let's look at some of these effects.

Frequency Response

Suppose you have an audio device—a mic, mixer, effects unit, recorder, or speaker. You send a musical signal through the device. Usually the music contains some high and some low frequencies.

The device might respond differently to different frequencies. It might amplify the low notes and weaken the high notes. You can graph how the device responds to different frequencies by plotting its output level versus frequency. This graph is called a frequency response (Figure 3.13). The level in the graph is measured in dB, while frequency is measured in Hz. Generally, 1 dB is the smallest change in loudness that we can hear.

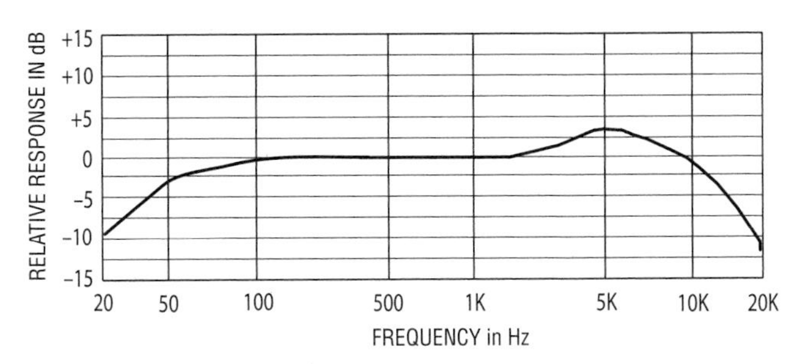

FIGURE 3.13
An example of a frequency response.

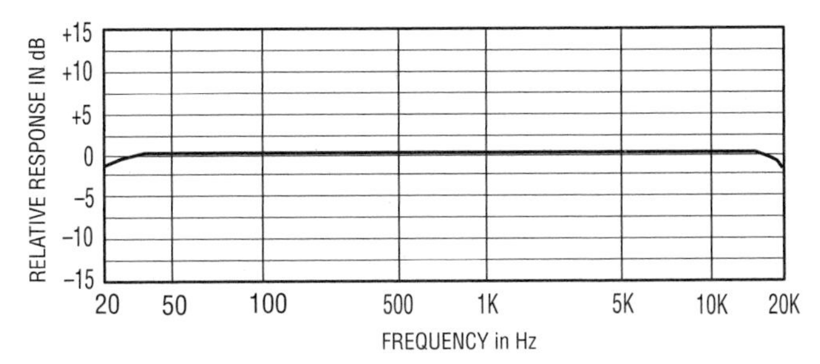

FIGURE 3.14
Flat frequency response.

Suppose the level is the same at all frequencies. Then the graph forms a horizontal straight line and is called a "flat frequency response" (Figure 3.14). All frequencies are reproduced at an equal level. In other words, the device passes all the frequencies without changing their relative levels. You get out the same amount of bass and treble that went in. A flat response doesn't affect the tonal balance of the incoming sound.

Many audio devices don't have a flat response across the audio band from 20 to 20,000 Hz. They have a limited range of frequencies that can be reproduced at an equal level (within a tolerance, such as ±3 dB). In Figure 3.14, the frequency response shown by the solid line is 50 to 12,000 Hz ± 3 dB. That means the audio device passes all frequencies from 50 to 12,000 Hz at a nearly equal level—within 3 dB. It reproduces low and high sounds equally well. The response is down 3 dB at 50 and 12,000 Hz, and is up 3 dB at 5000 Hz.

Usually, the more extended or "wide" the frequency range is, the more natural and real the recording sounds. A wide, flat response gives accurate reproduction. A frequency response of 200 to 8000 Hz (±3 dB) is narrow (poor fidelity); 80 to 12,000 Hz is wider (better fidelity), and 20 to 20,000 Hz is widest (best fidelity). *Play audio clip 7 at www.elsevierdirect.com/companions/9780240811444.*

Also, the flatter the frequency response, the greater the fidelity or accuracy. A response deviation of ±3 dB is good, ±2 dB is better, and ±1 dB is excellent. There are exceptions to this statement, which we'll look at in Chapter 10 in the section on equalizers.

When you turn a bass or treble knob on your guitar amp, mixer EQ, or stereo, you're changing the frequency response. If you turn up the bass, the low frequencies rise in level. If you turn up the treble, the high frequencies are emphasized. The ear interprets these effects as changes in tone quality—warmer, brighter, thinner, duller, and so on.

Figure 3.13 shows a non-flat frequency response. Toward the right side of this line, the response at high frequencies "rolls off" or declines. This shows that the upper harmonics are weak. The result is a dull sound. Toward the left side, the response at low frequencies rolls off. This means the fundamentals are weakened and the result is a thin sound.

The frequency response of an audio device might be made non-flat on purpose. For example, you might cut low frequencies with an equalizer to reduce breath pops from a microphone. Also, a microphone may sound best with a non-flat response, such as boosted high frequencies that add presence and sizzle.

Noise

Noise is another characteristic of audio signals. Every audio component produces a little noise—a rushing sound like wind in trees. Noise in a recording is undesirable unless it's part of the music.

You can make noise less audible by keeping the signal level in a device relatively high. If the level is low, you have to turn up the listening volume to hear the signal well. Turning up the volume of the signal also turns up the volume of the noise, so you hear noise along with the signal. But if the signal level is high, you don't have to turn up the listening level as much. Then the noise remains in the background.

Distortion

If you turn up the signal level too high, the signal distorts and you hear a gritty, grainy sound or clicks. This type of distortion is sometimes called "clipping" because the peaks of the signal are clipped off so they are flattened. To hear distortion, simply record a signal at a very high recording level (with the meters going well into the red area) and play it back. *Play audio clip 9 at www.elsevierdirect.com/companions/9780240811444*. Digital recorders also produce "quantization" distortion at very low signal levels.

Optimum Signal Level

You want the signal level high enough to cover up the noise, but low enough to avoid distortion. Every audio component works best at a certain optimum signal level, and this is usually indicated on a level meter built into the device. More on this in Appendix A.

Figure 3.15 shows the range of signal levels in an audio device. At the bottom is the noise floor of the device—the level of noise it produces with no signal. At the top is the distortion level—the point at which the signal distorts and sounds grungy. In between is a range in which the signal sounds clean. The idea is to maintain the signal around normal operation level on the average. Generally you want the signal level to be as high as possible without clipping or distorting.

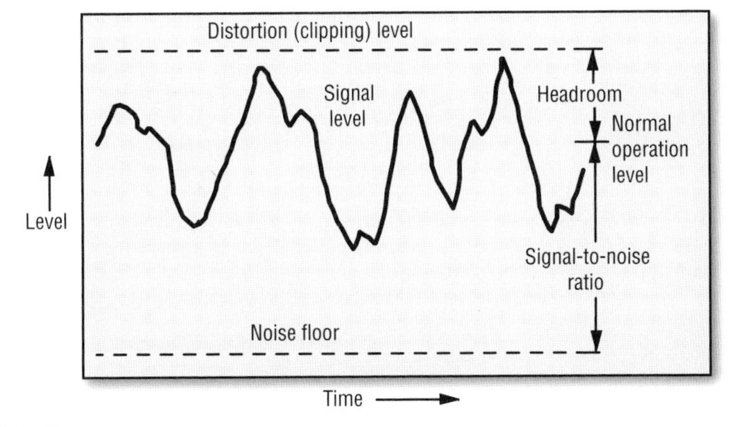

FIGURE 3.15
The range of signal levels in an audio device.

With digital recorders (such as computer recording software), "0 dBFS" (0 decibels full scale) on the meter is the maximum undistorted signal. In that case, an average level of −20 dBFS and a peak level of −6 dBFS is the normal operation level. In a digital audio file that is used to master a CD, an average level of −14 dBFS is a good goal to aim for. There's more on digital signal levels in Chapter 13 on Computer Recording.

Signal-to-Noise Ratio

The level difference in decibels between the signal level and the noise floor is called the signal-to-noise ratio or S/N (see Figure 3.15). The higher the S/N, the cleaner the sound. An S/N of 60 dB is fair, 70 dB is good, and 80 dB or greater is excellent. *Play audio clip 8 at www.elsevierdirect.com/companions/9780240811444.*

To illustrate S/N, imagine a person yelling a message over the sound of a train. The message yelled is the signal; the noise is the train. The louder the message, or the quieter the train, the greater the S/N. And the greater the S/N, the clearer the message.

Headroom

The level difference in dB between the normal signal level and the distortion level is called "headroom" (see Figure 3.15). The greater the headroom, the greater the signal level the device can pass without running into distortion. If an audio device has a lot of headroom, it can pass high-level peaks without clipping them.

You want to set your mixer controls so that the signal has some headroom, is well above the noise floor, and is below distortion. Tips on doing that are in Chapter 9, Digital Recording, under the heading Digital Recording Level. Also see Chapter 12, Mixer Operation, under the headings Set Recording Levels and Set Levels.

With these settings, the signal levels in the mixer should be just about right, with no audible noise or distortion. And the mixer should have enough headroom so that loud peaks won't distort.

CHAPTER 4

Equipping Your Studio

You want to set up a recording system that's affordable, easy to use, and sonically excellent. With today's wide array of user-friendly sound tools, you can do just that. This chapter is a guide to equipment for a recording studio: what it does, what it costs, and how to set it up.

In this chapter we'll examine:

- Equipment
- Cables and connectors
- Preventing hum and radio frequency interference

What is the bare-bones equipment you need to crank out a decent demo CD? How much does it cost? Thanks to the new breed of affordable equipment, you can put together a complete home recording studio for as little as $1200. That includes powered speakers, two mics and mic stands, recording software, a sound card, headphones, and cables.

EQUIPMENT

Let's examine each piece of equipment in a recording studio. You'll need some sort of recording device, headphones, cables, mics, direct box, monitor speakers, audio interface, and effects.

Recording Device

Five types of recording devices to choose from are a 2-track recorder, a recorder–mixer, a separate multitrack recorder and mixer, a computer, and a keyboard workstation. We'll look at each one.

PORTABLE 2-TRACK RECORDER

This device is great for on-location stereo recording of classical music: an orchestra, symphonic band, string quartet, pipe organ, or soloist. Sometimes it can be used to record folk music or jazz, or to capture a practice session or an audience perspective of a rock group. Two types of 2-track recorder are a Flash memory recorder (Figure 4.1), or laptop computer with recording software. Prices range from $200 to $1800. Details on these units are in Chapter 9. Some Flash recorders are the Sony PCM-D1 and D50 (sony.com/proaudio), Marantz PMD620 (d-mpro.com), Edirol R-1, R-09 and R-09HR (edirol.com), M-Audio MicroTrack II (m-audio.com), Zoom H4 and H2 (samsontech .com), Olympus LS-10 (olympus.com), Korg MR-1 and MR-1000 (korg.com), Yamaha PockeTrak 2G (yamaha.com), and TASCAM DR-1 and HD-P2 (tascam.com).

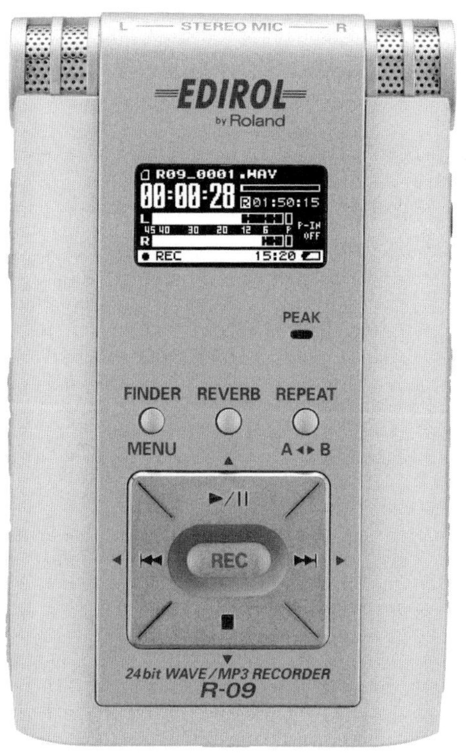

FIGURE 4.1
Edirol R-09W, an example of a portable 2-track recorder.

MINI STUDIO

Also called a portable studio, personal studio, or pocket studio, this piece of gear combines a 4- or 8-track recorder with a mixer—all in one portable chassis (Figure 4.2). A mini studio records in MP3 format to a Flash memory card. It lets you record several instruments and vocals, then mix them to stereo; that is, you "bounce" the tracks to a stereo file in the memory card. You can copy the stereo mix file to your computer via USB. (USB is a type of cable and ports for high-speed data transfer). The sound quality of Pocket Studios is good enough to make demos, or to use as a musical scratchpad for ideas, but it is not quite good enough to release commercial albums. Costing about $230 to $599, a mini studio can be a good choice for beginning recordists.

Some features include:

- An internal MIDI sound module (synthesizer) that plays back MIDI sequences. Includes MIDI files or rhythm patterns to jam with.
- Built-in effects.

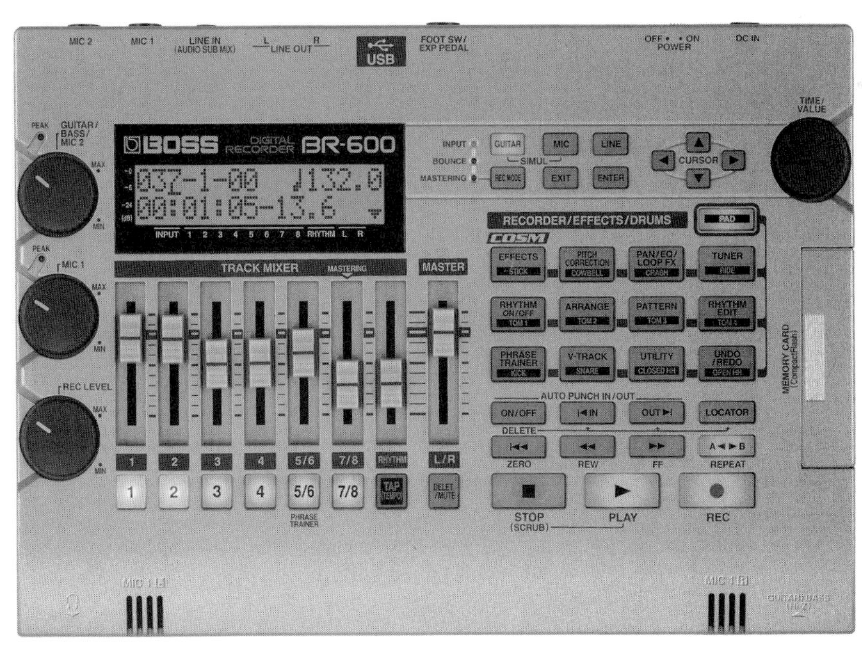

FIGURE 4.2
Boss BR-600, an example of a mini studio.

- Built-in mic (in some models).
- Autopunch. Automatically "punch" into and out of record mode to correct mistakes.
- Battery or AC-adapter powered.
- Virtual tracks let you record multiple takes of a single performance, then select your favorite take during mixdown.
- Guitar-amp modeling simulates various guitar amps; mic modeling simulates mic models.
- 2 mic inputs; records up to 2 tracks at a time. Plays back 4 or 8 tracks at once.

DIGITAL MULTITRACKER (RECORDER–MIXER)

Other names for this device are stand-alone Digital Audio Workstation (DAW), portable digital studio, or personal digital studio.

Like the mini studio, the digital multitracker combines a multitrack recorder with a mixer in a single chassis (Figure 4.3). It's convenient and portable. Offering better sound than a mini studio, the digital multitracker provides CD sound quality because it can record WAVE files. It records 8 to 32 tracks onto a built-in hard drive or Flash memory card. The mixer includes faders (volume controls) for mixing, EQ or tone controls, and aux sends for effects (such as reverb). An LCD screen displays recording levels, waveforms for editing, and other functions.

Price ranges from $700 to $3600. Some manufacturers of mini studios or digital multitrackers are TASCAM, Boss, Fostex, Korg, Roland, Yamaha, and Zoom (Samson).

Chapter 9, under the heading Recorder–Mixer (Personal Digital Studio), lists some features to look for in a recorder–mixer.

SEPARATE MULTITRACK RECORDER AND MIXER

Ideal for on-location recording, a multitrack hard-disk recorder (HD Recorder) records up to 24 tracks reliably on a built-in hard drive (Figure 4.4), just like the hard drive in your computer. The recorder doesn't have mic preamps, so it must be used with a mixer or mic preamps. Multiple recorders can be linked to get more tracks. Some examples are the Alesis ADAT HD24XR, TASCAM MX-2424 and X-48, Otari DR-100, iZ Technology RADAR, Fostex D-2424LV, and Mackie HDR 24/96. Prices range from $1500 to $20,000.

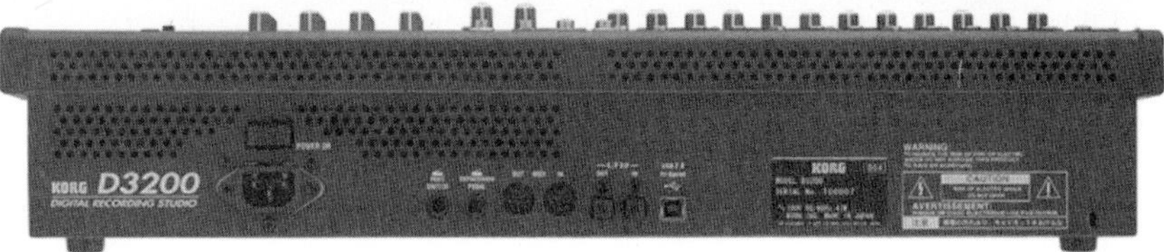

FIGURE 4.3
Korg D3200, an example of a digital multitracker (recorder–mixer).

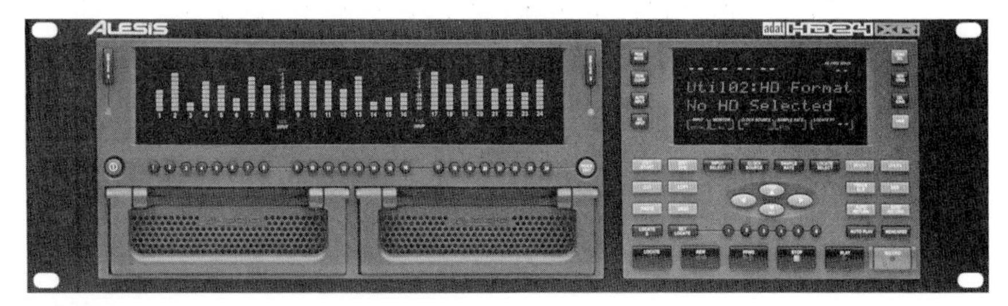

FIGURE 4.4
Alesis HD24XR, an example of a multitrack HD recorder.

Current units record with 24-bit resolution and up to 96 kHz sampling rate. Some record 24 tracks of 24-bit programs at sample rates of 44.1 and 48 kHz. They record 12 tracks of 24-bit recording at 88.2 and 96 kHz. (These terms are explained in Chapter 9). Some HD recorders allow track editing, either on a built-in LCD screen or on a plug-in computer monitor. A few models have built-in removable hard drives for backing up projects.

A mixer (Figure 4.5) is an electronic device with several mic preamplifiers, volume and tone controls for each mic, monitor volume controls, and outputs for effects devices. To use it, plug mics and electric instruments into the mixer, which amplifies their signals. Connect the output signal of each mic preamp to a multitrack recorder. While recording, use the mixer to send those signals to the desired recorder tracks, and to set recording levels. During mixdown, the mixer combines (mixes) the tracks to stereo. It also lets you adjust each track's volume, tone quality, effects, and stereo position. A large, complex mixer is called a

FIGURE 4.5
Mackie 1604-VLZ3, an example of a mixer.

mixing console (board or desk) and costs upwards of $3000. Chapter 11 explains mixing consoles in detail.

Price is about $180 to $1500. Mixers are made by such companies as TASCAM, Alesis, Yamaha, Mackie, Allen & Heath, Soundcraft, M-Audio, Tapco, Peavey, Samson, and many others.

COMPUTER DIGITAL AUDIO WORKSTATION (DAW)

This low-cost recording setup has three parts: a personal computer, an audio interface, and recording software (Figure 4.6). Some software records MIDI data (explained later) as well as audio. The audio interface converts audio from your mics, mic preamps, or mixer into a signal that is recorded on the computer's hard drive as a magnetic pattern. Dozens of tracks can be recorded with professional quality. You mix the tracks with a mouse by adjusting the "virtual" controls that appear on your computer screen. Then you record the mix on your computer's hard drive. Details are in Chapter 13, Computer Recording.

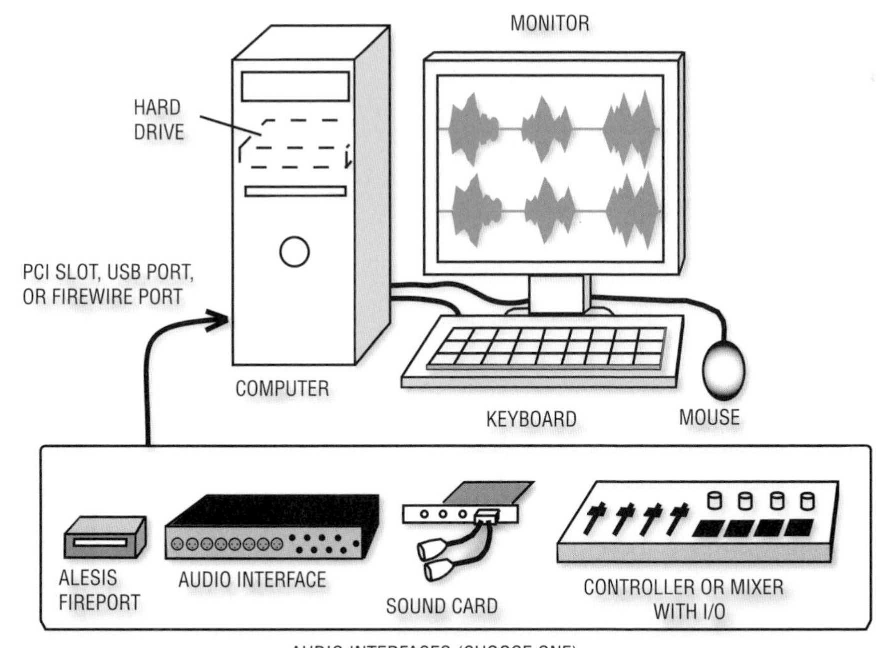

FIGURE 4.6
Computer with recording software and a choice of audio interface.

A computer studio costs about the same as a mini studio and is more powerful. It's a bargain. But since software requires computer skills, it's a little harder to learn and use than a hardware multitracker. Software recordings are at least CD quality—better than the MP3 recordings you get with a mini studio.

Recording software costs anywhere from $0 (freeware) to $1800, with $150 to $500 being typical prices. Software examples include Cakewalk Home Studio, Sonar Studio and Sonar Producer; Digidesign Pro Tools LE, M-Powered and HD; Apple Logic Studio and Logic Pro, Steinberg Cubase and Nuendo, Sony Vegas and Sound Forge, BIAS Deck, Pro Tracks Plus, Magix Samplitude and Sequoia, Mackie Tracktion, Reaper; N Track Studio, RML Labs SAW Studio, Adobe Audition, and MOTU Digital Performer (Mac only).

Free multitrack recording programs are available for download. Although they lack extensive features, they offer a chance to practice your skills at no cost. Examples are Audacity (http://audacity.sourceforge.net), Kristal (from Kreative.org), and Garage Band (supplied free with any new MacIntosh computer or $79 with a Mac software upgrade).

Related to recording software is music creation software, which comes with samples of musical instruments. You set up loops (repeating drum riffs and musical patterns), play along with them using a MIDI keyboard, and record that MIDI sequence. Examples include Ableton Live, Propellerhead Reason, Sony ACID Pro, and Spectrasonics Stylus RMX.

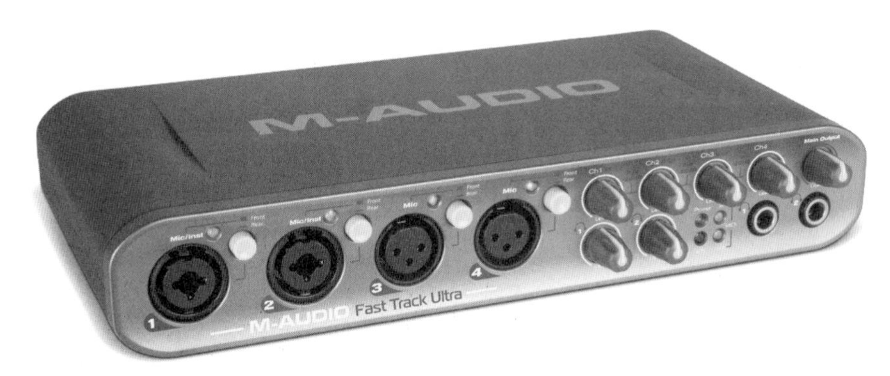

FIGURE 4.7
M-Audio Fast Track Ultra, an example of a USB 2.0 audio interface. It offers low-latency direct monitoring with reverb, and can create two headphone mixes.

A sound card (PCI audio interface) costs $100 to $1700, a USB audio interface (Figure 4.7) costs $100 to $500, and a FireWire audio interface costs $200 to $1950.

Some makers of sound cards include Echo, Event, Frontier Design, M-Audio, RME, Lynx, and E-MU. USB and FireWire Audio interface manufacturers include Digidesign, Metric Halo, MOTU, PreSonus, Apogee, Focusrite, M-Audio, TC Electronic, Mackie, E-MU, Alesis, Edirol, TASCAM, Lexicon, and many others.

Another type of audio interface is a mixer that connects to a computer by a FireWire or USB port. It sends the output signal of each mic preamp to a separate track in your recording software, and returns the stereo mix back to the mixer for monitoring. This can be a good system for recording a band because it offers easy setup of monitor mixes with effects. Some examples are the Mackie Onyx 1640 with the optional Onyx FireWire Card (16 recording channels), Alesis Multimix USB or FireWire series (up to 8 recording channels), Edirol M-16DX (16 recording channels, 4 mic inputs), and Behringer Xenyx 2442FX (4 recording channels). Generally these mixers are not controllers for recording software—you still need to use a mouse—but the Edirol M-16DX is also a control surface.

A few microphones have USB built in, so you just plug the mic directly into your computer's USB port. Examples are the Samson C01U, Blue Snowball, Rode Podcaster USB microphone; and MXL USB.009, 990 USB, and 990 USB stereo mics. The MXL MicMate plugs into any pro mic and converts its output to USB. Phantom power is included. Its digital-audio format is 16-bit, 44.1 or 48 kHz. The Centrance MicPort Pro is similar but records up to 24-bit/96 kHz audio.

Using a mouse can be fatiguing and can lead to repetitive stress syndrome. An alternative to the mouse is a control surface or controller. It looks like a mixer with faders, but it adjusts the virtual controls you see on the computer screen. That way you can use your computer for recording, and control the software with knobs and faders instead of a mouse. Two controllers with one fader are the Frontier Design AlphaTrack ($200, Figure 4.8) and PreSonus FaderPort ($150). Multi-fader controllers are made by such companies as Mackie, TASCAM, M-Audio, Digidesign, and Alesis ($350 to $1300).

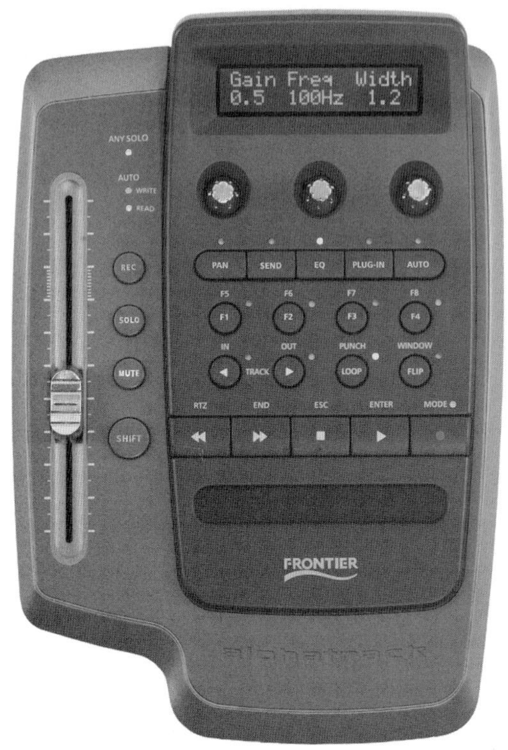

FIGURE 4.8
Frontier Design AlphaTrack, an example of a controller for recording software.

KEYBOARD WORKSTATION

Here's another device to record music. This is a synthesizer/sampler with a piano-style keyboard and a built-in sequencer (MIDI recorder) and effects. Examples include Alesis Fusion, Promega 2 and Andromeda; Korg K2600X; Yamaha Motif, Roland Fantom-X, Juno-G and GW-7. The workstation lets you create a tune with several instrument sounds. Then you record the audio from the keyboard into a computer. If you want to add a vocal, use MIDI/audio recording software. An arranger keyboard workstation automatically creates backing tracks (drums, bass, and chords) by following your left hand's notes and right-hand's melodies. Examples include Yamaha Tyros2 and PSR-S900.

A similar device is a beat box or groove box, which is a sample player with pads that you tap to generate sounds and grooves. It lets you assemble a stereo music track from drum and synth samples. Copy the music from the beat box into your computer and add vocals. Examples include Akai MPC series and MPD24, Roland sampling workstations, Zoom RT-223 and Streetboxx SB-246, Korg Electribe series.

LEGACY RECORDING DEVICES

Some recording equipment has become almost obsolete, but you may need to work with it if a client brings older gear into your studio to transfer audio tracks. Some of these formats are listed below.

Analog tape recorder: This records 2 to 24 tracks of analog audio on reels of magnetic tape from 1/4 to 2 inches wide. Compared to digital recorders, analog recorders have more noise, distortion, frequency-response errors, and unstable pitch. Their electronics and tape heads need frequent alignment. Still, many people love the sound of them. Some have mic inputs; all have analog line inputs and line outputs.

DAT recorder: This records 2 tracks of audio digitally on a small DAT-tape, a cassette about half the size of a standard analog cassette. The DAT also writes absolute time (hours, minutes, and seconds) on tape as it records. Inputs are analog mic/line level and digital; outputs are analog line level and digital.

Modular Digital Multitrack (MDM): This records 8 digital tracks on a videocassette, using a rotating drum like a DAT recorder. Two popular models are the Alesis ADAT which records on S-VHS tape, and the TASCAM DA-88, DA-78 or DA-38 which record on Hi-8 mm tape. ADAT records up to 40 minutes on a single tape; DA-88 records up to 1 hour 48 minutes. With both types, you can sync several 8-track units with a cable to add more tracks. Both have analog line inputs and outputs. ADAT has Lightpipe-format digital outputs; TASCAM models have TDIF-format digital outputs.

Microphone

So far we covered recording devices. Let's move on to other studio equipment.

A microphone converts the sound of a voice or instrument into an electrical signal that can be recorded. Microphone sound quality varies widely, so be sure to get one or more good mics costing at least $100 each. Your ears should tell you if a mic's fidelity is adequate for your purpose. Some people are happy to get any sound recorded; others settle for nothing less than professional sound quality.

Probably the most useful mic types for home recording are the cardioid condenser mic and cardioid dynamic mic. The cardioid pickup pattern helps reject room acoustics for a tighter sound. The condenser type is commonly used on cymbals, acoustic instruments, and studio vocals; dynamics are used typically on drums and guitar amps. You'll also need at least one mic cable, mic stand, and boom costing about $25 each. For more information on microphones, see Chapter 6.

Do you want to record solo instruments or classical musical ensembles in stereo with two mics at a distance? You'll need either a stereo mic or a matched pair of high-quality condenser or ribbon mics of the same model number, plus a stereo mic-stand adapter. See Chapter 18 for details.

Phantom-Power Supply

A phantom-power supply powers the circuits in condenser mics. It uses the same cable that carries the mic's audio signal. You don't need the supply if your condenser mic has a battery, or if your mixer or audio interface supplies phantom power at its mic connectors (most do). Some makers of phantom supplies are ART, Audio-Technica, Rolls, and Whirlwind. Prices range from $45 to $139.

Mic Preamp

This device amplifies a mic signal up to a higher voltage called "line level," which is required by mixers and recorders. A stand-alone mic preamp provides a slightly cleaner, smoother, or more colorful sound than a mic preamp built into a mixer or audio interface, but costs much more. A 2-channel preamp costs $120 to $2000, while 8-channel preamps range from $600 to $6000. Studios on a budget can do without a preamp.

Some manufacturers are Manley, True Systems, Focusrite, Universal Audio, GLM, Chandler, A Designs, Millennia, Avalon, Great River, John Hardy, Benchmark, AEA, Apogee, Vintech, Grace, Groove Tubes, Presonus, Mackie, Summit Audio, Studio Projects, Blue, Samson, dbx, Aphex, and M-Audio.

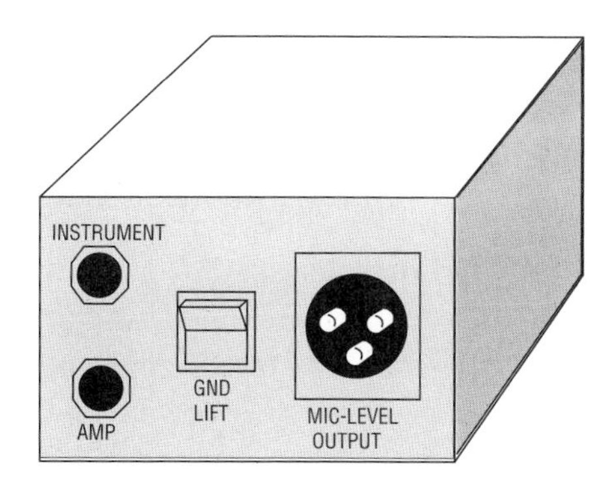

FIGURE 4.9
A direct box.

Direct Box (DI)

A direct box (Figure 4.9) is used to connect an electric instrument (guitar, bass, synth) to an XLR-type mic input of a mixer, recorder–mixer, or audio interface. It lets you record electric instruments directly into your mixer without a microphone. A direct box picks up a very clean sound, which may be undesirable for electric guitar. If you want to pick up the distortion of the guitar amp, use a microphone instead. Or use a guitar-amp modeling device or modeling plug-in.

Some recorder–mixers and audio interfaces have "high-impedance" or "instrument" inputs meant for electric guitars and synthesizers. In this case, simply use a short guitar cord between the instrument and the mixer high-impedance input. If the cable run is over 15 feet, you'll get higher quality with a direct box plugged into an XLR mic input.

Direct boxes sell for $50 to $700. Some manufacturers are Radial, Countryman, Whirlwind, Manley, Groove Tubes, Pro Co, BSS, Tapco, Aphex, and ART.

Monitor System

Another important part of your studio is the monitor system: a pair of quality headphones or loudspeakers. The monitor system lets you hear what you're recording and mixing—what you're doing to the recorded sound. The sound you hear over the monitors is approximately what the final listener will hear.

Very good headphones are available for $100 and up from Sony, AKG, Sennheiser, Beyerdynamic, Audio Technica, Ultrasone, Shure, and others. A headphone amp can power several headphones at once, and some units let the musician adjust the headphone mix.

If you're recording only yourself, one set of headphones is enough. But if you're recording another musician, you both need headphones. Many recorder–mixers have two headphone jacks for this purpose.

If you want to record or overdub several people at once, you need headphones for all of them. For example, if you're overdubbing three harmony vocalists, each one needs headphones to hear previously recorded tracks to sing with. To connect all these headphones, you could build a headphone junction box: an aluminum or plastic box that contains several headphone jacks. These are wired to a cable coming from your mixer's headphone jack. Or you could use a splitter cable, which makes two jacks out of one. Another option is to purchase an affordable Rolls or Behringer headphone amplifier. Feed the amp from aux sends or the stereo mix.

Powered speakers with built-in amplifiers provide accurate, high-fidelity monitoring. Described in Chapter 5, these Nearfield studio monitors are small, bookshelf-type speakers that are placed about 3 feet apart and 3 feet from you as you sit at your mixer. Some monitor companies include Genelec, Dynaudio, JBL, Event, Quested, Focal, Lipinski, KRK, Blue Sky, Yamaha, Adam, Mackie, Alesis, Tannoy, M-audio, Behringer, Edirol, and Samson. Prices range from $200 to $3860 a pair.

Effects

A recording without effects sounds rather dead and plain. Effects such as reverberation, echo, and chorus can add sonic excitement to a recording. They are produced by devices called signal processors (see Figure 4.10) or by plug-ins, which are software effects used in a computer recording program. See Chapter 10 for more information on effects.

Although effects are built into most recording programs and recorder–mixers, most analog mixers require external effects units. On the mixer is a set of connectors (labeled "aux send" and "aux return") for hooking up an external effects unit, such as a reverb or delay device. A unit with one effects send lets you add one type of effect; a unit with two effects sends lets you add two different effects to create more sonic interest.

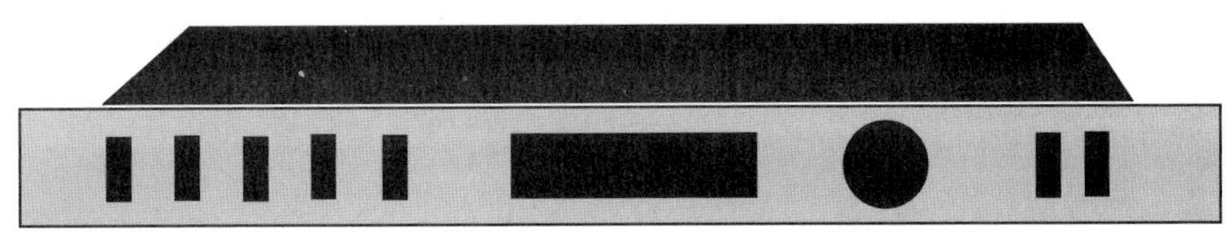

FIGURE 4.10
An effects unit.

Miscellaneous Equipment

Other equipment for your home studio includes audio cables, USB or FireWire cables, power outlet strips, lighting, tables or studio furniture, mic pop filters, masking tape and a pen to label inputs and cables, contact cleaning fluid, MIDI equipment stands, music stands, session forms, connector adapters, pen and paper, a flashlight, and so on.

MIDI Studio Equipment

MIDI is covered in detail in Chapter 16. It describes the components in a typical MIDI studio, such as shown in Figure 4.11. Also see Chapter 2 under the heading MIDI Sequencing.

We've looked at several types of recording setups. All can help you create quality demos. As you go higher in price you get more features and better sound. For example, if you want to record an entire band playing all at once, with each instrument having its own mic, you'll need a system with more microphones, more tracks, and more headphones.

As we've seen, putting together a home studio or project studio needn't cost much. As technology develops, better equipment is available at lower prices. That dream of owning your own studio is within reach.

SETTING UP YOUR STUDIO

Once you have your equipment, you need to connect it together with cables, and possibly install equipment racks and acoustic treatment. Let's consider each step.

Cables

Cables carry electric signals from one audio component to another. They are usually made of one or two insulated conductors (wires) surrounded by a fine-wire mesh shield that reduces hum. Outside

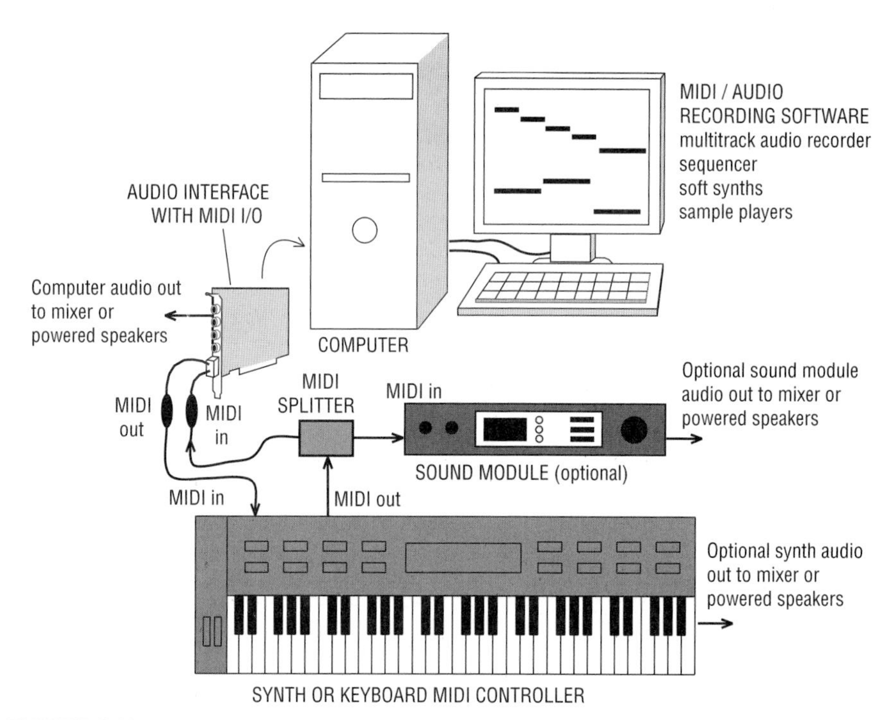

FIGURE 4.11
One type of MIDI studio.

the shield is a plastic or rubber insulating jacket. On both ends of the cable are various types of connectors.

Cables are either balanced or unbalanced. A balanced line is a cable that uses two wires (conductors) to carry the signal, surrounded by a shield (see Figure 4.12). Each wire has equal impedance to ground. An unbalanced line has a single conductor surrounded by a shield (see Figure 4.13). The conductor and shield carry the signal. A balanced line rejects hum better than an unbalanced line, but an unbalanced line less than 10 feet long usually provides adequate hum rejection and costs less.

A cable carries one of these five signal levels or voltages:

1. Mic level: About 2 mV (0.002 volt) to about 1 V depending on how loud the sound source is, and on how sensitive the mic is.
2. Instrument level: Typically 0.1 to 1 V for passive pickups; up to 1.75 V for active pickups.
3. Semi-pro or consumer line level: −10 dBV (0.316 volt).
4. Pro line level: +4 dBu (1.23 volts).
5. Speaker level (about 20 volts).

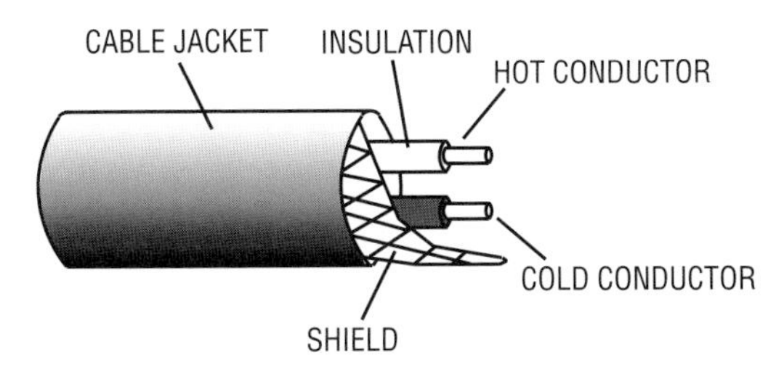

FIGURE 4.12
A 2-conductor shielded, balanced line.

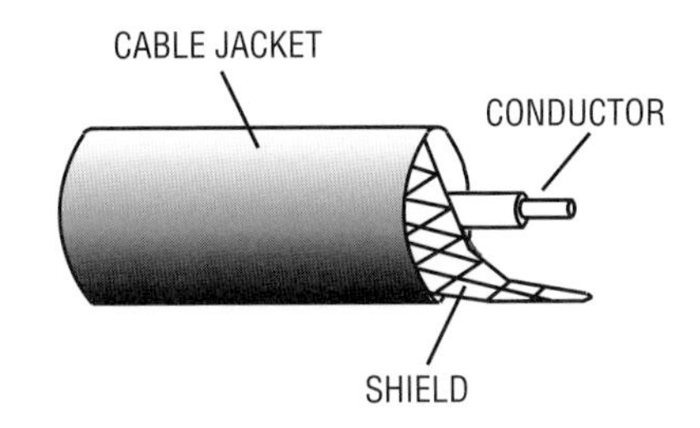

FIGURE 4.13
A 1-conductor shielded, unbalanced line.

Note that "instrument level" can overlap mic level and line level.

Some manufacturers use dBu instead of volts (for an explanation, see Appendix A):

Mic level: About −52 dBu to +2.2 dBu.
Instrument level: −17.7 dBu to +7 dBu.
Semi-pro or consumer line level: −7.8 dBu
Pro line level: +4 dBu.

Equipment Connectors

Recording equipment also has balanced or unbalanced connectors built into the chassis. Be sure your cable connectors match your equipment connectors.

Balanced equipment connectors:

- 3-pin (XLR-type) connector—Figure 4.14
- 1/4-inch TRS (tip-ring-sleeve) phone jack—Figure 4.15

Unbalanced equipment connectors:

- 1/4-inch TS (tip-sleeve) phone jack—Figure 4.15
- Phono jack (RCA connector)—Figure 4.16

A jack is a receptacle; a plug inserts into a jack.

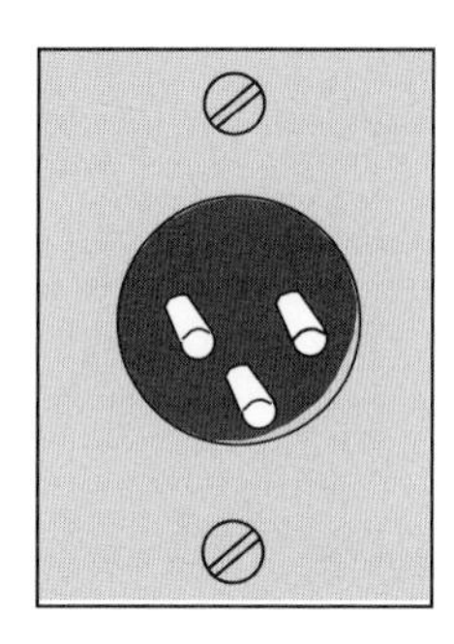

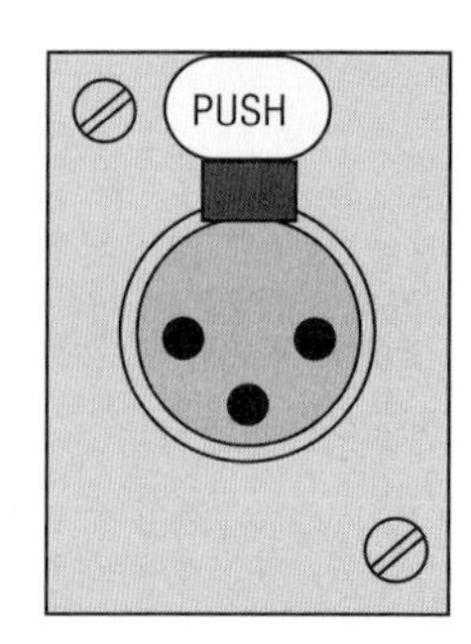

FIGURE 4.14
A 3-pin XLR-type connector used in balanced equipment. Left: male output connector. Right: female input connector.

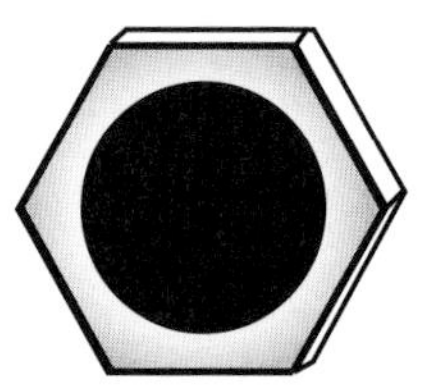

FIGURE 4.15
A 1/4-inch phone jack used in balanced and unbalanced equipment.

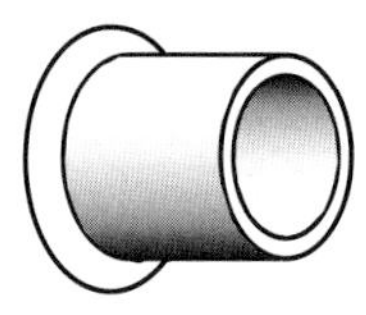

FIGURE 4.16
A phono (RCA) jack used in unbalanced equipment.

Connectors are confusing because a single connector can have several functions (usually not at the same time). Here are some examples:

- XLR: balanced line input at +4 dBu, balanced mic input at 2 mV to 1 V, or balanced line output at +4 dBu.
- TRS (stereo 1/4-inch phone jack): balanced mic input, insert send/return connectors (line-level), instrument input, or stereo headphones.
- TS (mono 1/4-inch phone jack): unbalanced mic input, unbalanced line-level input or output (+4 dBu or −10 dBV), instrument input, or low-cost speaker connector.
- Combi connector: an XLR mic input plus a TRS input (instrument-level or line-level).

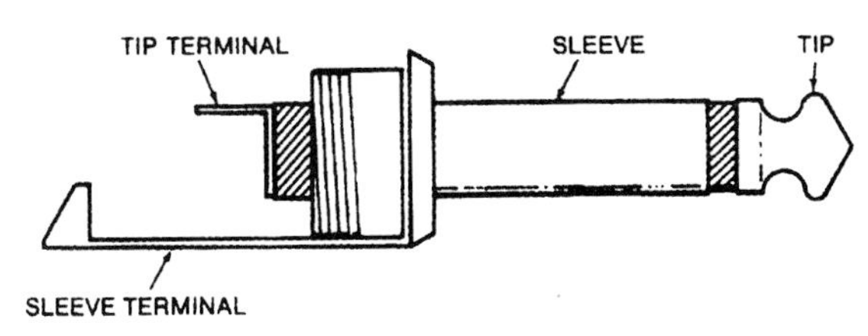

FIGURE 4.17
A mono (TS) 1/4-inch phone plug.

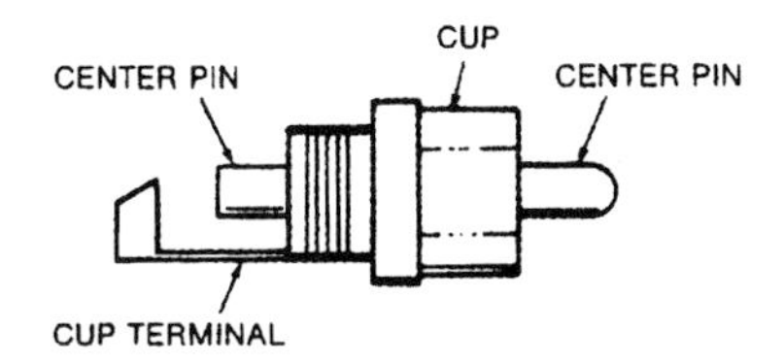

FIGURE 4.18
An RCA (phono) plug.

- RCA (phono): home stereo line-level input or output at −10 dBV, composite video input/output, or SPDIF digital-audio input/output.
- 1/8-inch (3.5 mm) mini phone jack: headphone jack, low-cost stereo mic input, line output in a portable recorder at −10 dBV; or a computer sound card's line in, line out, or speaker out.

Equipment connectors are labeled according to their function. If you see an XLR connector with the label "MIC," you know it's a balanced mic input. If it's an 1/8-inch connector on a sound card, look at the icon near the connector. It's either a mic input, line input, line output, or speaker output. You could download the manual for the sound card, which should describe the function of each connector.

Cable Connectors

Several types of cable connectors are used in audio. Figure 4.17 shows a 1/4-inch mono phone plug (or TS phone plug), used with cables for unbalanced microphones, synthesizers, and electric instruments. The tip terminal is soldered to the cable's center conductor; the sleeve terminal is soldered to the cable shield.

Figure 4.18 shows an RCA or phono plug, used to connect unbalanced line-level signals. The center pin is soldered to the cable's center conductor; the cup terminal is soldered to the cable shield.

Figure 4.19 shows a 3-pin pro audio connector (XLR-type). It is used with cables for balanced mics and balanced recording equipment. The female connector (with holes; Figure 4.19A) plugs into equipment outputs. The male connector (with pins; Figure 4.19B) plugs into equipment inputs. Pin 1 is soldered to the cable shield, pin 2 is soldered to the "hot" red or white lead, and pin 3 is soldered to the remaining lead. This wiring applies to both female and male connectors.

Figure 4.20 shows a stereo (TRS) phone plug, used with stereo headphones and with some balanced line-level cables. For headphones, the tip terminal is soldered to the left-channel lead; the ring terminal is soldered to the right-channel lead; and the sleeve terminal is soldered to the common lead. For balanced line-level cables, the sleeve terminal is soldered to the shield; the tip terminal is soldered to the hot red or white lead; and the ring terminal is soldered to the remaining lead.

Some mixers have insert jacks that are stereo phone jacks; each jack accepts a stereo phone plug. Tip is the send signal to an audio device input, ring is the return signal from the device output, and sleeve is ground.

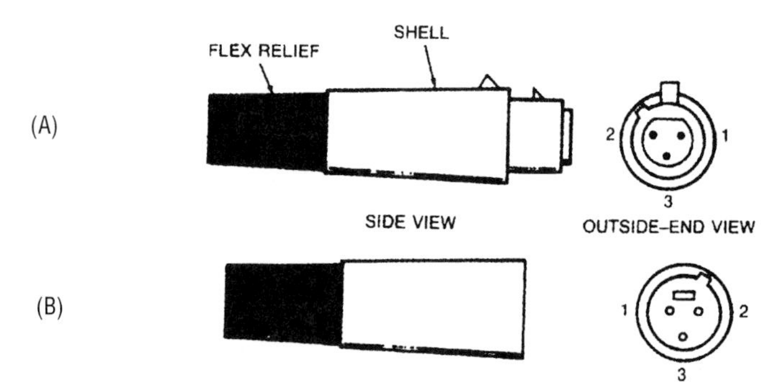

FIGURE 4.19
A 3-pin pro audio connector (XLR-type). (A) female; (B) male.

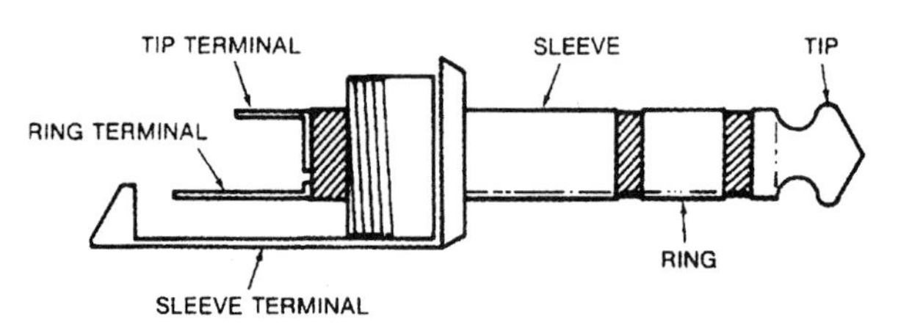

FIGURE 4.20
A stereo (TRS) phone plug.

Cable Types

Cables are also classified according to their function. In a studio, you'll use several types of cables: power, mic, MIDI, speaker, USB, FireWire, S/PDIF, TASCAM TDIF, Alesis Lightpipe, guitar cords, and patch cords.

A power cable, such as an AC extension cord or a power cord on a device, is made of three heavy-gauge wires surrounded by an insulating jacket. The wires are thick to handle high current without overheating.

A mic cable is usually 2-conductor shielded. It has two wires to carry the signal, surrounded by a fine-wire cylinder or shield that reduces hum pickup. On one end of the cable is a connector that plugs into the microphone, usually a female XLR-type. On the other end is either a 1/4-inch phone plug or a male XLR-type connector that plugs into your mixer or audio interface.

Rather than running several mic cables to your mixer or interface, you might consider using a snake: a box with multiple mic connectors, all wired to a thick multiconductor cable (Figure 4.21). A snake is especially convenient if you're running long cables to recording equipment from another room. It's essential for most on-location recording.

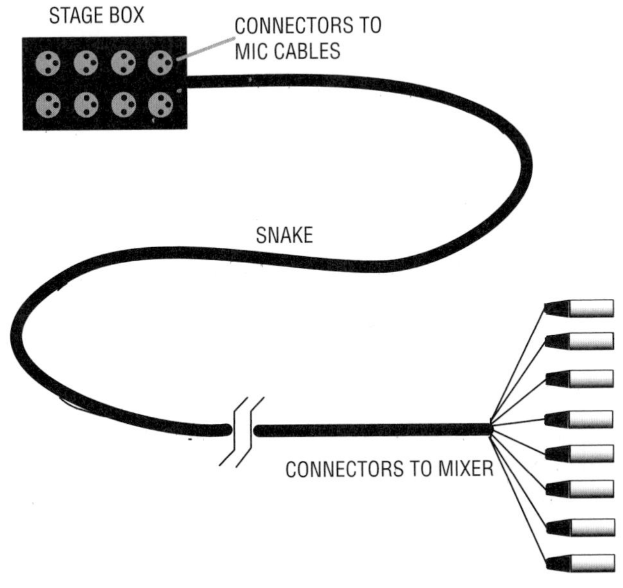

FIGURE 4.21
A stage box and snake.

Professional balanced equipment is interconnected with mic cable: 2-conductor shielded cable having a female XLR on one end and a male XLR on the other. Professional patch bays (described in the next section) use balanced cables with TRS phone plugs.

A MIDI cable uses a 5-pin DIN connector on each end of a 2-conductor shielded cable. The cable connects MIDI OUT to MIDI IN, or MIDI THRU to MIDI IN.

A speaker cable connects a power amp to each loudspeaker. Speaker cables are normally made of lamp cord (zip cord). To avoid wasting power, speaker cables should be as short as possible and should be heavy gauge (between 12 and 16 gauge). Number 12 gauge is thicker than 14; 14 is thicker than 16.

A USB cable or a FireWire cable connects a peripheral device (like an audio interface) to a computer. USB and FireWire are covered in detail in Chapter 13.

An S/PDIF cable transfers a digital signal from one device's S/PDIF output to another device's S/PDIF input. It uses a shielded unbalanced cable (ideally a 75-ohm RG59 video cable) with an RCA plug on each end.

A TASCAM TDIF cable is a multiconductor cable with a 25-pin D-sub connector on both ends. It's used to connect multiple digital-audio signals from TASCAM multitrack recorders to digital mixers or computer TDIF interfaces.

An Alesis Lightpipe cable is a fiber-optic cable with a TOSLINK connector on both ends. This cable is used to connect 8 channels of digital-audio signals from an Alesis multitrack recorder to a digital mixer or computer Lightpipe interface.

A guitar cord is constructed of a 1-conductor shielded cable with a 1/4-inch phone plug on each end. It connects a guitar amp to an electric instrument (electric guitar, electric bass, synthesizer, or an acoustic guitar with a pickup). Also, it connects an electric instrument to a direct box or to an instrument input.

Patch cords connect your recorder–mixer to external devices: an effects unit, a 2-track recorder, and a power amplifier. They also connect an analog mixer to the analog inputs and outputs of a multitrack recorder, usually as a snake that combines several cables. An unbalanced patch cord is made of 1-conductor shielded cable with either a 1/4-inch phone plug or a phono (RCA) connector on each end. A stereo patch cord is two patch cords joined side by side.

Rack/Patch Bay

You might want to mount your signal processors in a rack, an enclosure made of wood or metal with rails of screw holes for mounting equipment (Figure 4.22). You also might want to install a patch panel or patch bay: a group of connectors that are wired to equipment inputs and outputs. Using a patch bay and patch cords, you can change equipment connections easily. You also can bypass or patch around defective equipment. Note that a patch bay increases the chance of hum pickup slightly because of the additional cables and connectors. Racks and patch bays aren't essential, but they are convenient.

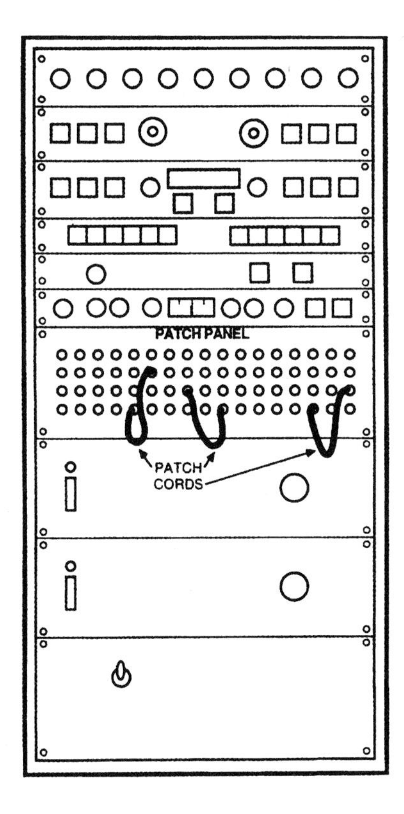

FIGURE 4.22
A rack and patch panel.

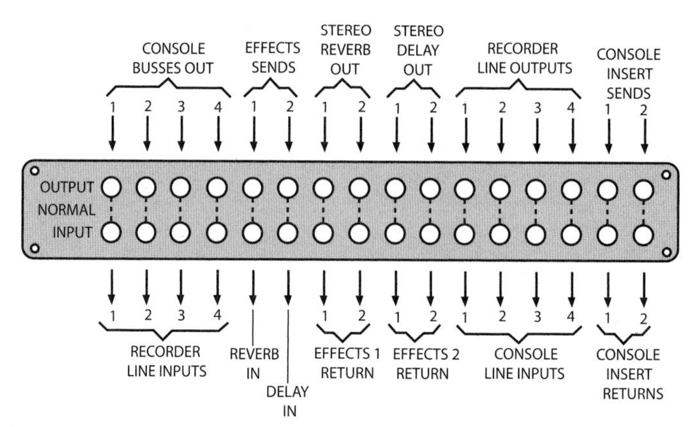

Figure 4.23
Some typical patch-bay assignments.

Figure 4.23 shows some typical patch-panel assignments.

Equipment Connections

The instruction manuals of your equipment tell how to connect each component to the others. In general, use cables that are as short as possible to reduce hum, but that are long enough to let you make changes.

Be sure to label all your cables on both ends according to what they plug into—for example, MIXER CH1 MONITOR OUT or ALESIS 3630 IN. If you change connections temporarily, or the cable becomes unplugged, you'll know where to plug it back in. A piece of masking tape folded over the end of the cable makes a stay-put label.

Typically, you follow this procedure to connect equipment (see Figure 4.24):

1. Plug the AC power cords of audio equipment and electric musical instruments into AC outlet strips fed from the same circuit breaker. Make sure that the sum of the equipment current ratings doesn't exceed the breaker's amp rating for that outlet. Plug the power amplifier or powered speakers into their own outlet on the same breaker so that they receive plenty of current.
2. Connect mic cables to mics.
3. Connect mic cables either to the snake junction box, or directly into mic inputs on a mixer, mic preamps, or audio

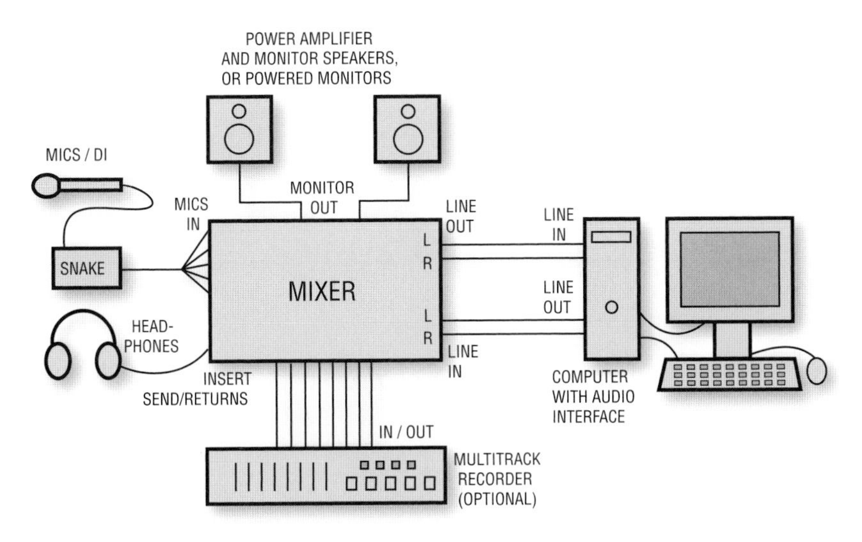

FIGURE 4.24
Typical connections for recording-studio equipment.

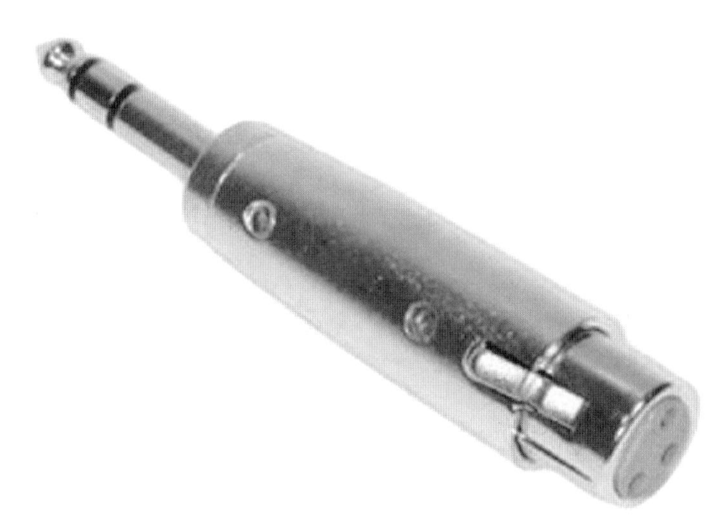

FIGURE 4.25
A female XLR to phone plug adapter.

interface. Plug the snake connectors into the mic inputs. If your mixer has phone-jack mic inputs, you may need to use an impedance-matching adapter (female XLR to phone) between the mic cable and the mic input jack (Figure 4.25).

4. Set the output volume of synthesizers and sound modules about three-quarters of the way up. Connect their audio outputs to instrument or line inputs on your mixer or audio interface. If this causes hum, use a direct box.

If you're recording a guitar direct, connect it either to (1) an instrument input on your mixer or audio interface, or (2) a direct box. Connect the XLR output of the direct box to a mic input on your mixer or audio interface.

5. If the mixer is a stand-alone unit (not part of a recorder–mixer), connect the mixer's stereo outputs to the inputs of an audio interface. If the mixer has a USB or FireWire connector, connect that to the mating connector in your computer.
6. Connect the audio interface audio outputs to the mixer's 2-track or tape inputs, or to a volume-control unit, or directly to powered speakers.
7. Connect the mixer's monitor outputs to the power-amp inputs. Connect the power-amp outputs to loudspeakers. Or if you're using powered (active) monitors, connect the mixer monitor outputs to the monitor-speaker inputs.
8. If the mixer doesn't have internal effects, connect the mixer aux-send connectors to effects inputs (not shown). Connect the effects outputs to the mixer aux-return or bus-in connectors.
9. If you're using a separate mixer and multitrack recorder, connect mixer bus 1 to recorder track 1 IN, connect bus 2 to track 2, and so on. Also connect the recorder's track 1 OUT to the mixer's line input 1, connect the track 2 OUT to the mixer's line input 2, and so on. As an alternative, connect insert jacks to multitrack inputs and outputs. At each insert plug, connect the tip (send) terminal to a track input; connect the ring (return) terminal to the same track's output.
10. If you have several headphones for musicians, connect the cue output to a small amplifier to drive their headphones. Or if the mixer's headphone signal is powerful enough, connect it to a box with several headphone jacks wired in parallel.

Semi-pro studio equipment with unbalanced connectors (usually phone or RCA) operates at a level called −10 or −10 dBV. Pro studio equipment with balanced connectors (XLR or TRS) works at +4 or +4 dBu. Check your equipment manuals to determine their input and output levels. When you connect devices that run at different

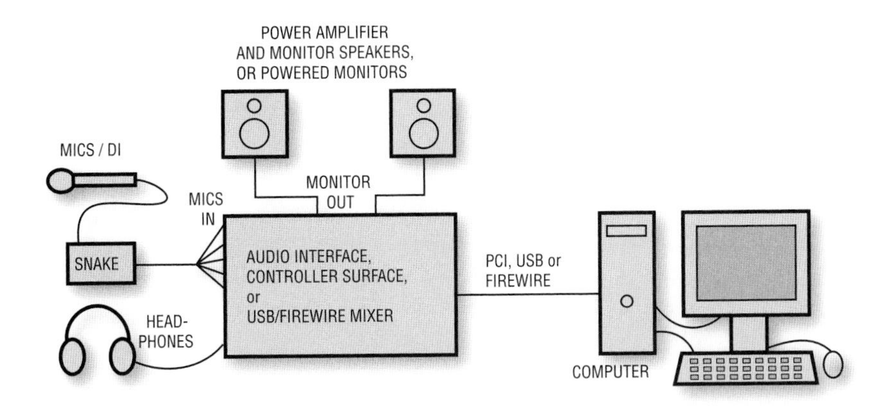

FIGURE 4.26
Typical connections in a DAW recording studio.

levels, set the +4/−10 switch on each unit to match the levels. If there is no such switch on either device, connect between them a +4/−10 converter box such as the Whirlwind LM2U line-level converter (www.whirlwindusa.com). Or try the cables shown in Appendix A (dB or Not dB), Figures A.3 and A.4.

Figure 4.26 shows typical connections in a DAW recording studio that uses an audio interface, controller surface, or USB/FireWire mixer (described in Chapter 13). Figure 4.27 shows a typical layout for a DAW recording studio.

A recorder–mixer studio can be quite simple. Plug mics into mic inputs, plug headphones into headphone jacks, and plug powered speakers into the mixer monitor outputs.

HUM PREVENTION

You patch in a piece of audio equipment, and there it is—HUM! It's a low-pitched tone or buzz. This annoying sound is a tone at 60 Hz (50 Hz in Europe) and multiples of that frequency.

Hum is caused mainly by

1. Cables picking up magnetic and electrostatic hum fields radiated by power wiring—especially if the cable shield connection is broken.
2. Ground loops. A ground loop is a closed loop of ground wires. It's made of two separated pieces of audio equipment each

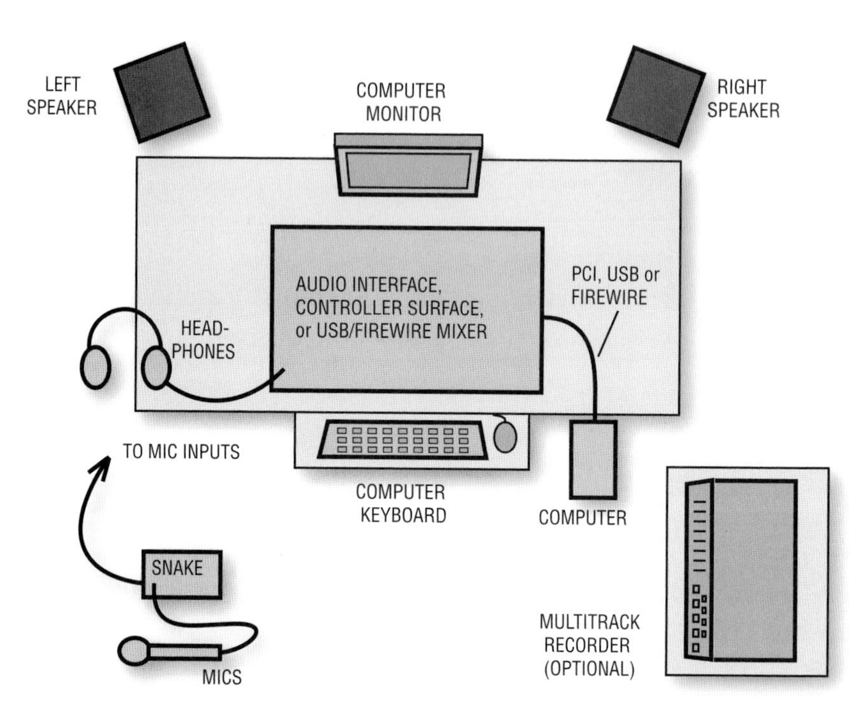

FIGURE 4.27
Typical layout of a DAW recording studio.

connected to power ground through a 3-prong power cord, and also connected to each other through a cable shield. The ground voltage may be slightly different at each piece of equipment, so a 50- or 60-Hz hum signal flows between the components along the cable shield.

These are the most important points to remember about hum prevention:

- To prevent ground loops, plug all equipment into outlet strips powered by the same AC outlet.
- Do not use an AC (electrical) 3-to-2 adapter to disconnect the power ground—this creates a safety hazard.
- Some power amps create hum if they don't get enough AC current. So connect the power amp (or powered speakers) AC plug to its own wall outlet socket—the same outlet that feeds the outlet strips for the recording equipment.
- If possible, use balanced cables going into balanced equipment. Balanced cables have XLR or TRS connectors and two conductors

surrounded by a shield. At both ends of the cable, connect the shield to a screw in the chassis, not to XLR pin 1. Or use audio gear whose XLR connectors are wired with pin 1 to chassis ground, not to signal ground.

- Transformer-isolate unbalanced connections. If that isn't an option, use the cable assemblies shown in Figures A.3 and A.4 in Appendix A.
- Don't use conventional SCR dimmers to change the studio lighting levels. Use Luxtrol® variable-transformer dimmers or multi-way incandescent bulbs instead.

Even if your system is wired properly, a hum or buzz may appear when you make a connection. Follow these tips to stop the hum:

- If the hum is coming from a direct box, flip its ground-lift switch.
- Check cables and connectors for broken leads and shields.
- Unplug all equipment from each other. Start by listening just to the powered monitor speakers. Connect a component to the system one at a time, and see when the hum starts.
- Remove audio cables from your devices and monitor each device by itself. It may be defective.
- Lower the volume on your power amp (or powered speakers), and feed them a higher level signal.
- Use a direct box instead of a guitar cord between instrument and mixer.
- To stop a ground loop when connecting two devices, connect between them a 1:1 isolation transformer, direct box or hum eliminator (such as Jensen CI-2RR, Behringer HD400, Rolls HE18 or Ebtech He2PKG). See Figures A.3 and A.4 in Appendix A.
- Make sure that the snake box isn't touching metal.
- To prevent accidental ground loops, don't connect XLR pin 1 to the connector shell except for permanent connections to equipment inputs and outputs.
- Try another mic.
- If you hear a hum or buzz from an electric guitar, have the player move to a different location or aim in a different direction. You might also attach a wire between the player's body and the guitar strings near the tailpiece to ground the player's body.
- Turn down the high-frequency EQ on a buzzing bass-guitar track.
- To reduce buzzing between notes on an electric-guitar track, apply a noise gate.

- Route mic cables and patch cords away from power cords; separate them vertically where they cross. Also keep recording equipment and cables away from computer monitors, power amplifiers, and power transformers.
- See Rane's excellent article on sound system interconnections at www.rane.com.

By following all these tips, you should be able to connect audio equipment without introducing any hum. Good luck!

REDUCING RADIO FREQUENCY INTERFERENCE (RFI)

RFI is heard as buzzing, clicks, radio programs, or "hash" in the audio signal. It's caused by CB transmitters, computers, lightning, radar, radio and TV transmitters, industrial machines, auto ignitions, stage lighting, and other sources. Many of the following techniques are the same used to reduce hum from other sources. To reduce RFI do the following:

- If you think that a speaker cable, mic cable, or patch cord is picking up RFI, wrap the cable several times around an RFI choke (available at Radio Shack). Put the choke near the device that is receiving audio.
- Install high-quality RFI filters in the AC power outlets. The cheap types available from local electronics shops are generally ineffective.
- If the shield is disconnected in a balanced cable connector, solder it to pin 1.
- If a mic is picking up RFI, solder a 0.047 microfarad capacitor between pin 1 and 2 and between pin 1 and 3 in the female XLR connector of the mic cable. Caution: this might cause high-frequency distortion with some mics.
- At equipment XLR mic inputs, use a cable connector with a tab that is connected to the metal shell. Use a wide strip of metal braid to connect this tab to pin 1. Make sure the connector makes metal-to-metal contact with the chassis.
- Periodically clean connector contacts with Caig Labs DeoxIT, or at least unplug and plug them in several times.

This chapter briefly covered the equipment and connectors for a recording studio. The rest of this book explains each piece of equipment in detail and tells how to use it for best results.

CHAPTER 5

Monitoring

One of the most exciting moments in recording comes when the finished mix is played over the studio monitor speakers. The sound is so clear you can hear every detail, and so powerful you can feel the deep bass throbbing in your chest.

You use the monitor system to listen to the output signals of the console, audio interface, or recorders. It consists of the console monitor mixer, the power amplifiers, loudspeakers, and the listening room. The power amplifier boosts the electrical power of the console signal to a sufficient level to drive a loudspeaker. The speaker converts the electrical signal into sound, and the listening-room acoustics affect the sound from the speaker.

A quality monitor system is a must if you want your mixes to sound good. The power amp and speakers tell you what you're doing to the recorded sound. According to what you hear, you adjust the mix and judge your mic techniques. Clearly, the monitor system affects the settings of many controls on your mixer, as well as your mic selection and placement. And all those settings affect the sound you're recording. So, using inadequate monitors can result in a poor-sounding product coming out of your studio.

It's important to use accurate speakers that have a flat frequency response. If your monitors are weak in the bass, you will tend to boost the bass in the mix until it sounds right over those monitors. But when that mix is played over speakers with a flatter response, it will sound too bassy because you boosted the bass on your mixer. So, using monitors with weak bass results in bassy recordings; using monitors with exaggerated treble results in dull recordings, and so on. In general, colorations in the monitors will be inverted in your mixdown recording.

That's why it's so important to use an accurate monitor system—one with a wide, smooth frequency response. Such a system lets you hear exactly what you recorded.

SPEAKER REQUIREMENTS

The requirements for an accurate studio monitor include:

- Wide, smooth frequency response. To ensure accurate tonal reproduction, the on-axis response of the direct sound should be ±4 dB or less from at least 40 Hz to 15 kHz. The low-frequency response of a small monitor speaker should extend to at least 70 Hz.
- Uniform off-axis response. The high-frequency output of a speaker tends to diminish off-axis. Ideally the response at 30 degrees off-axis should be only a few decibels down from the response on-axis. That way, a producer and engineer sitting side by side will hear the same amount of treble. Also, the treble won't change as the engineer moves around at the console.
- Good transient response. This is the ability of the speaker to accurately follow the attack and decay of musical sounds. If a speaker has good transient response, the bass guitar sounds tight, not boomy, and drum hits have sharp impact. Some speakers are designed so that the woofer and tweeter signals are aligned in time. This aids transient response.
- Clarity and detail. You should be able to hear small differences in the sonic character of instruments, and to sort them out in a complex musical passage.
- Low distortion. Low distortion is necessary because it lets you listen to the speaker for a long time without your ears hurting. A good spec might be: Total harmonic distortion under 3% from 40 Hz to 20 kHz at 90 dBSPL (sound pressure level).
- Sensitivity. This is the sound pressure level a speaker produces at 1 meter (m) when driven with 1 watt (W) of pink noise. Pink noise is random noise with equal energy per octave. This noise is either band-limited to the range of the speaker or is a one-third-octave band centered at 1 kHz. Sensitivity is measured in dB/W/m (dB sound pressure level per 1 W at 1 m). A spec of

93 dB/W/m is considered high; 85 dB/W/m is low. The higher the sensitivity, the less amplifier power you need to get adequate loudness.

- High output capability. This is the ability of a speaker to play loudly without burning out. You often need to monitor at high levels to hear quiet details in the music. Plus, when you record musicians who play loudly in the studio, it can be a letdown for them to hear a quiet playback. So you may need a maximum output of 110 dBSPL.

This formula calculates the maximum output of a speaker (how loud it can play):

$$\text{dBSPL} = 10\log(P) + S$$

where dBSPL is the sound pressure level at 1 m, P is the continuous power rating of the speaker in Watts, and S is the sensitivity rating in dB/W/m.

For example, if a speaker is rated at 100 W maximum continuous power, and its sensitivity is 94 dBSPL/W/m, its maximum output SPL is $10\log(100) + 94 = 114$ dBSPL (at 1 m from the speaker). The level at 2 m will be about 4 to 6 dB less.

NEARFIELD™ MONITORS

Many professional recording studios use large monitor speakers that have deep bass. However, they are expensive, heavy, and difficult to install, and they are affected by the acoustics of the control room. All studios, large or small, need a pair of Nearfield monitor speakers (Figure 5.1). A Nearfield monitor is a small, wide-range speaker typically using a cone woofer and dome-shaped tweeter. You place a pair of them about 3 or 4 feet apart, on stands just behind the console, about 3 or 4 feet from you. Nearfields are far more popular than large wall-mounted speakers.

This technique, developed by audio consultant Ed Long, is called Nearfield monitoring. Because the speakers are close to your ears, you hear mainly the direct sound of the speakers and tend to ignore the room acoustics. Plus, Nearfield monitors sound very clear and provide sharp stereo imaging. Some units have bass or treble tone

FIGURE 5.1
A Nearfield monitor speaker.

controls built in to compensate for the effects of speaker placement and room surfaces.

Nearfield monitors have enough bass to sound full when placed far from walls. Although most Nearfields lack deep bass, they can be supplemented with a subwoofer to reproduce the complete audio spectrum. Or you can check the mix occasionally with headphones that have deep bass.

Some Nearfields are in a satellite-subwoofer format. The two satellite speakers are small units, typically including a 4-inch woofer and 3/4-inch dome tweeter. The satellites are too small to produce deep bass, but that is handled by the subwoofer—a single cabinet with one or two large woofer cones. Typically, the subwoofer (sub) produces frequencies from 100 Hz down to 40 Hz or below. Because we don't localize sounds below about 100 Hz, all of the sound seems to come from the satellite speakers. The sub-satellite system is more complicated to set up than two larger speakers, but offers deeper bass.

FIGURE 5.2
Crown D-75A, an example of a power amplifier.

POWERED (ACTIVE) MONITORS

Most monitors have a power amplifier built in. You feed them a line-level signal (labeled MONITOR OUT) from your mixing console or audio interface. Most powered monitors are bi-amplified: they have one amplifier for the woofer and another for the tweeter. The advantages of bi-amplification include:

- Distortion frequencies caused by clipping the woofer power amplifier won't reach the tweeter, so there is less likelihood of tweeter burnout if the amplifier clips. In addition, clipping distortion in the woofer amplifier is made less audible.
- Intermodulation distortion is reduced.
- Peak power output is greater than that of a single amplifier of equivalent power.
- Direct coupling of amplifiers to speakers improves transient response—especially at low frequencies.
- Bi-amping reduces the inductive and capacitive loading of the power amplifier.
- The full power of the tweeter amp is available regardless of the power required by the woofer amp.

THE POWER AMPLIFIER

If your monitor speakers aren't powered, you need a power amplifier (Figure 5.2). It boosts your mixer's line-level signal to a higher power to drive the speakers.

How many watts of power do you need? The monitor speaker's data sheet gives this information. Look for the specification called "Recommended amplifier power." A power amp of 50 W per channel continuous is about the minimum for Nearfield monitors; 150 W is better. Too much power is better than too little, because an under-powered system is likely to clip or distort. This creates high frequencies that can damage tweeters.

A good monitor power amp has distortion under 0.05% at full power. It should have a high damping factor—at least 100—to keep the bass tight. The amp should be reliable. Look for separate level controls for left and right channels. The amplifier should have a clip or peak light that flashes when the amp is distorting.

SPEAKER CABLES AND POLARITY

When you connect the power amp to the speakers, use good wiring practice. Long or thin cables waste amplifier power by heating. So put the power amp(s) close to the speakers and use short cables with thick conductors—at least 16 gauge. The low resistance of these cables helps the power amplifier to damp the speaker motion and tighten the bass.

If you wire the two speakers in opposite polarity, one speaker's cone moves out while the other speaker's cone moves in. This causes vague stereo imaging, weak bass, and a strange sense of pressure on your ears. Be sure to wire the speakers in the same polarity as follows: In both channels, connect the amplifier positive (or red) terminal to the speaker positive (or red) terminal. Setting the correct polarity is also called "speaker phasing."

CONTROL-ROOM ACOUSTICS

The acoustics of the control room affect the sound of the speakers. At your ears, the sound waves reflected from the room surfaces combine with the direct sound from the speakers. Reflections that arrive within 20 to 65 milliseconds (msec) after the direct sound blend with the direct sound and affect the tonal balance you hear.

Chapter 3, under the heading Controlling Room Problems with Acoustic Treatments, described how to treat the acoustics of studios and control rooms to reduce excessive reverb, echo, standing waves, and noise.

Using Nearfield monitors makes the room acoustics less important, but it still helps to put absorbent material on the wall behind the monitors. This treatment improves the monitors' sound. Stereo imaging and depth are greatly improved, the sound is clearer, and the frequency response is flatter. The treatment will reduce boominess and ringing, and make transients sharper. Also, your recordings will translate better to other speakers.

While sitting in front of your mixer, have someone slide a mirror along the walls at eye height. Note the spots where you can see the monitor speakers in the mirror, and put some absorbent material there to absorb sound reflections. Repeat this procedure with the mirror on the ceiling between the monitors and your mix position.

If your control room is separate from the studio, the control room should be built to keep out sound from the studio. You want to hear only the sound from the monitors, not the live sound from the musicians. In a home studio, you might be able to achieve isolation simply by putting the control-room equipment in a room far removed from the studio, with the doors closed.

A control room built next to the studio needs good isolation. Use staggered-stud construction, seal holes between rooms, and install a double-pane window (mounted in rubber) between the control room and studio.

In some home or project studios, the control room is the same room as the studio. Because no isolation is used, the cost of building the studio is much less. You record while listening with headphones, and do the critical monitoring with speakers during playback and mixdown.

SPEAKER PLACEMENT

Once you have acquired the speakers and worked on the room acoustics, you can install the speakers. Mount them at ear height so the mixer doesn't block their sound. To prevent sound reflections off the mixing console, place the speakers on stands behind the console's meter bridge, rather than putting them on top. For best stereo imaging, align the speaker drivers vertically and mount the speakers symmetrically with respect to the side walls. Place the two speakers as far apart as you're sitting from them; aim them toward you, and sit exactly between them (Figure 5.3). To get the smoothest low-frequency response, put the speakers near the shorter wall, and sit forward of the halfway point in the room.

Try to position the monitors several feet from the nearest wall. Wall reflections can degrade the frequency response and stereo imaging. The closer to the wall the monitors are, the more bass you hear. In small rooms you might have to place the monitors against the

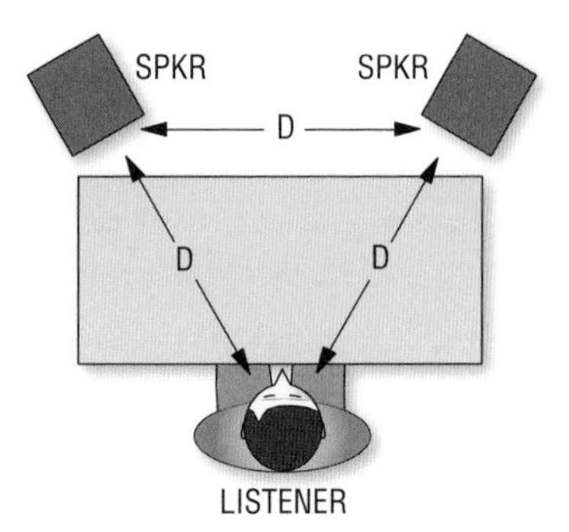

FIGURE 5.3
The recommended speaker/listener relationship for best stereo imaging.

wall, which will exaggerate the bass. But some monitors have a low-frequency attenuation switch to compensate.

USING THE MONITORS

You've treated the room acoustics, and you've connected and placed the speakers as described earlier. Now it's time to adjust the stereo balance.

1. Play a mono musical signal into an input channel on your mixer, and assign it to the stereo output channels 1 and 2.
2. Adjust the input channel's pan pot so that the signal reads the same on the stereo output channel 1 and 2 meters.
3. Place the two speakers the same distance from you.
4. Sit at the mixer exactly midway between the speakers. If you sit off-center, you will hear the image shifted toward one side. Listen to the image of the sound between the speaker pair. You should localize it midway between the monitors; that is, straight ahead.
5. If necessary, center the image by adjusting the left or right volume control on your power amp or powered monitors.

When you do a mixdown, try to keep the listening level around 85 dBSPL—a fairly loud home listening level. As discovered by Fletcher and Munson, we hear less bass in a program that is played quietly than in the same program played loudly. If you mix a program while monitoring at, say, 100 dBSPL, the same program will sound weak in the bass when heard at a lower listening level—which is likely in the home. So, programs meant to be heard at 85 dBSPL should be mixed and monitored at that level.

Loud monitoring also exaggerates the frequencies around 4 kHz. A recording mixed loud may sound punchy, but the same recording heard at a low volume will sound dull and lifeless.

Here's another reason to avoid extreme monitor levels: Loud sustained sound can damage your hearing or cause temporary hearing loss at certain frequencies. If you must do a loud playback for the musicians (who are used to high SPLs in the studio), protect your ears by wearing earplugs or leaving the room.

You can get a low-cost sound level meter from Radio Shack. Play a musical program at 0 VU or 0 dB on the mixer meters and adjust the monitor level to obtain an average reading of 85 dBSPL on the sound level meter. Mark the monitor-level setting.

Before doing a mix, you may want to play some familiar commercial CDs over your monitors to remind yourself what a good tonal balance sounds like. Listen to the amount of bass, midrange and treble, and try to match those in your mixes. But listen to several CDs because they vary.

While mixing, monitor the program alternately in stereo and mono to make sure there are no out-of-phase signals that cancel certain frequencies in mono. Also beware of center-channel buildup: instruments or vocals that are panned to center in the stereo mix sound 3 dB louder when monitored in mono than they do in stereo. That is, the balance changes in mono—the center instruments are a little too loud. To prevent this, don't pan tracks hard left and hard right. Bring in the side images a little so they will be louder in mono.

You'll mix the tracks to sound good on your accurate monitors. But also check the mix on small inexpensive speakers to see whether anything is missing or whether the mix changes drastically. Make sure that bass instruments are recorded with enough edge or harmonics to be audible on the smaller speakers. It's a good idea to make a CD copy of the mix for listening in a car, boom box, or compact stereo.

HEADPHONES

Compared to speakers, headphones have several advantages:

- They cost much less.
- There is no coloration from room acoustics.
- The tone quality is the same in different environments.
- They are convenient for on-location monitoring.
- It's easy to hear small changes in the mix.
- Transients are sharper due to the absence of room reflections.

Headphones have several disadvantages:

- They become uncomfortable after long listening sessions.
- Cheap headphones have inaccurate tone quality.
- Headphones don't project bass notes through your body.
- The bass response varies due to changing headphone pressure.
- The sound is in your head rather than out front.
- You hear no room reverberation, so you may add too much or too little artificial reverb.
- It's difficult to judge the stereo spread. Over headphones, panned signals tend to sound closer to center than the same signals heard over speakers. The same is true of stereo recordings made with a coincident pair of mics.

Because speakers sound different from headphones, it's best to do mixes over speakers. But check your mixes on headphones too because so many listeners use them with MP3 players.

If your monitor speakers are in the same room as your microphones, the mics pick up the sound of the speakers. This causes feedback or a muddy sound. In this case you must monitor only with headphones while recording or overdubbing, then monitor with speakers during playback or mixdown.

If you monitor with headphones as the musicians are playing, external sounds leak through the headphones and mask the sounds you're trying to monitor. That makes it hard to judge the sound quality you're picking up. Check it during playback. Closed-cup headphones or in-the-ear earphones provide the best isolation from outside sounds.

THE CUE SYSTEM

The cue system is a monitor system for musicians to use as they're recording. It includes some of the aux knobs in your mixer, a small power amplifier or headphone amp, a headphone connector box, and headphones. Musicians often can't hear each other well in the studio, but they can listen over headphones to hear each other in a good balance. Also, they can listen to previously recorded tracks while overdubbing.

Headphones for a cue system should be durable and comfortable. They should be closed-cup to avoid leakage into microphones. This is an ideal situation; open-air phones may work well enough. Also, the cue "phones" should have a smooth response to reduce listening fatigue, and should play loud without burning out. Make sure they are all the same model so each musician hears the same thing. A built-in volume control is convenient.

A suggested cue system is shown in Figure 5.4. Connect a power amp to an aux-send or monitor output of your mixer. The amp drives several resistor-isolated headphones, which are in parallel; that is, all

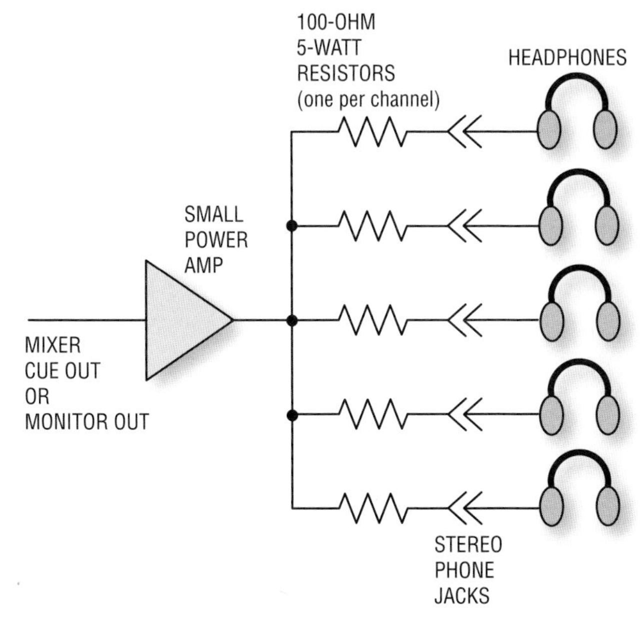

FIGURE 5.4
A cue system.

the tip (left-channel) terminals on the jacks are wired together, and all the ring (right-channel) terminals on the jacks are wired together.

If your mixer has a strong signal at its headphone jack, you can get by with four headphone jacks in a small metal box wired in parallel and connected to the mixer's headphone jack.

Although some consoles can provide several independent cue mixes, the ideal situation is to set up individual cue mixers near each musician. Then they can set their own cue mix and listening level. The inputs of these mixers are fed from the console output busses.

Suppose a vocalist sings into a microphone and hears that mic's signal over the cue headphones. If the singer's voice and the headphone's sound are opposite in polarity, the voice partially cancels or sounds funny in the headphones. Make sure that the voice and headphones are the same polarity.

Here's how. While talking into a mic and listening to it on headphones, reverse the ground and signal leads to the headphones connector. The position that gives the fullest, most solid sound in the headphones is correct.

All the headphones in your studio should be the same model, so that everyone will hear with correct polarity.

CONCLUSION

Ultimately, what you hear from the monitors influences your recording techniques and affects the quality of your recordings. So take the time to plan and adjust the control-room acoustics. Choose and place the speakers carefully. Monitor at proper levels and listen on several systems. You'll be rewarded with a monitor system you can trust.

CHAPTER 6

Microphones

What microphone is best for recording an orchestra? What's a good snare mic? Should the microphone be a condenser or dynamic, omni or cardioid?

You can answer these questions more easily once you know the types of microphones and understand their specs. First, it always pays to get a high-quality microphone. The mic is a source of your recorded signal. If that signal is noisy, distorted, or tonally colored, you'll be stuck with those flaws through the whole recording process. Better to get it right up front.

Even if you have a MIDI studio and get your sounds from samples or synthesizers, you still might need a good microphone for sampling, or to record vocals, sax, acoustic guitar, and so on.

A microphone is a transducer—a device that changes one form of energy into another. Specifically, a mic changes sound into an electrical signal. Your mixer amplifies and modifies this signal.

TRANSDUCER TYPES

Mics for recording can be grouped into three types depending on how they convert sound to electricity: dynamic, ribbon, or condenser.

A dynamic mic capsule, or transducer, is shown in Figure 6.1. A coil of wire attached to a diaphragm is suspended in a magnetic field. When sound waves vibrate the diaphragm, the coil vibrates in the magnetic field and generates an electrical signal similar to the

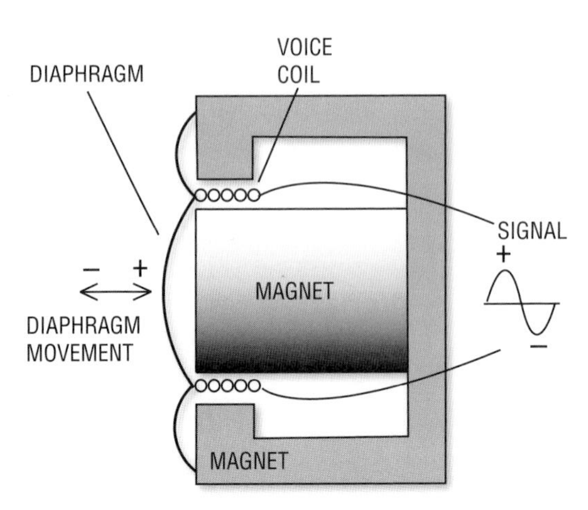

FIGURE 6.1
A dynamic transducer.

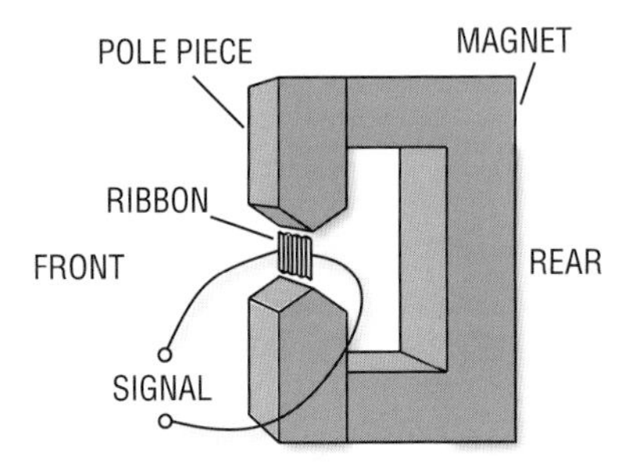

FIGURE 6.2
A ribbon transducer.

incoming sound wave. Another name for a dynamic mic is moving-coil mic, but this term is seldom used.

In a ribbon mic capsule, a thin metal foil or ribbon is suspended in a magnetic field (Figure 6.2). Sound waves vibrate the ribbon in the field and generate an electrical signal.

A condenser or capacitor mic capsule has a conductive diaphragm and a metal backplate placed very close together (Figure 6.3). They are charged with static electricity to form two plates of a capacitor. When sound waves strike the diaphragm, it vibrates. This varies the spacing between the plates. In turn, this varies the capacitance and generates a signal similar to the incoming sound wave. Because of

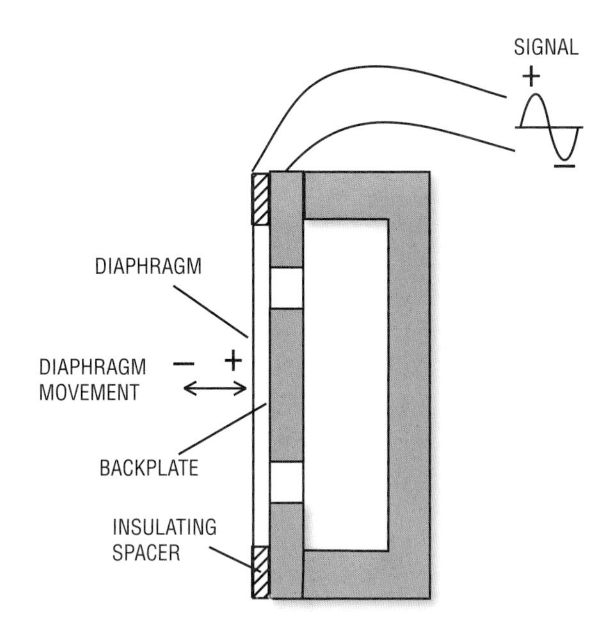

FIGURE 6.3
A condenser transducer.

its lower diaphragm mass and higher damping, a condenser mic responds faster than a dynamic mic to rapidly changing sound waves (transients).

Two types of condenser mic are true condenser and electret condenser. In a true condenser mic (externally biased mic), the diaphragm and backplate are charged with a voltage from a circuit built into the mic. In an electret condenser mic, the diaphragm and backplate are charged by an electret material, which is in the diaphragm or on the backplate. Electrets and true condensers can sound equally good, although some engineers prefer true condensers, which tend to cost more.

A condenser mic needs a power supply to operate, such as a battery or phantom power supply. Phantom power is 12 to 48 volts DC applied to pins 2 and 3 of the mic connector through two equal resistors. The microphone receives phantom power and sends audio signals on the same two conductors. Ground for the phantom power supply is through the cable shield. Nearly all mixing consoles and audio interfaces supply phantom power at their mic input connectors. You simply plug the mic into the mixer to power it.

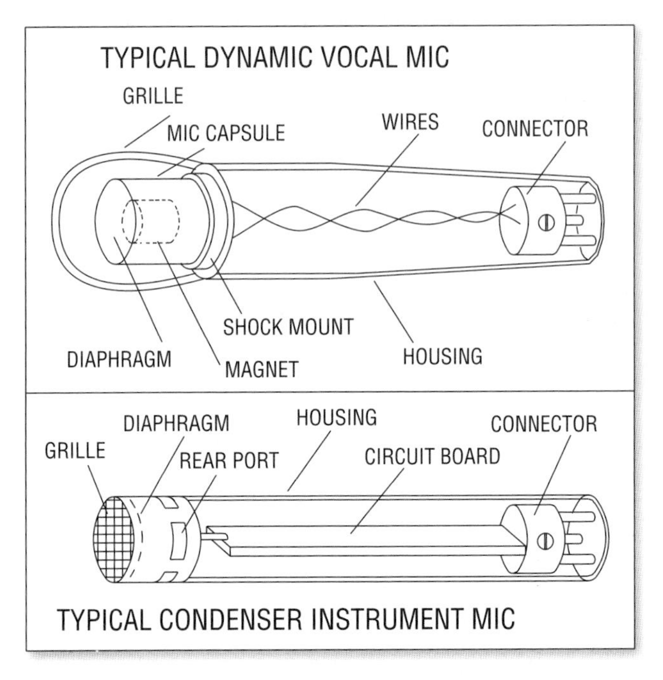

FIGURE 6.4
Inside a typical dynamic vocal mic and condenser instrument mic.

Dynamics and ribbons need no power supply. You can plug these types of mics into a phantom supply without damage because the voice coil or ribbon isn't connected to ground (unless they are accidentally shorted to the mic housing). Some newer ribbon mics have a built-in preamp that is phantom powered.

Figure 6.4 shows a cutaway view of a typical dynamic vocal mic and condenser instrument mic.

General Traits of Each Transducer Type

Condenser

- Wide, smooth frequency response
- Detailed sound, extended highs
- Omni type has excellent low-frequency response
- Transient attacks sound sharp and clear
- Preferred for acoustic instruments, cymbals, studio vocals
- Can be miniaturized

Dynamic

- Tends to have rougher response, but still quite usable
- Rugged and reliable
- Handles heat, cold, and high humidity
- Handles high volume without distortion
- Preferred for guitar amps and drums
- If flat response, can take the "edge" off woodwinds and brass

Ribbon

- Prized for its warm, smooth tone quality
- Delicate
- Complements digital recording

There are exceptions to the tendencies listed above. Some dynamics have a smooth, wide-range frequency response. Some condensers are rugged and handle high SPLs. It depends on the specs of the particular mic. *Audio clip 13 at www.elsevierdirect.com/companions/9780240811444 demonstrates the sound of each transducer type.*

POLAR PATTERN

Microphones also differ in the way they respond to sounds coming from different directions. An omnidirectional microphone is equally sensitive to sounds arriving from all directions. A unidirectional mic is most sensitive to sound arriving from one direction—in front of the mic—but softens sounds entering the sides or rear of the mic. A bidirectional mic is most sensitive to sounds arriving from two directions—in front of and behind the mic—but rejects sounds entering the sides.

There are three types of unidirectional patterns: cardioid, supercardioid, and hypercardioid. A mic with a cardioid pattern is sensitive to sounds arriving from a broad angle in front of the mic. It's about 6 dB less sensitive at the sides, and about 15 to 25 dB less sensitive in the rear. The supercardioid pattern is 8.7 dB less sensitive at the sides and has two areas of least pickup at 125 degrees away from the front. The hypercardioid pattern is 12 dB less sensitive at the sides and has two areas of least pickup at 110 degrees away from the front.

To hear how a cardioid pickup pattern works, talk into a cardioid mic from all sides while listening to its output. Your reproduced voice is loudest when you talk into the front of the mic, and softest when you talk into the rear. *Play audio clip 14 at www.elsevierdirect.com/companions/9780240811444.*

The super- and hypercardioid mics reject sound from the sides more than the cardioid. They are more directional, but they pick up more sound from the rear than the cardioid.

A microphone's polar pattern is a graph of its sensitivity versus the angle at which sound approaches the mic. The polar pattern is plotted on polar graph paper. Sensitivity is plotted as distance from the origin. Figure 6.5 shows various polar patterns.

Traits of Each Polar Pattern

Omnidirectional

- All-around pickup
- Most pickup of room reverberation (*play audio clip 14 at www.elsevierdirect.com/companions/9780240811444*)
- Not much isolation unless you mike close
- Low sensitivity to pops (explosive breath sounds)
- Low handling noise
- No up-close bass boost (proximity effect)
- Extended low-frequency response in condenser mics—great for pipe organ or bass drum in an orchestra or symphonic band
- Lower cost in general

Unidirectional (cardioid, supercardioid, hypercardioid)

- Selective pickup
- Rejection of room acoustics, background noise, and leakage
- Good isolation—good separation between tracks
- Up-close bass boost (except in mics that have holes in the handle)
- Better gain-before-feedback in a sound-reinforcement system
- Coincident or near-coincident stereo miking (explained in Chapter 7)

Cardioid

- Broad-angle pickup of sources in front of the mic
- Maximum rejection of sound approaching the rear of the mic
- Most popular pattern

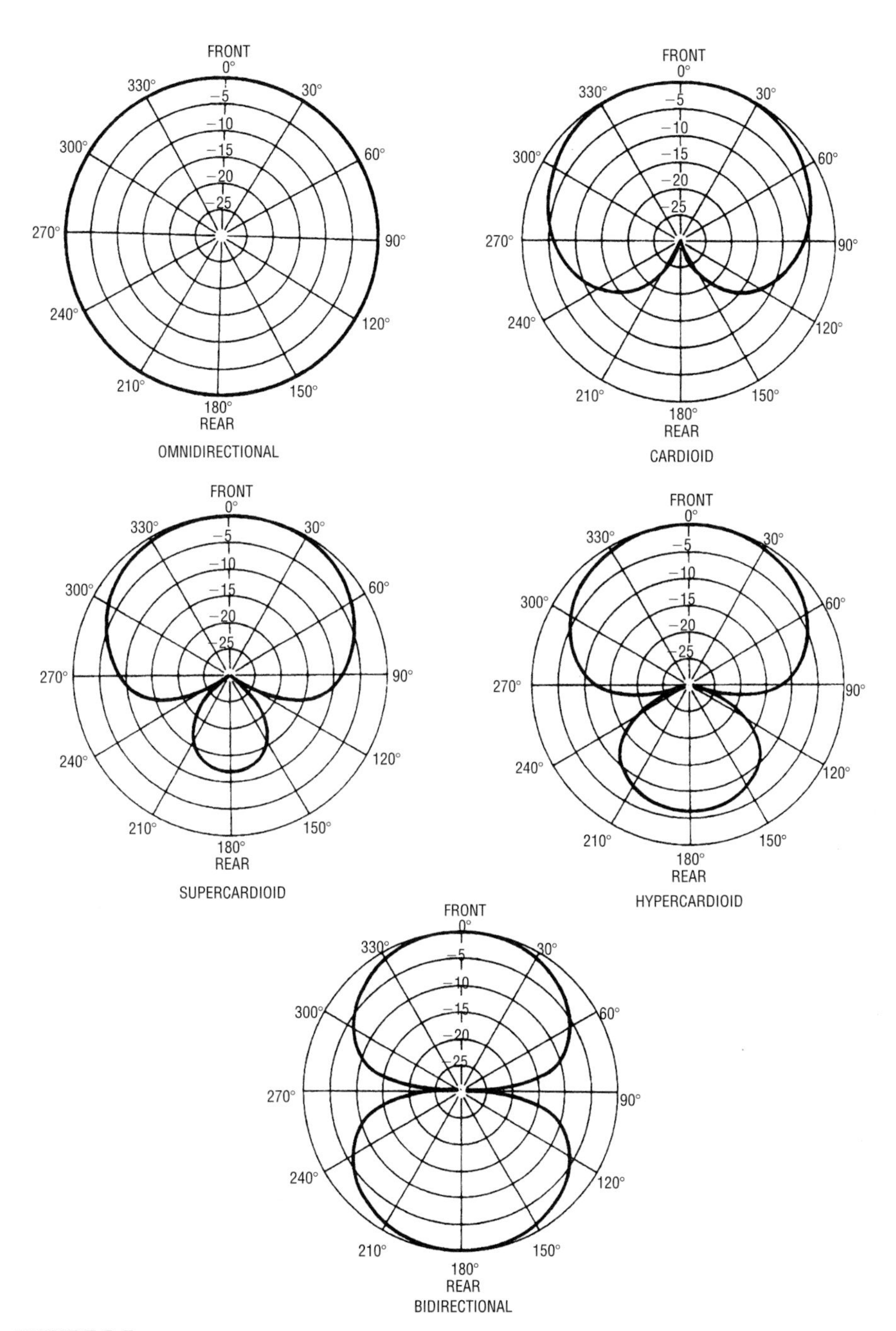

FIGURE 6.5
Various polar patterns. Sensitivity is plotted vs. angle of sound incidence.

Supercardioid

- Maximum difference between front hemisphere and rear hemisphere pickup (good for stage-floor miking)
- More isolation than a cardioid
- Less reverb pickup than a cardioid

Hypercardioid

- Maximum side rejection in a unidirectional mic
- Maximum isolation—maximum rejection of reverberation, leakage, feedback, and background noise

Bidirectional

- Front and rear pickup, with side sounds rejected (for across-table interviews or two-part vocal groups, for example)
- Maximum isolation of an orchestral section when miked overhead
- Blumlein stereo miking (two bidirectional mics crossed at 90 degrees)

In a good mic, the polar pattern should be about the same from 200 Hz to 10 kHz. If not, you'll hear off-axis coloration: the mic will have a different tone quality on and off axis. Small-diaphragm mics tend to have less off-axis coloration than large-diaphragm mics.

You can get either the condenser or dynamic type with any kind of polar pattern (except bidirectional dynamic). Ribbon mics are either bidirectional or hypercardioid. Some condenser mics come with switchable patterns. Note that the shape of a mic doesn't indicate its polar pattern.

If a mic is end-addressed, you aim the end of the mic at the sound source. If a mic is side-addressed, you aim the side of the mic at the sound source. Figure 6.6 shows a typical side-addressed condenser mic with switchable polar patterns.

Boundary mics that mount on a surface have a pattern that is half-omni (hemispherical), half-supercardioid, or half-cardioid (like an apple sliced in half through its stem). The boundary mounting makes the mic more directional so it picks up less room acoustics.

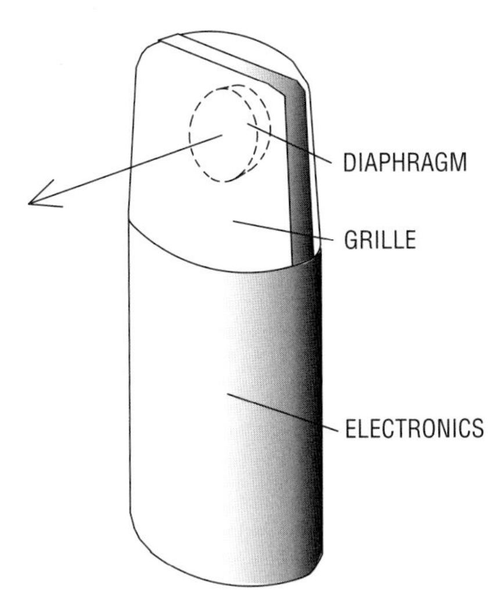

FIGURE 6.6
A typical multi-pattern mic that is side-addressed.

FREQUENCY RESPONSE

As with other audio components, a microphone's frequency response is the range of frequencies that it will reproduce at an equal level (within a tolerance, such as ±3 dB).

The following is a list of sound sources and the microphone frequency response that is adequate to record the source with high fidelity. A wider range response works, too.

- Most instruments: 80 Hz to 15 kHz
- Bass instruments: 40 Hz to 9 kHz
- Brass and voice: 80 Hz to 12 kHz
- Piano: 40 Hz to 12 kHz
- Cymbals and some percussion: 300 Hz to 15 or 20 kHz
- Orchestra or symphonic band: 40 Hz to 15 kHz

If possible, use a mic with a response that rolls off below the lowest fundamental frequency of the instrument you're recording. For example, the frequency of the low-E string on an acoustic guitar is about 82 Hz. A mic used on the acoustic guitar should roll off below that frequency to avoid picking up low-frequency noise such as rumble from trucks and air conditioning. Some mics have a built-in low-cut switch for this purpose. Or you can filter out the unneeded lows at your mixer.

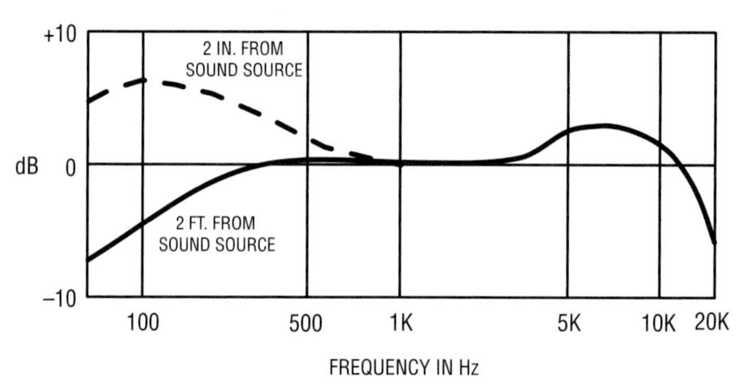

FIGURE 6.7
An example of the frequency response of a microphone with proximity effect and a presence peak around 5 kHz.

A frequency-response curve is a graph of the mic's output level in dB at various frequencies. The output level at 1 kHz is placed at the 0 dB line on the graph, and the levels at other frequencies are so many decibels above or below that reference level.

The shape of the response curve suggests how the mic sounds at a certain distance from the sound source. (If the distance isn't specified, it's probably 2 to 3 feet.) For example, a mic with a wide, flat response reproduces the fundamental frequencies and harmonics in the same proportion as the sound source. So a flat-response mic tends to provide accurate, natural reproduction at that distance.

A rising high end or a "presence peak" around 5 to 10 kHz sounds more crisp and articulate because it emphasizes the higher harmonics (Figure 6.7). *Play audio clip 15 at www.elsevierdirect.com/companions/9780240811444.* Sometimes this type of response is called tailored or contoured. It's popular for guitar amps and drums because it adds punch and emphasizes attack. Some microphones have switches that alter the frequency response.

Most uni- and bidirectional mics boost the bass when used within a few inches of a sound source. You've heard how the sound gets bassy when a vocalist sings right into the mic. This low-frequency boost related to close mic placement is called the proximity effect, and it's often plotted on the frequency-response graph. Omni mics have no proximity effect; they sound tonally the same at any distance.

The warmth created by proximity effect adds a pleasing fullness to drums. In most recording situations, though, the proximity effect lends an unnatural boomy or bassy sound to the instrument or voice picked up by the mic. Some mics—multiple-D or variable-D types—are designed to reduce it. These types have holes or slots in the mic handle. Some mics have a bass-rolloff switch to compensate for the bass boost. Or you can roll off the excess bass with your mixer's equalizer until the sound is natural. By doing so, you also reduce low-frequency leakage picked up by the microphone.

Note that mic placement can greatly affect the recorded tone quality. A flat-response mic doesn't always guarantee a natural sound because mic placement has such a strong influence. Tonal effects of mic placement are covered in Chapter 7.

IMPEDANCE (Z)

This spec is the mic's effective output resistance at 1 kHz. A mic impedance between 150 and 600 ohms is low; 1000 to 4000 ohms is medium; and above 25 kilohms is high.

Always use low-impedance mics. If you do, you can run long mic cables without picking up hum or losing high frequencies. The input impedance of a mixer mic input is about 1500 ohms. If it were the same impedance as the mic, about 250 ohms, the mic would "load down" when you plug it in. Loading down a mic makes it lose level, distort, or sound thin. To prevent this, a mic input has an impedance much higher than that of the microphone. But it's still called a low-Z input.

More information on impedance is in Appendix C.

MAXIMUM SPL

To understand this spec, first we need to understand sound pressure level (SPL). It is a measure of the intensity of a sound. The quietest sound we can hear, the threshold of hearing, is 0 dBSPL. Normal conversation at 1 foot measures about 70 dBSPL; painfully loud sound is above 120 dBSPL.

If the maximum SPL spec is 125 dBSPL, the mic starts to distort when the instrument miked is putting out 125 dBSPL at the mic. A maximum SPL spec of 120 dB is good, 135 dB is very good, and 150 dB is excellent.

Dynamic mics tend not to distort, even with very loud sounds. Some condensers are just as good. Some have a pad you can switch in to prevent distortion in the mic circuitry. Because a mic pad reduces signal-to-noise ratio (S/N), use it only if the mic distorts.

SENSITIVITY

This spec tells how much output voltage a mic produces when driven by a certain SPL. A high-sensitivity mic puts out a stronger signal (higher voltage) than a low-sensitivity mic when both are exposed to an equally loud sound.

A low-sensitivity mic needs more mixer gain than a high-sensitivity mic. More gain usually results in more noise. When you record quiet music at a distance (classical guitar, string quartet), use a mic of high sensitivity to override mixer noise. When you record loud music or mike close, sensitivity matters little because the mic signal level is well above the mixer noise floor; that is, the S/N is high. Listed below are typical sensitivity specs for three transducer types:

- Condenser: 5.6 mV/Pa (high sensitivity)
- Dynamic: 1.8 mV/Pa (medium sensitivity)
- Ribbon or small dynamic: 1.1 mV/Pa (low sensitivity)

The louder the sound source, the higher the signal voltage the mic puts out. A very loud instrument, such as a kick drum or guitar amp, can cause a microphone to generate a signal strong enough to overload the mic preamp in your mixer. That's why most mixers have pads or input-gain controls—to prevent preamp overload from hot mic signals.

SELF-NOISE

Self-noise or equivalent noise level is the electrical noise or hiss a mic produces. It's the dBSPL of a sound source that would produce the same output voltage that the noise does.

Usually the self-noise spec is A-weighted. That means the noise was measured through a filter that makes the measurement correlate more closely with the annoyance value. The filter rolls off low and high frequencies to simulate the frequency response of the ear.

An A-weighted self-noise spec of 14 dBSPL or less is excellent (quiet); 21 dB is very good, 28 dB is good; and 35 dB is fair—not good enough for quality recording.

Because a dynamic mic has no active electronics to generate noise, it has very low self-noise (hiss). So most spec sheets for dynamic mics don't specify self-noise.

S/N (SIGNAL-TO-NOISE RATIO)

This is the difference in decibels between the mic's sensitivity and its self-noise. The higher the SPL of the sound source at the mic, the higher the S/N. Given an SPL of 94 dB, an S/N spec of 74 dB is excellent; 64 dB is good. The higher the S/N ratio, the cleaner (more noise-free) the signal, and the greater the "reach" of the microphone.

Reach is the clear pickup of quiet, distant sounds due to high S/N. Reach isn't specified in data sheets because any mic can pick up a source at any distance if the source is loud enough. For example, even a cheap mic can reach several miles if the sound source is a thunderclap.

POLARITY

The polarity spec relates the polarity of the electrical output signal to the acoustic input signal. The standard is "pin 2 hot." That is, the mic produces a positive voltage at pin 2 with respect to pin 3 when the sound pressure pushes the diaphragm in (positive pressure).

Be sure that your mic cables don't reverse polarity. On both ends of each cable, the wiring should be pin 1 shield, pin 2 red, pin 3 white or black. Or the wiring on both ends should be pin 1 shield, pin 2 white, pin 3 black. If some mic cables are correct polarity and some are reversed, and you mix their mics to mono, the bass may cancel.

MICROPHONE TYPES

The following sections describe several types of recording mics.

Large-Diaphragm Condenser Microphone

This is a condenser microphone, usually side-addressed, with a diaphragm 1 inch or larger in diameter (Figure 6.6). It generally has very good low-frequency response and low self-noise. Common uses are studio vocals and acoustic instruments. Examples include: AKG C12VR, C414, Perception 100 and 200, C2000B and C3000B; Audio-Technica AT2020/3035/4040, Audix SCX25, Blue Blueberry,

CAD Equitek Series and M177, DPA 4041, Lawson L47MP MKII and L251, Manley Gold Reference, Neumann U87, U47 and TLM 103; Soundelux Elux 251, Sterling Audio ST51, ADK various models, SE Electronics various models, Shure KSM Series, MXL V67G, V69, 900, 2001, 2003 and 2006; Rode NT1A, Studio Projects B and C Series, Samson CL7 and C01, Nady SCM 950 and 100, Violet Flamingo, M-Audio Solaris, Sputnik, Luna and Nova; and Behringer B1 and B2.

Small-Diaphragm Condenser Microphone

This is a stick-shaped or "pencil" condenser microphone, usually cardioid and end-addressed, with a diaphragm under 1 inch in diameter (Figure 6.4). It generally has very good transient response and detail, making it a fine choice for close miking acoustic instruments—especially cymbals, acoustic guitar, and piano. Examples include: AKG C 451 B; Audio-Technica AT 3031 and AT 4051a; Audix SCX1, ADX50, and ADX51; Berliner CM-33, CAD Equitek e60; M-Audio Pulsar; Samson C02; Crown CM-700; DPA 4006; Mojave MA-100, Neumann KM 184; Sennheiser e614 and MKH50; Shure KSM109/SL, KSM137/Sl and SM81; MXL 600 and 603S; Behringer B5 with cardioid capsule; and Studio Projects C4.

Dynamic Instrument Microphone

This is a stick-shaped dynamic microphone, end-addressed (Figure 6.4). Although it may have a flat response, it generally has a presence peak and some low-frequency rolloff to prevent boominess when used up close. It's often used on drums and guitar amps. Examples include: Shure SM57, AKG D112 (kick drum), Audio-Technica AT AE2500 (kick), Electro-Voice N/D868 (kick), Heil PR40 (kick), Audix D1 through D6 and I-5, and Sennheiser MD421, e604 and e602 (kick).

Live-Vocal Microphone

This unidirectional mic is shaped like an ice-cream cone because of its large grille used to reduce breath pops. It can be a condenser, dynamic, or ribbon type, and it usually has a presence peak and some low-frequency rolloff. Examples include: Shure SM58, Beta 58, SM85, SM86, SM87A and KSM9; AKG D3800, C 535 EB, C5, ELLE C, and D 5/D 5 S; Audix OM5, Beyerdynamic M88 TG, Crown CM-200A, EV N-Dym Series, and Neumann KMS104 and KMS105.

Ribbon Microphone

This mic can be side- or end-addressed. It generally is used wherever you want a warm, smooth tone quality (sometimes with reduced highs). Examples include: models by Beyerdynamic, Coles, Royer, Cascade, Blue, and AEA.

Boundary Microphone

Boundary mics are designed to be used on surfaces. Tape them to the underside of a piano lid, or tape them to the wall for pickup of room ambience. They can be used on hard baffles between instruments, or on panels to make the mics directional. A boundary mic uses a mini condenser mic capsule mounted very near a sound-reflecting plate or boundary (Figure 6.8). Because of this construction, the mic picks up direct sound and reflected sound at the same time, in-phase at all frequencies. So you get a smooth response free of phase cancellations. A conventional mic near a surface sounds colored; a boundary mic on a surface sounds natural. Examples include: AKG C 562 BL, Audio-Technica AT 841a, Beyerdynamic MPC 22, Crown PZM-30D and PZM-6D, and Shure Beta 91.

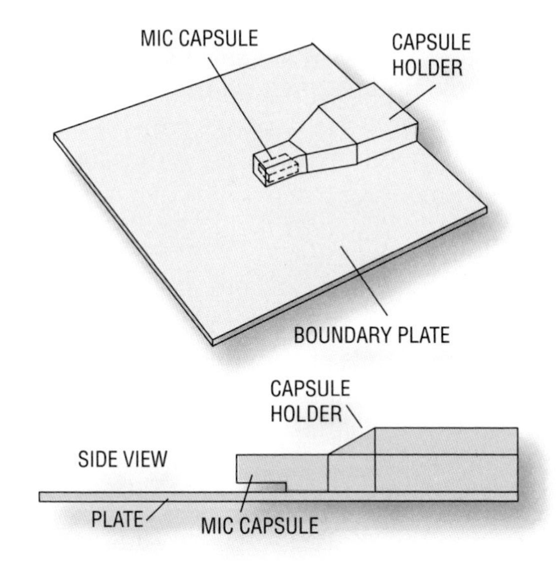

FIGURE 6.8
Typical PZM construction.

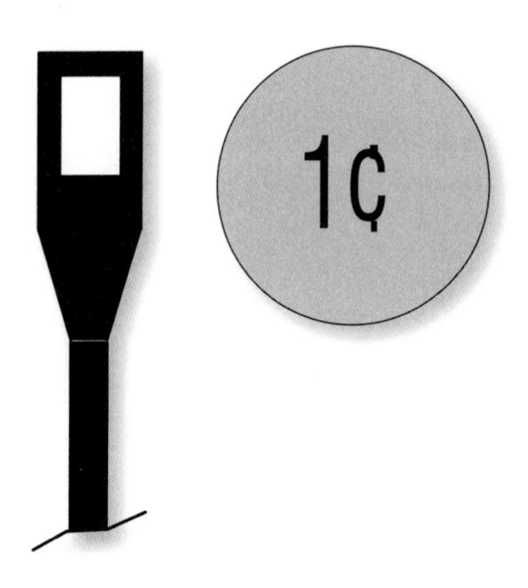

FIGURE 6.9
A mini mic is the size of a penny.

Other benefits are a wide, smooth frequency response free of phase cancellations, excellent clarity and reach, and the same tone quality anywhere around the mic. The polar pattern is half-omni or hemispherical. Some boundary mics have a half-cardioid or half-supercardioid polar pattern. They work great on a conference table, or near the front edge of a stage floor to pick up drama or musicals.

Miniature Microphone

Mini condenser mics can be attached to drum rims, flutes, horns, acoustic guitars, fiddles, and so on. Their tone quality is about as good as larger studio microphones and the price is relatively low. With these tiny units you can mike a band in concert without cluttering the stage with boom stands (Figure 6.9), or you can mike a whole drum set with two or three of these. Although you lose individual control of each drum in the mix, the cost is low and the sound is quite good with some bass and treble boost. Compared to large mics, mini mics tend to have more noise (hiss) in distant-miking applications. A lavalier mic is a mini mic worn on the chest to pick up speech from a newscaster or a wandering lecturer. Examples include: AKG Micro Mic Series, Shure Beta 98S, Audix M1245 and Micro-D, Countryman Isomax B6, Crown GLM-100, DPA 4060, and Sennheiser e608.

FIGURE 6.10
Audio-Technica AT825, an example of a stereo microphone.

Stereo Microphone

A stereo microphone combines two directional mic capsules in a single housing for convenient stereo recording (Figure 6.10). Simply place the mic a suitable distance and height from the sound source, and you'll get a stereo recording with little fuss. Examples include: AKG C426 B Comb, Audio-Technica AT825 and AT822, Crown SASS-P MKII, Neumann SM69, Shure VP88, Nady RSM-2, AEA R88, and Royer SF-12.

Because there is no spacing between the mic capsules, there also is no delay or phase shift between their signals. Coincident stereo microphones are mono-compatible—the frequency response is the same in mono and stereo—because there are no phase cancellations if the two channels are combined.

Digital Microphone

This condenser microphone has a built-in A/D converter. It is usually side-addressed, has a large diaphragm, has a flat response, and very low self-noise. Its output is a digital signal, which is immune to picking up hum. Examples include: Beyerdynamic MCD 100 and Neumann Solution-D.

Headworn Microphone

This microphone is used for a live performance that might be recorded. It's a small condenser mic worn on the head, either omni- or unidirectional. The headworn mic allows the performer freedom of movement on stage. Some models provide excellent gain before feedback and isolation. Examples include: AKG C 420, Audio-Technica ATM73, Countryman Isomax E6, DPA 4065, and Crown CM-311A.

MICROPHONE SELECTION

Table 6.1 is a guide to choosing a mic based on your requirements.

Suppose you want to record a grand piano playing with several other instruments. You need the microphone to reduce leakage. Table 6.1 recommends a unidirectional mic or an omni mic up close. For this particular piano, you also want a natural sound, for which the table suggests a mic with a flat response. You want a detailed sound, so a condenser mic is the choice. A microphone with all these characteristics is a flat-response, unidirectional condenser mic. If you're miking close to a surface (the piano lid), a boundary mic is recommended.

Now suppose you're recording an acoustic guitar on stage, and the guitarist roams around. This is a moving sound source, for which the table recommends a mini mic attached to the guitar. Feedback and leakage might not be a problem because you're miking close, so you can use an omni mic. Thus, an omni condenser mic is a good choice for this application.

For a home studio, a suggested first choice is a cardioid condenser mic with a flat frequency response. This type of mic is especially good for studio vocals, cymbals, percussion, and acoustic instruments. Remember that the mic needs a power supply to operate, such as a battery or phantom power supply.

Your second choice of microphone for a home studio is a cardioid dynamic microphone with a presence peak in the frequency response. This type is good for drums and guitar amps. I recommend cardioid over omni for a home studio. The cardioid pattern rejects the leakage, background noise, and room reverb often found in home studios. An omni mic, however, can do that, too, if you mike

Table 6.1 Mic Application Guide

Requirement	Characteristic
Natural, smooth tone quality	Flat frequency response
Bright, present tone quality	Presence peak (emphasis around 5 kHz)
Extended lows	Omni condenser or dynamic with good low-frequency response
Extended highs (detailed sound)	Condenser
Reduced "edge" or detail	Dynamic or ribbon
Boosted bass up close	Directional mic
Flat bass response up close	Omni mic, or directional mic with sound holes in the handle
Reduced pickup of leakage, feedback, and room acoustics	Directional mics
Enhanced pickup of room acoustics	Omni mics
Miking close to a surface, even coverage of moving sources or large sources, inconspicuous mic	Boundary mic
Coincident or near-coincident stereo (see Chapter 18)	Stereo mic
Extra ruggedness	Dynamic mic
Reduced handling noise	Omni mic, or unidirectional with shock mount
Reduced breath popping	Omni mic, or unidirectional with pop filter
Distortion-free pickup of very loud sounds	Condenser with high maximum SPL spec, or dynamic
Low self-noise, high sensitivity, noise-free pickup of quiet sounds	Large-diaphragm condenser mic

close enough. Also, omni mics tend to provide a more natural sound at lower cost, and they have no proximity effect.

MIC ACCESSORIES

There are many devices used with microphones to route their signals or to make them more useful. These include pop filters, stands and booms, shock mounts, cables and connectors, stage boxes and snakes, and splitters.

Pop Filter

A much needed accessory for a vocalist's microphone is a pop filter or windscreen. It usually is a foam "sock" that you put over the mic. Some microphones have pop filters or ball-shaped grilles built in.

Why is it needed? When a vocalist sings a word starting with "p," "b," or "t" sounds, a turbulent puff of air is forced from the mouth. A microphone placed close to the mouth is hit by this air puff, resulting in a thump or little explosion called a pop. The windscreen reduces this problem.

The best type of pop filter is a nylon screen in a hoop, or a perforated-metal disk, placed a few inches from the mic.

You can also reduce pop by placing the mic above or to the side of the mouth, or by using an omni mic. *Audio clip 16 at www.elsevierdirect.com/companions/9780240811444 demonstrates how a pop filter or mic placement can prevent breath pops.*

Stands and Booms

Stands and booms hold the microphones and let you position them as desired. A mic stand has a heavy metal base that supports a vertical pipe. At the top of the pipe is a rotating clutch that lets you adjust the height of a smaller telescoping pipe inside the large one. The top of the small pipe has a standard 5/8-inch 27 thread, which screws into a mic stand adapter.

A boom is a long horizontal pipe that attaches to the vertical pipe. The angle and length of the boom are adjustable. The end of the boom is threaded to accept a mic stand adapter, and the opposite end is weighted to balance the weight of the microphone.

Shock Mount

A shock mount holds a mic in a resilient suspension to isolate the mic from mechanical vibrations, such as floor thumps and mic-stand bumps. Many mics have an internal shock mount which isolates the mic capsule from its housing; this reduces handling noise as well as stand thumps.

Cables and Connectors

Mic cables carry the electrical signal from the mic to the mixing console, mic preamp, or recorder. With low-impedance mics, you can use hundreds of feet of cable without hum pickup or high-frequency loss. Some mics have a permanently attached cable for convenience and low cost; others have a connector in the handle to accept a separate mic cable. The second method is preferred for serious recording because if the cable breaks, you have to repair or replace only the cable, not the whole microphone.

Mic cables are made of one or two insulated conductors surrounded by a fine-wire mesh shield to keep out electrostatic hum. If you hear a loud buzz when you plug in a microphone, check that the shield is securely soldered to XLR pin 1 on both ends of the cable.

After acquiring a microphone, you may need to wire its 2-conductor shielded cable to a 3-pin XLR audio connector. Here are the solder connections:

- Pin 1: Shield
- Pin 2: "Hot" or "in-polarity" lead (usually red or white)
- Pin 3: "Cold" or "out-of-polarity" lead (usually black)

If the mic output is a 3-pin XLR, but your recorder or mixer mic input is an unbalanced phone jack, a different wiring is needed:

- Phone-plug tip (the short terminal): Hot lead
- Phone-plug sleeve (the long terminal): Shield and cold lead

Wind your mic cables onto a large spool, which can be found in the electrical section of hardware stores. Plug the cables together as you wind them.

Snake

It's messy and time-consuming to run mic cables from several mics all the way to a mixer. Instead, you can plug all your mics into a stage box with several connectors (Figure 6.11). The snake—a thick multi-conductor cable—carries the signals to the mixer. At the mixer end, the cable divides into several mic connectors that plug into the mixer.

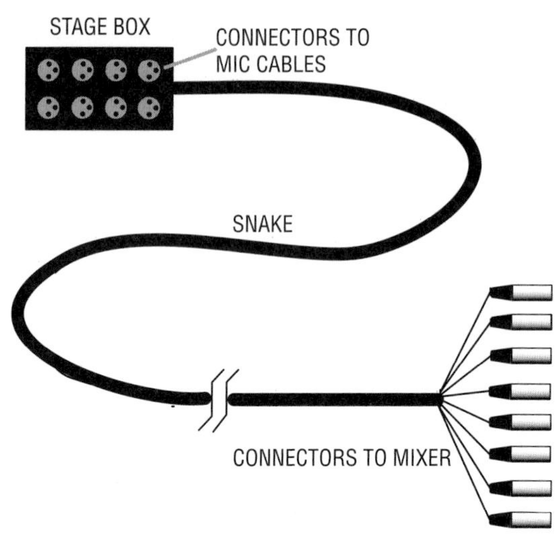

FIGURE 6.11
A stage box and snake.

Splitter

When you record a band in concert, you might want to feed each mic's signal to your recording mixer and to the band's PA and monitor mixers. A mic splitter does the job. For each microphone channel, it has one XLR input for a microphone, a "direct" XLR output wired to the input, and one or more transformer-isolated XLR outputs with a ground-lift switch. The mixer that provides phantom power must be connected to the direct XLR output. There's more on splitters in Chapter 17 under the heading Splitting the Mic Signals.

SUMMARY

We talked about some mic types, specs, and accessories. You should have a better idea about what kind of microphone to choose for your own applications.

Mic manufacturers are happy to send you free catalogs and application notes, which are also available on mic company Web sites. Mic dealers also may have this literature.

Remember, you can use any microphone on any instrument if it sounds good to you. Just try it and see if you like it. To make high-quality recordings, though, you need good mics with a smooth, wide-range frequency response, low noise, and low distortion.

CHAPTER 7

Microphone Technique Basics

Suppose you're going to mike a singer, a sax, or a guitar. Which mic should you choose? Where should you place it?

Your mic technique has a powerful effect on the sound of your recordings. In this chapter we'll look at some general principles of miking that apply to all situations. Chapter 8 covers common mic techniques for specific instruments.

WHICH MIC SHOULD I USE?

Is there a "right" mic to use on a piano, a kick drum, or a guitar amp? No. Every microphone sounds different, and you choose the one that gives you the sound you want. Still, it helps to know about two main characteristics of mics that affect the sound: frequency response and polar pattern.

Most condenser mics have an extended high-frequency response—they reproduce sounds up to 15 or 20 kHz. This makes them great for cymbals or other instruments that need a detailed sound, such as acoustic guitar, strings, piano, and voice. Dynamic moving-coil microphones have a response good enough for drums, guitar amps, horns, and woodwinds. Loud drums and guitar amps sound dull if recorded with a flat-response mic; a mic with a presence peak (a boost around 5 kHz) gives more edge or punch.

Suppose you're choosing a microphone for a particular instrument. In general, the frequency response of the mic should cover at least the frequencies produced by that instrument. For example, an acoustic guitar produces fundamental frequencies from 82 Hz to about

1 kHz, and produces harmonics from about 1 to 15 kHz. So a mic used on an acoustic guitar should have a frequency response of at least 82 Hz to 15 kHz if you want to record the guitar accurately. Table 7.1 shows the frequency ranges of various instruments.

The polar pattern of a mic affects how much leakage and ambience it picks up. Leakage is unwanted sound from instruments other than the one at which the mic is aimed. Ambience is the acoustics of the

Table 7.1 Frequency Ranges of Various Musical Instruments

Instrument	Fundamentals (Hz)	Harmonics (kHz)
Flute	261–2349	3–8
Oboe	261–1568	2–12
Clarinet	165–1568	2–10
Bassoon	62–587	1–7
Trumpet	165–988	1–7.5
French horn	87–880	1–6
Trombone	73–587	1–7.5
Tuba	49–587	1–4
Snare drum	100–200	1–20
Kick drum	30–147	1–6
Cymbals	300–587	1–15
Violin	196–3136	4–15
Viola	131–1175	2–8.5
Cello	65–698	1–6.5
Acoustic bass	41–294	1–5
Electric bass	41–300	1–7
Acoustic guitar	82–988	1–15
Electric guitar	82–1319	1–3.5 (through amp)
Electric guitar	82–1319	1–15 (direct)
Piano	28–4196	5–8
Bass (voice)	87–392	1–12
Tenor (voice)	131–494	1–12
Alto (voice)	175–698	2–12
Soprano (voice)	247–1175	2–12

recording room—its early reflections and reverb. The more leakage and ambience you pick up, the more distant the instrument sounds.

An omni mic picks up more ambience and leakage than a directional mic when both are the same distance from an instrument. So an omni tends to sound more distant. To compensate, you have to mike closer with an omni.

HOW MANY MICS?

The number of mics you need varies with what you're recording. If you want to record an overall acoustic blend of the instruments and room ambience, use just two microphones or a stereo mic (Figure 7.1). This method works great on an orchestra, symphonic band, choir, string quartet, pipe organ, small folk group, or a piano/voice recital. Stereo miking is covered in detail later in this chapter under the heading Stereo Mic Techniques.

To record a pop-music group, you mike each instrument or instrumental section. Then you adjust the mixer volume control for each mic to control the balance between instruments (Figure 7.2).

To get the clearest sound, don't use two mics when one will do the job. Sometimes you can pick up two or more sound sources with one mic (Figure 7.3). You could mike a brass section of four players with

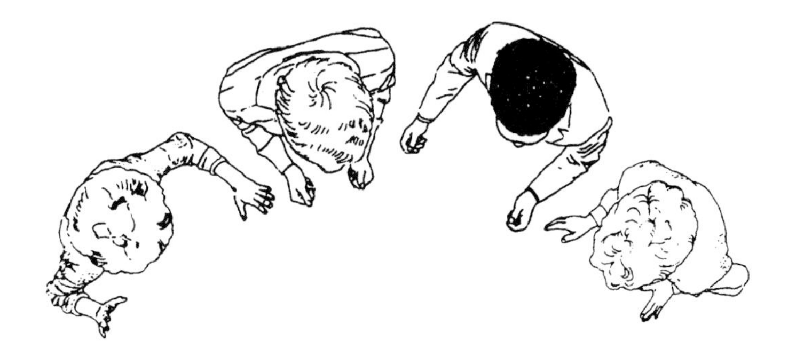

FIGURE 7.1
Overall miking of a musical ensemble with two distant microphones.

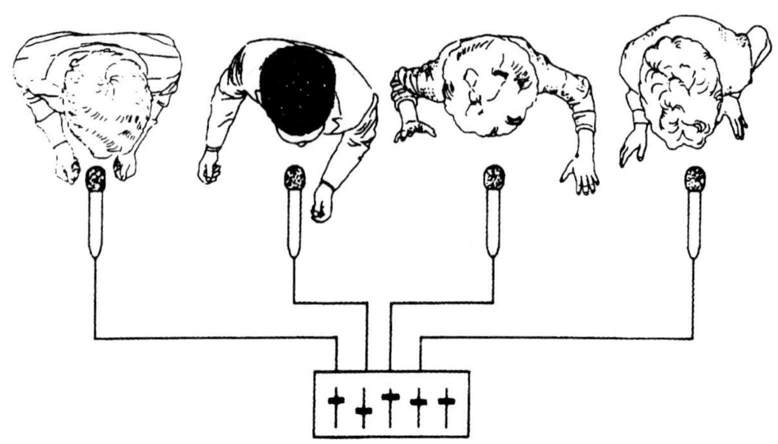

FIGURE 7.2
Individual miking with multiple close mics and a mixer.

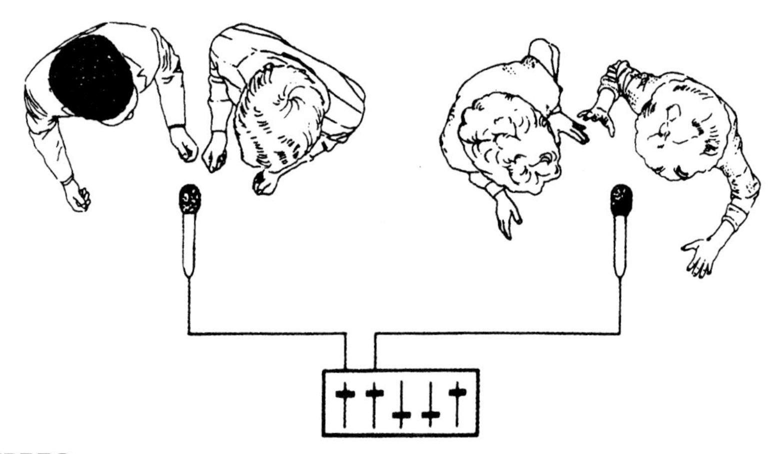

FIGURE 7.3
Multiple miking with several sound sources on each microphone.

one mic on four players, or with two mics on every two players. Or mike a choir in a studio in four groups: put one mic on the basses, one on the sopranos, and so on.

Picking up more than one instrument with one mic has a problem: during mixdown, you can't adjust the balance among instruments recorded on the same track. You have to balance the instruments before recording them. Monitor the mic, and listen to see if any instrument is too quiet. If so, move it closer to the mic.

HOW CLOSE SHOULD I PLACE THE MIC?

Once you've chosen a mic for an instrument, how close should the mic be? Mike a few inches away to get a tight, present sound; mike farther away for a distant, spacious sound. (Try it to hear the effect.) *Play audio clip 17 at www.elsevierdirect.com/companions/9780240811444.* The farther a mic is from the instrument, the more ambience, leakage, and background noise it picks up. So mike close to reject these unwanted sounds. Mike farther away to add a live, loose, airy feel to overdubs of drums, lead-guitar solos, horns, etc.

Close miking sounds close; distant miking sounds distant. Here's why. If you put a mic close to an instrument, the sound at the mic is loud. So you need to turn up the mic gain on your mixer only a little to get a full recording level. And because the gain is low, you pick up very little reverb, leakage, and background noise (Figure 7.4A).

If you put a mic far from an instrument, the sound at the mic is quiet. You'll need to turn up the mic gain a lot to get a full recording level. And because the gain is high, you pick up a lot of reverb, leakage, and background noise (Figure 7.4B).

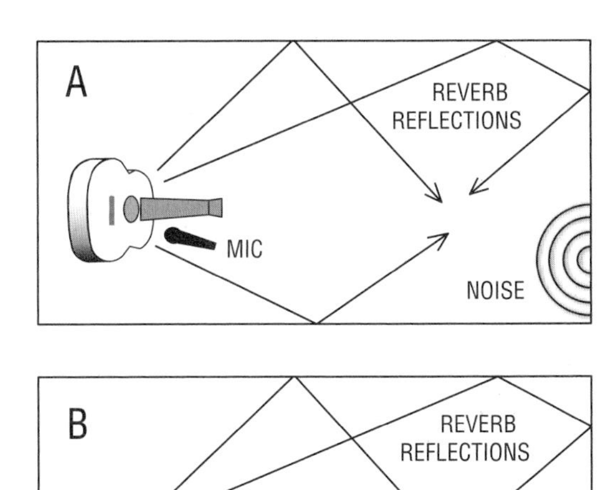

FIGURE 7.4
(A) A close microphone picks up mainly direct sound, which results in a close sound quality. (B) A distant microphone picks up mainly reflected sound, which results in a distant sound quality.

If the mic is very far away—maybe 10 feet—it's called an ambience mic or room mic. It picks up mostly room reverb. A popular mic for ambience is a boundary microphone taped to the wall. You mix it with the usual close mics to add a sense of space. Use two for stereo. When you record a live concert, you might want to place ambience mics over the audience, aiming at them from the front of the hall, to pick up the crowd reaction and the hall acoustics.

Classical music is always recorded at a distance (about 4 to 20 feet away) so that the mics will pick up reverb from the concert hall. It's a desirable part of the sound.

Leakage (Bleed or Spill)

Suppose you're close-miking a drum set and a piano at the same time (Figure 7.5). When you listen to the drum mics alone, you hear a close, clear sound. But when you mix in the piano mic, that nice, tight drum sound degrades into a distant, muddy sound. That's because the drum sound leaked into the piano mic, which picked up a distant drum sound from across the room. *Audio clip 11 at www.elsevierdirect.com/companions/9780240811444 is an example of leakage.*

There are many ways to reduce leakage:

- Mike each instrument closely. That way the sound level at each mic is high. Then you can turn down the mixer gain of each mic, which reduces leakage at the same time.
- Overdub each instrument one at a time.

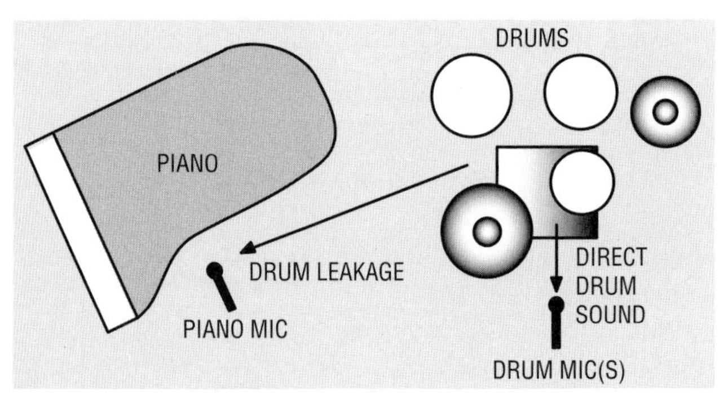

FIGURE 7.5
Example of leakage. The piano mic picks up leakage from the drums, which changes the close drum sound to distant.

- Record direct. Record an acoustic guitar off its pickup during tracking, then overdub the guitar with a mic. Record an electric guitar off its pickup during tracking, then play the guitar signal through a guitar-amp modeling plug-in during mixdown. Or record the electric guitar through a Line 6 Pod, which is a guitar-amp emulator.
- Filter out frequencies above and below the range of each instrument.
- Use directional mics (cardioid, etc.) instead of omni mics.
- Record in a large, fairly dead studio. In such a room, leakage reflected from the walls is weak.
- Put portable walls (goboes) between instruments.
- Use noise gates on drum tracks (see Chapter 10 under the heading Noise Gate.)

Don't Mike Too Close

Miking too close can color the recorded tone quality of an instrument. If you mike very close, you might hear a bassy or honky tone instead of a natural sound. Why? Most musical instruments are designed to sound best at a distance, at least 1-1/2 feet away. The sound of an instrument needs some space to develop. A mic placed a foot or two away tends to pick up a well-balanced, natural sound. That is, it picks up a blend of all the parts of the instrument that contribute to its character or timbre.

Think of a musical instrument as a loudspeaker with a woofer, midrange, and tweeter. If you place a mic a few feet away, it will pick up the sound of the loudspeaker accurately. But if you place the mic close to the woofer, the sound will be bassy. Similarly, if you mike close to an instrument, you emphasize the part of the instrument that the microphone is near. The tone quality picked up very close may not reflect the tone quality of the entire instrument.

Suppose you place a mic next to the sound hole of an acoustic guitar, which resonates around 80 to 100 Hz. A microphone placed there hears this bassy resonance, giving a boomy recorded timbre that doesn't exist at a greater miking distance. To make the guitar sound more natural when miked close to the sound hole, you need to roll off the excess bass on your mixer, or use a mic with a bass rolloff in its frequency response.

The sax projects highs from the bell, but projects mids and lows from the tone holes. So if you mike close to the bell, you miss the warmth and body from the tone holes. All that's left at the bell is a harsh tone quality. You might like that sound, but if not, move the mic out and up to pick up the entire instrument. If leakage forces you to mike close, change the mic or use equalization (EQ).

Usually, you get a natural sound if you put the mic as far from the source as the source is big. That way, the mic picks up all the sound-radiating parts of the instrument about equally. For example, if the body of an acoustic guitar is 18 inches long, place the mic 18 inches away to get a natural tonal balance. If this sounds too distant or hollow, move in a little closer.

WHERE SHOULD I PLACE THE MIC?

Suppose you have a mic placed a certain distance from an instrument. If you move the mic left, right, up, or down, you change the recorded tone quality. In one spot, the instrument might sound bassy; in another spot, it might sound natural, and so on. So, to find a good mic position, simply place the mic in different locations—and monitor the results—until you find one that sounds good to you.

Here's another way to do the same thing. Close one ear with your finger, listen to the instrument with the other ear, and move around until you find a spot that sounds good. Put the mic there. Then make a recording and see if it sounds the same as what you heard live. Don't try this with kick drums or screaming guitar amps! You could also move a mic around while monitoring its signal with good headphones.

Why does moving the mic change the tone quality? A musical instrument radiates a different tone quality in each direction. Also, each part of the instrument produces a different tone quality. For example, Figure 7.6 shows the tonal balances picked up at various spots near a guitar. *Audio clip 18 at www.elsevierdirect.com/companions/9780240811444 illustrates the effect of mic placement on guitar tonal balance. Audio clip 19 demonstrates close and distant stereo miking of the acoustic guitar.*

Other instruments work the same way. A trumpet radiates strong highs directly out of the bell, but doesn't project them to the sides.

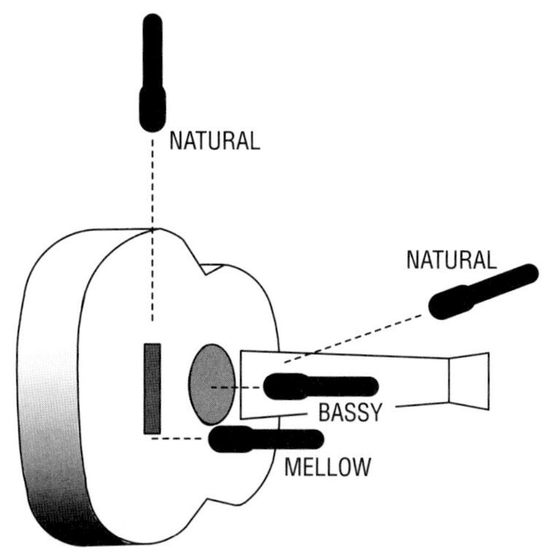

FIGURE 7.6
Microphone placement affects the recorded tonal balance.

So a trumpet sounds bright when miked on-axis to the bell and sounds more natural or mellow when miked off to one side. A grand piano miked one foot over the middle strings sounds fairly natural, under the soundboard sounds bassy and dull, and in a sound hole it sounds mid-rangey.

It pays to experiment with all sorts of mic positions until you find a sound you like. There is no one right way to place the mics because you place them to get the tonal balance you want.

ON-SURFACE TECHNIQUES

Sometimes you're forced to place a mic near a hard reflecting surface.

Examples:

- Recording drama or opera with the mics near the stage floor
- Recording an instrument that has hard surfaces around it
- Recording a piano with the mic close to the lid

In these cases, you'll often pick up an unnatural, filtered tone quality. Here's why. Sound travels to the microphone via two paths: directly from the sound source, and reflected off the nearby surface (Figure 7.7). Because of its longer travel path, the reflected sound is delayed compared to the direct sound. The direct and delayed sound

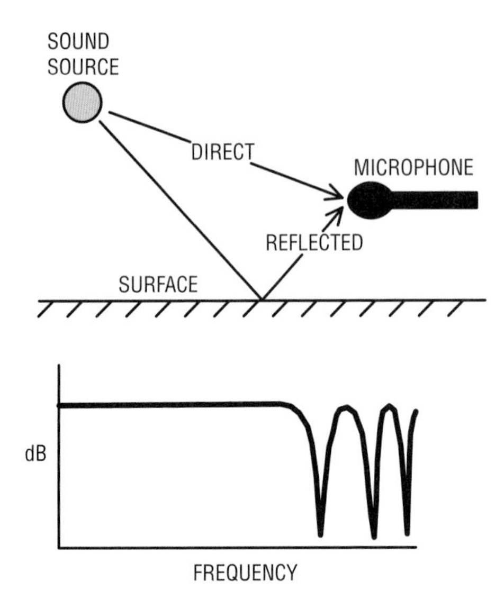

FIGURE 7.7
A mic placed near a surface picks up direct sound and a delayed reflection, which gives a comb-filter frequency response.

waves combine at the mic, which causes phase cancellations of various frequencies. The series of peaks and dips in the response is called a comb-filter effect, and it sounds like mild flanging.

Boundary mics solve the problem. In a boundary mic, the diaphragm is very close to the reflecting surface so that there is no delay in the reflected sound. Direct and reflected sounds add in-phase over the audible range of frequencies, resulting in a flat response (Figure 7.8). *Play audio clip 20 at www.elsevierdirect.com/companions/9780240811444.*

You might tape an omni boundary mic to the underside of a piano lid, to a hard-surfaced panel, or to a wall for ambience pickup. A unidirectional boundary mic works great on a stage floor to pick up drama. A group of these mics will clearly pick up people at a conference table.

THE 3:1 RULE

Suppose you're recording a singer/guitarist. There's a mic on the singer and a mic on the acoustic guitar. When you monitor the mix, something's wrong: the singer's voice sounds hollow or filtered. You're hearing the effect of phase interference.

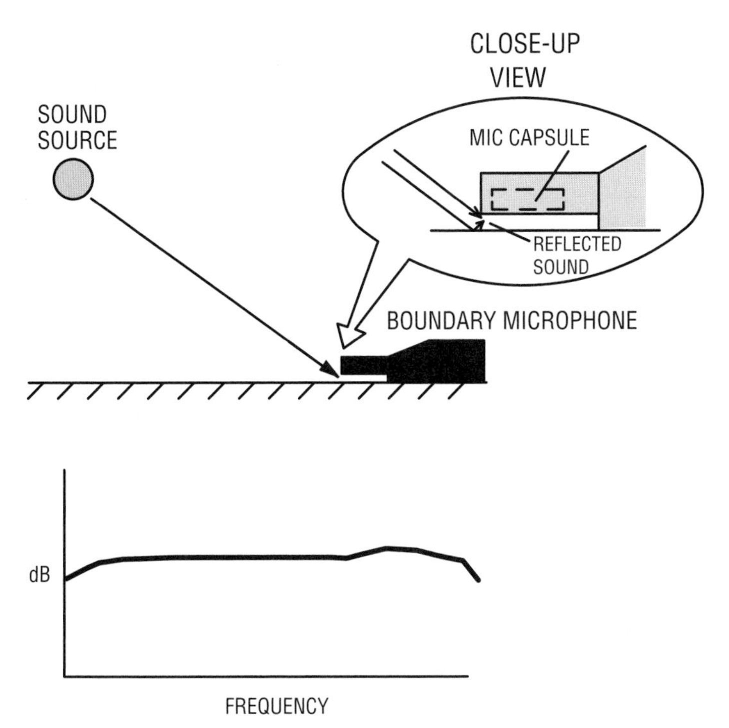

FIGURE 7.8
A boundary mic on the surface picks up direct and reflected sounds in phase.

In general, if two mics pick up the same sound source at different distances, and their signals are mixed to the same channel, this might cause phase cancellations. These are peaks and dips in the frequency response—a comb filter—caused by some frequencies combining out of phase. The result is a colored, filtered tone quality that sounds like mild flanging.

To prevent this problem, follow the 3:1 rule: Space the mics at least three times the mic-to-source distance (as in Figure 7.9). For example, if two mics are 12 inches apart, they should be less than 4 inches from their sound sources to prevent phase cancellations.

Play audio clip 21 at www.elsevierdirect.com/companions/9780240811444. The mics can be closer together than 3:1 if you use two cardioid mics aiming in opposite directions. The goal is to get at least 9 dB of separation between recorded tracks.

What if you pick up an instrument with two mics that are panned left and right? You don't get phase interference. Instead you get stereo imaging.

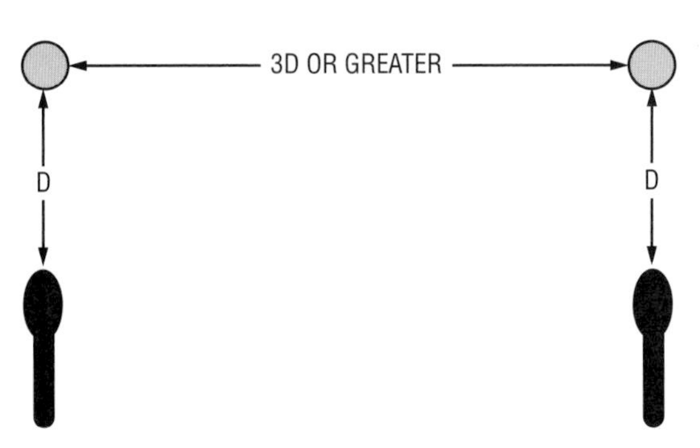

FIGURE 7.9
The 3:1 rule of microphone placement avoids phase interference between microphone signals.

OFF-AXIS COLORATION

Some mics have off-axis coloration—a dull or colored effect on sound sources that aren't directly in front of the mic. Try to aim the mic at sound sources that put out high frequencies, such as cymbals. When you pick up a large source such as an orchestra, use a mic that has the same response over a wide angle. Such a mic has similar polar patterns at middle and high frequencies. Most large-diaphragm mics have more off-axis coloration than smaller mics (under 1 inch).

STEREO MIC TECHNIQUES

Stereo mic techniques capture the sound of a musical group as a whole, using only two or three microphones. When you play back a stereo recording, you hear phantom images of the instruments in various spots between the speakers. These image locations—left to right, front to back—correspond to the instrument locations during the recording session.

Stereo miking is the preferred way to record classical-music ensembles and soloists. In the studio, you can stereo-mike a piano, drum set cymbals, vibraphone, harmony singers, or other large sound sources.

Goals of Stereo Miking

One goal is accurate localization; that is, instruments in the center of the group are reproduced midway between the two speakers. Instruments at the sides of the group are heard from the left or right

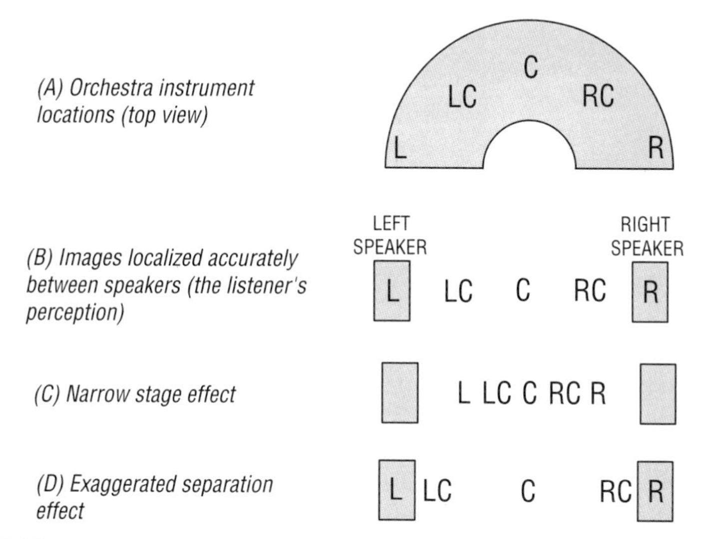

FIGURE 7.10
Stereo localization effects. (A) Orchestra instrument locations (top view). (B) Images localized accurately between speakers (the listener's perception). (C) Narrow stage effect. (D) Exaggerated separation effect.

speaker. Instruments halfway to one side are heard halfway to one side, and so on.

Figure 7.10 shows three stereo localization effects. Figure 7.10A shows some instrument positions in an orchestra: left, left-center, center, right-center, right. In Figure 7.10B, the reproduced images of these instruments are accurately localized between the speakers. The stereo spread, or stage width, extends from speaker to speaker. (You might want to record a string quartet with a narrower spread.)

If you space or angle the mics too close together, you get a narrow stage effect (Figure 7.10C). If you space or angle the mics too far apart, you hear exaggerated separation (Figure 7.10D); that is, instruments halfway to one side are heard near the left or right speaker.

To judge stereo effects, you have to sit exactly between your monitor speakers (the same distance from each). Sit as far from the speakers as the spacing between them. Then the speakers appear to be 60 degrees apart. This is about the same angle an orchestra fills when viewed from a typical ideal seat in the audience (say, tenth row center). If you sit off-center, the images shift toward the side on which you're sitting and are less sharp.

Types of Stereo Mic Techniques

To make a stereo recording, you use one of these basic techniques:

- Coincident pair (XY or MS)
- Spaced pair (AB)
- Near-coincident pair (ORTF, etc.)
- Baffled pair (sphere, OSS, SASS, PZM wedge, etc.)

Let's look at each technique.

COINCIDENT PAIR

With this method, you mount two directional mics with grilles touching, diaphragms one above the other, and angled apart (Figure 7.11). For example, mount two cardioid mics with one grille above the other, and angle them 120 degrees apart. You can use other patterns too: supercardioid, hypercardioid, or bidirectional. The wider the angle between mics, the wider the stereo spread. If the angle is too wide, center images will be weak (there will be a "hole in the middle").

How does this technique make images you can localize? A directional mic is most sensitive to sounds in front of the mic (on-axis) and progressively less sensitive to sounds arriving off-axis. That is, a directional mic puts out a high-level signal from the sound source it's aimed at, and produces lower-level signals from other sound sources.

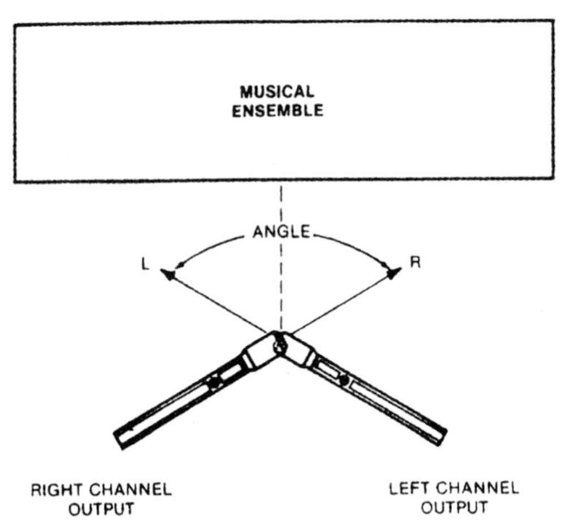

FIGURE 7.11
Coincident-pair technique.

The coincident pair uses two directional mics that are angled symmetrically from the center line (Figure 7.11). Instruments in the center of the group make the same signal from each mic. During playback, you hear a phantom image of the center instruments midway between your speakers. That's because identical signals in each channel produce an image in the center.

If an instrument is off-center to the right, it is more on-axis to the right-aiming mic than to the left-aiming mic. So the right mic will produce a higher level signal than the left mic. During playback of this recording, the right speaker will play at a higher level than the left speaker. This reproduces the image off-center to the right—where the instrument was during recording.

The coincident pair codes instrument positions into level differences between channels. During playback, the brain decodes these level differences back into corresponding image locations. A pan pot in a mixing console works on the same principle. If one channel is 15 to 20 dB louder than the other, the image shifts all the way to the louder speaker.

Suppose we want the right side of the orchestra to be reproduced at the right speaker. That means the far-right musicians must produce a signal level 20 dB higher from the right mic than from the left mic. This happens when the mics are angled far enough apart. The correct angle depends on the polar pattern.

Instruments partway off center produce interchannel level differences less than 20 dB, so you hear them partway off center.

Listening tests have shown that coincident cardioid mics tend to reproduce the musical group with a narrow stereo spread; that is, the group doesn't spread all the way between speakers.

A coincident-pair method with excellent localization is the Blumlein array. It uses two bidirectional mics angled 90 degrees apart and facing the left and right sides of the group.

A special form of the coincident-pair technique is mid-side or MS (Figure 7.12). In this method, a cardioid or omni mic faces the middle of the orchestra. A matrix circuit sums and differences the cardioid mic with a bidirectional mic aiming to the sides. This produces left- and right-channel signals. You can remotely control the stereo

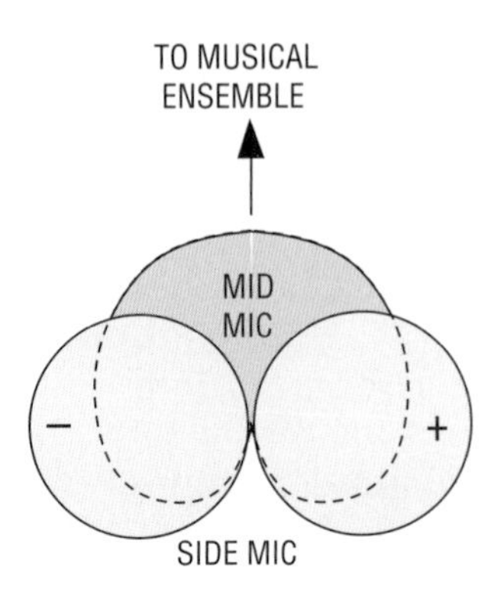

FIGURE 7.12
Mid-side (MS) technique.

spread by changing the ratio of the mid signal to the side signal. This remote control is useful at live concerts, where you can't physically adjust the mics during the concert. MS localization can be accurate. Chapter 18, under the heading Stereo-Spread Control, tells how to replace the matrix with a computer DAW.

To make coincident recordings sound more spacious, boost the bass 4 dB (+2 dB at 600 Hz) in the L–R or side signal.

A recording made with coincident mics is mono-compatible; that is, the frequency response is the same in mono or stereo. Because the mics occupy almost the same point in space, there is no time or phase difference between their signals. And when you combine them to mono, there are no phase cancellations to degrade the frequency response. If you expect that your recordings will be heard in mono (say, on TV), then you'll probably want to use coincident methods.

SPACED PAIR

Here, you mount two identical mics several feet apart and aim them straight ahead (Figure 7.13). The mics can have any polar pattern, but omni is most popular for this method. The greater the spacing between mics, the greater the stereo spread.

How does this method work? Instruments in the center of the group make the same signal from each mic. When you play back this

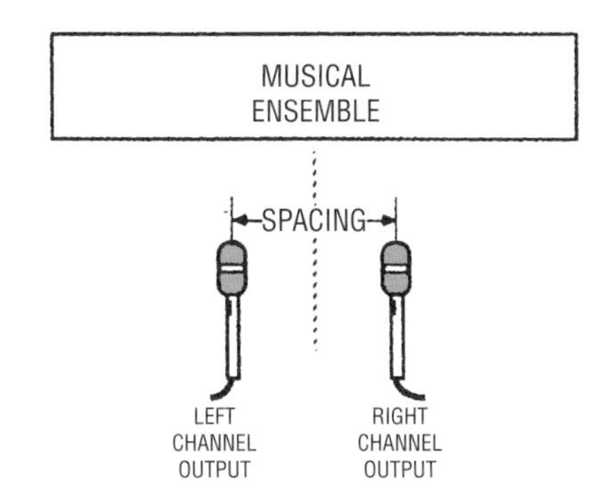

FIGURE 7.13
Spaced-pair technique.

recording, you hear a phantom image of the center instruments midway between your speakers. If an instrument is off-center, it is closer to one mic than the other, so its sound reaches the closer microphone before it reaches the other one. Both mics make about the same signal, except that one mic signal is delayed compared with the other.

If you send a signal to two speakers with one channel delayed, the sound image shifts off center. With a spaced-pair recording, off-center instruments produce a delay in one mic channel, so they are reproduced off center.

The spaced pair codes instrument positions into time differences between channels. During playback, the brain decodes these time differences back into corresponding image locations.

A delay of 1.2 msec is enough to shift an image all the way to one speaker. You can use this fact when you set up the mics. Suppose you want to hear the right side of the orchestra from the right speaker. The sound from the right-side musicians must reach the right mic about 1.2 msec before it reaches the left mic. To make this happen, space the mics about 2 to 3 feet apart. This spacing makes the correct delay to place right-side instruments at the right speaker. Instruments partway off center make interchannel delays less than 1.2 msec, so they are reproduced partway off center.

If the spacing between mics is, say, 12 feet, then instruments that are slightly off center produce delays between channels that are greater than 1.2 msec. This places their images at the left or right speaker. I call this "exaggerated separation" or a "ping-pong" effect (Figure 7.10D).

On the other hand, if the mics are too close together, the delays produced will be too small to provide much stereo spread. Also, the mics will tend to emphasize instruments in the center because the mics are closest to them.

To record a good musical balance of an orchestra, you might need to space the mics about 10 or 12 feet apart. But then you get too much separation. You could place a third mic midway between the outer pair and mix its output to both channels. That way, you pick up a good balance, and you hear an accurate stereo spread.

The spaced-pair method tends to make off-center images unfocused or hard to localize. Why? Spaced-pair recordings have time differences between channels. Stereo images produced solely by time differences are unfocused. You still hear the center instruments clearly in the center, but off-center instruments are hard to pinpoint. Spaced-pair miking is a good choice if you want the sonic images to be diffuse or blended, instead of sharply focused.

Another flaw of spaced mics: If you mix both mics to mono, you may get phase cancellations of various frequencies. This may or may not be audible.

Spaced mics, however, give a "warm" sense of ambience, in which the concert-hall reverb seems to surround the instruments and, sometimes, the listener. Here's why: The two channels of recorded reverb are incoherent; that is, they have random phase relationships. Incoherent signals from stereo speakers sound diffuse and spacious.

Because spaced mics pick up reverb incoherently, it sounds diffuse and spacious. The simulated spaciousness caused by this phasiness isn't necessarily realistic, but it's pleasant to many listeners.

Another advantage of the spaced pair is that you can use omni mics. An omni condenser mic has deeper bass than a uni condenser mic.

NEAR-COINCIDENT PAIR

In this method, you angle apart two directional mics, and space their grilles a few inches apart horizontally (Figure 7.14). Even a few inches of spacing increases the stereo spread and adds a sense of ambient warmth or air to the recording. The greater the angle or spacing between mics, the greater the stereo spread. If the angle is too wide, center images will be weak (there will be a "hole in the middle").

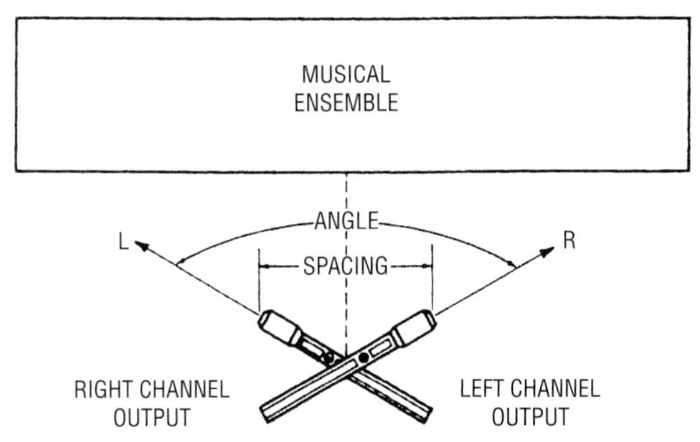

FIGURE 7.14
Near-coincident pair technique.

How does this method work? Angling directional mics produces level differences between channels. Spacing mics produces time differences. The level differences and time differences combine to create the stereo effect.

If the angling or spacing is too great, you get exaggerated separation. If the angling or spacing is too small, you'll hear a narrow stereo spread.

A common near-coincident method is the ORTF system, which uses two cardioids angled 110 degrees apart and spaced 7 inches (17 cm) horizontally. Usually this method gives accurate localization; that is, instruments at the sides of the orchestra are reproduced at or very near the speakers, and instruments halfway to one side are reproduced about halfway to one side.

BAFFLED OMNI PAIR

This method uses two omni mics, usually ear-spaced, and separated by either a hard or soft baffle (Figure 7.15). To create stereo, it uses time differences at low frequencies and level differences at high frequencies. The spacing between mics creates time differences. The baffle creates a sound shadow (reduced high frequencies) at the mic farthest from the source. Between the two channels, there are spectral differences—differences in frequency response.

Some examples of baffled-omni pairs are the Schoeps or Neumann sphere microphones, the Jecklin Disk, and the Crown SASS-P MKII stereo microphone.

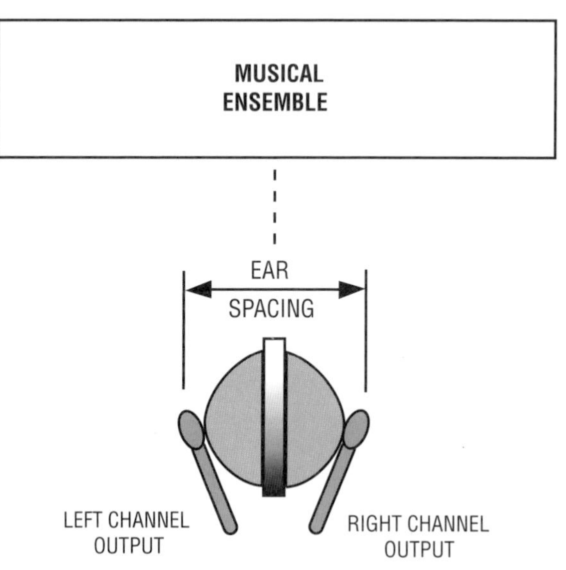

FIGURE 7.15
Baffled-omni technique.

Comparing the Four Techniques

Coincident pair:

- Uses two directional mics angled apart with grilles touching.
- Level differences between channels produce the stereo effect.
- Images are sharp.
- Stereo spread ranges from narrow to accurate.
- Signals are mono-compatible.

Spaced pair:

- Uses two mics spaced several feet apart, aiming straight ahead.
- Time differences between channels produce the stereo effect.
- Off-center images are diffuse.
- Stereo spread tends to be exaggerated unless a third center mic is used, or unless spacing is under 2 to 3 feet.
- Provides a warm sense of ambience.
- Tends not to be mono-compatible, but there are exceptions.
- Good low-frequency response if you use omni condensers.

Near-coincident pair:

- Uses two directional mics angled apart and spaced a few inches apart horizontally.

- Level and time differences between channels produce the stereo effect.
- Images are sharp.
- Stereo spread tends to be accurate. Provides a greater sense of air than coincident methods.
- Tends not to be mono-compatible.

Play audio clip 22 at www.elsevierdirect.com/companions/9780240811444 to hear a comparison of the coincident, near-coincident and spaced-pair techniques.

Baffled omni pair:

- Uses two omni mics, usually ear-spaced, with a baffle between them.
- Level, time, and spectral differences produce the stereo effect.
- Images are sharp.
- Stereo spread tends to be accurate but is not adjustable (except partly by panning).
- Good low-frequency response.
- Good imaging with headphones.
- Provides more air than coincident methods.
- Tends not to be mono-compatible, but there are exceptions.

HARDWARE

A handy device is a stereo mic adapter or stereo bar (Figure 7.16). It mounts two mics on a single stand, and lets you adjust the angle and spacing. You might prefer to use a stereo mic instead of two mics. It has two mic capsules in a single housing for convenience.

HOW TO TEST IMAGING

Here's a way to check the stereo imaging of a mic technique.

1. Set up the stereo mic array in front of a stage.
2. Record yourself speaking from various locations on stage where the instruments will be—center, half-right, far right, half-left, far left. Announce your position.
3. Play back the recording over speakers.

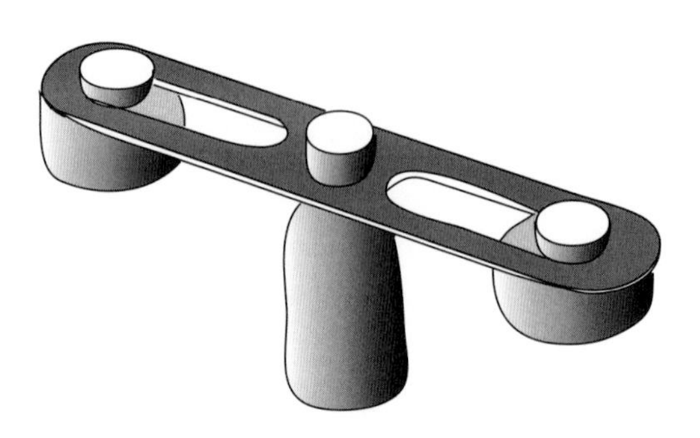

FIGURE 7.16
Stereo mic adapter.

You'll hear how accurately the technique translated your positions, and you'll hear how sharp the images are.

We looked at several mic arrays to record in stereo. Each has its pros and cons. Which method you choose depends on the sonic compromises you're willing to make.

CHAPTER 8
Microphone Techniques

This chapter describes some ways to select and place mics for musical instruments and vocals. These techniques are popular, but they're just suggested starting points. Feel free to experiment.

Before you mike an instrument, listen to it live in the studio, so you know what sound you're starting with. You might want to duplicate that sound through your monitor speakers.

ELECTRIC GUITAR

Let's start by looking at the chain of guitar, effects, amplifier, and speaker. At each point in the chain where you record, you'll get a different sound (Figure 8.1).

1. The electric guitar puts out an electrical signal that sounds clean and clear.
2. This signal might go through some effects boxes, such as distortion, wah wah, compression, chorus, or stereo effects.
3. Then the signal goes through a guitar amp, which boosts the signal and adds distortion. At the amplifier output (preamp out or external speaker jack), the sound is very bright and edgy.
4. The distorted amp signal is played by the speaker in the amp. Because the speaker rolls off above 4 kHz, it takes the edge off the distortion and makes it more pleasant.

You can record the electric guitar in many ways (Figure 8.1):

- With a mic in front of the guitar amp
- With a direct box

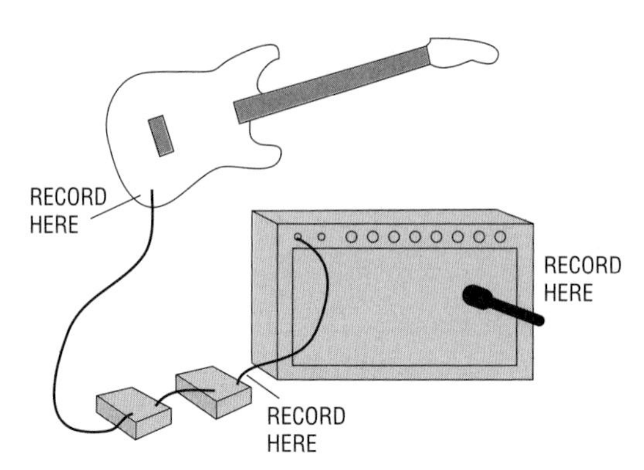

FIGURE 8.1
Three places to record the electric guitar.

- Both miked and direct
- Through a signal processor or stomp box

The song you're recording will tell you what method it wants. Just mike the amp when you want a rough, raw sound with tube distortion and speaker coloration. Rock 'n' roll or heavy metal usually sounds best with a miked amp. If you record through a direct box, the sound is clean and clear, with crisp highs and deep lows. That might work for quiet jazz or R&B. Use whatever sounds right for the particular song you're recording.

First, try to kill any hum you hear from the guitar amp. Turn up the guitar's volume and treble controls so that the guitar signal overrides hum and noise picked up by the guitar cable. Ask the guitarist to move around, or rotate, to find a spot in the room where hum disappears. Flip the polarity switch on the amp to the lowest hum position. To remove buzzes between guitar notes, try a noise gate, or ask the player to keep his hands on the strings.

Miking the Amp

Small practice amplifiers tend to be better for recording than large, noisy stage amps. If you use a small one, place it on a chair to avoid picking up sound reflections from the floor (unless you like that effect).

A common mic for the guitar amp is a cardioid dynamic type with a "presence peak" in its frequency response (a boost around 5 kHz).

The cardioid pattern reduces leakage (off-mic sounds from other instruments). The dynamic type handles loud sounds without distorting, and the presence peak adds "bite." Of course, you can use any mic that sounds good to you.

As a starting point, try miking the amp about an inch from the grille near a speaker cone, slightly off-center—where the cone meets the dome.

The closer you mike the amp, the bassier the tone. The farther off-center the mic is, the duller the tone. Often, distant miking sounds great when you overdub a lead guitar solo played through a stack of speakers in a live room. Try a boundary mic on the floor or on the wall several feet away.

Recording Direct

Now let's look at recording direct (also known as direct injection or DI). The electric guitar produces an electrical signal that you can plug into your mixer. You bypass the mic and guitar amp, so the sound is clean and clear. Just remember that amp distortion is desirable in some songs.

Mixer mic inputs tend to have an impedance (Z) around 1500 ohms. But a guitar pickup is several thousand ohms. So if you plug a high-Z electric guitar directly into a mic input, the input will load down the pickup and give a thin or dull sound.

To get around this loading problem, use a direct box between the guitar and your mixer (Figure 8.2). The DI box has a high-Z input and low-Z output, thanks to a built-in transformer or circuit. Some mixers and audio interfaces have a high-Z (instrument) input jack built in, so you can plug the electric guitar or bass directly into this jack.

The direct box should have a ground-lift switch to prevent ground loops and hum. Set it to the position where you monitor the least hum. You might try a mix of direct sound and miked sound. *Play audio clip 23 at www.elsevierdirect.com/companions/9780240811444 to hear demonstrations of electric-guitar recording methods.*

It's a good idea to record the guitar direct on its own track even if you mike the amp. Later during mixdown, you can run the DI track through a guitar-amp simulator plug-in, which might sound better than the real guitar amp did.

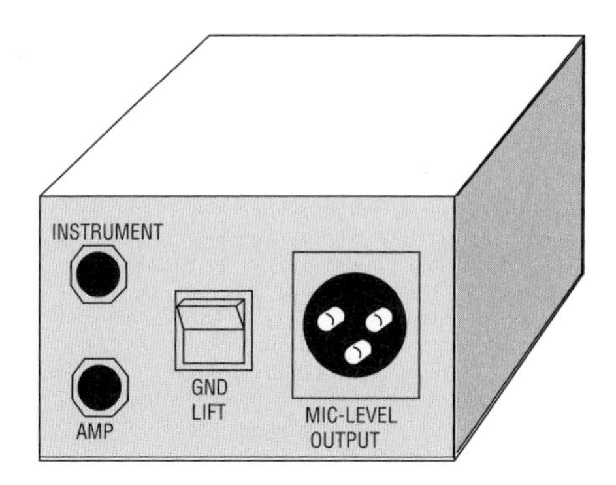

FIGURE 8.2
Typical direct box.

Electric Guitar Effects

If you want to record the guitarist's effects, connect the output of the effects boxes into the direct-box input. Many players have a rack of signal processors that creates their unique sound, and they just give you their direct feed. Be open to their suggestions, and be diplomatic about changing the sound. If they are studio players, they often have a better handle on effects than you might as the engineer.

You might want a "fat" or spacious lead-guitar sound. Here are some ways to get it:

- Send the guitar signal through a digital delay set to 20 to 30 msec. Pan guitar left, delay right. Adjust levels for nearly equal loudness from each speaker. (Watch out for phase cancellations in mono.)
- Send the guitar signal through a pitch-shifter, set for about 10 cents of pitch bending. Pan guitar left, pitch-shifted guitar right. (A cent is 1/100 of an equal-temperament semitone. There are 100 cents in a half-tone or semitone interval of pitch).
- Record two guitarists playing identical parts, and pan them left and right. This works great for rhythm-guitar parts in heavy metal.
- Double the guitar. Have the player re-record the same part on an unused track while listening to the original part. Pan the original part left and pan the new part right.
- Add stereo reverb or stereo chorus.

Some guitar processors add many effects to an electric guitar, such as distortion, EQ, chorus, and compression. An example is the Line 6 Pod.

You simply plug the electric guitar into the processor, adjust it for the desired sound, and record the signal direct. You wind up with a fully produced sound with a minimum of effort.

Re-amping is a technique that lets you work on the amp's sound during mixdown rather than during recording. Record the guitar direct, then feed that track's signal into a guitar processor or miked guitar amp during mixdown. Use a low- to high-Z transformer between the track output and the processor or amp input. Record the processor or amp on an open track. In a digital audio workstation (DAW), you can start with a track of a direct-recorded guitar, then insert a guitar-amp modeling plug-in.

ELECTRIC BASS

BWAM, dik diddy bum. Do your bass tracks sound that clear? Or are they more muddy, like, "Bwuh, dip dubba duh"? Here's how to record the electric bass so it's clean and easy to hear in a mix.

As always, first you work on the sound of the instrument itself. Put on new strings if the old ones sound dull. Adjust the pickup screws (if any) for equal output from each string. Also adjust the intonation and tuning.

Usually, you record the electric bass direct for the cleanest possible sound. A direct pickup gives deeper lows than a miked amp, but the amp gives more midrange punch. You might want to mix the direct and miked sound. Use a condenser or dynamic mic with a good low-frequency response, placed 1 to 6 inches from the speaker.

When mixing a direct signal and a mic signal, make sure they are in-phase with each other. To do this, set them to equal levels and reverse the polarity of the direct signal or the mic signal. The polarity that gives the most bass is correct.

Have the musician play some scales to see if any notes are louder than the rest. You might set a parametric equalizer to soften these notes, or use a compressor.

The bass guitar should be fairly constant in level (a dynamic range of about 6 dB) to be audible throughout the song, and to avoid clipping the recording on loud peaks. To do this, run the bass guitar through a compressor. Set the compression ratio to about 4:1; set the attack time fairly slow (8 to 20 msec) to preserve the attack transient; and set the release time fairly fast (1/2 second). If the release time is too fast, you get harmonic distortion.

EQ can make the bass guitar clearer. Try cutting around 60 to 80 Hz, or at 400 Hz. A boost at 2 to 2.5 kHz adds edge or slap, and a boost at 700 to 900 Hz adds "growl" and harmonic clarity. If you boost the lows around 100 Hz, try boosting at a lower frequency in the kick drum's EQ to keep their sounds distinct. A fretless bass will probably need different EQ or less EQ than a fretted bass.

Here are some ways to make the bass sound clean and well defined:

- Record the bass direct.
- Use no reverb or echo on the bass.
- Have the bass player turn down the bass amp in the studio, just loud enough to play adequately. This reduces muddy-sounding bass leakage into other mics.
- Better yet, don't use the amp. Instead, have the musicians monitor the bass (and each other) with headphones.
- Have the bass player try new strings or a different bass. Some basses are better for recording than others. Use roundwound strings for a bright tone or flatwounds for a rounder tone.
- Ask the bass player to use the treble pickup near the bridge.
- Be sure to record the bass with enough edge or harmonics so the bass will be audible on small, cheap speakers.
- Try a bass-guitar signal processor such as the Line 6 Bass Pod, Zoom B1, or DigiTech BP80.

If the bass part is full and sustained, it's probably best to go for a mellow sound without much pluck. Let the kick drum define the rhythmic pattern. But if both the bass and kick are rhythmic and work independently, then you should hear the plucks. Listen to the song first, then get a bass sound appropriate for the music. A sharp, twangy timbre is seldom right for a ballad; a full, round tone will get lost in a fusion piece.

Often, a musician plays bass lines on a synth or sound module. The module is triggered from a keyboard, a sequencer, or a bass guitar plugged into a pitch-to-MIDI converter. Connect the module output to your mixer or audio interface line input.

Two effects boxes for the electric bass are the octave box and the bass chorus. The octave box takes the bass signal and drops it an octave in pitch; that is, it divides the bass signal's fundamental frequency in two. You put 82 Hz in; you get 41 Hz out. This gives an extra deep, growly sound. So does a 5-string bass.

A bass chorus gives a wavy, shimmering effect. Like a conventional chorus box, it detunes the signal and combines the detuned signal with the direct signal. Also, it removes the lowest frequencies from the detuned signal, so that the chorus effect doesn't thin out the sound.

SYNTHESIZER, DRUM MACHINE, AND ELECTRIC PIANO

For the most clarity, you usually DI a synth, MIDI sound module, drum machine, or electric piano. Set the volume on the instrument about three-quarters up to get a strong signal. Try to get the sound you want from patch settings rather than EQ.

Plug the instrument into a phone jack input on your mixer, or use a direct box. If you connect to a phone jack and hear hum, you probably have a ground loop. Here are some fixes:

- Power your mixer and the instrument from the same outlet strip. If necessary, use a thick extension cord between the outlet strip and the instrument.
- Use a direct box instead of a guitar cord, and set the ground-lift switch to the position where you monitor the least hum.
- To reduce hum from a low-cost synth, use battery power instead of an AC adapter.

A synth can sound dry and sterile. To get a livelier, funkier sound, you might run the synth signal into a power amp and speakers, and mike the speakers a few feet away.

If the keyboard player has several keyboards plugged into a keyboard mixer, you may want to record a premixed signal from that mixer's output. Record both outputs of stereo keyboards.

LESLIE ORGAN SPEAKER

This glorious device has a rotating dual-horn on top for highs and a woofer on the bottom for lows. Only one horn of the two makes sound; the other is for weight balance. The swirling, grungy sound comes from the phasiness and Doppler effect of the rotating horn,

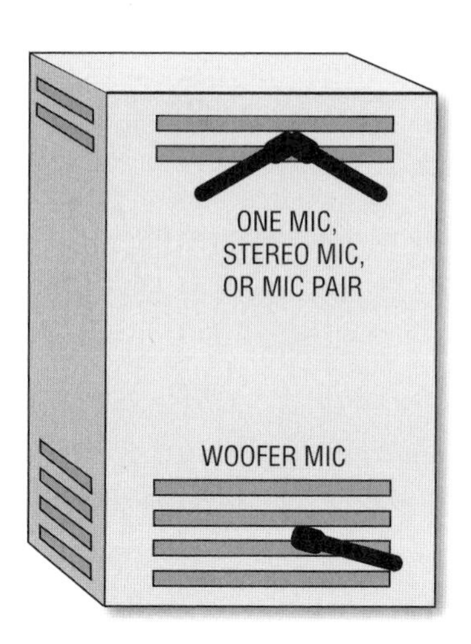

FIGURE 8.3
Miking a Leslie organ speaker.

and from the distorted tube electronics that drive the speaker. Here are a few ways to record it (Figure 8.3):

- In mono: Mike the top and bottom separately, 3 inches to 1 foot away. Aim the mics into the louvers. In the top mic's signal, roll off the lows below 150 Hz.
- In stereo: Record the top horn with a stereo mic or a pair of mics out front. Put a mic with a good low end on the bottom speaker, and pan it to center.

When you record the Leslie, watch out for wind noise from the rotating horn and buzz from the motor. Mike farther away if you monitor these noises.

Rather than recording an actual Hammond B3 organ and Leslie speaker, you might prefer to use a software emulation of those instruments: an organ soft synth or sample and a Leslie speaker plug-in. Trigger the synth or sample by a MIDI sequencer or MIDI controller (covered in Chapter 16). You can automate the horn rotation speed in the Leslie speaker plug-in.

DRUM SET

The first step is to make the drums sound good live in the studio. If the set sounds poor, you'll have a hard time making it sound great in the control room! You might put the drum set on a riser 1–1/2 feet high to reduce bass leakage and to provide better eye contact between the drummer and the rest of the band. To reduce drum leakage into other mics, you could surround the set with goboes—padded thick-wood panels about 4 feet tall. For more isolation, place the set in a drum booth, a small padded room with windows. It's also common to overdub the set in a live room.

Tuning

One secret to creating a good drum sound lies in careful tuning. It's easier to record a killer sound if you tune the set to sound right in the studio before miking it.

First let's consider drum heads. Plain heads have the most ring or sustain, while heads with sound dots or hydraulic heads dampen the ring. Thin heads are best for recording because they have crisp attack and long sustain. Old heads become dull, so use new heads.

When you tune the toms, first take off the heads and remove the damping mechanism, which can rattle. Put just the top head on and hand-tighten the lugs. Then, using a drum key, tighten opposite pairs of lugs one at a time, one full turn. After you tighten all the lugs, repeat the process, tightening one-half turn. Then press on the head to stretch it. Continue tightening a half-turn at a time until you reach the pitch you want. You'll get the most pleasing tone when the heads are tuned within the range of the shell resonance.

To reduce ugly overtones, try to keep the tension the same around the head. While touching the center of the head, tap with a drum-stick on the head near each lug. Adjust tension for equal pitch around the drum. If you want a downward pitch bend after the head is struck, loosen one lug.

Keep the bottom head off the drum for the most projection and the broadest range of tuning. In this case, pack the bottom lugs with felt to prevent rattles. But you may want to add the bottom head for extra control of the sound. Projection is best if the bottom head is tighter than the top head—say, tuned a fourth above the top head. There will be a muted attack, an "open" tone, and some note bend-ing. If you tune the bottom head looser than the top, the tone will be more "closed," with good attack.

With the kick drum (bass drum), a loose head gives lots of slap and attack, and almost no tone. The opposite is true for a tight head. Tune the head to complement the style of music. For more attack or click, use a hard beater.

Tune the snare drum with the snares off. A loose batter head or top head gives a deep, fat sound. A tight batter head sounds bright and crisp. With the snare head or bottom head loose, the tone is deep with little snare buzz, while a tight snare head yields a crisp snare response. Set the snare tension just to the point where the snare wires begin to "choke" the sound, then back off a little.

Damping and Noise Control

Usually the heads should ring without any damping. But if the toms or snare drum rings too much, put some plastic damping rings on them. Or tape some gauze pads, tissues, or folded handkerchiefs to the edge of the heads. Put masking tape on three sides of the pad so that the untaped edge is free to vibrate and dampen the head motion. Don't overdo the damping, or the drum set will sound like cardboard boxes.

Oil the kick drum pedal to prevent squeaks. Tape rattling hardware in place.

Sometimes a snare drum buzzes in sympathetic vibration with a bass-guitar passage or a tom-tom fill. Try to control the buzz by wedging a thick cotton wad between the snares and the drum stand. Or tune the snare to a different pitch than the toms.

Drum Miking

Now you're ready to mike the set. For a tight sound, place a mic near each drum head. For a more open, airy sound, use fewer mics or mix in some room mics placed several feet away. Typical room mics are omni condensers or boundary mics. Figure 8.4 shows typical mic placements for a rock drum set. Let's look at each part of the kit.

SNARE

The most popular type of mic for the snare is a cardioid dynamic with a presence peak. The cardioid pattern reduces leakage; its proximity effect boosts the bass for a fatter sound. The presence peak adds attack. You might prefer a cardioid condenser for its sharp transient response.

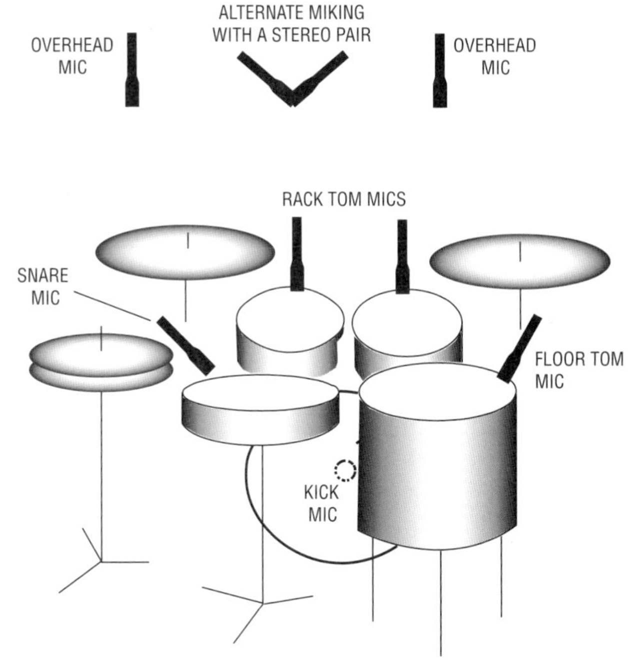

FIGURE 8.4
Typical mic placements for a rock drum set.

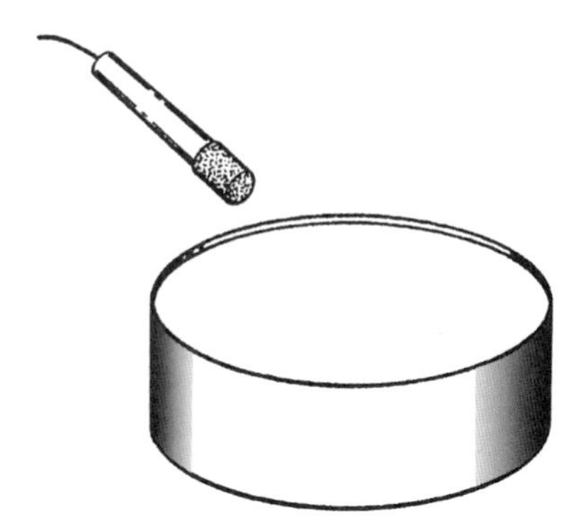

FIGURE 8.5
Snare-drum miking.

Bring the mic in from the front of the set on a boom. Place the mic even with the rim, 2 inches above the head (Figure 8.5). Angle the mic down to aim where the drummer hits, or attach a mini condenser mic to the side of the snare drum so it "looks at" the top head over the rim.

Some engineers mike both the top and bottom heads of the snare drum, with the microphones in opposite polarity. A mic under the snare drum gives a zippy sound; a mic over the snare drum gives a fuller sound. You might prefer to use just a top mic, and move it around until it picks up both the top head and snares. The sound is full with the mic near the top head, and thins out and becomes brighter as you move the mic toward the rim and down the side of the drum.

Whenever the hi-hat closes, it makes a puff of air that can "pop" the snare-drum mic. Place the snare mic so the air puff doesn't hit it. To prevent hi-hat leakage into the snare mic:

- Mike the snare closely.
- Bring the snare boom in under the hi-hat, and aim the snare mic away from the hi-hat.
- Use a piece of foam or pillow to block sound from the hi-hat.
- Use a de-esser on the snare.
- Don't play the hi-hat during tracking—overdub it later.

HI-HAT

Try a cardioid condenser mic about 6 inches over the cymbal edge that's farthest from the drummer (Figure 8.6). To avoid the air puff

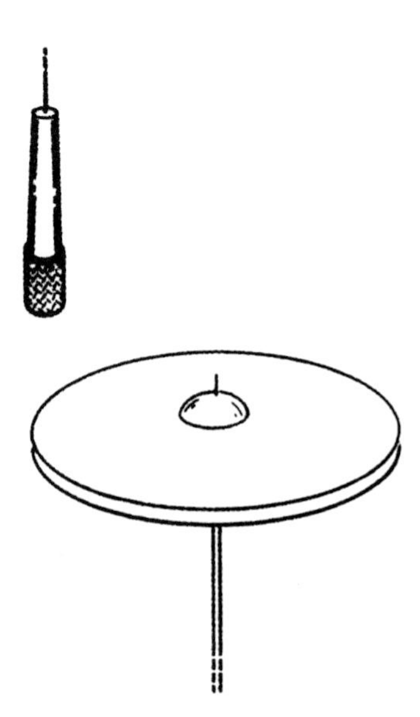

FIGURE 8.6
Hi-hat miking.

just mentioned, don't mike the hi-hat off its side; mike it from above aiming down. This also reduces snare leakage. Filter out the lows below about 500 Hz. You may not need a hi-hat microphone, especially if you use room mics. Usually the overhead mics pick up enough hi-hat.

TOM-TOMS

You can mike the toms individually, or put a mic between each pair of toms. The first option sounds more bassy. Place a cardioid dynamic about 2 inches over the drumhead and 1 inch in from the rim, angled down about 45 degrees toward the head (Figure 8.7). Again, the cardioid's proximity effect gives a full sound. Another way is to clip mini condenser mics to the toms, peeking over the top rim of each drum.

If the tom mics pick up too much of the cymbals, aim the "dead" rear of the tom mics at the cymbals. If you use a supercardioid or hypercardioid mic, aim the null of best rejection at the cymbals.

FIGURE 8.7
Tom-tom miking.

KICK DRUM

Place a blanket or folded towel inside the drum, pressing against the beater head to dampen the vibration and tighten the beat. The blanket shortens the decay portion of the kick-drum envelope. To emphasize the attack, use a wood or plastic beater—not felt—and tune the drum low.

A popular mic for kick drum is a large-diameter, cardioid dynamic type with an extended low-frequency response. Some mics are designed specifically for the kick drum, such as the AKG D112, Audio-Technica AT AE2500, Electro-Voice N/D868, and Shure Beta 52A.

For starters, place the kick mic inside on a boom, a few inches from where the beater hits (Figure 8.8). Mic placement close to the beater picks up a hard beater sound; off-center placement picks up more skin tone, and farther away picks up a boomier shell sound.

How should the recorded kick drum sound? Well, they don't call it a kick drum for nothing. THUNK! You should hear a powerful low-end thump plus an attack transient.

FIGURE 8.8
Kick-drum miking.

CYMBALS

To capture all the crisp "ping" of the cymbals, a good mic choice is a cardioid condenser with an extended high-frequency response, flat or rising at high frequencies. Place the overhead mics about 2 to 3 feet above the cymbal edges; closer miking picks up a low-frequency ring. The cymbal edges radiate the most highs. Place the cymbal mics to pick up all the cymbals equally. Try to place them the same distance from the snare. If your recording will be heard in mono, or for sharper imaging, you might want to mount the mic grilles together and angle the mics apart (Figure 8.4). This results in a narrow stereo spread. Another option is a stereo mic overhead. For wide stereo and sharp imaging, try a near-coincident pair aimed at the high-hat and floor tom.

Recorded cymbals should sound crisp and smooth, not muffled or harsh.

ROOM MICS

Besides the close-up drum mics, you might want to use a distant pair of room mics when you record drum overdubs. Place the mics about 10 or 20 feet from the set to pick up room reverb. When mixed with the close-up mics, the room mics give an open, airy sound to the drums. Popular room mics are omni condensers or boundary mics taped to the control-room window. You might compress the room mics for special effect. If you don't have enough tracks for room microphones, try raising the overhead mics.

BOUNDARY MIC TECHNIQUES

Boundary mics let you pick up the set in unusual ways. You can strap one on the drummer's chest to pick up the set as the drummer hears it. Tape them to hard-surfaced goboes surrounding the drummer. Put them on the floor under the toms and near the kick drum, or hang a pair over the cymbals. Try a supercardioid boundary mic in the kick drum.

RECORDING WITH TWO TO FOUR MICS

Sometimes you can mike the set simply. Place a stereo mic (or two mics) overhead and put another mic in the kick. If necessary,

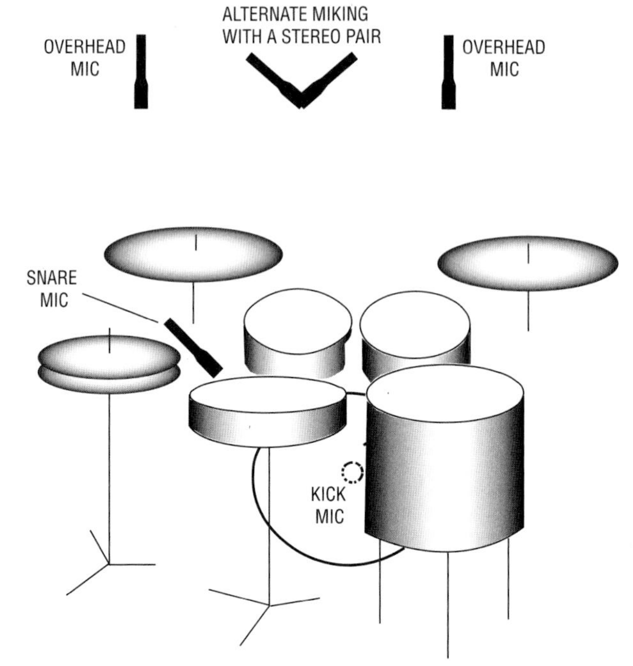

FIGURE 8.9
Miking a drum set with four mics.

add a snare-drum mic (Figure 8.9). This method works well for acoustic jazz, and often for rock. If you want the toms to sound fuller, boost the lows in the overhead mics. As an alternative try two mics about 18 inches apart angled down at the set from just over the drummer's head.

Another setup is shown in Figure 8.10. It uses only one mini omni condenser mic and one kick-drum mic. This method sometimes works well on small drum sets. Clip a mini omni condenser mic to the snare-drum rim about 4 inches above the rim, in the center of the set, aiming at the hi-hat. Or place a small-diaphragm omni there with a mic boom. Also mike the kick drum.

The mini mic will pick up the snare, hi-hat, and toms all around it, and will pick up the cymbals from underneath. Move the mic closer or farther from the toms, and raise or lower the cymbals, until you hear a pleasing balance. Add a little bass and treble. You'll be surprised at the good sound and even coverage you can get with this simple setup.

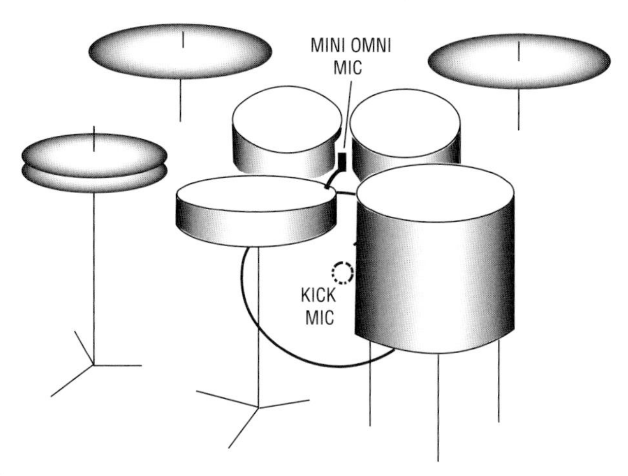

FIGURE 8.10
Miking a drum set with a mini omni mic.

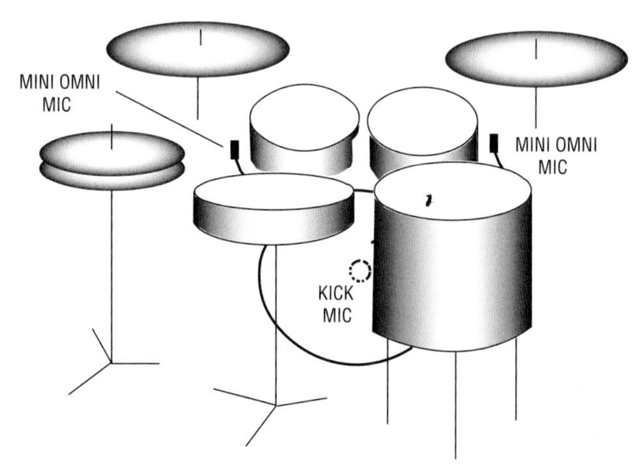

FIGURE 8.11
Miking a drum set with two mini omni mics.

Want a stereo effect? Mount one mic 4 inches above the snare drum rim between the hi-hat, snare drum, and rack tom. Adjust position for best balance. Mount another mic four inches above the floor-tom rim, on the side farthest from the drummer (Figure 8.11). Pan the mini mics left and right.

Audio clip 24 at www.elsevierdirect.com/companions/9780240811444 demonstrates several methods of miking a drum set.

Drum Recording Tips

After you set up all the mics, ask the drummer to play. Listen for rattles and leakage by soloing each microphone. Try not to spend much time getting a sound; otherwise you waste the other musicians' time and wear out the drummer.

To keep the drum sound tight during mixdown, mute or delete drum tracks that aren't in use in a particular tune, use a noise gate on the kick and toms, or overdub the drums.

One dated effect for the snare drum is gated reverb. It's a short splash of bright-sounding reverberation, which is rapidly cut off by a noise gate or expander. Many effects units have a gated-reverb program.

Another trick is recording "hot." Using an analog multitrack (or its plug-in equivalent), record the drums at a high level so they distort just a little. It's also common to compress the kick and snare.

A drummer might use drum pads, or drum triggers, fed into a sound module. Record directly off the module. You might want to mike the cymbals anyway for best sound.

If you're recording a drum machine and it sounds too mechanical, add some real drums. The machine can play a steady background while the drummer plays fills.

When miking drums on stage for PA, you don't need a forest of unsightly mic stands and booms. Instead, you can use short mic holders that clip onto drum rims and cymbal stands (Mic-Eze at www.ac-cetera.com), or use mini condenser mics.

In a typical rock mix, the drums either are the loudest element, or are slightly quieter than the lead vocal. The kick drum is almost as loud as the snare. If you don't want a wimpy mix, keep those drums up front!

Try these EQ settings to enhance the recorded sound of the drums:

- Snare: Fat at 200 Hz, crack at 5 kHz, sizzle at 10 or 12 kHz. Some snare drums ring a lot at one note. To fix it, set up an equalizer on the snare-drum track (equalization, EQ, is explained in Chapter 10.) Set a narrow, high-Q boost around 500 Hz and sweep it in frequency until you amplify the frequency that is ringing. Apply a narrow, high-Q cut there to remove the ringing.

- Toms: Cut around 600 to 800 Hz to reduce the papery sound. Then if necessary, boost around 100 to 200 Hz for more fullness on rack toms or 80 to 100 Hz on floor toms. Boost around 5 kHz for more attack.
- Cymbals: Sizzle at 10 kHz or higher. If you're close-miking the toms, you might roll off the lows in the cymbal mics below 500 Hz to reduce low-frequency leakage.
- Kick drum: To remove the "cardboard" sound, cut at 300 to 600 Hz. Then if necessary, boost at 3 to 5 kHz for more click. Don't overdo the high-frequency boost; usually you don't want too much "point" on the kick sound. Boost 60 to 80 Hz if the kick sounds thin. Filter out highs above 9 kHz to reduce leakage from cymbals.

Try these tricks to come up with unusual drum sounds:

- Record with a cheap dynamic or crystal mic, maybe in a can.
- Run the drums through extreme processing: compression, gating, distortion, pitch shifting, tremolo, and so on.
- Substitute other objects for drums, cymbals, drumsticks, and brushes.
- Move a mic around a cymbal or drumhead while recording it.
- Put the drums in a reverberant room or hallway.
- Try the preverb effect described in Chapter 10.

Instead of recording an acoustic drum set, you might use an electronic drum set or CD of drum samples. Copy the samples into a sampler or sampling software, then trigger them with a MIDI sequencer or MIDI controller with drum pads.

PERCUSSION

Let's move on to percussion, such as the cowbell, triangle, tambourine, or bell tree. A good mic for metal percussion is a condenser type because it has sharp transient response. Mike at least 1 foot away so the mic doesn't distort.

You can pick up congas, bongos, and timbales with a single mic between the pair, a few inches over the top rim, aimed at the heads. Or put a mic on each drum. It often helps to mike these drums top and bottom, with the bottom mic in opposite polarity. A cardioid dynamic with a presence peak gives a full sound with a clear attack.

For xylophones and vibraphones, place two cardioid mics 1–1/2 feet above the instrument, aiming down. Cross the mics 135 degrees apart or place them about 2 feet apart. You'll get a balanced pickup of the whole instrument.

ACOUSTIC GUITAR

The acoustic guitar has a delicate timbre that you can capture through careful mic selection and placement. First prepare the acoustic guitar for recording. To reduce finger squeaks, try commercial string lubricant, a household cleaner/waxer, talcum powder on fingers, or smooth-wound strings. Ask the guitarist to play louder; this increases the "music-to-squeak" ratio!

Replace old strings with new ones a few days before the session. Experiment with different kinds of guitars, picks, and finger picking to get a sound that's right for the song.

For acoustic guitar, a popular mic is a pencil-type condenser with a smooth, extended frequency response from 80 Hz up. This kind of mic has a clear, detailed sound. You can hear each string plucked in a strummed chord. Usually the sound picked up is as crisp as the real thing.

Now let's look at some mic positions. To record a classical guitar solo in a recital hall, mike about 3 to 6 feet away to pick up room reverb. Try a stereo pair (Figure 8.12A), such as XY, ORTF, MS, or spaced pair (described in Chapter 7). If you record a classical guitar solo in a dead studio, mike about 1.5 to 2 feet away and add artificial reverb.

When you record pop, folk, or rock music, try a spot about 6 to 12 inches from where the fingerboard joins the guitar body (Figure 8.12B). That's a good starting point for capturing the acoustic guitar accurately. Still, you need to experiment and use your ears. Close to the bridge, the sound is woody and mellow.

In general, close miking gives more isolation, but tends to sound harsh and aggressive. Distant miking lets the instrument "breathe"; you hear a gentler, more open sound.

Another spot to try: Tape a mini omni mic onto the body near the bottom of the sound hole, and roll off the excess bass. This spot gives good isolation (Figure 8.12C).

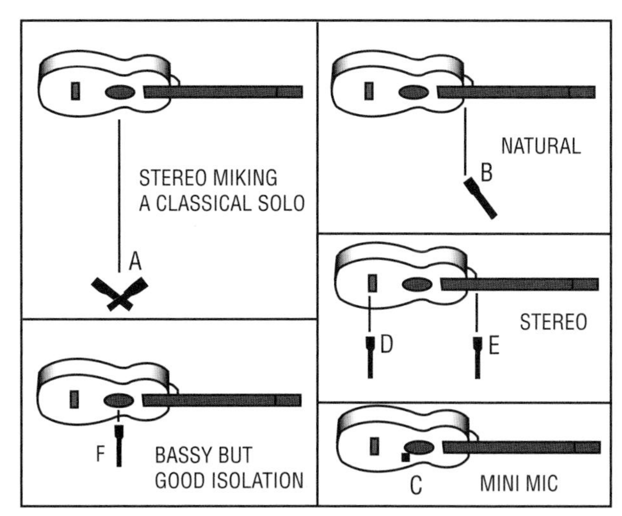

FIGURE 8.12
Some mic techniques for acoustic guitar.

The guitar will sound more real if you record in stereo. Try one mic near the bridge, and another near the 12th fret (Figures 8.12D and E). Pan part-way left and right. Another way to record stereo is with an XY pair of cardioid mics about 6 inches from the 12th or 16th fret, mixed with a 3-foot-spaced pair of omni mics about 3 feet away.

Is feedback or leakage a problem? Mike close to the sound hole (Figure 8.12F). The tone there is very bassy, so turn down the low-frequency EQ on your mixer until the sound is natural. Also cut a few decibels around 3 kHz to reduce harshness.

You get the most isolation with a contact pickup. It attaches to the guitar, usually under the bridge. The sound of a pickup is something like an electric guitar. You can mix a mic with a pickup to add air and string noise to the sound of the pickup. That way, you get good isolation and good tone quality. You might record the acoustic guitar off its pickup while tracking to prevent leakage, then overdub the guitar later with a microphone for its better sound quality.

SINGER/GUITARIST

Normally you overdub the guitar and vocal separately. But if you have to record both at once, the vocal might sound filtered or hollow because of phase cancellations between the vocal mic and guitar mic.

This can happen whenever two mics pick up the same source at approximately equal levels, at different distances, and both mixed to the same channel. Try one of these methods to solve the problem:

- Angle the vocal mic up and angle the guitar mic down to isolate the two sources. Follow the 3:1 rule described in Chapter 7.
- Mike the voice and guitar very close. Roll off the excess bass with your mixer's EQ.
- Use a pickup on the guitar instead of a mic.
- Place two bidirectional mics so the tops of their grilles touch. This gets rid of any delay between their signals. Aim the "dead" side of the vocal mic at the guitar; aim the dead side of the guitar mic at the mouth.
- Use just one mic, or a stereo mic, midway between the mouth and guitar about 1 foot out front. Adjust the balance between voice and guitar by changing the mic's height.
- Delay the vocal mic signal by about 1 msec. Then the signals of the two mics will be more in-phase, preventing phase cancellations when they are mixed to the same channel. Some multitrack recorders have a track-delay feature for this purpose.

GRAND PIANO

This magnificent instrument is a challenge to record well. First have the piano tuned, and oil the pedals to reduce squeaks. You can prevent thumps by stuffing some foam or cloth under the pedal mechanism.

For a classical-music solo, record in a reverberant room such as a recital hall or concert hall. Reverb is part of the sound. Set the piano lid on the long stick. Use condenser mics with a flat response. Place a stereo mic, or a stereo pair of cardioid mics, about 7 feet away and 7 feet high, up to 9 feet away and 9 feet high (Figure 8.13). Move the mics closer to reduce reverb, farther to increase it. When using a pair of omni mics, place them 1.3 to 2 feet apart, 3 to 6.5 feet from the piano, and 4 to 5 feet high (Figure 8.14). You might need to mix in a pair of hall mics: try cardioids aiming away from the piano about 25 feet away.

When recording a piano concerto, give the piano a spot mic about 1 to 3 feet away. Put the mic in a shock mount.

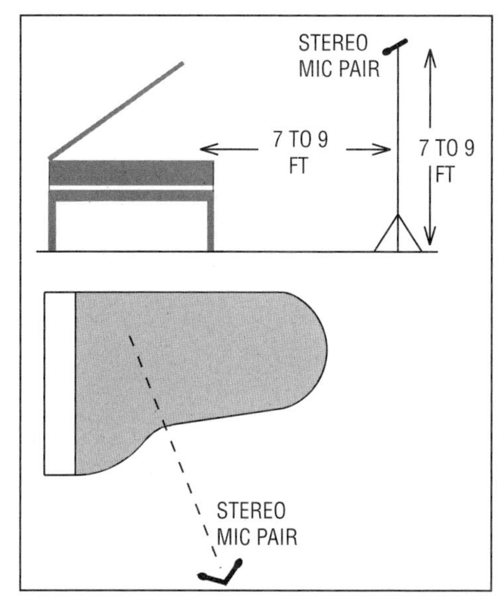

FIGURE 8.13
Suggested grand-piano miking for classical music (using cardioid mics).

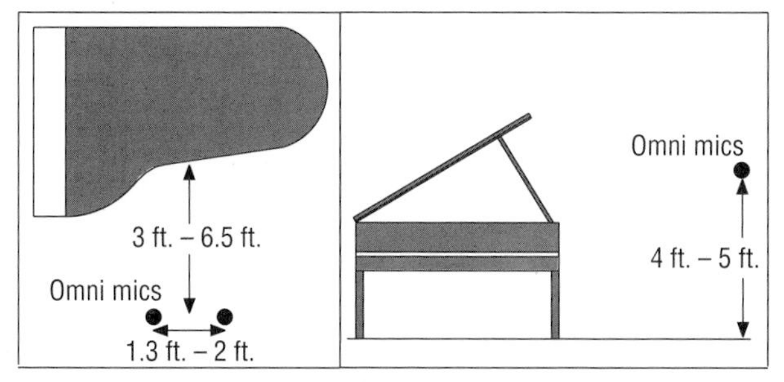

FIGURE 8.14
Suggested grand-piano miking for classical music (using omnidirectional mics).

Pop music demands close miking. Close mics pick up less room acoustics and leakage, and give a clear sound that cuts through the mix. Try not to mike the strings closer than 8 inches, or else you'll emphasize the strings closest to the mics. You want equal coverage of all the notes the pianist plays.

One popular method uses two spaced mics inside the piano. Use omni or cardioid condensers, ideally in shock mounts. Put the lid on the long stick. If you can, remove the lid to reduce boominess. Center one mic over the treble strings and one over the bass strings.

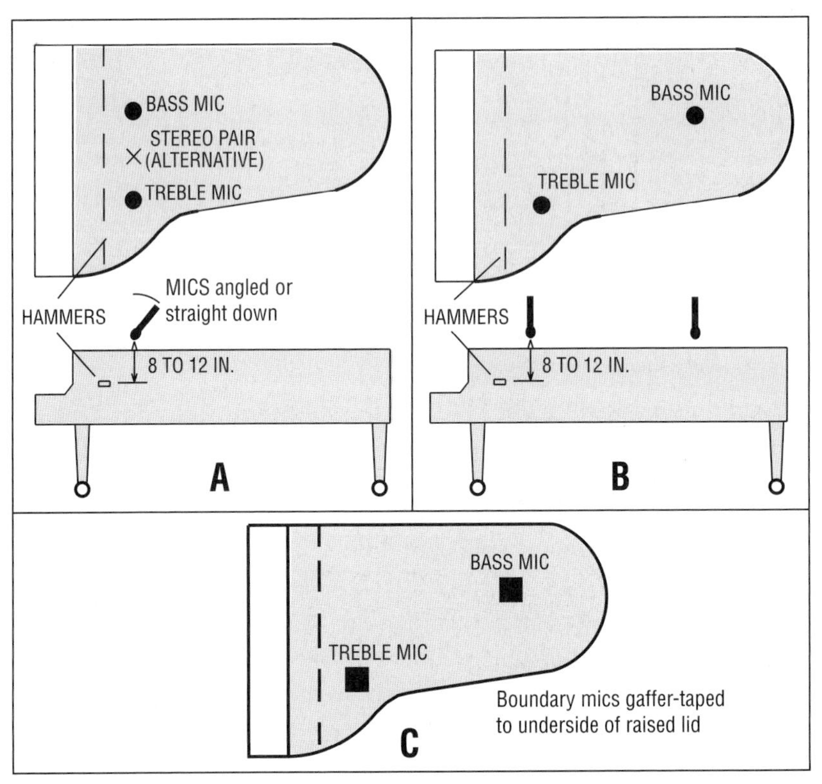

FIGURE 8.15
Suggested grand-piano miking for popular music.

Typically, both mics are 8 to 12 inches over the strings and 8 inches horizontally from the hammers (Figure 8.15A, bass and treble mics). Aim the mics straight down or angle them to aim at the hammers. Pan the mics partly left and right for stereo.

One alternative is to put the treble mic near the hammers, and put the bass mic about 2 feet toward the tail (Figure 8.15B). Another method uses two ear-spaced omni condensers or an ORTF pair about 12 to 18 inches above the strings. *Audio clip 25 at www.elsevierdirect.com/companions/9780240811444 demonstrates some mic techniques for grand piano.*

The spaced mics might have phase cancellations when mixed to mono, so you might want to try coincident miking (Figure 8.15A, stereo pair). Boom-mount a stereo mic, or an XY pair of cardioids crossed at 120 degrees. Miking close to the hammers sounds percussive; toward the tail has more tone.

For more clarity and attack, boost EQ around 10 kHz or use a mic with a rising high-frequency response.

Boundary mics work well, too. If you want to pick up the piano in mono, tape a boundary mic to the underside of the raised lid, in the center of the strings, near the hammers. Use two for stereo over the bass and treble strings. Put the bass mic near the tail of the piano to equalize the mic distances to the hammers (Figure 8.15C). If leakage is a problem, close the lid and cut EQ a little around 250 Hz to reduce boominess.

If your studio lacks a piano, consider using a software emulation of a piano. Some programs provide high-quality samples of piano notes that can be played with a sequencer or a MIDI controller. Examples include: Steinberg Grand VST 2.0 ($199 at www.steinberg.net) and Maxim Digital Audio Piano (freeware at www.mda-vst.com under VST synths).

UPRIGHT PIANO

Here are some ways to mike an upright piano:

(Figure 8.16A) Remove the panel in front of the piano to expose the strings over the keyboard. Place one mic near the bass strings and one near the treble strings about 8 inches away. Record in stereo and pan the signals left and right for the desired piano width. If you can spare only one mic for the piano, just cover the treble strings.

(Not shown) Remove the top lid and upper panel. Put a stereo pair of mics about 1 foot in front and 1 foot over the top. If the piano is against a wall, angle the piano about 17 degrees from the wall to reduce tubby resonances.

(Figure 8.16B) Aim the soundboard into the room. Mike the bass and treble sides of the soundboard a few inches away. In this spot, the mics pick up less pedal thumps and other noises. Try cardioid dynamic mics with a presence peak.

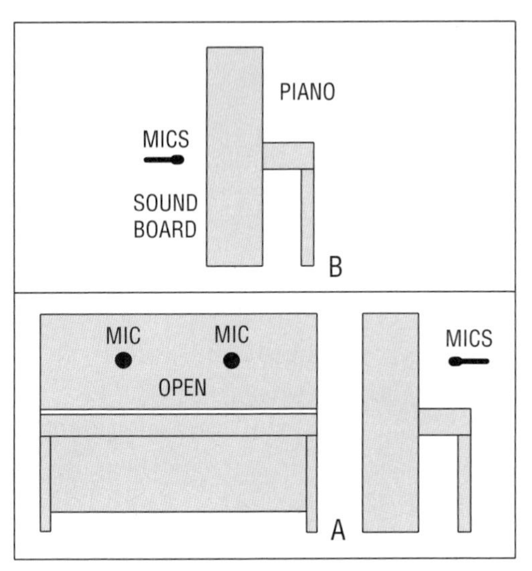

FIGURE 8.16
Some mic techniques for upright piano.

ACOUSTIC BASS

The acoustic bass (string bass, double bass, upright bass) puts out frequencies as low as 41 Hz, so use a mic with an extended low-frequency response such as a large-diaphragm condenser mic or ribbon mic. As always, closer miking improves isolation, while distant miking tends to sound more natural but can pick up too much room sound. Try these techniques (Figure 8.17):

- 4 to 8 inches in front of the bridge, a few inches above the bridge.
- 4 to 6 inches under the bridge, a few inches from the strings. This mic will pick up a deep sound with good definition. You might mix in a second mic near the plucking fingers for clarity.
- Mix a pickup with a mic, or use a pickup alone and EQ it to sound good.

Here are some methods that isolate the bass and let the player move around. They work well for PA:

- Wrap a mini omni condenser mic in foam rubber (or in a foam windscreen) and mount it in the bridge aiming up (Figure 8.17).
- Tape a mini omni mic to the bridge, or wedge it into a slot in the bridge.

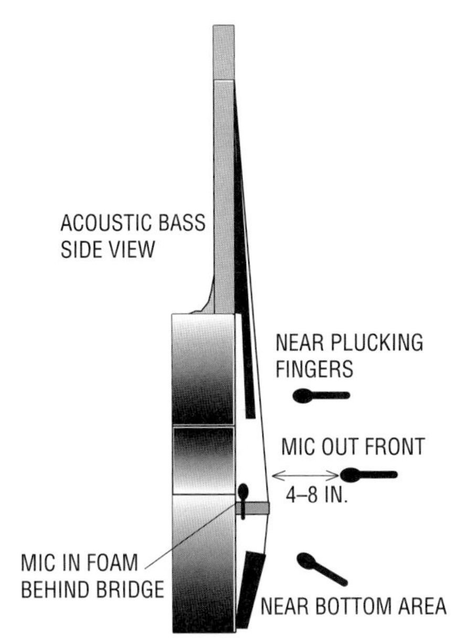

FIGURE 8.17
Some mic techniques for the acoustic bass.

- Wrap a regular cardioid mic in foam padding (except the front grille) and squeeze it behind the bridge (Figure 8.17) or tailpiece.
- For best isolation, try a direct feed from a pickup. This method adds clarity and deep bass, but probably will need some EQ. You might mix the pickup with a microphone.

BANJO

Try a flat-response condenser mic about 1 foot away from the center (Figure 8.18). Or try a cardioid dynamic mic 6 inches from the lower right area or 6 inches from the lower rim, aiming at the bridge. If you need more isolation, mike closer and roll off some bass. The banjo sounds pleasantly mellow when miked toward the edge of the head, near the resonator holes (if the banjo has them). Cloth stuffed inside will tighten the sound and will reduce feedback in PA situations.

For the most isolation, tape a mini omni condenser mic to the head about 1 inch in from the bottom edge, or on the tailpiece, or on the bridge. You can wedge a pickup between the strings below the bridge and the banjo head. Put the pickup flat against the head surface.

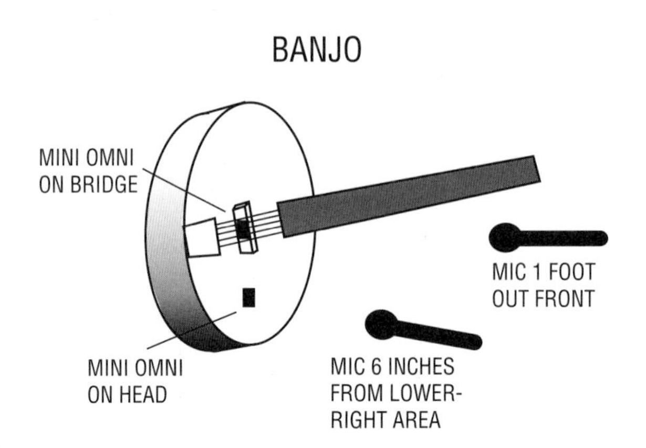

FIGURE 8.18
Four methods for miking a banjo.

MANDOLIN, DOBRO, BOUZOUKI, AND LAP DULCIMER

Mike these about 6 to 8 inches away from the hole with a small-diaphragm condenser mic. Some engineers also mix in a mic near the neck or dobro resonator. If you need more lows and more isolation, tape a mini omni condenser mic near a hole and tweak EQ for the best sound.

HAMMERED DULCIMER

Place a flat-response condenser mic about 2 feet over the center of the soundboard (Figure 8.19A). On stage, place a cardioid dynamic or condenser 6 to 12 inches over the middle of the top end (Figure 8.19B). For the best gain-before-feedback in a PA system, mix in a mini omni condenser mic (or a cardioid with bass rolloff) very near the sound hole (Figure 8.19C).

FIDDLE (VIOLIN)

Listen to the fiddle itself to make sure it sounds good. Correct any instrument problems before miking.

First try a flat-response condenser mic (omni or cardioid) about 2 feet over the bridge. This distant miking gives an airy, silky sound. Close miking (about 6 to 12 inches, Figure 8.20) sounds more aggressive, which is desirable in old-time or bluegrass music. Aim the

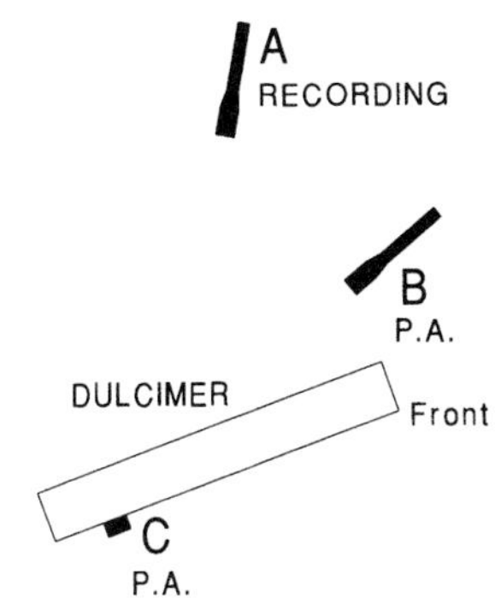

FIGURE 8.19
Some mic techniques for hammered dulcimer.

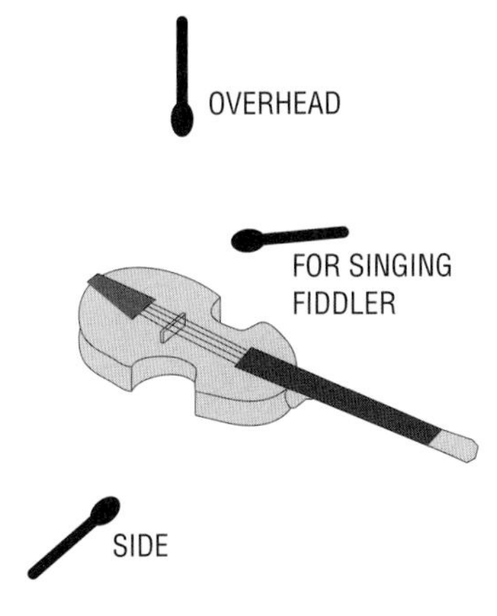

FIGURE 8.20
Three fiddle-miking methods.

mic toward the f-holes for warmth or toward the fingerboard for clarity. A fiddle that sounds too bright and scratchy can be tamed using a ribbon mic, or by miking it from the side or even underneath. If the ceiling over the fiddle is low, nail a square yard of acoustic foam up there to prevent reflections.

If you have to mike close—say, for a singing fiddler—aim the mic at the player's chin from 6 to 12 inches above the end of the fingerboard. The mic will pick up both the singer and the fiddle.

If you need more isolation, try a mini omni mic. Wrap its cable in foam rubber (or a windscreen) 1–1/2 inches from the capsule. Wedge the foam under the tailpiece, and position the mic capsule

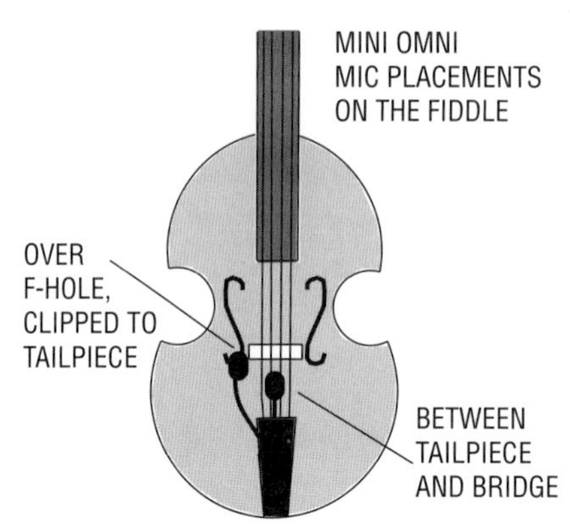

FIGURE 8.21
Two ways to close-mike a fiddle for isolation.

halfway between the tailpiece and bridge, a half inch over the body (Figure 8.21). If necessary, cut a little at 3 kHz to reduce harshness and boost around 200 Hz for warmth. Another option is to clip the mic to the tailpiece and mount it over an f-hole.

A good spot for a pickup is on the left side of the top (player's view), on the player's side of the bridge.

To record a classical violin solo, try a stereo mic (or a stereo pair) 5 to 15 feet away in a reverberant room.

STRING SECTION

Place the strings in a large, live room and mike them at a distance to pick up a natural acoustic sound. A common mic choice is a condenser with a flat response. First try a stereo mic or stereo pair of mics about 4 to 20 feet behind the conductor, raised about 13 to 15 feet.

If the room is noisy or too dead, or the balance is poor, you'll need to mike close and add digital reverb. Try one mic on every two to four violins, 6 feet off the floor, aiming down. Same for the violas. Mike the cello about 1 to 2 feet from the bridge, to the right side between the bridge and f-hole. When you mix the strings to stereo, pan them evenly between the monitor speakers. Spread them left, center, and right to make a "curtain of sound." If you can spare only one track for the strings, use a stereoizer effect during mixdown.

STRING QUARTET

Record a quartet in stereo using a stereo mic or a pair of mics. Place them about 6 to 10 feet away to capture the room ambience. The monitored instruments should not spread all the way between speakers. If you want to narrow the stereo stage, angle or space the mics closer together.

BLUEGRASS BAND AND OLD-TIME STRING BAND

Suppose you're recording a group that has a good acoustic balance. Try a stereo mic or stereo pair of mics about 2 to 3 feet away and 5 feet high (lower if the group is seated). Move the players toward or away from the mics to adjust their balance.

You'll have more control if you mike all the instruments up close and mix them. This also gives a more "commercial" sound. The production style aims for a natural timbre on all the instruments, either with no effects or with slight reverb.

HARP

Use a condenser mic with a flat response. If the harp is playing with an orchestra, mike the harp about 18 inches from the front of the soundboard, or 18 inches from the player's left hand. With a Celtic group, try a mic 1/3 of the way up the sound board about 6 to 12 inches away.

Tape a mini omni condenser mic to the soundboard if you need isolation. A mic on the inside of the soundboard has more isolation; a mic on the outside sounds more natural. Also try a cardioid condenser wrapped in foam, stuck into the center hole from the rear.

HORNS

"Horns" in studio parlance refers to the brass instruments: trumpets, cornets, trombones, baritones, French horns, and tubas.

All the brass radiate strong highs straight out from the bell, but do not project them to the sides. A mic close to and in front of the bell picks up a bright, edgy tone. To mellow out the tone, mike the bell off-axis with a flat-response mic (Figure 8.22). The sound on-axis to

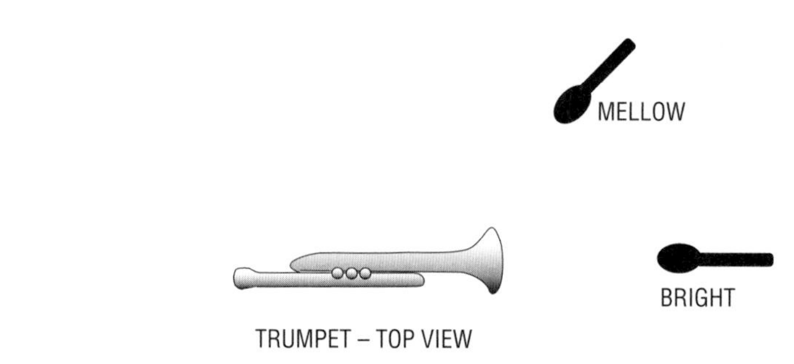

FIGURE 8.22
Miking for trumpet tone control.

the bell has a lot of spiky high harmonics that can overload a condenser mic, mixer input, or analog tape. That's another reason to mike off-axis.

Mike the trumpet with a dynamic or ribbon mic to take the edge off the sound. Use a condenser mic if you want a lot of sizzle. Mike about 1 foot away for a tight sound; mike several feet away for a fuller, more dramatic sound.

You can pick up two or more horns with one microphone. Several players can be grouped around a single omni mic, or around a stereo pair of mics. The musicians can play to a pair of boundary mics taped on the control-room window or on a large panel.

Record a classical brass quartet in a reverberant room. Use a stereo mic, or a stereo pair of mics, about 6 to 12 feet away.

SAXOPHONE

A sax miked very near the bell sounds bright, breathy, and rather hard (Figure 8.23). Mike it there for best isolation. To get a warm, natural sound, mike the sax about 1–1/2 feet away, halfway down the wind column. Don't mike too close, or else the level varies when the player moves. A compromise position for a close-up mic is just above the bell, aiming at the holes. You can group a sax section around one mic.

Figure 8.24 shows a typical miking setup for big-band jazz. It uses the techniques already described for the drums, bass, piano, electric guitar, trumpet, and sax.

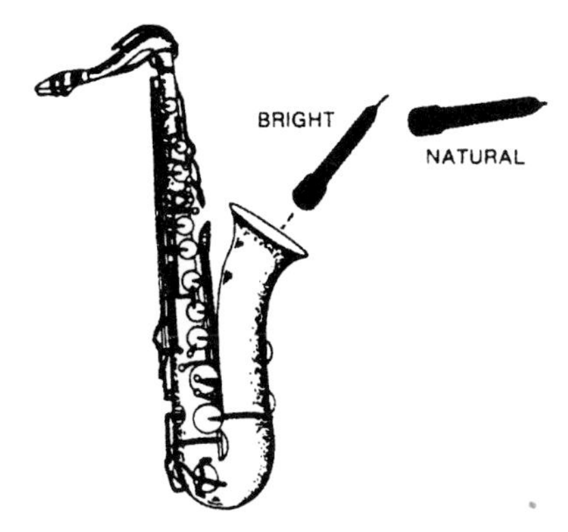

FIGURE 8.23
Two ways to mike a saxophone.

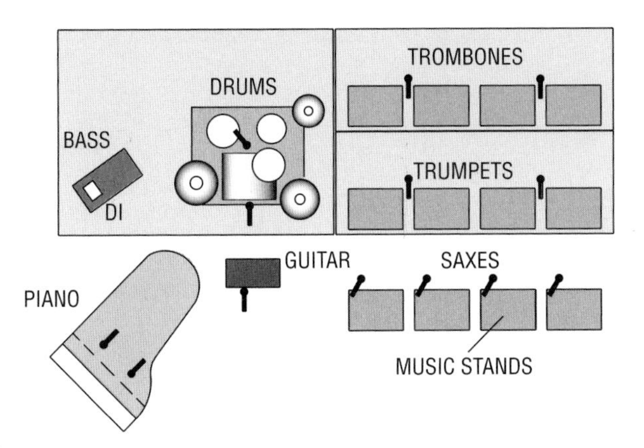

FIGURE 8.24
Typical miking setup for big-band jazz.

WOODWINDS

With woodwinds, most of the sound radiates not from the bell, but from the holes. So aim a flat-response mic at the holes about 1 foot away (Figure 8.25).

When miking a woodwind section within an orchestra, you need to reject nearby leakage from other instruments. To do that, try aiming a bidirectional mic down over the woodwind section. The side nulls of the mic cut down on leakage.

To pick up a flute in a pop-music group, try miking a few inches from the area between the mouthpiece and the first set of finger

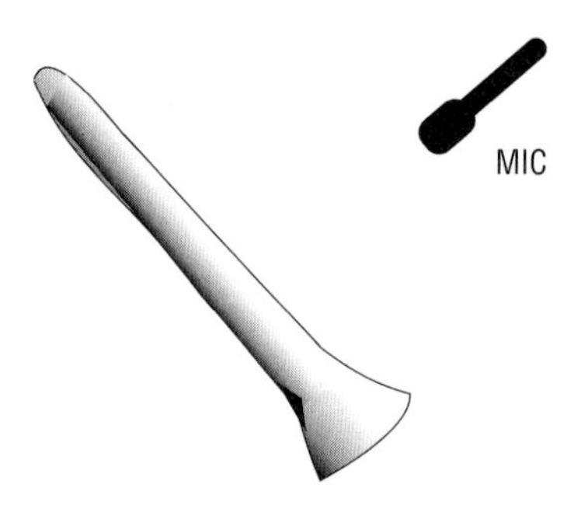

FIGURE 8.25
Miking a clarinet from the side.

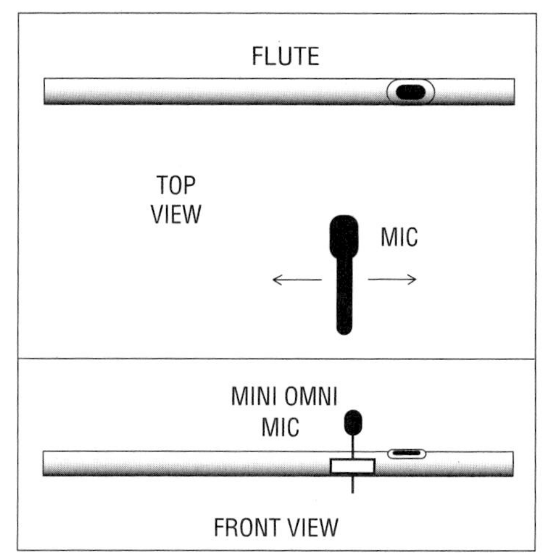

FIGURE 8.26
Two methods of miking a flute.

holes (Figure 8.26). You may need a pop filter. If you want to reduce breath noise, roll off high frequencies or mike farther away. You also can attach a mini omni mic to the flute a few inches above the body, between the mouthpiece and finger holes.

For classical music solos, try a stereo pair 4 to 12 feet away.

HARMONICA, ACCORDION, AND BAGPIPE

One way to mike a harmonica (harp) is to use a cardioid dynamic mic with a ball grille. Place the mic very close to the harmonica or have the player hold it. A condenser mic about 1 foot away gives a natural sound. To get a bluesy, dirty sound, use a "bullet"-type harmonica mic or play the harmonica through a miked guitar amp.

For an accordion, try a mic about 6 to 12 inches from the sound holes near the keyboard. Some accordions have sound holes on both sides, so you'll need two mics. Follow the 3:1 rule. The distance between mics should be at least three times the mic-to-source distance. One end of the accordion is in constant motion, so you might want to attach a mini omni mic to that end. A solo accordion or concertina could be miked with a stereo pair of flat-response cardioid condenser mics about 3 to 6 feet in front.

A bagpipe has two main sound sources: the chanter, which the musician plays with the fingers, and the drone pipes, which make a steady tone. Mike the chanter about a foot away from the side, and mike the drone pipes a foot from the end. Again, follow the 3:1 rule. You could also mike the bagpipe a few feet away with one mic.

LEAD VOCAL

The lead vocal is the most important part of a pop song, so it's critical to record it right. First set up a comfortable environment for the singer. Put down a rug, add some flowers or candles, dim the lights. Set up a good cue mix with effects to help the singer get into the mood of the song.

You might want to turn off the reverb in the singer's headphones; this makes it easier to hear pitch. If the vocalist is singing flat, reduce their headphone volume, and vice versa.

With any vocal recording, there are some problems to overcome, but we can deal with them. Among these are proximity effect, breath pops, wide dynamic range, sibilance, and sound reflections from the music stand. Let's look at these in detail.

Miking Distance

When you sing or talk close to most directional mics, the microphone boosts the bass in your voice. This is called the proximity effect. We've come to accept this bassy sound as normal in a PA system, but the effect just sounds boomy in a recording.

FIGURE 8.27
Typical miking technique for a lead vocal.

To prevent boomy bass, mike the singer at a distance, about 8 inches away (Figure 8.27). A popular mic choice is a flat-response condenser mic with a large diaphragm (1–1/4-inch diameter). As always, you can use any mic that sounds good to you. If the mic has a bass rolloff switch, set it to "flat."

Singers should maintain their distance to the mic. I ask the singer to spread the fingers, touch lips with the thumb, and touch the mic with the pinky. The hand forms a spacer for keeping a constant distance.

If the singer's voice is too harsh or edgy, try miking partly to the side of the mouth, aiming at the mouth. Or try a ribbon mic, or a multiband compressor set to compress from 2000 Hz and up.

Some singers feel more comfortable singing into a handheld mic. You can give them a handheld mic but also mike them several inches away with a good condenser microphone. Record both mics on separate tracks and choose whatever sounds the best.

If you must record the singer and the band at the same time—as in a concert—you'll have to mike close to avoid picking up the instruments with the vocal mic. Try a cardioid or supercardioid mic with a foam pop filter. The sound will be bassy because of proximity effect, so roll off the excess lows at your mixer. For starters, try −6 dB at 100 to 200 Hz. Some mics have a bass filter switch for this purpose. Aim the mic partly toward the singer's nose to prevent a nasal or closed-nose effect. This close-up method works well if you want an intimate, breathy sound.

When recording a classical-music singer who is accompanied by an orchestra, place the mic about 1 to 2 feet away. If the singer is a soloist (maybe accompanied by piano), use a stereo pair about 8 to 15 feet away to pick up room reverb.

Breath Pops

When you sing a word with "p" or "t" sounds, a turbulent puff of air shoots out of the mouth. The puff hits the mic and makes a thump or small explosion called a pop. To reduce it, put a foam-plastic pop filter on the mic. Some mics have a ball grille screen to cut pops, but foam works better. The pop filter should be made of special open-cell foam to pass high frequencies. For best pop rejection, allow a little air space between the foam and the front of the mic grille.

Foam pop filters reduce the highs a little. So they should be left off instrument mics, except for outdoor recording or dust protection. Pop filters don't reduce breathing sounds or lip noises. To get rid of these problems, mike farther away or roll off some highs.

The most effective pop filter is a hoop with a nylon stocking stretched over it (Figure 8.27) or a disk of perforated metal. You can buy those, or make one with a coat hanger and a crochet hoop. Place the filter a few inches from the mic.

Another way to get rid of pop is to put the mic at forehead height, aiming at the mouth. This way the puffs of air shoot under the mic and miss it. Make sure the vocalist sings straight ahead, not up at the mic, or the mic will pop.

Wide Dynamic Range

During a song, vocalists often sing too loud or too soft. They blast the listener or get buried in the mix; that is, many singers have a wider dynamic range than their instrumental backup. To even out these extreme level variations, ask the singer to use proper mic technique. Back away from the mic on loud notes; come in closer for soft ones. Or you can ride gain on singers: gently turn them down as they get louder, and vice versa. Usually this is done during mixdown with automation or volume envelopes.

Another solution is to pass the vocal signal through a compressor, which acts like an automatic volume control. Plug the compressor into the vocal channel's insert jacks or insert a compressor plug-in in the vocal track during mixdown. A typical compressor setting for vocals is a 2:1 ratio, −10 dB threshold, and about 3 to 6 dB of gain reduction. Of course, you should use whatever settings are needed for the particular singer. *Audio clip 26 at www.elsevierdirect.com/companions/9780240811444 demonstrates vocal compression, as well as breath pops and the effects of miking distance on recorded vocals.*

If the singer moves toward and away from the mic while singing, the average level will go up and down. Try to mike the singer at least 8 inches away, so that small movements of the singer won't affect the level.

If you must mike close to prevent leakage or feedback, ask the vocalist to sing with lips touching the foam windscreen to keep the same distance to the mic. Turn down the excess bass using your mixer's low-frequency EQ (typically −6 dB at 100 to 200 Hz).

Sibilance

Sibilance is the emphasis of "s" or "sh" sounds, which are strongest around 3 to 10 kHz. They help intelligibility. In fact, many producers like sizzly "s" sounds, which add a bright splash to the vocal reverb. But the sibilance shouldn't be piercing or strident.

If you want to reduce sibilance, use a mic with a flat response—rather than one with a presence peak—or cut the highs little around 8 kHz on your mixer. Better yet, use a de-esser signal processor or plug-in, which cuts the highs only when the singer makes sibilant sounds. You can set a multiband compressor to act as a de-esser: see Chapter 10 under the heading Suggested "Ballpark" Compressor Settings.

Reflections from the Music Stand and Ceiling

Suppose that a lyric sheet or music stand is near the singer's mic. Some sound waves from the singer go directly into the mic. Other sound waves reflect off the lyric sheet or music stand into the mic (Figure 8.28, top). The delayed reflections will interfere with the direct sound, making a colored tone quality like mild flanging.

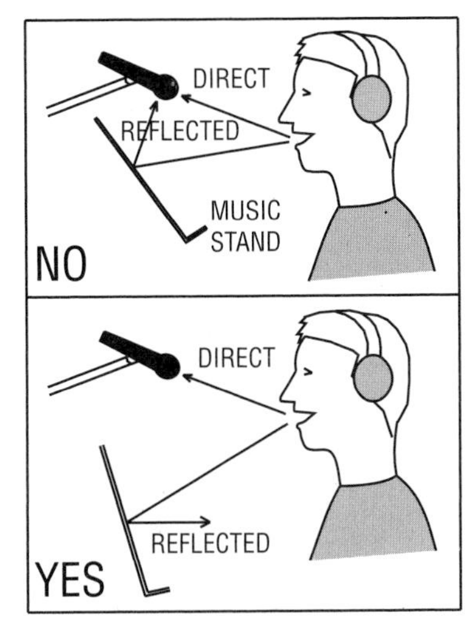

FIGURE 8.28
Preventing reflections from a music stand.

To prevent this, lower the music stand and tilt it almost vertically (Figure 8.28, bottom). This way, the sound reflections miss the mic.

If your studio has a low ceiling, the recorded vocal might have a colored tone quality due to phase cancellations from ceiling reflections. Try putting the mic lower and use a hoop-type pop filter. Also put a 3-foot square of acoustic foam or pressed-fiberglass panel on the ceiling over the singer and mic.

Vocal Effects

Some popular vocal effects are stereo reverb, echo, and doubling. You can record real room reverb by miking the singer at a distance in a hard-surfaced room. Slap echo provides a 1950s rock 'n' roll effect. Often a vocal is mixed dry, with no reverb. A little distortion might even be effective on some songs. You might try a vocal processor, which offers a variety of effects. Try different EQ or different effects on each section of a song.

Doubling a vocal gives a fuller sound than a single vocal track. Overdub a second take of the vocal on an empty track, in sync with the original take. During mixdown, mix the second vocal take with the original, at a slightly lower level than the original. You can double a vocal track by running it through a digital delay set at 15 to 35 msec, or through a pitch shifter that is detuned 10 to 15 cents.

BACKGROUND VOCALS

When you overdub background vocals (harmony vocals), you can group two or three singers in front of a mic. The farther they are from the mic, the more distant they will sound in the recording. Pan the singers left and right for a stereo effect. Because massed harmonies can sound bassy, roll off some lows in the background vocals.

If you want independent control of each background singer, give each one a close-up mic and record them with separate mixer channels or separate tracks.

Barbershop, vocal jazz, or gospel quartets with a good natural blend can be recorded with a stereo mic or stereo pair of mics about 2 to 4 feet away. If their balance is poor, close-mike each singer about 8 inches away, and balance them with your mixer. This also gives a more "commercial" sound. If you close-mike, spread the singers at least 2 feet apart in a circle to prevent phase cancellations.

SPOKEN WORD (SUCH AS PODCASTS AND AUDIO BOOKS)

The tips given earlier for a lead vocalist also apply to recording the spoken word. Be sure to keep the miking distance constant and use a hoop-type pop filter. To prevent sound reflections into the mic, put the script on a padded music stand that is angled almost vertically, and put the mic in the plane of the stand near the top edge. Fold up a corner of each script page to form a handle for turning pages silently.

The engineer and announcer should both have the same script. Mark the beginning of each misread sentence. The announcer should re-read each misread sentence from the beginning to make editing easier.

CHOIR AND ORCHESTRA

Figure 8.29 shows three ways of miking a choir. If the mics will also be used for PA, or if the venue is noisy or sounds bad, try miking close, pan the mic as desired, and add artificial reverb (Figure 8.29A).

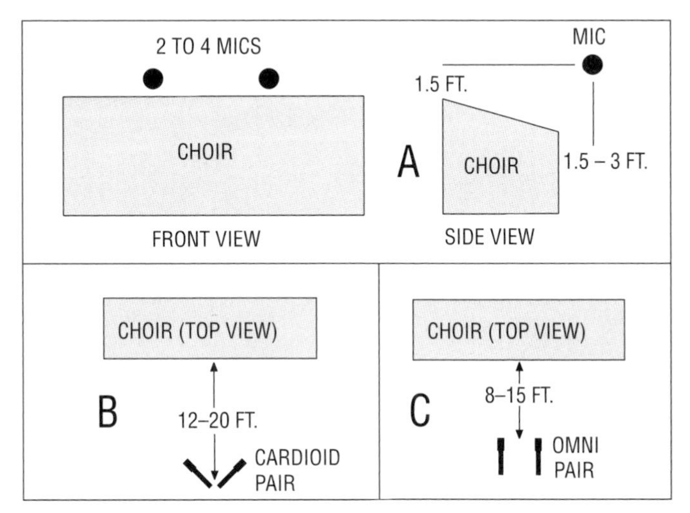

FIGURE 8.29
Choir miking suggestions. (A) Close-up panned mics. (B) Near-coincident stereo pair. (C) Spaced stereo pair.

Otherwise, try a near-coincident pair of cardioid mics (Figure 8.29B) or a pair of omni mics spaced about 2 feet apart (Figure 8.29C). Adjust the mic-to-choir distance until you hear the desired amount of hall acoustics in your monitors.

See Chapters 18 and 19 for suggestions on miking an orchestra.

SUMMARY

We can sum up mic placement like this: If leakage or feedback are problems, place the mic near the loudest part of the instrument, and add EQ to get a natural sound. Otherwise, place the mic in various spots until you find a position that sounds good over your monitors. There is no single "correct" mic technique for any instrument. Just place the mic where you hear the desired tonal balance and amount of room reverb.

Try the techniques described here as a starting point, then explore your own ideas. Trust your ears! If you capture the power and excitement of electric guitars and drums, if you capture the beautiful timbre of acoustic instruments and vocals, you've made a successful recording.

CHAPTER 9

Digital Recording

Before 1976, all music was recorded on analog tape recorders. They record the audio signal as magnetic patterns that rise and fall in strength as the signal waves do. Digital recorders, on the other hand, store the audio signal as a numerical code of ones and zeros.

Let's venture into the world of digital audio. We'll overview how digital recording works, explore 2-track digital recorders, and explain multitrack digital recorders. Chapter 13 covers computer recording in detail.

DIFFERENCES BETWEEN ANALOG AND DIGITAL

Both analog and digital recorders can sound accurate—the playback signal sounds like the input signal. But there are subtle differences. A digital recording usually is called "clean" because it adds almost no noise or distortion to the input signal. Some analog tape recorders add a little "warmth" to the sound. It's likely due to slight third harmonic distortion, head bumps (bass boost), and tape compression at high recording levels. Compared to analog tape recordings, digital recordings have measurably less hiss, frequency-response errors, modulation noise, distortion, unsteady pitch, and print-through. However, those characteristics are inaudible in the best analog tape recorders when they are properly maintained and aligned. Although older digital recorders sounded harsh compared to analog, they improved with each generation.

Compared to analog tape recorders, digital recorders and their media tend to be smaller and lower cost. They can record timing information.

Also, they have random-access so you can locate a particular section of the recording very quickly.

DIGITAL RECORDING

Like an analog tape deck, a digital recorder puts audio on a magnetic medium, but in a different way. Here's what happens in the most common digital recording method—pulse code modulation or PCM:

1. The signal from your mixer, preamp, or audio interface (Figure 9.1A) is run through a lowpass filter (anti-aliasing filter), which removes all frequencies above 20 kHz.
2. Next, the filtered signal passes through an analog-to-digital (A/D) converter. This converter measures (samples) the voltage of the audio waveform several thousand times a second (Figure 9.1B).
3. Each time the waveform is measured, a binary number (made of 1's and 0's) is generated that represents the voltage of the waveform at the instant it's measured (Figure 9.1C). This process is called quantization. Each 1 and 0 is called a bit, which stands for binary digit. The more bits that are used to represent each measurement (the higher the bit depth), the more accurate the measurement.
4. These binary numbers are stored on the recording medium as a modulated square wave recorded at maximum level (Figure 9.1D). For example, the numbers can be stored magnetically on a hard disk.

The playback process is the reverse:

1. The binary numbers are read from the recording medium, such as a hard disk (Figure 9.2A).
2. A digital-to-analog (D/A) converter translates the numbers back into an analog signal made of voltage steps (Figure 9.2B).
3. An anti-imaging filter (lowpass filter, smoothing filter, reconstruction filter) smoothes the steps in the analog signal, resulting in the original analog signal (Figure 9.2C).

As a simplified analogy, suppose you want to reproduce the circular shape of a wooden disk.

1. You lay a ruler across the disk every 1/4 inch and measure how wide the disk is at each point (Figure 9.3A). That's like sampling.

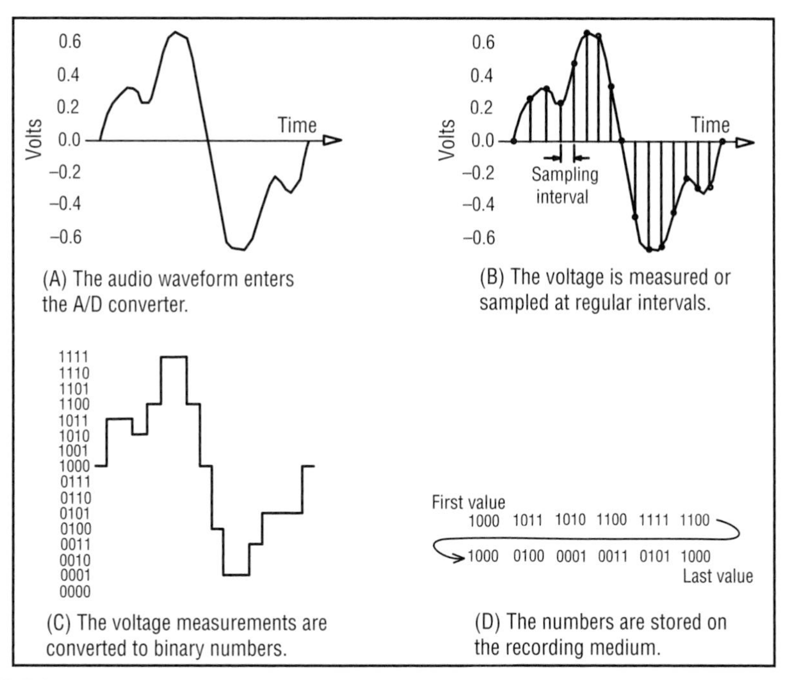

FIGURE 9.1
A/D conversion. See text. 4-bit measurements are shown for clarity, but 16- or 24-bit measurements are standard.

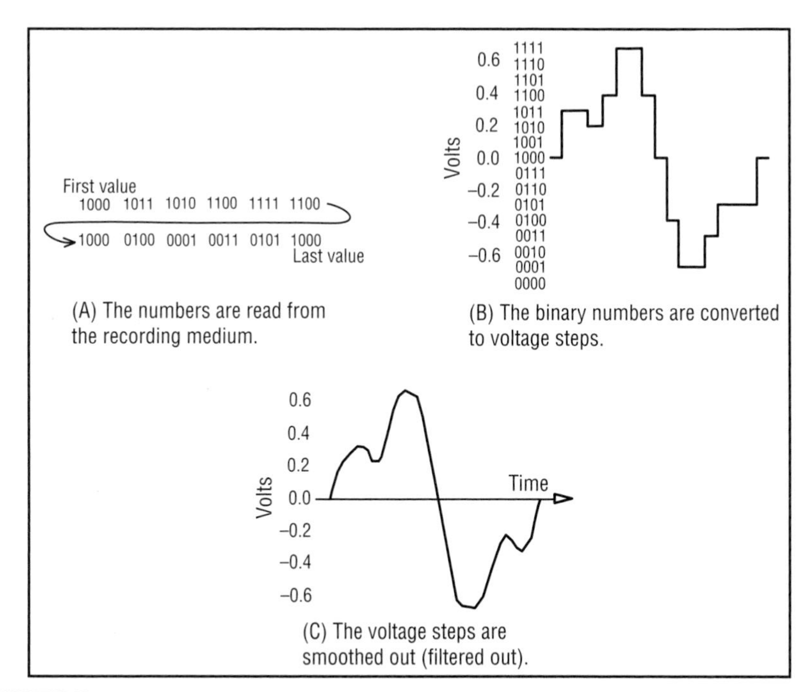

FIGURE 9.2
D/A conversion. See text. 4-bit measurements are shown for clarity, but 16- or 24-bit measurements are standard.

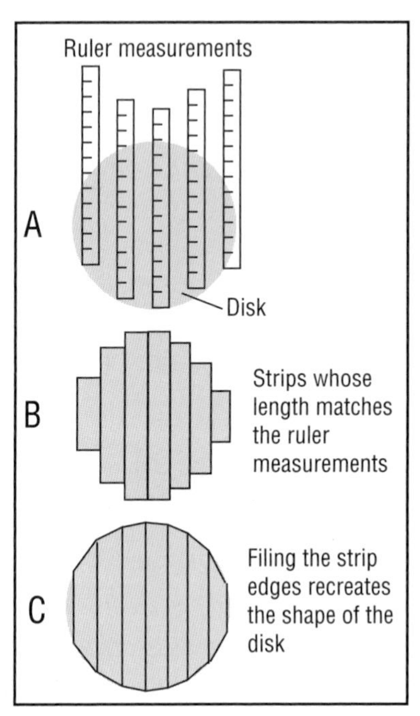

FIGURE 9.3
An analogy of digital recording and playback: recreating a disk shape using wooden strips.

2. You cut a bunch of wood strips with the same lengths as those measurements, and put the strips side-by-side. They look like stair steps going up and down (Figure 9.3B).
3. You file the raggedy edges of the wood strips in an arc so they join together smoothly. That's like the anti-imaging filter. You've just reproduced the shape of the original wooden disk (Figure 9.3C).

As we said, digital recording reduces noise, distortion, speed variations, and data errors. Because the digital playback head reads only 1's and 0's, it is insensitive to the magnetic medium's noise and distortion. During recording and playback, numbers are read into a buffer memory and read out at a constant rate, eliminating speed variations in the rotating media. Reed-Solomon coding during recording, and decoding during playback, corrects for missing bits by using redundant data.

If a digital recording is on a defective medium such as a scratched compact disc, errors (missing samples) can occur. Usually these errors can be corrected by interpolation. This algorithm looks at data before and after the blank sample and "guesses" what its value should be. If the errors are too extended to correct, the resulting audio has a silent spot or a burst of noise.

Almost all digital recording devices employ the same A/D, D/A conversion process, but use different storage media: a hard-disk drive records on magnetic hard disk, a compact disc and DVD recorder record on an optical disc, a memory recorder records onto a Flash memory card, and a sampler records into computer memory. The sound quality of any of these devices depends mainly on its A/D and D/A converters.

Digital audio is recorded on a computer hard drive as a wave file or Audio Interchange File Format (AIFF) file. Both are standard formats for audio files. Wave (.wav) is for PC and AIFF is for Mac. Both formats use linear PCM encoding, with no data compression (explained in Chapter 20). Two wave formats are Riff and Broadcast wave. The latter facilitates interchange of program material between audio workstations.

Bit Depth

As we said, the audio signal is measured many thousand times a second to generate a string of binary numbers (called words). The longer each word is (the more bits it has), the greater the accuracy of each measurement. Short words give poor resolution of the signal voltage (high distortion); long words give good resolution (low distortion). Bit depth or resolution are other terms for word length.

In our wooden disk analogy, bit depth (word length) corresponds to the accuracy of the ruler measurement: whether we measured to the closest 1/8", 1/16", 1/32", etc.

A word length of 16 bits is adequate (but not optimum) for hi-fi reproduction. It is the current standard for CDs. Most digital recorders offer 24-bit word lengths. More bits sound smoother and more transparent, but need more disk storage space.

Even though CDs are a 16-bit format, they sound better when made from 24-bit recordings. During mastering, you add dither (low-level noise) to the 24-bit recording, then export or save it as a 16-bit recording. You copy the 16-bit recording to CD. The dither helps the 16-bit recording sound more like the 24-bit recording.

In other words, dither lets you retain most of the quality and resolution that you recorded at 24 bits, even though the recording ends up on a 16-bit CD. We explain dither later in this chapter.

Suppose you record with a 24-bit word length. All signal levels in that recording have 24-bit resolution. Even low-level signals use all the bits—but most of the bits are zeroes.

Sampling Rate

Sampling rate or sampling frequency is the rate at which the A/D converter samples or measures the analog signal while recording. For example, a rate of 48 kHz is 48,000 samples per second; that is, 48,000 measurements are generated for each second of sound. The higher the sampling rate, the wider the frequency response of the recording. According to the Nyquist-Shannon theorem, the upper frequency limit of a digital recording is one-half the sampling rate. Compact discs use a 44.1 kHz sampling rate, so their frequency response extends to 22.05 kHz.

In our wooden disk analogy, sampling rate corresponds to the number of wood strips you cut to form the disk. If you measure the disk every 1/2″ across, that is like using a low sampling rate. If you measure the disk every 1/8″ across, that is like using a high sampling rate.

Sampling rate for high-quality audio can be 44.1, 48, 88.2, 96, or 192 kHz. CD quality is 44.1 kHz/16 bits. A 96 kHz sampling rate can be used on the DVD. State-of-the-art is Super Audio CD or linear PCM at 192 kHz/24 bits (but you're more likely to see 96 kHz/24 bits).

According to some sources, higher sampling rates sound smoother and more transparent but need more disk storage space and a faster hard drive. One study in the September 2007 issue of the *Journal of the Audio Engineering Society* has shown that rates higher than 44.1 kHz yield no audible improvement over 44.1 kHz.

Digital recording is based on the Nyquist-Shannon theorem. It defines the "Nyquist frequency" as half the sampling rate. If the sampling rate is 44.1 kHz, the Nyquist frequency is 22.05 kHz.

The theorem proves that the original analog signal can be reconstructed exactly from its samples—if the highest frequency in the signal is less than the Nyquist frequency. That happens when the original signal is sent through a lowpass (high-cut) filter to remove frequencies above the Nyquist frequency.

If the analog signal is not filtered, aliasing occurs: high frequencies above the Nyquist frequency appear as lower frequencies. In other words, inaudible ultrasonic frequencies are converted to audible tones. That's why an anti-aliasing filter is needed.

If you lowpass filter the analog signal to the Nyquist frequency before it's sampled, there is only one waveform that can pass through the sample points, and that is the original analog waveform. Figure 9.1B shows the sample points and the waveform that passes through them.

Current A/D and D/A converters use a process called oversampling to improve the sound. Oversampling is sampling an audio signal at a higher rate than needed to reproduce the highest frequency in the signal. For example, sampling a 20 kHz audio signal at 8 times 44.1 kHz is called "8x oversampling." This process is followed by a digital lowpass filter and a gentle-slope analog anti-alias filter. The result is less phase shift and less harshness compared to a steep, "brick-wall" analog filter used alone.

In summary, a digital audio system samples the analog signal several thousand times a second, and quantizes (assigns a value to) each sample. Sampling rate affects the high-frequency response. Bit depth affects the dynamic range, noise, and distortion.

An alternative to PCM is 1-bit (bitstream) encoding with a 2.8224 MHz or higher sampling rate. This is the Direct Stream Digital (DSD) process used in the Super Audio CD and in some Korg portable digital recorders. DSD offers a frequency response from DC to 100 kHz with 120-dB dynamic range and a very smooth, analog-like sound.

Most recent digital devices use delta-sigma A/D converters, which produce 1-bit streams at a very high sample rate (typically using 64x oversampling). Then a digital lowpass filter removes the aliasing and quantizing noise. Finally, a decimation circuit downsamples (converts) the bitstream to 16- or 24-bit PCM samples at the desired sample rate (44.1 kHz, 96 kHz, and so on).

With delta-sigma A/D conversion, recorded sample rates above 44.1 kHz don't improve the sound quality because the analog input signal is 64x oversampled, no matter what the recorded sample rate is.

In a digital transmission, the two channels of a stereo program are multiplexed. That is, one word from channel 1 is followed by one word from channel 2, which is followed by one word from channel 1, and so on.

Data Rate and Storage Requirements

The data rate of digital audio (in bytes per second) is

$$\text{Bit depth}/8 \times \text{Sampling rate} \times \text{Number of tracks.}$$

Divide by 1,048,576 to get megabytes (MB) per second. For example, the data rate of a 24-bit/44.1 kHz recording of 16 mono tracks is

$$(24/8 \times 44{,}100 \times 16)/1{,}048{,}576 = 2\ \text{MB/sec.}$$

Recording digital audio on a hard drive consumes a lot of space. The storage required is

$$\text{Bit depth}/8 \times \text{Sampling rate} \times \text{Number of tracks} \times 60 \times \text{Number of minutes}$$

Divide by 1,048,576 to get MB. Divide MB by 1024 to get gigabytes (GB). For example, suppose you record a concert at 24 bits, 44.1 kHz, 16 tracks, for 2 hours. The hard-drive space needed is

$$(24/8 \times 44{,}100 \times 16 \times 60 \times 120)/1{,}048{,}576 = 14{,}534.9\ \text{MB or } 14.2\ \text{GB}$$

Note that the operating system might report the capacity of the disk drive up to 7% lower than it really is.

Table 13.1 shows the amount of storage needed for a 1-hour recording. Not shown in Table 13.1, a 1-bit stereo recording at 2.8 MHz consumes about 42.3 MB/minute or 2.3 GB per hour.

Digital Recording Level

In a digital recorder, the record-level meter is a peak-reading LED or LCD bar graph meter (Figure 9.4) that reads up to 0 dBFS (FS means full scale). In a 16-bit digital recorder, 0 dBFS means all 16 bits

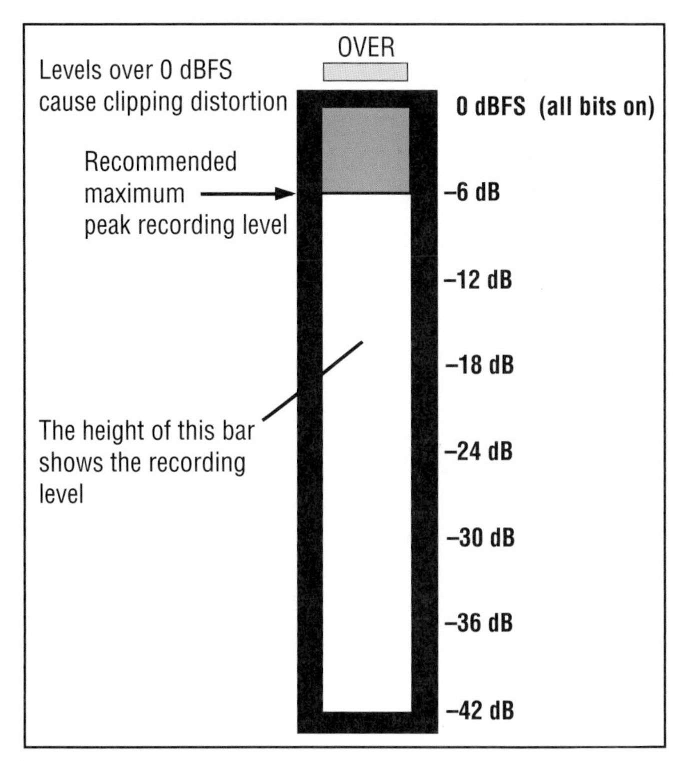

FIGURE 9.4
Digital record-level meter.

are on (1) at the waveform peak, and all 16 bits are off (0) at the waveform trough. In a 24-bit digital recorder, 0 dBFS means that all 24 bits are on or off.

The OVER indication means that the input level exceeded the voltage needed to produce 0 dBFS, and there is some short-duration clipping of the output analog waveform. This clipping can sound really nasty. Some manufacturers calibrate their meters so that 0 dBFS is less than 16 or 24 bits on; this allows a little headroom. When you set the recording level, it's a good idea to aim for −6 dB maximum so that unexpected peaks don't exceed 0 dBFS. If you're making a 24-bit recording, the recording level isn't very critical because a 16-bit signal is at −48 dBFS!

THE CLOCK

Each digital audio device has a clock or internal oscillator that sets the timing of its samples. The clock signal is a series of pulses running at the sampling rate. When you transfer digital audio from one device to another, their clocks must be synchronized. One device must provide the master clock and the other must be the slave. This clock comes embedded in the digital signal, or comes on a separate wire or connector as a word clock signal.

In your digital devices or DAW, look for a switch or software menu that selects the clock source. The three choices are internal clock, external clock, or word clock. Internal clock is the A/D converter's clock. External clock (also called digital input) is the clock signal embedded in another device's digital audio signal. Word clock is a separate cable with timing signals. Set the master device to internal clock and set the slave devices to external clock (digital input). If you're using a master word-clock generator feeding the word-clock connectors of other devices, set the slave devices to word clock.

Some receiving devices synchronize automatically to the embedded clock from the source device. In that case you don't need to set the clock source.

Your system's A/D converter or interface is usually the best master clock. Connect its word clock output (if any) to the input of a word-clock splitter or distribution unit. Connect a cable from each output of the distribution unit to the word-clock input of each of the other devices. That way all the devices will be synchronized.

Word-clock cables under 25 feet should be unbalanced 75-ohm cables with BNC connectors. Word-clock cables over 25 feet should be balanced 110-ohm AES digital cables with XLR connectors, although regular mic cables can work.

In large recording systems where you mix or assemble digital audio signals from a variety of sources, it's convenient to use a separate master clock such as Apogee's Big Ben, Mutec's iClock, or Drawmer's M-Clock. Most master clocks have several word-clock outputs. Connect each one to each slave device in your studio to keep them in sync. The receiving device's word-clock connector must be terminated properly—refer to the user's manual. If you daisy-chain the word-clock connectors between audio devices (not recommended), only the last device in the chain should be terminated.

DIGITAL AUDIO SIGNAL FORMATS

There are seven major formats for carrying digital-audio signals between various devices: AES3, MADI, S/PDIF, ADAT Lightpipe, TDIF, USB, and FireWire. With all these formats you connect a single cable between digital out and digital in. Let's look at each format.

- AES3 (also called AES/EBU or IEC-60958 Type I): This professional format transfers 2 channels of digital audio over a 110-ohm shielded twisted-pair cable with XLR connectors. That is not the same as a mic cable. An AES3 cable can be up to 100 meters long. AES3id uses 75 ohm coaxial cable and BNC connectors for runs up to 1000 meters. You need one AES3 interface for each pair of channels.
- MADI (Multichannel Audio Digital Interface or AES10-2003): This professional format transports 28, 56, or 64 channels up to 24 bits/96 kHz on 75-ohm coaxial cable or fiber-optic cable. It's often used for linking large mixing consoles to digital multitrack recorders. The cable may be up to 3000 meters long.
- S/PDIF (Sony/Philips Digital Interface; also called IRC-0958 Type II): This is a 2-channel consumer or semi-pro format. It uses a single 75-ohm coaxial cable up to 10 meters long with RCA or BNC connectors, or a fiber-optic cable with TOSLINK connectors. Optical interfaces prevent ground loops and cable losses.
- ADAT Lightpipe: The Alesis ADAT digital multitracks use a Lightpipe, which sends 8 channels of digital audio on a single optical cable with TOSLINK connectors. Every 8 channels of transfer require a separate Lightpipe cable. Data transfer is up to 24 bit/48 kHz or 24 bit/96 kHz with half the number of channels.
- TASCAM TDIF (TASCAM digital interface): The TASCAM DA-88 and similar modular digital multitracks use a multiconductor cable with standard DB-25 connectors. TDIF sends 8 channels of digital audio in and out on a single cable, which can be up to 5 m long.
- USB (Universal Serial Bus) and FireWire (IEEE 1394) are standard Mac/PC protocols for high-speed serial data transfer between a computer and an external device, such as a hard drive, USB drive, MIDI interface, or audio interface. More details are in Chapter 13 under the heading FireWire or USB Audio Interface.

AES3, S/PDIF, and ADAT Lightpipe are self-clocking systems: the clock is embedded in the signal and marks the start time of each sample. MADI, TDIF, and ADAT Sync carry a separate word-clock signal on a separate connector or wire. Note that FireWire and USB signals don't have clock information; they are reclocked at the receiving device.

Converting Signal Formats

AES and S/PDIF signals are similar but not necessarily compatible. You can convert one to the other using a format converter. Some sound cards and digital mixers do this conversion. Lightpipe and TDIF signals can be converted as well.

Some digital audio devices don't implement AES or S/PDIF correctly, so they don't interface with some other devices.

DITHER

Your HD recorder, software, and digital mixer may run at 24 bits, but the result ends up on a 16-bit CD. When you save a 24-bit audio file as a 16-bit file to transfer to CD, those last 8 bits are truncated or cut off. The result may be a grainy static sound at very low levels. This distortion can be prevented by adding low-level random noise (dither) to the signal. Let's explain.

A 24-bit resolution can accurately capture the quietest parts of a musical program: very low-level signals such as the end of long fades and reverb tails. But truncation of that signal to 16 bits makes those low-level signals sound grainy or fuzzy, because 16 bits is a less accurate measurement of the analog waveform than 24 bits. This fuzzy sound, called quantization distortion, doesn't exist at normal high levels.

What causes this distortion? Each digital word is made of a certain number of bits. During quantization, the A/D converter assigns the closest possible digital number to represent the measured voltage of each sample. The last or rightmost bit (least significant bit or LSB) switches on or off depending on whether the converter rounds the word value up or down. If this switching occurs in the 16th bit, it may be faintly audible as a fuzzy noise during quiet passages.

Also, a 24-bit recording has 256 possible levels in the lower 8 bits. But after the signal is truncated to 16 bits, that resolution is lost.

To solve this problem, dithering adds random noise (random 1's and 0's) to the lowest 8 bits of the 24-bit signal (at about −100 dB) before they are truncated to 16 bits. That noise modulates the 16th bit with some 24-bit information (bits 17 through 24) in the form of pulse-density modulation. The average value of that modulated square wave is recovered by a lowpass filter. Then most of the 24-bit sound quality is restored, and the quantization distortion changes to a smooth hiss.

To make that added hiss less obvious, noise shaping is used. Noise shaping applies an oversampling filter to the noise, which reduces its level in the midrange where our ears are most sensitive, and increases its level in the high frequencies where it's less audible.

Compared to a truncated signal, a truncated-and-dithered signal sounds slightly more transparent. Fades and reverb tails sound smoother, and there's more sonic detail. Signals below the noise floor become audible.

For best sound quality, apply dither only once when you convert a high bit-depth source to its 16-bit CD format. For example, record at 24 bits and stay there through the entire project. Don't re-dither material that has already been dithered—switch off any dithering. Add dither as the very last step when you export your stereo mix to 16 bits.

To hear the effects of dithering, start with a clean 24-bit recording, reduce its level 50 dB in your editing software, and export it as a 16-bit file. Export it in three ways: without dither, with dither, and with noise shaping added. Next, normalize the exported recordings so that the highest peak hits 0 dBFS. Then listen to the resulting 16-bit files at high level over headphones. Compare the processing and use whatever sounds best. Generally, a signal that is truncated with no dither is accompanied by a rough, grainy noise. A signal that is truncated and dithered is accompanied by smooth, quiet hiss or silence.

With good dithering algorithms, it's possible to preserve most of the 24-bit quality (ultra-low distortion, fine detail, and ambience) when converting to 16 bits. One such system is the POW-r™ Psychoacoustically Optimized Word-Length Reduction algorithm from the Pow-r Consortium LLC. It reduces the high-resolution, higher word lengths (20 to 32 bit) to a CD-standard, 16-bit format while retaining the transparency of the high-resolution recording. In other words, the 16-bit CD sounds like the original 24-bit recording.

JITTER

Accurate A/D and D/A conversions rely on the clock precisely sampling the analog signal at equal time intervals. Jitter is the unstable timing of samples that occurs in A/D and D/A conversion. Any change between the sample times, even nanoseconds, creates amplitude errors—small changes in the audio waveform's shape—resulting in a slight veiling of the sound (low-level distortion or noise). Some researchers claim that a jitter spec under 250 nanoseconds is inaudible; others say 20 nanoseconds; others say 250 picoseconds.

During A/D and D/A conversion, one cause of jitter is analog noise and crosstalk in the recording system. They affect the switching times and switching threshold of the clock, causing frequency modulation of the clock. They also affect the analog filters and oscillators used in the clock's phase-locked loops (PLLs).

Jitter is also caused by inadequate cables. They pick up hum and noise, and introduce phase shift and high-frequency attenuation, which degrade the timing of the digital signal. However, the AES3 and ADAT interfaces can handle a lot of timing variation, so cable-induced jitter is almost never a problem.

In theory, jitter doesn't occur during real-time data transmission over FireWire or USB because they are clockless systems. In other words, FireWire and USB connections don't introduce jitter on their own. FireWire data transfers are "isochronous," which means these data must be delivered at a certain minimum data rate. Digital audio and video signal streams require isochronous data flow to ensure that audio data are delivered as fast as they are played and recorded.

However, there's a potential problem. Part of the FireWire and USB signals are sync-message packets. FireWire and USB interfaces can slightly vary the timing of those messages. Then, when the receiving device extracts the clock with a PLL, jitter can be created.

To reduce jitter:

- Use high-quality clock sources with low jitter specs (under 1 nanosecond). Some audio interfaces and A/D converters are specially designed to reduce jitter, and usually include a PLL.
- Use high-quality, well-shielded cables designed for digital signals, and as short as possible. Don't use mic cables for AES3; don't use hi-fi cables for S/PDIF.

- Keep analog and digital cabling separate.
- Use your audio interface or A/D converter as the master clock. Don't drive it from an external clock source.

DIGITAL TRANSFERS AND COPIES

When you send a digital audio signal in real time from one device to another, they must be set to the same sampling rate. For example, if the digital signal from a multitrack recorder is 48 kHz, the digital mixer it feeds must also be set to 48 kHz. In a digital transfer, the sending device usually is the master, and the receiving device is the slave. Normally you can set the slave to automatically lock onto the sampling rate of the master. Set the master device to Internal Clock, and set the slave device to External Clock. Or set the slave's clock source to whatever device the master is.

The word length (number of bits) of digital transfers is less critical. As long as the two machines handle the same digital format (AES, S/PDIF, etc.), transfers of any word length can be done.

It's okay to send a lower bit signal to a higher bit device; for instance, feed a 16-bit signal from a DAT into a 24-bit digital mixer. The mixer will add more zeroes to fill out the digital word, with no effect on sound quality. It's also no problem to feed a device more bits than it can hold. The receiving device ignores the extra bits. For example, if you send a 24-bit signal to a 16-bit recorder, the last 8 bits will be truncated or cut off.

Truncation adds slight distortion, but this can be reduced by adding dither—low-level noise—to the signal before truncating. So it's best to keep the word length as long as possible as you're working on a project. Always apply dither before you reduce the word length, and only then.

What if you have a 48 K recording that you want to release on CD? The CD mastering engineer can (1) convert the 48 K digital signal to 44.1 K, which might degrade the sound, or (2) use the analog output of his playback machine, and convert the analog signal to digital at 44.1 K. This may or may not sound better.

Digital audio transfers to or from DAT or MDM tape, and to or from CD-Rs, may or may not create perfect clones because of inadequate error correction or interpolation errors.

When you copy a WAVE or AIFF file rather than a digital audio signal, the copy is a perfect clone of the original file. The file transfers much faster than the real-time playback of the signal. Flawless file copies can be made in these ways:

- Inside a computer from one hard drive location to another
- From one computer's hard drive to another computer's hard drive
- Via Ethernet, USB, FireWire, or the Internet
- Between computers via data CD-R, Jaz, or Zip drive; floppy disks; or Flash memory cards.

The following data compression algorithms don't lose any data, so they don't degrade audio quality:

- Emagic's Zap
- Waves' TrackPac
- PKWare's PKZIP
- WinZip Computing's WinZip
- Aladdin Systems' StuffIt
- Meridian Lossless Packing (MLP).

2-TRACK DIGITAL RECORDERS

Now that digital recording has been overviewed, let's look at digital recorders. They come in 2-track and multitrack formats. The 2-track formats are

- Portable hard-drive recorder
- Digital Audio Workstation; this is a computer (usually a laptop) with a sound card or other audio interface, plus recording software
- CD-R recorder
- Flash memory recorder.

We'll look at each one. No matter which format you use, the sound quality depends on the A/D converter and mic preamp (if any).

Portable Hard-Drive Recorder

This device records onto a built-in hard drive. For example, the Edirol R4 records up to 4 tracks simultaneously onto a 40 GB drive, either MP3 or WAV files up to 24 bit/96 kHz. The unit has XLR/phone inputs with phantom power, built-in stereo speakers, USB 2.0 interface to a computer, waveform editing, limiting, and a compact-flash slot. See CD www.edirol.com.

The Digital Audio Workstation

The Digital Audio Workstation (DAW) is a computer running recording software with a connected audio interface such as a sound card (Figure 9.5). A DAW allows you to record, edit, and mix audio programs entirely in digital form. It can store up to several hours of digital audio or MIDI data. You can edit this data with great precision on a computer monitor screen. What's more, you can add digital effects and perform automated mixdowns. We'll cover DAWs in detail in Chapter 13.

A laptop computer can make a great portable 2-track digital recorder. Just add some recording software and an audio interface connected by USB or FireWire.

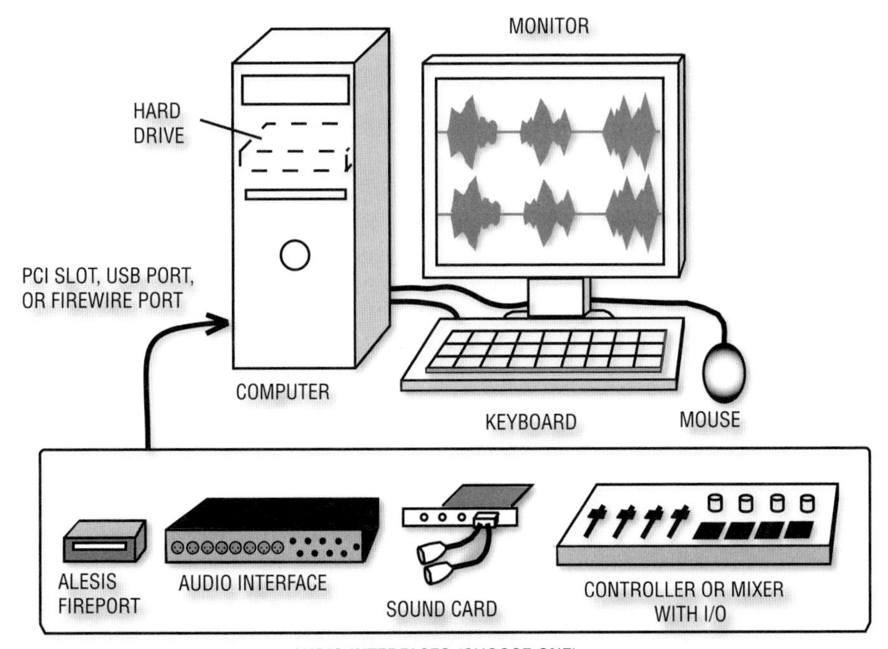

FIGURE 9.5
A computer Digital Audio Workstation (DAW).

CD Recordable (CD-R)

Another form of digital recording is the compact disc. On your own desktop, you can cut a CD by using a CD-R recorder (CD burner). It's exciting to hear one of these CDs playing your music with the purity of digital sound. The sound quality meets or exceeds CD standards.

CD-R stands for compact disc recordable. This optical medium is a write-once (nonerasable) format. CD-RW stands for compact disc re-writable; you can erase it and record a new program. Although any modern CD player or CD-ROM drive will play a CD-R, many CD players—especially older ones—can't play CD-RW discs. Most new CD-ROM drives support CD-RW, but not all will read CD-RW discs at full speed. CD-RW blank discs cost more than CD-R blank discs.

How can you use the CD-R format? You could make demos for your own band, or make a one-off copy of your stereo mixes for clients. Use the CD-R as a pre-master to send to a CD replicator. Another function is to compile sound libraries of production music, samples, and sound effects. In addition to recording audio, CD-Rs can store data files such as WAVE, AIFF, or MP3 files to send to other studios. If handled and stored with care, the CD format can archive your recordings (but always use another medium as a backup).

CD-R FORMATS

First you'll be faced with two basic choices of CD-R recorder:

1. A stand-alone CD-R recorder, sometimes called a consumer CD recorder
2. A computer peripheral CD-R recorder, also called a computer CD burner; you plug it into your computer system, or buy a computer with a CD burner built in.

Both types produce discs that sound equally good. Both types of discs will play on any audio CD player.

The stand-alone CD recorder (Figure 9.6) has everything you need in one chassis. Inside is a CD transport, laser, and microprocessor. On the back are analog and digital ins and outs. On the front are the level meters, record-level knob, display, and keypad. Because the

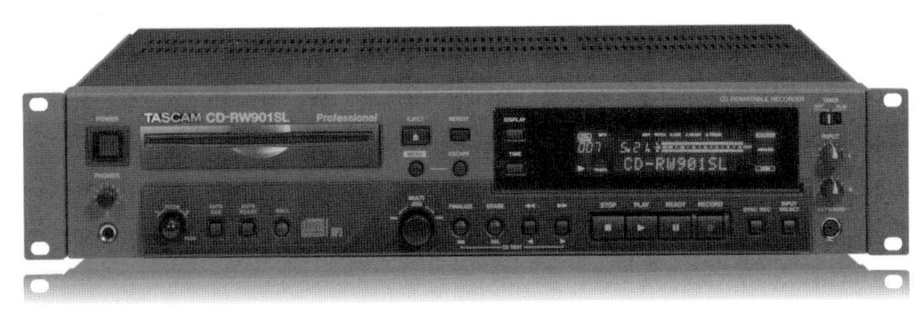

FIGURE 9.6
TASCAM CD-RW901SL, an example of a stand-alone CD recorder.

stand-alone unit needs no external computer, it's user-friendly. Just connect your audio source such as a mixer output or an analog or digital recording. Set the recording level and start recording.

The stand-alone unit can write audio but not data. The 63-minute blank discs it uses are the "Music CD-R" or "Digital Audio CD-R" format. Prices for stand-alone CD-R writers start around $250.

A computer CD recorder costs less: about $35 and up. The unit plugs into an EIDE or SCSI connector in your computer. It can write audio or computer data. Its blank discs, called "CD-R" or "Data CD-R," cost less than 35 cents in quantity. (Music CD-Rs will work in a computer CD burner as well.) Disc length is 74 minutes (650 MB) or 80 minutes (700 MB). Some discs permit recording at up to 52 times real time if your computer and hard drive are fast enough.

The computer CD-R recorder also requires a CD recording program, usually packaged with the recorder. You don't have to use that program; other ones are available that you might prefer. Be sure the program is compatible with your recorder. You'll also need a sound card and some software to record audio onto hard disk.

Multi-drive CD burners let you record or copy several CDs at once.

CD-R TECHNOLOGY

While conventional CD players follow the Sony-Philips Red Book standard, CD-Rs conform to the Orange Book part II standard. Once recorded, a CD-R disc meets the Red Book standard.

A recordable CD is the same size as a standard CD, but it is more colorful. On top is a metal reflective layer; on the bottom is a recording layer made of blue cyanine dye or yellow (gold) phthalocyanine dye. The blue layer appears green because of the gold layer behind it. Yellow dye lasts a little longer in accelerated aging tests, and it may work better with high-speed CD-R drives. A few other colors are available as well.

A blank CD-R is made of four layers (Figure 9.7):

1. Clear plastic (protects the metal layer)
2. Metal layer (gold, silver, or silver alloy, which reflects the laser light)
3. Dye (for the recording)
4. Clear plastic (protects the dye layer).

The dye fills a spiral groove that is etched in the bottom clear-plastic layer. This groove guides the laser.

To record data on a disc, the laser melts holes in the dye layer. The plastic layer flows into the holes to form pits. During playback,

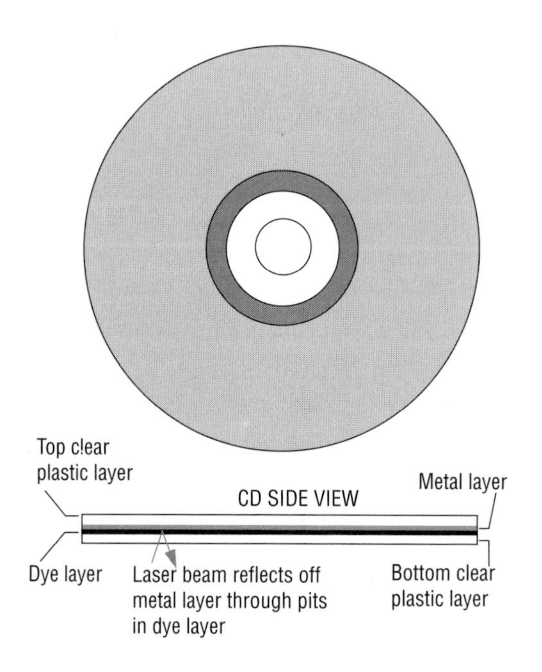

FIGURE 9.7
CD-R construction.

the same laser reads the disc at lower power. At each pit, laser light reflects off the metal layer. The reflected light enters the laser reader, which detects the varying reflectance as the pits go by.

In contrast with a standard CD, a CD-R disc has two more data areas:

- The program calibration area (PCA). The CD recorder uses this area to make a test recording, which determines the right amount of laser power to burn the disc (4 to 8 milliwatts).
- The program memory area (PMA). This area stores a temporary table of contents (TOC) as the CD-R tracks are assembled. The TOC is a list of the tracks, their start times, and the total program time. The recorder uses the PMA for this information until it writes the final TOC.

According to CD-R manufacturers, the expected lifetime of a CD-R is about 70 to 100 years with careful handling. Avoid sunlight and high temperatures. Be sure not to damage the label side of the disc by writing on it with a ballpoint pen or pencil—use a soft felt-tip pen that is water-based, such as a CD-marking Sharpie. Store CDs vertically and keep them in their cases. Avoid labels because the adhesive can attack the protective plastic layer, and the paper might warp over time. CD-RW life expectancy is claimed to be about 25 years.

CD-R SESSIONS, DISC-AT-ONCE, AND TRACK-AT-ONCE

A session on disc is made up of a lead-in, program area, and lead-out. Each session has its own TOC. Each lead-in and lead-out consumes 13 MB of disc space.

With the multisession feature, you can write several sessions on a disc at different times. This feature comes in handy when you need to add information to a disc a little at a time. Only the first session on the disc will play on an audio CD player, so the discs are just for your own use and not for distribution. On a multisession disc, you lose about 23 MB of space when you close the first session and about 14 MB for each following session.

CD-R recorders also permit Disc-At-Once recording, in which the entire disc must be recorded nonstop. You can't add new material once you write to the disc. With some software, Disc-At-Once lets you set the length of silence between tracks (down to 0 seconds), and lets you control how the tracks are laid out on disc. Use Disc-At-Once for all pro audio work.

Most CD-R recorders allow Track-At-Once recording. They can record one track (or a few tracks) at a time—up to 99 tracks. You can play a partly recorded disc on a CD-R recorder, but the disc won't play on a regular CD player until the final TOC is written. Track-At-Once isn't recommended for audio because it puts 2-second spaces and clicks between audio tracks.

If you want no pauses between tracks (as on a live album), get a CD-R writer with Disc-At-Once. Also get some software that can adjust the pause length down to zero, or that can set the start ID of each song anywhere in the program. Note that a stand-alone CD-R recorder will copy your edited program as it is, with or without pauses.

A good CD-R burner has a buffer of at least 2 MB. Some burners include SCMS copy code. Be sure that you can return the CD-R burner if it proves to be unreliable.

HOW TO USE A STAND-ALONE CD-R WRITER

Let's say that you have an edited recording of song mixes on a 2-track recorder, and you want to copy them onto a CD by using a self-contained CD-R writer.

Connect the 2-track recorder's output to the CD-R writer's input. Use a Music CD-R or Digital Audio CD-R. Set your levels and begin recording. Your program will copy to disc in real time.

Depending on the CD-R writer, the recording's start IDs may or may not convert to CD track numbers. If not, you can use a converter box, or add the track numbers manually while you record.

Using a computer CD-R burner is covered in Chapter 15. For more information on CD-R technology, see www.cdrfaq.org and www.harmony-central.com/Features/CDRecorder/.

Flash Memory Recorder

This is a portable 2-track digital recorder that records onto a Flash memory card such as a Compact Flash or Secure Digital (SD) card. It records MP3 or uncompressed PCM WAV files (which are CD quality or better). You can copy the recorded files to a computer hard drive through a USB cable.

Available mic connectors are XLR, 1/4″ phone or 1/8″ phone, with or without 48V phantom power or plug-in power for the mics.

Options are sample rate (44.1 to 96 kHz), battery life (4 to 19 hours), battery type (alkaline, lithium, or NiMH), battery rechargeable or standard, memory size (up to 4 GB), USB version (1 or 2), mics (built-in or not), and extra features such as a guitar tuner, vocal cancellation, pitch change, overdubbing, CD burner, and real-time audio interfacing to a computer. The Korg MR-1 and MR-1000 record in 1-bit DSD format rather than PCM.

Examples of models are in Chapter 4 under the heading Portable 2-Track Recorder, and one is shown in Figure 9.8.

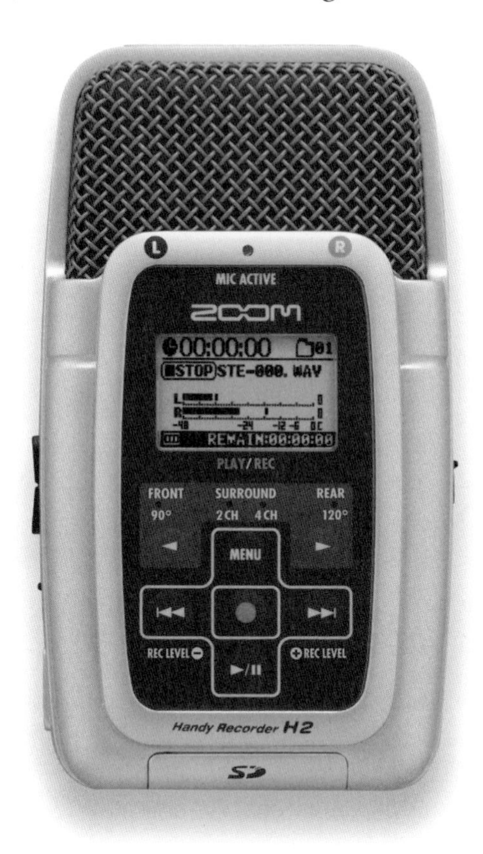

FIGURE 9.8
Zoom H2 Flash memory recorder.

MULTITRACK DIGITAL RECORDERS

So far we've covered 2-track digital recorders. Now we'll consider the multitrack digital recorders listed below:

- Computer DAW (covered earlier)
- Stand-alone hard-disk recorder
- Recorder–mixer
- DVD recorder (see Chapter 19).

Hard-Disk (HD) Recorder

This digital multitrack recorder was described in Chapter 4 under the heading Separate Multitrack Recorder and Mixer.

Recorder–Mixer (Digital Personal Studio)

This device was explained in Chapter 4 under the heading Digital Multitracker (Recorder–Mixer). An example is shown in Figure 9.9. Listed below are some features to look for.

FIGURE 9.9
TASCAM DP-02, an example of a digital personal studio.

- Type of analog inputs and outputs. Balanced XLR or TRS phone-jack mic connectors are preferred over unbalanced (TS) phone-jack connectors. Balanced connections let you run longer mic cables without picking up hum.
- Type of digital inputs and outputs. S/PDIF is a common format.
- Number of mic inputs: 2 and up. If you're recording an entire band or drum set at once, you'll need at least 8 mic inputs.
- Number of mixer channels: 8 and up.
- Number of tracks: 4 to 36. The more tracks you have, the more instruments and vocals you can record on individual tracks.
- Number of tracks that can be recorded simultaneously: 8 to 24. If you want to record an entire band at once, you need to record many tracks at the same time.
- A/D converter's bit depth: 16 to 24 bit. The more bits, the higher the sound quality.
- A/D converter's sampling rate: 44.1 to 96 kHz. The higher the sampling rate, the higher the sound quality, but 44.1 kHz is often good enough for projects that will end up on a CD. The latest units record with 24-bit resolution and up to 96 kHz sampling rate. Some record 24 tracks of 24-bit programs at sample rates of 44.1 and 48 kHz. They record 12 tracks of 24-bit recording at 88.2 and 96 kHz.
- Number of built-in digital effects: Up to 8 stereo or 16 mono. Examples of effects are reverberation, echo, flanging, chorus, and compression. They add sonic excitement and a professional touch to your productions. Some recorder–mixers have an expansion card that accepts downloadable plug-in effects.
- Types of EQ: 2- or 3-band, fixed or parametric. EQ means tone control. A 2-band EQ adjusts bass and treble, and a 3-band adjusts bass, midrange, and treble. A fixed-frequency EQ is less flexible than a parametric or sweepable EQ, which lets you adjust the frequency range you want to work on.
- Phantom power for condenser mics.
- Backlit LCD display (bigger is better) with a waveform display.
- Built-in CD burner.
- Backup or transfer recordings to computer via USB.
- Data compression: Recordings can be made in MP3 format as well as WAV format. The WAV format is preferred for higher sound quality.

- Footswitch so you can record over mistakes (punch-in) with your foot while playing your instrument.
- Pitch control. Adjust the recording speed up or down so you can match the pitch of recorded tracks to the pitch of new instruments to be recorded.
- Backup options: Removable hard drive or CD-R.
- Levels of undo: Up to 1000. The Undo function lets you undo an editing change you just did. If you don't like how an edit sounded, you can press Undo and go back to the way things were before you made the change.
- Number of locate points: Up to 1000. A locate point is a point in time in the recording that you can have the recorder memorize and locate later. For example, you might want to mark the time where the beginning of a song is, so you can tap a button and go there instantly.
- Expansion ports for SMPTE and MTC sync, MDM interface, and extra ins and outs. Beginners probably don't need these features, but they let you connect your multitracker to other devices.
- MIDI implementation: MIDI Machine Code (MMC), Control Change (CC). These features let you control the mixer via MIDI commands.
- Types of synchronization: SMPTE, Word Clock, MTC, MIDI Clock with Song Position Pointer, ADAT, DA-88, and RS-422. These are various formats that you might want your multitracker to sync with.
- Jog/shuttle wheel to "scrub" audio–play it slowly forward and backward to locate an edit point.
- Remote control.
- Automated mixing. A hard-drive or memory recording in the multitracker remembers your song-mix settings and mix changes. The next time you play the song, the mixer resets to those settings automatically. Some multitrackers have motorized faders, so that the faders move up and down just as they did when you adjusted them while mixing.
- Editing (cut and paste, etc.). This lets you do all sorts of things with the song arrangement. You might remove parts of songs that you don't want to keep, loop a drum part so that it repeats continuously, or copy a chorus to several points in a song.
- Video output that plugs into a computer monitor screen. During editing, it's a lot easier to see details in the sound waveform if you can plug a large monitor into your multitracker.

- Mastering tools and CD-burning ability. In mastering you compile a playlist of song mixes from which you can create an album. Some multitrackers include programs that let you record your songs on a CD-R recorder, either external or built-in.
- Number of virtual tracks (recordings of separate takes of the same instrument): 8 and up.

A virtual track is a 1-channel recording of a single take or performance on a random-access medium, such as a hard drive. You can record several virtual tracks or takes of a single instrument, then select which take you want to hear during mixdown. For example, suppose you have a 16-track hard-disk recorder. You might record 8 or more virtual tracks (takes) of a vocal on hard disk. Then during mixdown, choose which virtual track will play as one of those 16 tracks. Some recorders can be set up to select parts of different virtual tracks during playback—a process called comping. For example, play vocal take 15 during the verse, play vocal take 3 during the chorus, and so on.

In other words, you can associate several virtual tracks, or takes, with a single "real" track. You might play 8 tracks at once, and you can choose which virtual track (take) plays on each track. For example, you can record several takes of a guitar solo—keeping each one—and choose the best take during mixdown. You also can create a composite track, which is made of the best parts of several takes. All HD recorder–mixers let you record virtual tracks.

Pros and Cons of Three Multitrack Recording Systems

Let's compare a hard-drive recorder, recorder–mixer, and computer DAW.

Multitrack Hard-Drive Recorder and Separate Mixer

Pros:

- The recorder is portable—great for on-location recording.
- Real faders and knobs.
- Can export audio data to a computer for editing, or can use its own editing firmware.
- Stable. Less likely to crash than a computer recording system. This is a big advantage both in on-location recording and studio recording.
- The mixer allows easy set up of monitoring and cue mixes.
- The recorder might have higher quality converters than a computer recording system.

Cons:

- If you want to do automated mixes, this system requires a mixing console with automation, or a computer recording program with automation.
- If you mix with a hardware mixer rather than a computer, you'll also need some hardware effects.
- Lacks the advantages of a computer recording system.

Recorder–Mixer

Pros:

- Easy to use—hands on.
- Simple. No need for a computer, external multitrack recorder, or outboard effects.
- Portable—good for on-location recording, but bigger than an HD recorder.
- Less likely to crash than a computer recording system.
- Can export audio data to a computer for editing, or use its own editing firmware.
- All-in-one system. No external cables needed between recorder and mixer or between mixer and effects.
- Low-cost CD backup.
- Automated mixing.

Cons:

- Lacks the advantages of a computer recording system.
- Less flexible than recording software. Not as upgradable. However, some recorder–mixers have expansion cards that accept downloadable plug-in effects.
- May require copying data to a computer for sophisticated editing and plug-ins.
- If you already have a mixer, a recorder–mixer may be more than you need. Just get a recorder, unless you need automation.

Computer Recording System (DAW)

Pros:

- Can be low cost if you already have a fast computer (except for the top-level Pro Tools systems). Software effects (plug-ins) cost much less than hardware effects.

- Flexible (due to software upgrades and downloadable plug-ins).
- Everything can be "in the box." The mixer, recorder, effects, and CD burner are all in the computer. Fewer cables are needed.
- Fast backup of files to CD or external hard drive.
- Automated mixing.
- When you quit a project, the computer remembers all your mixer settings and edits. It's easy to recall a project.
- Sophisticated editing.
- Can be used with your existing analog or digital mixer.
- You record tracks directly to the computer. You don't need to copy external recordings to the computer before editing and mastering them.
- It's easy to find parts of a song by looking at the tracks on screen.

Cons:

- Computer may crash.
- Not as "hands on" as a recorder–mixer, but a controller surface can help with this.
- Not easily portable except in laptop systems.
- Requires a fast computer with a large hard drive optimized for digital audio.
- Requires a lot of computer tweaks and strong computer skills (but this is a plus for some users).

As we've seen, there is a wide variety of digital recording formats. Learn all you can about digital technology, then choose the formats that meet your needs.

BACKUP

The hard drive on which you store your projects will eventually crash. So it's vital to back up your recordings' WAV files and session files after each session. That is, copy them to another storage medium.

Some backup media are CD-R, CD-RW, hard drive (internal or external FireWire), DVD-R, and DVD-RW. Some engineers prefer to archive audio programs on analog tape as well as digital media.

Clone your system drive using a program such as Casper. If your hard drive crashes during a session, you can replace it with the clone drive, reboot, and be back in business.

CHAPTER 10

Effects and Signal Processors

With effects, your mix sounds more like a real "production" and less like a bland home recording. You might simulate a concert hall with reverb. Put a guitar in space with stereo chorus. Make a kick drum punchy by adding compression. Used on all pop-music records, effects can enhance plain tracks by adding spaciousness and excitement. They are essential if you want to produce a commercial sound. But many jazz, folk, and classical groups sound fine without any effects.

This chapter describes the most popular signal processors and effects, and suggests how to use them.

Effects are available both as hardware and software (called plug-ins). To add a hardware effect to a track, you feed the track's signal from your mixer's aux send to an effects device, or signal processor (Figure 10.1). It modifies the signal in a controlled way. Then the modified signal returns to your mixer, where it blends with the dry, unprocessed signal.

SOFTWARE EFFECTS (PLUG-INS)

Most recording programs include plug-ins: software effects that you control on your computer screen. Each effect is an algorithm (small program) that runs either in your computer's CPU or in a DSP card. Some plug-ins come already installed with the recording software; others you can download or purchase on CD, then install them on your hard drive. Each plug-in becomes part of your recording program (called the host), and can be called up from within the host.

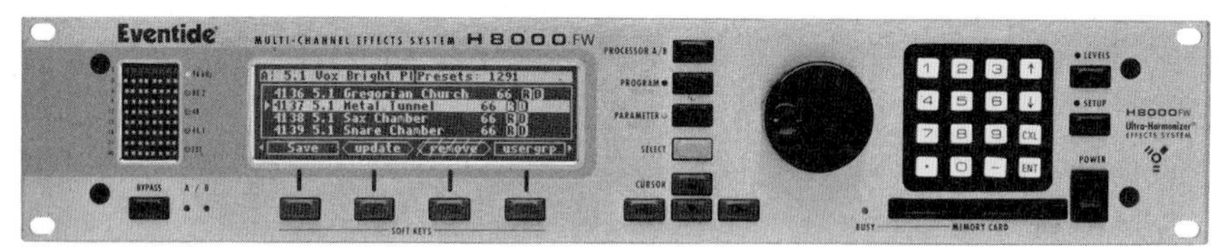

FIGURE 10.1
Eventide H8000FW Multichannel Effects System, an example of a signal processor.

You can use plug-ins made by your recording software company or by others. Some manufacturers make plug-in bundles, which are a variety of effects in a single package.

Plug-ins are the usual way to create effects in a DAW. Some DAWs let you configure your audio interface to produce an aux-send signal, which you feed to an external hardware processor. The processed signal returns to the interface and blends with the dry signal.

All the effects described below are available as plug-ins as well as hardware.

EQUALIZER

Recall from Chapter 2 that an equalizer (usually in the mixer) is a sophisticated tone control, something like the bass and treble controls in a stereo system. Equalization (EQ) lets you improve on reality: add crispness to dull cymbals or add bite to a wimpy electric guitar. EQ also can make a track sound more natural; for instance, remove tubbiness from a close-miked vocal.

To understand how EQ works, we need to know the meaning of a spectrum. Each instrument or voice produces a wide range of frequencies called its spectrum—the fundamentals and harmonics. The spectrum gives each instrument its distinctive tone quality or timbre.

If you boost or cut certain frequencies in the spectrum, you change the tone quality of the recorded instrument. EQ adjusts the bass, treble, and midrange of a sound by turning up or down certain frequency ranges; that is, it alters the frequency response. For example, a boost (a level increase) in the range centered at 10 kHz makes percussion sound bright and crisp. A cut at the same frequency dulls the sound.

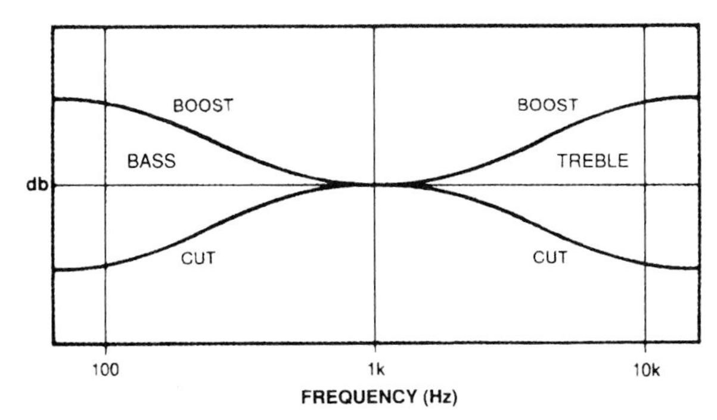

FIGURE 10.2
The effect of the bass and treble control.

Types of EQ

Equalizers range from simple to complex. The most basic type is a **bass and treble control** (labeled LF EQ and HF EQ). Figure 10.2 shows its effect on frequency response. Typically, this type has up to 15 dB of boost or cut at 100 Hz (for the low-frequency EQ knob) and at 10 kHz (for the high-frequency EQ knob).

With **3-band EQ** you can boost or cut the lows, mids, and highs at fixed frequencies (Figure 10.3). Sweepable EQ is more flexible because you can "tune in" the exact frequency range needing adjustment (Figure 10.4). If your mixer has **sweepable EQ**, one knob sets the center frequency while another sets the amount of boost or cut.

Parametric EQ lets you set the frequency, amount of boost/cut, and bandwidth—the range of frequencies affected. Figure 10.5 shows how a parametric equalizer varies the bandwidth of the boosted part of the spectrum. The "Q" or quality factor of an equalizer is the center frequency divided by the bandwidth. A boost or cut with a low-Q setting (like 1.5) affects a wide range of frequencies; a high-Q setting (like 10) makes a narrow peak or dip.

A **graphic equalizer** (Figure 10.6) is usually outside the mixing console. This type has a row of slide pots that work on 5 to 31 frequency bands. When the controls are adjusted, their positions graphically show the resulting frequency response. Usually, a graphic equalizer is used for monitor-speaker EQ, or is patched into a channel for sophisticated tonal tweaking.

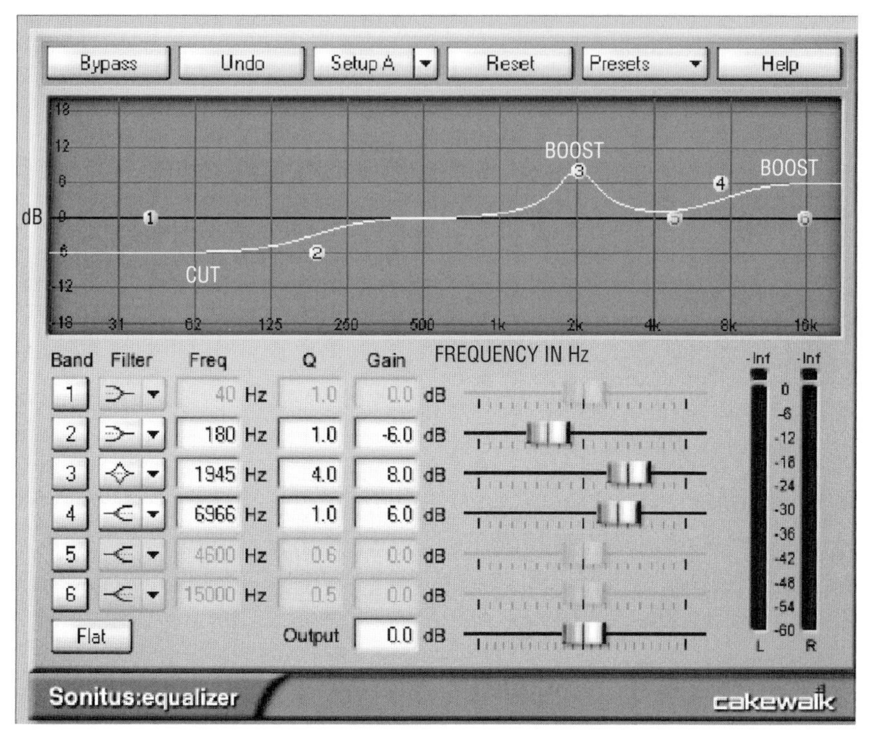

FIGURE 10.3
The effect of 3-band EQ.

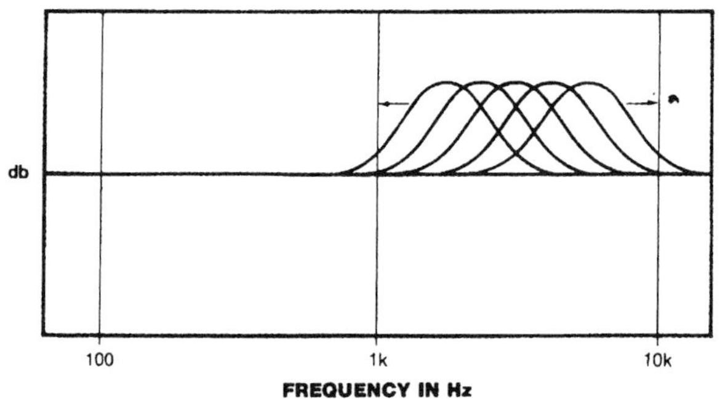

FIGURE 10.4
The effect of sweepable EQ.

Equalizers can also be classified by the shape of their frequency response. **Peaking EQ** shapes the response like a hill or peak when set for a boost (Figure 10.7). With **shelving EQ**, the shape of the frequency response resembles a shelf (Figure 10.8). *Audio clip 27 at*

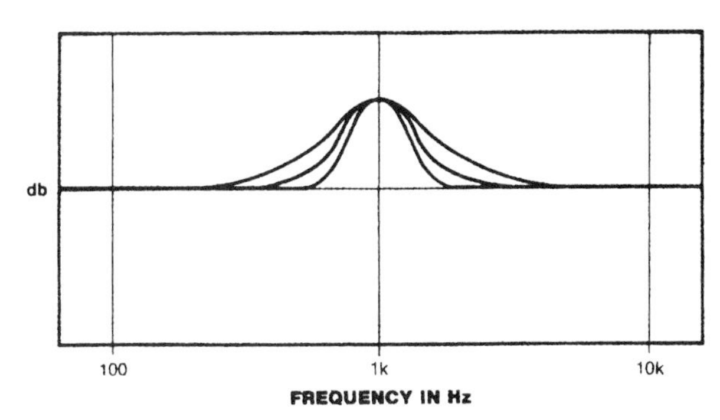

FIGURE 10.5
Curves that illustrate varying the bandwidth of a parametric equalizer.

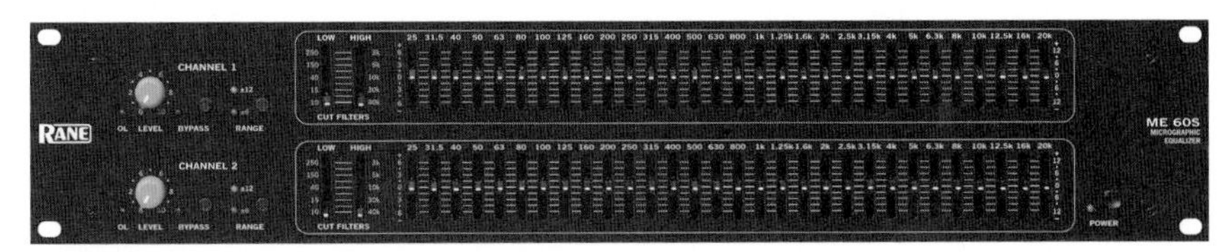

FIGURE 10.6
Rane ME60S, an example of a graphic equalizer.

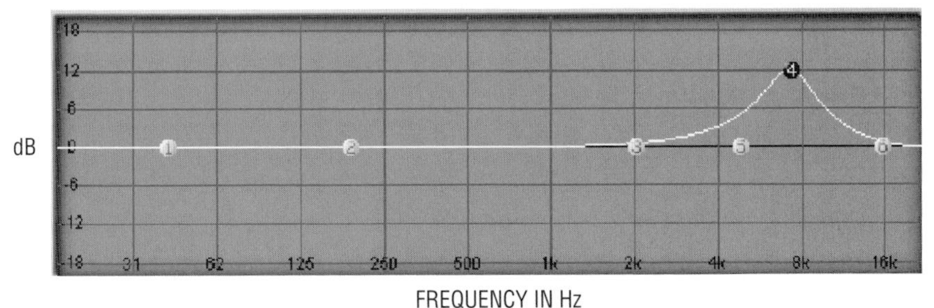

FIGURE 10.7
Peaking EQ at 7 kHz.

www.elsevierdirect.com/companions/9780240811444 demonstrates various types of EQ.

A **filter** causes a rolloff at the frequency extremes. It sharply rejects (attenuates) frequencies above or below a certain frequency. Figure 10.9 shows three types of filters: lowpass, highpass, and bandpass. For example, a 10-kHz lowpass filter (high-cut filter) removes frequencies above 10 kHz. Its response is down 3 dB at 10 kHz and more above

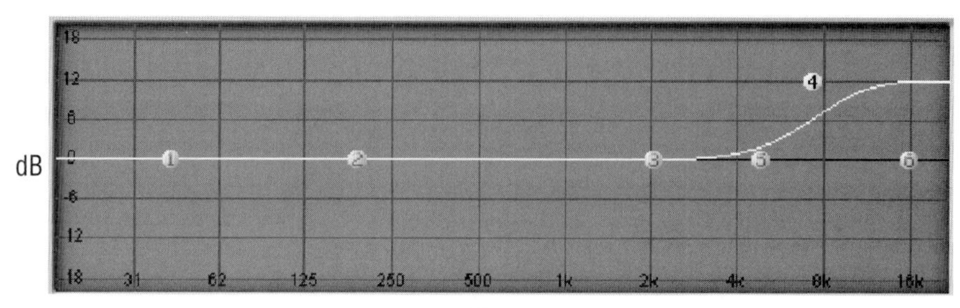

FIGURE 10.8
Shelving EQ at 7 kHz.

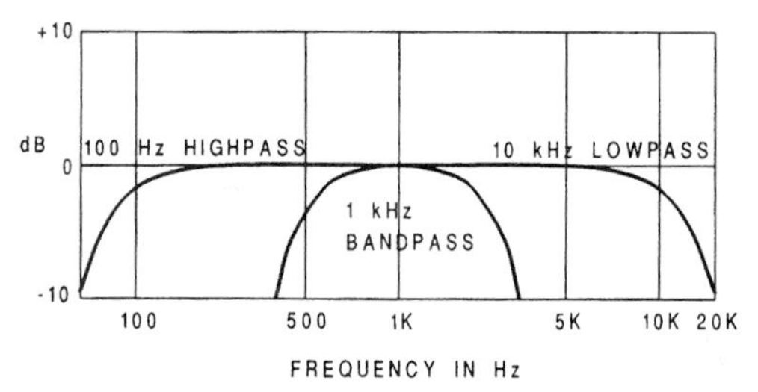

FIGURE 10.9
Lowpass, highpass, and bandpass filters.

that. This reduces hiss-type noise without affecting tone quality as much as a gradual treble rolloff would. A 100-Hz highpass filter (low-cut filter) attenuates frequencies below 100 Hz. Its response is down 3 dB at 100 Hz and more below that. This removes low-pitched noises such as air handler rumble or breath pops. A 1-kHz bandpass filter cuts frequencies above and below a frequency band centered at 1 kHz.

The crossover filter in some monitor speakers consists of lowpass, highpass, and bandpass filters. They send the lows to the woofer, mids to the midrange, and highs to the tweeter.

A filter is named for the steepness of its rolloff: 6 dB per octave (first order), 12 dB per octave (second order), 18 dB per octave (third order), and so on.

How to Use EQ

If your mixer has bass and treble controls, their frequencies are pre-set (usually at 100 Hz and 10 kHz). Set the EQ knob at 0 to have

Table 10.1 Frequency Ranges of Musical Instruments and Voices

Instrument	Fundamentals	Harmonics
Flute	261–2349 Hz	3–8 kHz
Oboe	261–1568 Hz	2–12 kHz
Clarinet	165–1568 Hz	2–10 kHz
Bassoon	62–587 Hz	1–7 kHz
Trumpet	165–988 Hz	1–7.5 kHz
French horn	87–880 Hz	1–6 kHz
Trombone	73–587 Hz	1–7.5 kHz
Tuba	49–587 Hz	1–4 kHz
Snare drum	100–200 Hz	1–20 kHz
Kick drum	30–147 Hz	1–6 kHz
Cymbals	300–587 Hz	1–15 kHz
Violin	196–3136 Hz	4–15 kHz
Viola	131–1175 Hz	2–8.5 kHz
Cello	65–698 Hz	1–6.5 kHz
Acoustic bass	41–294 Hz	700 Hz–5 kHz
Electric bass	41–294 Hz	700 Hz–7 kHz
Acoustic guitar	82–988 Hz	1500 Hz–15 kHz
Electric guitar	82–1319 Hz	1–15 kHz (direct)
Elec. guitar amp	82–1319 Hz	1–4 kHz
Piano	28–4196 Hz	5–8 kHz
Bass (voice)	87–392 Hz	1–12 kHz
Tenor (voice)	131–494 Hz	1–12 kHz
Alto (voice)	175–698 Hz	2–12 kHz
Soprano (voice)	247–1175 Hz	2–12 kHz

no effect (flat setting). Turn it clockwise for a boost; turn it counterclockwise for a cut. If your mixer has multiple-frequency EQ or sweepable EQ, one knob sets the frequency range and another sets the amount of boost or cut.

Table 10.1 shows the fundamentals and harmonics of musical instruments and voices. The harmonics given represent an approximate range. Percussion, cymbals, and muted trumpet actually have some energy up to 80 to 100 kHz. For each instrument, turn up the lower end of the fundamentals to get warmth and fullness. Turn down the fundamentals if the tone is too bassy or tubby. Turn up the harmonics for presence and definition; turn down the harmonics if the tone is too harsh or sizzly.

Here are some suggested frequencies to adjust for specific instruments. If you want the effects described below, apply boost. If you don't, apply cut. Try these suggestions and accept only the sounds you like:

- Bass: Full and deep at 60 to 100Hz, growl at 600 Hz, presence at 2.5 kHz, string noise at 3 kHz and up. Cut around 200 to 500 Hz for clarity.
- Electric guitar: Thumpy at 60 Hz, full at 100 Hz, puffy at 500 Hz, presence or bite at 2 to 3 kHz, sizzly and raspy above 6 kHz.
- Drums: Full at 100 to 200 Hz, wooly at 250 to 800 Hz (try cutting in that range), trashy snare at 1 to 3 kHz, attack at 5 kHz. Try cutting toms around 600 Hz to reduce boxiness. Dull cymbals sound more sizzly and crisp with a boost at 10 to 12 kHz.
- Kick drum: Full and powerful below 60 Hz, papery at 300 to 800 Hz (cut at 400 to 600 Hz for better tone), click or attack at 2 to 6 kHz.
- Sax: Warm at 500 Hz, harsh at 3 kHz, key noise above 10 kHz.
- Acoustic guitar: Full or thumpy at 80 Hz, presence at 5 kHz, pick noise above 10 kHz.
- Acoustic guitar pickup: To make the guitar sound more "acoustic," try a narrow cut at 1.2 to 1.5 kHz and maybe some high-frequency cut.
- Voice: Full at 100 to 200 Hz (males), full at 200 to 400 Hz (females), honky or nasal at 500 Hz to 1 kHz, presence at 5 kHz, sibilance ("s" and "sh "sounds) around 3 to 10 k Hz.

Example: Suppose a vocal track sounds too full or bassy. Reach for the LF EQ knob (say, 100 Hz) and turn it down until the voice sounds natural. Or suppose a snare drum sounds dull or muffled. Grab the mid-frequency EQ knob, set it to 5 to 10kHz, and turn it up until the snare sounds clear and crisp.

Set EQ to the approximate frequency range you need to work on. Then apply full boost or cut so the effect is easily audible. Finally, fine-tune the frequency and amount of boost or cut until the tonal balance is the way you like it.

What if an instrument sounds honky, tubby, or harsh, and you don't know what frequency to tweak? Set a sweepable EQ for extreme boost. Then sweep the frequencies until you find the frequency range matching the coloration. Cut that range by the amount that sounds right. For example, a piano miked with the lid closed might have a tubby coloration—maybe too much output around 300 Hz. Set your low-frequency EQ for boost, and vary the center frequency until the tubbiness is exaggerated. Then cut at that frequency until the piano sounds natural.

In general, avoid excessive boost because it can distort the signal. Try cutting the lows instead of boosting the highs. To reduce muddiness or enhance clarity, cut 1 to 2 dB around 300 Hz—either on individual instruments or on the entire mix. Don't boost everything at the same frequency.

When to Use EQ

Before using EQ, try to get the desired tone quality by changing the mic or its placement. This gives a more natural effect than EQ. Many purists shun the use of EQ, complaining of excessive phase shift or ringing caused by the equalizer—a "strained" sound. Instead, they use carefully placed, high-quality microphones to get a natural tonal balance without EQ.

The usual practice is to record flat (without EQ) and then equalize the track during mixdown.

Sometimes the instruments need a lot of EQ to sound good. If so, you might want to record with EQ so that the playback for the musicians will sound good. When you play the multitrack recording through your monitor mixer, the recording may not sound right unless the tracks are already equalized. (That's assuming the monitor mixer in your board has no EQ.)

Uses of EQ

Here are some applications for EQ:

Improve tone quality. The main use for EQ is to make an instrument sound better tonally. For example, you might use a high-frequency rolloff on a singer to reduce sibilance, or on a direct-recorded electric guitar to take the "edge" off the sound.

You could boost 100 Hz on a floor tom to get a fuller sound, or cut around 250 Hz on a bass guitar for clarity. Cut around 100 Hz to reduce bass buildup on massed harmony vocals. The frequency response and placement of each mic affect tone quality as well.

Although you can set the EQ for each track when it is soloed, a better way is to set the equalizers when the entire mix is playing. That's because one instrument can mask or hide certain frequencies in another instrument. For example, the cymbals might mask the "s" sounds in the vocal, making the vocal sound dull—even though it might sound fine when soloed.

Create an effect. Extreme EQ reduces fidelity, but it also can make interesting sound effects. Sharply rolling off the lows and highs on a voice, for instance, gives it a "telephone" sound. A 1-kHz bandpass filter does the same thing. To make a mono keyboard track sound stereo, send it to two mixer channels. Boost lows and cut highs in one channel panned left; cut lows and boost highs in the other channel panned right.

Reduce noise and leakage. You can reduce low-frequency noises—bass leakage, air-conditioner rumble, mic-stand thumps—by turning down the lows below the fundamental-frequency range of the instrument you're recording. This information is shown in Table 10.1.

For example, a fiddle's lowest frequency is about 200 Hz, so you'd use a low-cut filter (highpass filter) set to 200 Hz (if possible). This low-cut filter won't change the fiddle's tone quality because the filtered-out frequencies are below the fiddle's lowest frequency. Similarly, a kick drum has little or no output above 9 kHz, so you can filter out highs above 9 kHz on the kick drum to reduce cymbal leakage. Filtering out frequencies below 100 Hz on most instruments reduces air-conditioning rumble and breath pops. Try rolling off the lows on audience mics to prevent muddy bass. To reduce hum, set a parametric EQ for a 24-dB cut, Q of 30, at these frequencies: 60, 120, and 180 Hz (in the United States) or 50, 100, and 150 Hz (in Europe).

Compensate for the Fletcher-Munson effect. As discovered by Fletcher and Munson, the ear is less sensitive to bass and treble at low volumes than at high volumes. So, when you record a very loud instrument and play it back at a lower level, it might lack bass and treble. To restore these, you may need to boost the lows (around 100 Hz) and the highs (around 4 kHz) when recording loud rock groups. The louder the group, the more boost you need. It also helps to use cardioid mics with proximity effect (for bass boost) and a presence peak (for treble boost).

Make a pleasing blend. If you mix two instruments that sound alike, such as lead guitar and rhythm guitar, they tend to mush together—it's hard to tell what each is playing. You can make them more distinct by equalizing them differently. For example, make the lead guitar edgy by boosting 3 kHz, and make the rhythm guitar mellow by cutting 3 kHz. Then you'll hear a more pleasing blend and a clearer mix. The same philosophy applies to bass guitar and kick drum. Because they occupy about the same low-frequency range, they tend to mask or cover each other. To make them distinct, either fatten the bass and thin out the kick a little, or vice versa. The idea is to give each instrument its own space in the frequency spectrum; for example, the bass fills in the lows, synth chords emphasize mid-bass, lead guitar adds edge in the upper mids, and cymbals add sparkle in the highs.

Compensate for mic placement. Sometimes you're forced to mike very close to reject background sounds and leakage. But a close mic emphasizes the part of the instrument that the mic is near. This gives a colored tone quality, but EQ can partly compensate for it. Suppose you had to record an acoustic guitar with a mic near the sound hole. The guitar track will sound bassy because the sound hole radiates strong low frequencies. But you can turn down the lows on your mixer to restore a natural tonal balance.

This use of EQ can save the day by fixing poorly recorded tracks in live concert recordings. During a concert, the stage monitors might be blaring into your recording/PA microphones, so you're forced to mike close to reject monitor leakage and feedback. This close placement, or the monitor leakage itself, can give the recording an unnatural tone quality. In this case, EQ is the only way to get usable tracks.

"Re-mix" a single track. If a track contains two different instruments, sometimes you can change the mix within that track by using EQ. Imagine a track that has both bass and synth. By using LF EQ, you can bring the bass up or down without affecting the synth very much. Mixing with EQ is more effective when the two instruments are far apart in their frequency ranges.

Improve the tonal balance of an entire mix. During mastering, you can EQ the stereo mix of each song to make it better, to make the songs on an album sound more similar, or to make the album sound more like commercial albums. An effective tool for this purpose is Harmonic Balancer (www.har-bal.com).

Whenever you record, the ideal situation is to use the right mic in the right position, and in a good-sounding room. Then you don't need or want EQ. Otherwise, though, your recordings will sound better with EQ than without it.

COMPRESSOR

A compressor acts like an automatic volume control, turning down the volume when the signal gets too loud. Here's why it's necessary.

Suppose you're recording a female vocalist. Sometimes she sings too softly and gets buried in the mix; other times she hits loud notes and blasts the listener. Or she may move toward and away from the mic while singing, so that her average recording level changes.

To control this problem, you can ride gain—turn her down when she gets too loud; turn her up when she gets too quiet. But it's hard to anticipate these changes. You might prefer to use a compressor, which does the same thing automatically. It reduces the gain (amplification) when the input signal exceeds a preset level (called the threshold). The greater the input level, the less the gain. As a result, loud notes are made softer, so the dynamic range is reduced (Figure 10.10). *Play audio clip 28 at www.elsevierdirect.com/companions/9780240811444.*

Compression keeps the level of vocals or instruments more constant, so they are easier to hear throughout the mix, and it prevents loud notes that might clip. Also, it can be used for special effects—to make drums sound fatter, or to increase the sustain on a bass guitar. In pro studios, compression is used almost always on vocals; often on bass guitar, kick drum, and acoustic guitar; and sometimes on other instruments.

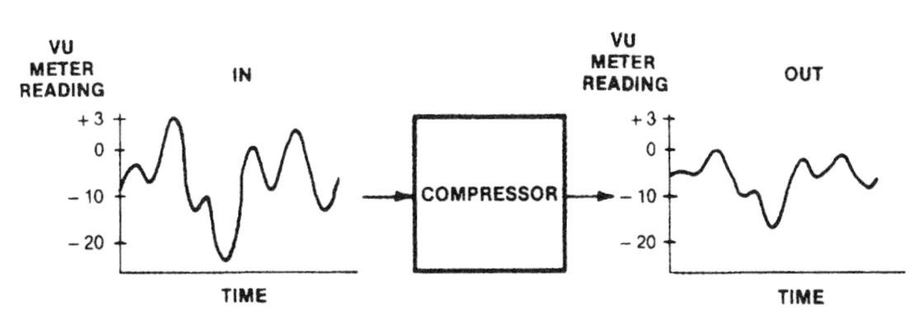

FIGURE 10.10
Compression.

Doesn't compression rob the music of its expressive dynamics? Yes, if overdone. But a vocal that gets too loud and soft is annoying. You need to tame it with a compressor. Even then, you can tell when the vocalist is singing loudly by the tone of the voice. It also helps to compress the bass and kick drum to ensure a uniform, driving beat.

You can avoid vocal compression if the singer uses proper mic technique. He or she should back away from the mic on loud notes, and come in close on soft notes. To tell whether you need a compressor, listen to your finished mix. If you can understand all the words, and no notes are too loud, omit the compressor.

Using a Compressor

Normally, you compress individual tracks or instruments, not the entire mix. You want to compress only the stuff that needs it. To compress a stereo mix, you need a 2-channel compressor with a stereo link, which keeps the left–right balance from changing during compression.

Multiband compression (covered later) is usually a better choice for compressing the stereo mix.

Let's describe the controls on the compressor. Some compressors have few controls; most of their settings are preset at the factory. Figure 10.11 shows a compressor plug-in.

COMPRESSION RATIO OR SLOPE

This is the ratio of the change in input level to the change in output level. For example, a 2:1 ratio means that for every 2 dB change in input level, the output of the compressor changes 1 dB. A 20-dB change in input level results in a 10-dB change in the output, and so on.

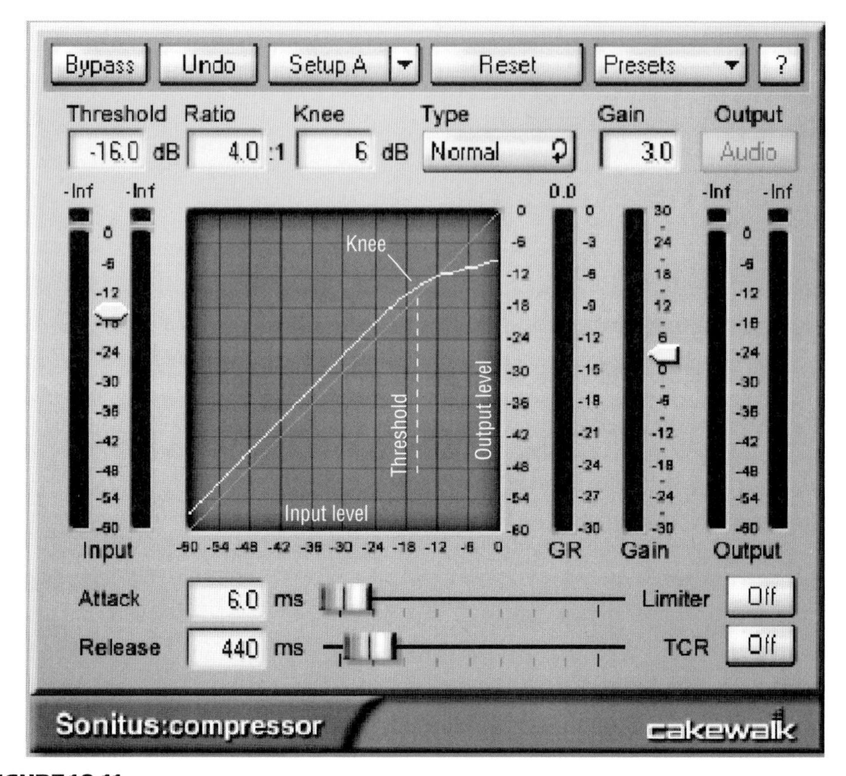

FIGURE 10.11
Sonitus Compressor, an example of a compressor plug-in. (The graph labels are by the author.)

Typical ratio settings are 2:1 to 4:1. A "soft knee" or "over easy" characteristic is a low compression ratio for low-level signals and a high ratio for high-level signals. Some manufacturers say that this characteristic sounds more natural than a fixed compression ratio.

THRESHOLD

This is the input level above which compression occurs. Set the threshold high (about –5 dB) to compress only the loudest notes; set it low (−10 or −20 dB) to compress a broader range of notes. A setting of −10 is typical. If the compressor has a fixed threshold, adjust the amount of compression with the input level control.

GAIN REDUCTION

This is the number of dB that the gain is reduced by the compressor. It varies with the input level. You set the ratio and threshold controls so that the gain is reduced on loud notes by an amount that sounds right. The amount of gain reduction shows up on a meter—3 to 10 dB is typical.

COMPRESSION GRAPH

Most compressor plug-ins display a compression graph of output level versus input level in dB (Figure 10.11). As the input level increases from zero up to the threshold level, the graph is an upward-sloping straight line in which the change in output level matches the change in input level. Where the input level is above the threshold, the graph transitions to a flatter slope where compression occurs: the change in output level is less than the change in input level. The transition point is called the knee of the compression curve. The transition from no compression to compression can be abrupt (hard knee) or gradual (soft knee).

The slope of the compressed part of the curve (the part on the right) indicates the compression ratio. The more horizontal this part is, the greater the ratio. A horizontal line indicates limiting (explained later in this chapter under the heading Limiter). As you vary the compression ratio and threshold, the compression curve changes accordingly.

ATTACK TIME

This is how fast the compressor reduces the gain when it's hit by a musical attack. Typical attack times range from 0.25 to 10 msec. Some compressors adjust the attack time automatically to suit the music; others have a factory-set attack time. The longer the attack time, the larger the peaks that are passed before gain reduction occurs. So, a long attack time sounds punchy; a short attack time reduces punch by softening the attack.

RELEASE TIME

This is how fast the gain returns to normal after a loud passage ends. It's the time the compressor takes to reach 63% of its normal gain. You can set the release time from about 50 msec to several seconds. One-half second to 0.2 second is typical. For bass instruments, the release time must be longer than about 0.4 second to prevent harmonic distortion.

Short release times make the compressor follow rapid volume changes in the music, and keep the average level higher. But because the noise rises along with the gain, short release times can give a pumping or breathing sound. Long release times sound more natural. If the release time is too long, though, a loud passage will reduce the gain during a subsequent quiet passage. In some units, the release time varies automatically, or is factory-set to a useful value.

Some compressors disable the attack and release settings when the compressor is set to RMS or average mode. Those settings are adjusted automatically.

Figure 10.12 shows the effects of compressor attack time and release time on the envelope of a waveform. The gray block represents a musical note fed into the compressor. As Figure 10.12 shows, a long

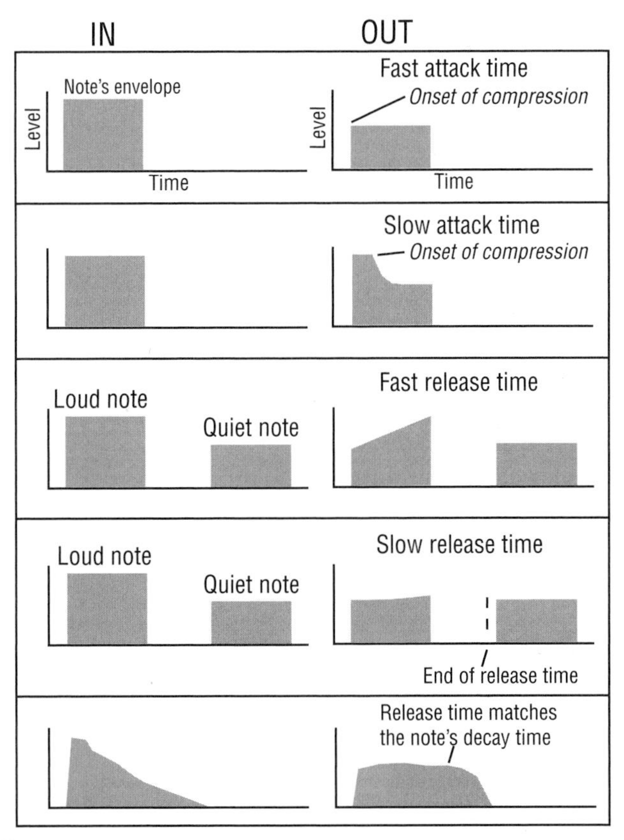

FIGURE 10.12
Effects of attack time and release time on the envelope of a note.

attack time tends to increase the attack portion of the envelope, giving a sound with a sharper attack or "edge." A short release time tends to bring up the level in between notes, giving a louder but more fatiguing sound. A release time similar to the note's decay time increases sustain.

OUTPUT-LEVEL CONTROL

Also called make-up gain, this control is used to increase the output level of the compressor by the amount of gain reduction. For example, if a compressor is causing 6 dB of level reduction, set the make-up gain to 6 dB to achieve unity gain. This also brings up the quiet parts of the track by 6 dB. Some compressors keep the output level constant when other controls are varied.

Spend some free time playing with all the settings so you learn how they affect the sound. Play various instruments and vocals through a compressor, vary the settings, and take notes on what you hear.

Some compressors have a side chain. This is a pair of in/out jacks for connecting an equalizer. To compress only the sibilant sounds on a vocal track, boost the side-chain EQ around 3 to 10 kHz. To compress only the breath pops on a vocal track, boost the side-chain EQ around 20 Hz.

A multiband or split-frequency compressor divides the audio band into three to five bands (bass, mids, treble) and compresses each band separately. That way, the compressor can squash a loud bass note, or soften "s" sounds, without bringing down the overall level. Multiband compression is sometimes applied to the final mix of each song during mixdown or mastering.

Connecting a Compressor

To compress one track in DAW, select a compressor plug-in for that track. Don't use an aux send for compression. Connect a hardware compressor in line with the signal you want to compress in one of the following ways:

- To compress one instrument or voice while recording: Locate the input module of the instrument you want to compress. On the back of that module, connect the insert send jack to compressor in; connect compressor out to the insert return jack.

(Chapter 11 explains these terms.) Or, take a signal from the input module's direct out. Feed that into the compressor, and feed the compressor output to the recorder track input.

- To compress a group of instruments while recording: Locate the bus output of the instruments you want to compress. Go from bus out to compressor in, and go from compressor out to recorder-track in. If the bus has insert jacks, you could connect to them instead.
- To compress one track during mixdown: Go from track out to compressor in, and go from compressor out to mixer channel in. Or locate the mixer input module for that track, then find the insert send and return jacks in that module. (There might be a single insert jack with send and return terminals.) Connect the insert send to compressor in; connect compressor out to insert return.

Suggested "Ballpark" Compressor Settings

- Vocals: Ratio 2:1 to 3:1, fast attack, 1/2 second release, set threshold for 3 to 6 dB of gain reduction. Singers with extreme dynamic range might need 12 dB of gain reduction and a ratio of 4:1.
- Bass and drums: Ratio 4:1, slow release, set threshold for 3 to 6 dB of gain reduction on loud bass "pops." Adjust attack time depending on how much you want to soften the attack. Short attack time = soft attack, long attack time = loud attack.
- Electric guitar: 4:1 to 8:1 ratio, 10-dB gain reduction, 400 msec release.
- To reduce breath pops: Use a multiband compressor. Enable only the lowest frequency band. Try these settings: ratio 30:1, upper frequency 100 Hz, make-up gain 0 dB, attack 1 msec, release 100 msec, threshold −18 dB. Experiment with the threshold setting.
- To reduce sibilance (de-ess mode): First, set an equalizer for a narrow boost around 6 kHz. While playing the vocal track, sweep the frequency up and down until you find the frequency area where sibilance ("s" and "sh" sounds) is exaggerated the most. Note that frequency. Remove the equalizer and insert a multiband compressor. Disable all bands except a band from about

3 to 10kHz. Try to set up a narrower band around the frequency you noted earlier. Suggested settings: ratio 10:1, attack 1 msec, release 100 msec, threshold −32 dB. Experiment with the threshold setting to get the desired amount of de-essing.
- To compress the stereo mix: Try these settings: 2:1, soft knee, attack 20 msec, release 200 msec. Set the threshold to get 5 to 10 dB of gain reduction. If you can hear the compressor squashing the sound, back off the gain reduction unless that's the sound you want.

Should you compress while tracking or mixing? If you compress while tracking, it will be difficult or impossible to change the amount of compression during mixdown. If you compress tracks during mixdown, you can change the settings at will.

When applying low-cut EQ to a track, insert the EQ before the compressor. Often there is too much bass on a track, and that extra bass will trigger the compressor unless you EQ it out first. When you apply a boost, put the EQ after the compressor so that the boost isn't compressed.

LIMITER

A limiter keeps signal peaks from exceeding a preset level. While a compressor reduces the overall dynamic range of the music, a limiter affects only the highest peaks (Figure 10.13). To act on these rapid peaks, limiters have a very fast attack time—1 microsecond to 1 millisecond. The compression ratio in a limiter is very high—10:1 or greater—and the threshold is set high, say at 0 dB. For input levels up to 0 dB, the output level matches the input. For input levels above 0 dB, the output level stays at 0 dB. This prevents overload in the device following the limiter.

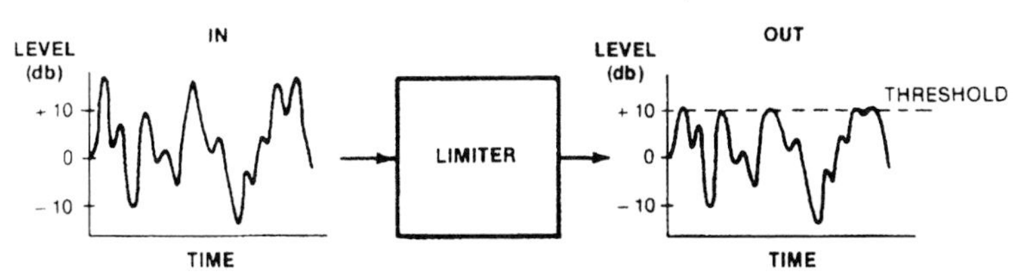

FIGURE 10.13
Limiting.

A compressor/limiter carries out both of the functions in its name. It compresses the average signal levels over a wide range, and limits its peaks to prevent overload. It has two thresholds: one low for the compressor and one high for the limiter.

Limiters can be used to prevent recorder overload during field recording, or to prevent PA power amps from clipping. When you master a program of several mixed songs in a DAW, you might use limiting to reduce the level of signal peaks in the program. Set the threshold about 6 dB below the highest peak level. Then apply normalization, which raises the level of the entire program until the highest peak in the program reaches maximum level. Limiting and normalization create a louder program on your finished CD without compressing the music's dynamics.

A limiter plug-in with the "look-ahead" feature checks the upcoming audio for peaks. Audio enters a look-ahead buffer, and the limiter measures that audio signal and reacts quickly to reduce the incoming peaks.

NOISE GATE

A noise gate (expander) acts like an on-off switch that removes noises during pauses in an audio signal. It reduces the gain when the input level falls below a preset threshold. That is, when an instrument stops playing for a moment, the noise gate drops the volume, which removes any noise and leakage during the pause (Figure 10.14).

Note: The gate doesn't remove noise while the instrument is playing.

Where is it used? The noise gate helps to clean up drum tracks by removing leakage between beats. It can shorten the decay time of the

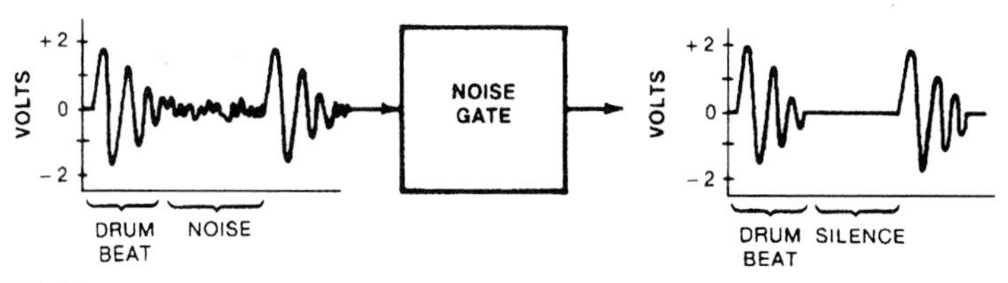

FIGURE 10.14
Gating.

drums, giving a very tight sound. If you're recording a noisy guitar amp, try a gate to cut out the buzz and hiss between phrases.

How do you use a noise gate? Patch it between a recorder-track output and a mixer line input, or use a gate plug-in in a DAW. Solo the track that you want to gate. Set the gate's threshold so that noise and leakage go away during pauses. If the gate chops off each note, the threshold is set too high—turn it down. Set the release time as short as possible, but long enough so that you hear the entire note (or tom hit) before the gate cuts off the sound. To fix a boomy kick drum, adjust the threshold until the kick sounds as "tight" as you want. That is, use the gate to shorten the decay portion of the kick-drum's envelope.

Excellent recordings can be made without gating. But if you want a tighter sound, gates come in handy. Some signal processors have compression, limiting, and noise gating in a single package.

Some gates have a side-chain input or key input. It's an input for an external signal that controls the gating action. The control signal triggers the output of the gate's main audio path. For example, you could feed a bass guitar through the noise gate, and gate the bass with a kick-drum signal fed into the side chain. Then the bass will follow the kick drum's envelope.

DELAY: ECHO, DOUBLING, CHORUS, AND FLANGING

A digital delay (or a delay plug-in) takes an input signal, holds it in memory, then plays it back after a short delay—about 1 msec to 1 second (Figure 10.15). Delay is the time interval between the input signal and its repetition at the output of the delay device.

If you listen to the delayed signal by itself, it sounds the same as the undelayed (dry) signal. But if you combine the delayed and dry signals, you may hear two distinct sounds: the signal and its repetition.

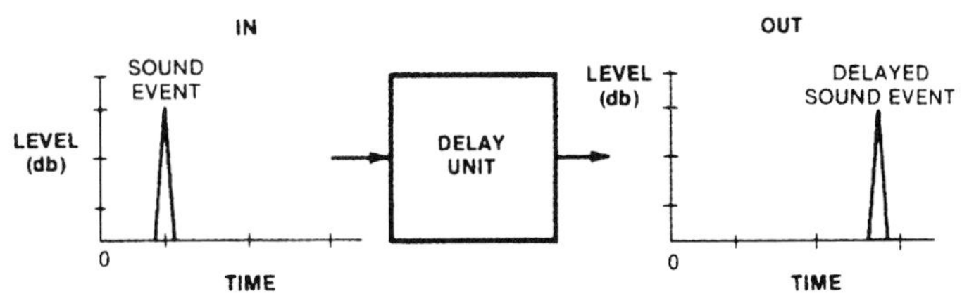

FIGURE 10.15
Delaying the signal.

By delaying a signal, a processor can create several effects such as echo, repeating echo, doubling, chorus, and flanging.

Echo

If the delay is about 50 msec to 1 second, the delayed repetition of a sound is called an echo. This is shown in Figure 10.16 by the two pulses. Echoes occur naturally when sound waves travel to a distant room surface, bounce off, and return later to the listener—repeating the original sound. A delay unit can mimic this effect. Many people use the term delay to mean echo.

In setting up a mix with echo, you want to hear both the dry sound and its echo. You do this by creating an effects loop: from the mixer, to the effects box, back to the mixer. Here's how to set up echo with a hardware mixer and effects unit:

1. On the delay unit, set the dry/wet mix control all the way to wet or 100% mix. Then the output of the delay unit will be only the delayed signal.
2. Suppose you want to use aux1 as the echo control. Connect aux1 send to delay unit IN. Connect delay unit OUT to Bus 1 and 2 IN (or to the effects-return jacks).
3. Find the mixer module for the instrument you want to add echo to.
4. Assign the instrument to busses 1 and 2. Monitor busses 1 and 2.
5. Find the knobs labeled Bus 1 IN and Bus 2 IN. They might be called Aux Return or Effects Return. Turn them up to 0, about three-fourths of the way up.
6. Turn up the aux1 send knob, and there's your echo.

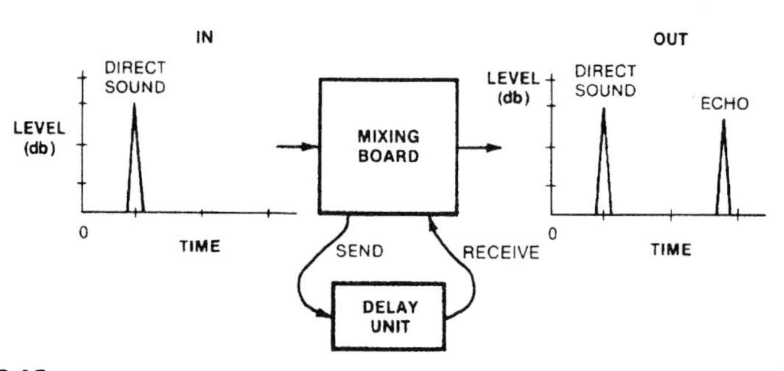

FIGURE 10.16
Echo.

The delayed sound mixes with the dry sound in busses 1 and 2. You hear both sounds, which together make an echo. Each aux knob controls the amount of echo on each track, while the effect-return knobs control the overall amount of echo on all tracks that are feeding the echo unit.

To set up echo in a DAW, follow this procedure:

1. Create or use a stereo aux bus that has an Echo or Delay plug-in enabled.
2. In a track that you want to have echo, enable an aux send to that bus.
3. Open the Echo plug-in. Set its dry/wet mix control all the way to wet or 100% mix.
4. In the track that you want to have echo, gradually turn up the aux-send until you hear the desired amount of echo. In the Echo plug-in, adjust the delay parameters for the desired effect.

SLAP ECHO

A delay from about 100 to 130 msec is called a slap echo or slapback echo. It was often used in 1950s rock 'n' roll and rockabilly tunes, and still is used today.

REPEATING ECHO

Most delay units can be made to feed the output signal back into the input, internally. Then the signal is re-delayed many times. This creates a repeating echo—several echoes that are evenly spaced in time (Figure 10.17, and *audio clips 29 and 30 at www.elsevierdirect.com/companions/9780240811444*). The regeneration (feedback) control sets the number of repeats.

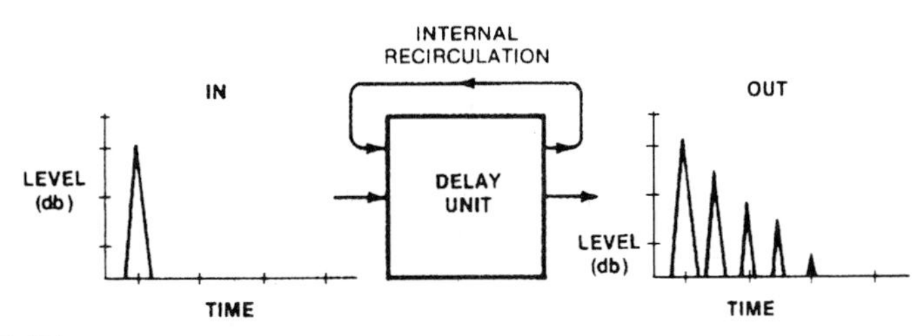

FIGURE 10.17
Repeating echo.

Repeating echo is most musical if you set the delay time to create an echo rhythm that fits the tempo of the song. The formula is

$$\text{Delay in seconds} = 60/\text{tempo}$$

So if the tempo is 120 bpm, the delay is 0.5 sec (500 msec). That's one echo per quarter note. Use half that delay to get one echo per eighth note. Use one-third that delay for triplets. A slow repeating echo—0.5 second between repeats; for example, gives an outer-space or haunted-house effect. It often sounds good on lead vocal in a ballad.

Doubling

If you set the delay around 23 to 30 msec, the effect is called doubling or automatic double tracking (ADT). It gives an instrument or voice a fuller sound, especially if the dry and delayed signals are panned to opposite sides. The short delays used in doubling sound like early sound reflections in a studio, so they add some "air" or ambience.

Doubling a vocal can be done without a delay unit. Record a vocal part, then overdub another performance of the same vocal part. Mix the parts, pan them both to center, or pan them left and right.

Chorus

This is a wavy or shimmering effect. The delay is 15 to 35 msec, and the delay varies at a slow rate. Sweeping the delay time causes the delayed signal to bend up and down in pitch, or to detune. When you combine the detuned signal with the original signal, you get chorusing.

STEREO CHORUS

This is a beautiful effect. In one channel, the delayed signal is combined with the dry signal in the same polarity. In the other channel, the delayed signal is inverted in polarity, then combined with the dry signal. Thus, the right channel has a series of peaks in the frequency response where the left channel has dips, and vice versa. The delay is slowly varied or modulated. *Hear a demonstration with audio clip 33 at www.elsevierdirect.com/companions/9780240811444.*

BASS CHORUS

This is chorus with a highpass filter so that low frequencies aren't chorused, but higher harmonics are. It gives an ethereal quality to the bass guitar.

Flanging

If you set the delay around 0 to 20 msec, you usually can't resolve the direct and delayed signals into two separate sounds. Instead, you hear a single sound with a strange frequency response. The direct and delayed signals combine and have phase interference, which puts a series of peaks and dips in the frequency response. This is called a comb-filter effect (Figure 10.18). It gives a very colored, filtered tone quality. The shorter the delay, the farther apart the peaks and dips are spaced in frequency.

The flanging effect varies or sweeps the delay between about 0 and 20 msec. This makes the comb-filter nulls sweep up and down the spectrum. As a result, the sound is hollow, swishing, and ethereal, as if the music were playing through a pipe. Flanging is easiest to hear with broadband signals such as cymbals but can be used on any instrument, even voices. *Hear a demonstration with audio clip 34 at www.elsevierdirect.com/companions/9780240811444.*

Some examples of flanging are on many Jimi Hendrix records, and on the oldies "Itchycoo Park" by the Small Faces and "Listen to the Music" by the Doobie Brothers. The first use of flanging was on "The Big Hurt" sung by Toni Fisher.

Positive flanging refers to flanging in which the delayed signal is the same polarity as the direct signal (Figure 10.18). With negative flanging, the delayed signal is opposite in polarity to the direct

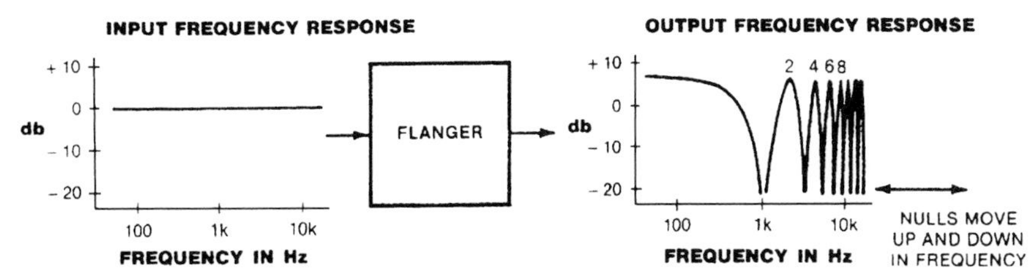

FIGURE 10.18
Flanging (or positive flanging).

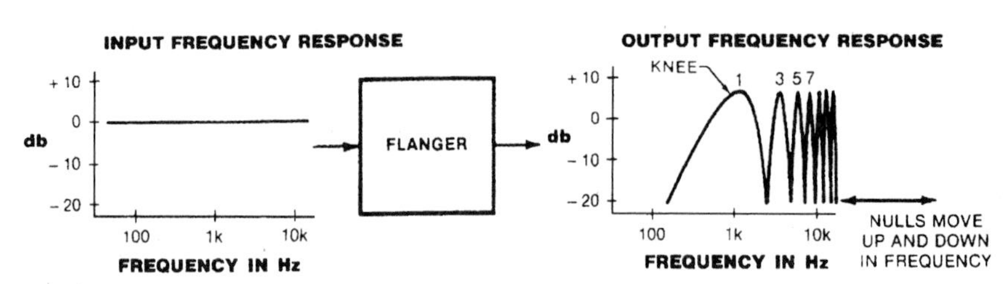

FIGURE 10.19
Negative flanging.

signal, which makes a stronger effect. The low frequencies are canceled (the bass rolls off), and the "knee" of the bass rolloff moves up and down the spectrum as the delay is varied. The high frequencies are still comb-filtered (Figure 10.19). Negative flanging makes the music sound like it's turning inside out.

When the flanger feeds some of the output signal back into the input, the peaks and dips get bigger. It's a powerful "science fiction" effect called resonant flanging.

REVERBERATION

This effect adds a sense of room acoustics, ambience, or space to instruments and voices. To know how it works, we need to understand how reverb happens in a real room. Natural reverberation in a room is a series of multiple sound reflections that make the original sound persist and gradually die away or decay. These reflections tell the ear that you're listening in a large or hard-surfaced room. For example, reverberation is the sound you hear just after you shout in an empty gymnasium.

A reverb effect simulates the sound of a room—a club, auditorium, or concert hall—by generating random multiple echoes that are too numerous and rapid for the ear to resolve (Figure 10.20). Digital reverb is available either in a dedicated reverb unit, as part of a multi-effects processor, or as a plug-in.

The most natural sounding digital reverb is a sampling reverb or convolution reverb, which creates the reverb from impulse-response samples (WAVE files) of real acoustic spaces, rather than from algorithms. One convolution reverb plug-in is SIR at www.knufinke.de/sir/. Free impulse-response samples are at www.noisevault.com and www.cksde.com/p_6_250.htm.

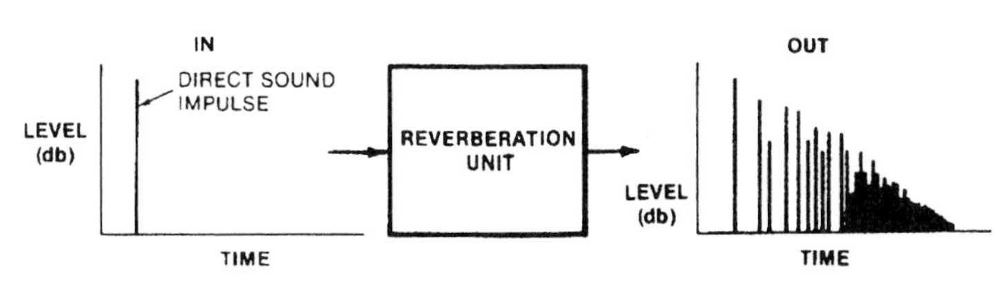

FIGURE 10.20
Reverberation.

Reverb and echo aren't the same thing. Echo is a repetition of a sound (HELLO hello hello); reverb is a smooth decay of sound (HELLO-OO-oo-oo).

Multichannel digital reverbs are available for surround sound, both as hardware and software. Examples include: Eventide's Orville, Sony's DRE-S777, TC Electronics' System 6000, Lexicon's 960L, and Kind of Loud's RealVerb 5.1 Pro Tools plug-in. Surround reverb plug-ins for Steinberg's Nuendo platform include Steinberg's Surround Edition plug-in bundle and TC Works SurroundVerb plug-in.

Reverb Parameters

Here are some controls in a reverb unit or plug-in (Figure 10.21):

- Reverb Time (RT60): The time it takes for reverberation to decay 60 dB below its original level. Set it long (1–1/2 to 2 seconds) to simulate a large room; set it short (under 1 second) to simulate a small room. Generally you use short reverbs (or no reverb) for fast songs, and long reverbs for slow songs.
- Pre-delay (pre-reverb delay): A short delay (10 to 100 msec) before the onset of reverb to simulate the delay that happens in real rooms before reverb starts. The longer the pre-delay, the bigger the room sounds. Using pre-delay on an instrument's reverb often helps to clarify the sound by removing the onset of reverb from the direct sound of the instrument. If your reverb unit doesn't have pre-delay, you can create it by connecting a delay unit between your mixer's aux-send and the reverb input. If your reverb plug-in lacks pre-delay, insert a delay plug-in before the reverb plug-in. *Audio clip 31 at www.elsevierdirect.com/companions/9780240811444 demonstrates reverb: short reverb time, long reverb time, and pre-delay.*
- Density or diffusion: A high density setting produces many echoes spaced close together. It gives a smooth decay but

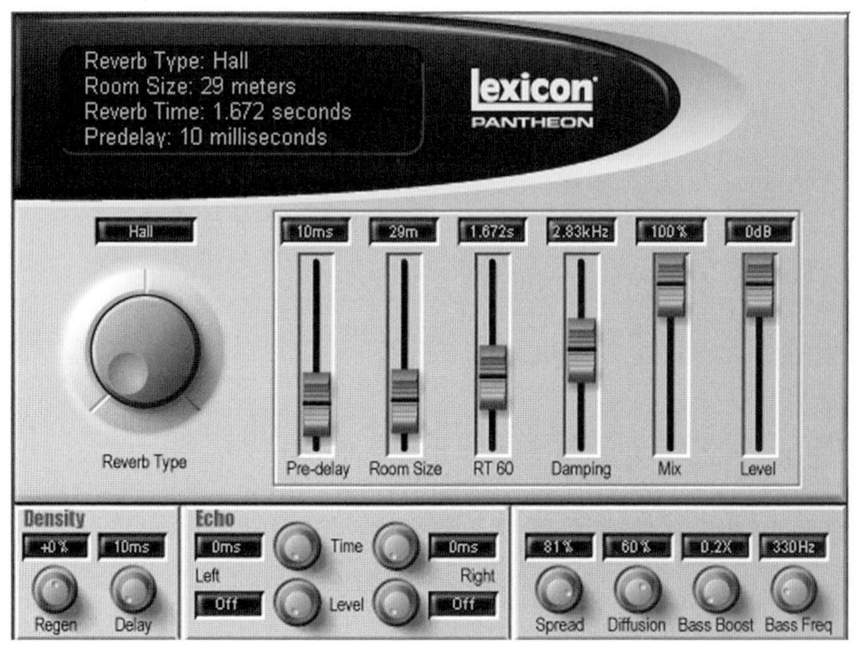

FIGURE 10.21
Lexicon Pantheon Reverb plug-in.

increases the load on the CPU. A low-density setting produces fewer echoes spaced farther apart, and may be adequate for vocals, synth pads, and organ. Use high density for percussive sounds to prevent grainy-sounding reverb.

- Damping: Adjusts the reverb time or decay at high frequencies. Set the damping frequency high (say, 7 kHz) to simulate a hard-surfaced room; set it low (say, 2 kHz) to simulate a soft-surfaced room. The latter is also called a "warm room" reverb.
- Presets: Factory-supplied reverb settings of small rooms, auditoriums, halls, and so on. A plate reverb setting duplicates the bright sound of a metal-foil plate, which used to be the most popular type of reverb in pro studios. Unnatural effects are available, such as nonlinear decay, reverse reverb that builds up before decaying, or gated reverb. With gated reverb, the reverb cuts off suddenly shortly after a note is hit. In the 1980s it was often used on a snare drum. A good example is the oldie "You Can Call Me Al" on Paul Simon's album Graceland.

Reverb Connections

To connect a reverb unit to a hardware mixer, connect a cable from the mixer aux-send to the reverb input. Connect a cable (two for stereo) from the reverb outputs to the mixer aux returns (effects returns or bus inputs). Set the mix control on the reverb unit all the way to wet or reverb. Turn the mixer's aux-return or bus-in knobs (if any) about two-thirds of the way up, and adjust the amount of reverb on each track with the aux-send knobs. Try to get an overall reverb-send level near 0 on the meter; then fine-tune the aux return level for the desired amount of reverb.

To enable reverb in a DAW, use the same procedure as for setting up echo, but choose a reverb plug-in.

PREVERB

Preverb is reverb that precedes a note rather than follows it. The reverb starts from silence and builds up to a note's attack (*audio clip 32 at www.elsevierdirect.com/companions/9780240811444*). When used on a snare drum, preverb gives a whip-cracking kind of sound, like "shSHK!"

Here's how to add preverb to a snare drum track in a DAW:

1. Set up an aux1 bus with reverb. Set the reverb all the way to wet or 100% mix.
2. Solo the drum track.
3. On the drum track, turn up the aux1 send and set it to pre-fader. Find a good snare hit and play it.
4. Turn down the drum-track fader and play the snare hit. You should hear only the reverb from the aux1 bus.
5. Highlight the snare hit up until the next hit. Export or save the mix (the snare hit) as Drum reverb.wav.
6. Select a blank track and call it Drum Reverb. Import Drum reverb.wav into that track.
7. Select the drum-reverb clip, then select the Reverse processing in your DAW. This reverses the drum-reverb clip.
8. On the original snare-drum track, reset the aux1 send to post-fader and turn it back down. Turn up the track fader where it used to be.

9. Slide in time the reversed drum-reverb clip so that it ends just as a snare-drum hit starts (check the waveforms).
10. Play the reversed drum-reverb track along with the snare-drum track. You should hear preverb.

Some signal processors have a reverse reverb effect in which the reverb comes after the note that produced it, but builds up before it fades out. This isn't quite the same as preverb. Reverse reverb can upset the musical timing; preverb doesn't.

ENHANCER

If a track or a mix sounds dull and muffled, you can run it through an enhancer to add brilliance and clarity. An enhancer works by adding slight distortion (as in the Aphex Aural Exciter).

OCTAVE DIVIDER

This unit takes a signal from a bass guitar and provides deep, growling bass notes one or two octaves below the pitch of the bass guitar. It does this by dividing the incoming frequency by 2 or 4: If you put 82 Hz in, you get 41 Hz out. Some MIDI sound modules have bass patches with extra-deep sound, and 5-string bass guitars have an extra string tuned especially low.

HARMONIZER

Basically a delay unit with delay modulation, a harmonizer makes a variety of pitch-shifting effects. It can create harmonies, change pitch without changing the duration of the program, change duration without changing pitch, and many other oddities. You've heard harmonizers on radio-station spots when the announcer's voice sounds like a Munchkin or Darth Vader. *Play audio clip 35 at www.elsevierdirect.com/companions/9780240811444*.

VOCAL PROCESSOR

This device or plug-in can affect the vocal's inflection, add growls or whispers, correct the pitch, add vibrato, make the voice more-or-less nasal or chesty, and so on. The latest vocal formant-corrected pitch-shifters maintain the voice formant structure when they shift pitch;

this prevents the "chipmunk" effect. An example includes TC-Helicon VoiceModeler plug-in.

Another type of vocal processor is called a channel strip. It includes a high-quality mic preamp or two, plus EQ, compression, gating, de-esser, and perhaps some tube saturation distortion. An example is Focusrite VoiceMaster Pro.

PITCH CORRECTION

This plug-in provides automatic or manual pitch correction of a single track (but not of chords). In automatic mode, it corrects flat or sharp notes by changing their pitch to match a musical scale of your choice. In manual mode, you see a graph of the notes' pitches on your monitor screen, and slide certain notes up or down to correct their pitch. Manual mode is less obvious than automatic. You also can use this effect as a "robotic" effect where the sung notes change pitch in a step-wise, jerky way rather than smoothly. Some pitch-correction plug-ins are Antares Auto-Tune, Celemony Melodyne, TC Helicon Intonator, and Roland's V-Vocal plug-in for Cakewalk Sonar Producer.

Melodyne Direct Note Access lets you edit individual notes within chords. See www.celemony.com.

TUBE PROCESSOR

This device or plug-in uses a vacuum tube or a simulation of one. Tubes have euphonic even-order harmonic distortion, which is claimed to add "richness" or "warmth" when the tube distorts (*audio clip 36 at www.elsevierdirect.com/companions/9780240811444*). There are tube mics, tube mic preamps, tube compressors, and stand-alone tube processors.

ROTARY SPEAKER SIMULATOR

This effect simulates the sound of a Leslie organ speaker, which plays music through rotating horns. It's a complex sound effect of pitch shifting, tremolo, and phase shifting. The speed and depth of the effect are adjustable.

ANALOG TAPE SIMULATOR

Analog tape saturation is mainly third-harmonic distortion and compression. An analog tape simulator adds this distortion to digital recordings in an attempt to smear or warm up the sound in a pleasant way (*Audio clip 37 at www.elsevierdirect.com/companions/9780240811444*).

SPATIAL PROCESSOR

Spatial processors enhance the stereo imaging or spatial aspects of a mix heard over two speakers. Some units have joystick-type pan pots, which move the image of each track anywhere around the listener. Other units make the stereo stage wider, so that images can be placed to the left of the left speaker, and to the right of the right speaker. The listener might hear images toward the sides of the listening room. In 5.1 surround systems, this spatial processing is done by surround panning and surround reverbs.

MICROPHONE MODELER

Antares and Roland offer a microphone modeler or simulator. You tell it which mic you're using and which mic you want it to sound like. A wide variety of vintage and current mic simulations are available. Mic modeling comes in three forms: hardware device, plug-in, and firmware (programmed into a chip) in a recorder–mixer.

GUITAR AMPLIFIER MODELER

Another simulator takes the sound of a direct-recorded guitar, and makes the guitar sound like it is played through a guitar amp (*audio clip 38 at www.elsevierdirect.com/companions/9780240811444*). Several amp models can be simulated, as well as effects, tone, drive, the mic used to pick up the amp, and the mic's position.

Two hardware examples of amp modelers are the Line 6 Pod and the Johnson J Station. Amp Farm is a guitar modeling plug-in for Pro Tools. Roland's DAWs offer COSM mic modeling and guitar-amp modeling.

Guitar processors or guitar stomp boxes can be used on any instrument or vocal to add distortion.

FIGURE 10.22
CamelCrusher distortion plug-in.

DISTORTION

Some plug-ins add intentional distortion of various types to generally "shred" the sound for a low-fi effect. Two examples are Izotope Trash and Camel Audio CamelCrusher (Figure 10.22).

DE-CLICK AND DE-NOISE

Also called "Audio restoration programs," these are plug-ins—or stand-alone programs—that can remove the clicks and pops from LP records, or remove hiss and hum from noisy recordings. Examples include: Dart Pro 24, iZotope RX, Bias SoundSoap, and Waves Native Restoration Bundle.

SURROUND SOUND

Recent plug-ins for surround sound are surround panning, surround reverb, and surround encoding/decoding.

MULTI-EFFECTS PROCESSOR

This provides several effects in a single device or plug-in. Some units let you combine up to four effects in any order. Others have several channels, so you can put a different effect on different instruments. With most processors, you can edit the sounds and save them in memory as new programs. *At www.elsevierdirect.com/companions/9780240811444 audio clips 29–38 demonstrate various effects on voice.*

Some processors offer 100 or more programmable presets with MIDI control over any parameter. For example, with some units you can place an instrument in a simulated room, and use a MIDI controller to continuously change the size of that room. Many signal processors can be controlled by MIDI program-change commands. You can quickly change the type of effect, or effect parameters, by entering certain program changes into a sequencer.

Suppose you want each tom-tom hit in a drum fill to have a different size room added to it. For example, put the high-rack tom in a small room; put the low-rack tom in a concert hall; and put the floor tom in a cave. To do this, first assign a different program number (patch or preset number) to each effects parameter. You do this with the effects device. Then, using the sequencer, punch in the appropriate program number for each note.

A MIDI program-change footswitch lets guitarists call up different effects on MIDI signal processors. By tapping a footswitch, they can get fuzz, flanging, wah-wah, spring reverb, and so on.

A MIDI mapper lets you control some effects parameters with any controller. For example, vary reverb decay time with a pitch wheel, or vary a filter with key velocity.

LOOKING BACK

We've come a long way with effects. Looking back over the past few decades, each era had its own "sound" related to the effects used at the time. The 1950s had tube distortion and slap echo; the 1960s used fuzz, wah-wah, and flanging. Much of the early 1970s sounded dry, and the early 1990s emphasized synthesizers, drum machines, and gated reverb. Now vacuum tubes and acoustic instruments are back, along with occasional low-fi (tinny, distorted, or noisy) sounds and dry vocals. Whatever effects you choose, they can enhance your music if used with taste.

SOUND-QUALITY GLOSSARY

The sound of mic techniques, effects, and EQ in a recording can be hard to translate into engineering terms. For example, what EQ should you use to get a "fat" sound or a "thin" sound? The glossary below may help. It's based on conversations with producers, musicians, and reviewers over 30 years. Not everyone agrees on these definitions, but they are common. This glossary doesn't suggest the cause of the sound quality or how to change it; that's up to you to determine.

AIRY—Spacious. The instruments sound like they are surrounded by a large reflective space full of air. A pleasant amount of reverb or early reflections. High-frequency response that extends to 15 or 20 kHz.

BALLSY *or* BASSY—Emphasized low frequencies below about 200 Hz.

BLOATED—Excessive mid-bass around 250 Hz. Poorly damped low frequencies, low-frequency resonances.

BLOOM—Adequate low frequencies. Spacious. Good reproduction of dynamics and reverberation. Early reflections or a sense of "air" around each instrument in an orchestra.

BOOMY—Excessive bass around 125 Hz. Poorly damped low frequencies or low-frequency resonances.

BOXY—Having resonances as if the music were enclosed in a box. Speaker cabinet diffraction or vibration. An emphasis around 250 to 600 Hz.

BREATHY—Audible breath sounds in vocals, flute, or sax. Good high-frequency response.

BRIGHT—High-frequency emphasis. Harmonics are strong relative to fundamentals.

BRITTLE—High-frequency peaks, or weak fundamentals. Slightly distorted or harsh highs. Opposite of round or mellow. See *Thin*.

Objects that are physically thin and brittle emphasize highs over lows when you crack them.

CHESTY—A vocal signal with a bump in the low-frequency response around 125 to 250 Hz.

CLEAN—Free of noise, distortion, and leakage.

CLEAR—See *Transparent*.

CLINICAL—Too clean or analytical. Emphasized high-frequency response, sharp transient response. Not warm.

COLORED—Having timbres that aren't true to life. Non-flat response, peaks, or dips.

CLOUDY—See *Wooly*.

CONSTRICTED—Poor reproduction of dynamics. Dynamic compression. Distortion at high levels. Also see *Pinched*.

CRISP—Extended high-frequency response. Like a crispy potato chip or crisp bacon frying. Often referring to cymbals.

CRUNCH—Pleasant guitar-amp distortion.

DARK—Opposite of bright. Weak high frequencies.

DELICATE—High frequencies extending to 15 or 20 kHz without peaks. A sweet, airy, open sound with strings or acoustic guitar.

DEPTH—A sense of closeness and farness of instruments, caused by miking them at different distances. Good transient response that reveals the direct/reflected sound ratio in the recording.

DETAILED—Easy to hear tiny details in the music; articulate. Adequate high-frequency response, sharp transient response.

DISTANT—Too much leakage. Low direct-to-reverb ratio.

DRY—Without effects. Not spacious. Reverb tends toward mono instead of spreading out. Overdamped transient response.

DULL—See *Dark*.

ECHOEY—Having audible echoes or reverberation.

EDGY—Too much high frequencies. Trebley. Harmonics are too strong relative to the fundamentals. When you view the waveform on an oscilloscope, it even looks edgy or jagged, because of excessive high frequencies. Distorted, having unwanted harmonics that add an edge or raspiness to the sound.

EFFORTLESS—Low distortion, usually coupled with flat response.

ETCHED—Clear but verging on edgy. Emphasis around 10 kHz or higher.

FAT—See *Full* and *Warm*. Also, a diffuse spatial effect. Also, smeared out in time, with some reverberant decay. Also, the sound of a snare drum tuned low.

FOCUSED—Referring to the image of a musical instrument which is easy to localize, pinpointed, having a small spatial spread.

FORWARD—Sounding close to the listener.. Emphasis around 2 to 5 kHz.

FULL—Opposite of thin. Strong fundamentals relative to harmonics. Good low-frequency response, not necessarily extended, but with adequate level around 100 to 300 Hz.

GENTLE—Opposite of edgy. The harmonics—highs and upper mids—aren't exaggerated, or may even be weak.

GLARE, GLASSY—A little less extreme than edgy. A little too bright or trebley.

GRAINY—The music sounds like it's segmented into little grains, rather than flowing in one continuous piece. Not liquid or fluid. Suffering from distortion. Some early A/D converters sounded grainy, as do current ones of inferior design. "Powdery" is finer than "grainy"!

GRUNGY—Lots of distortion.

HARD—Too much upper midrange, usually around 3 kHz. Or, good transient response, as if the sound is hitting you hard.

HARSH—Too much upper midrange. Peaks in the frequency response from 2 to 6 kHz. Or, excessive phase shift.

HEAVY—Good low-frequency response below about 50 Hz. Suggesting great weight or power, like a diesel locomotive or thunder.

HOLLOW—See *Honky*. Or, too much reverberation, a mid-frequency dip, or comb filtering.

HONKY—The music sounds the way your voice sounds when you cup your hands around your mouth. A bump in the response around 500 to 700 Hz.

IN YOUR FACE—Dry (without effects, without reverb), possibly with compression.

LIQUID—Opposite of grainy. A sense of seamless flowing of the music. Flat response and low distortion. High frequencies are flat or reduced relative to mids and lows.

LOW-FI (low fidelity)—"Trashy" sounding. Tinny, distorted, noisy, or muddy.

MELLOW—Reduced high frequencies, not edgy.

MUDDY—Not clear. Weak harmonics, smeared time response, distortion. Too much reverb at low frequencies. Too much emphasis around 200 to 350 Hz. Too much leakage.

MUFFLED—The music sounds covered up. Weak highs or weak upper mids.

MUSICAL—Conveying emotion. Flat response, low distortion, no edginess.

NASAL—The vocalist sounds as if she is singing with the nose closed. Also applies to strings. Bump in the response around 500 to 1000 Hz. See *Honky*.

NEUTRAL—Accurate tonal reproduction. No obvious colorations. No serious peaks or dips in the frequency response.

PAPERY—Referring to a kick drum that has too much output around 400 to 600 Hz.

PHASEY—Having phase interference (comb filtering). The sound of a direct signal and its delayed repetition mixed to the same channel (delay usually under 20 msec). Might be due to multiple mics picking up the same source, or one mic picking up direct sound and delayed reflected sound. Or a delayed signal mixed with itself undelayed. Or, some opposite-polarity crosstalk between stereo channels. Or one monitor speaker is reversed in polarity.

PIERCING—Strident, hard on the ears, screechy. Having sharp, narrow peaks in the response around 3 to 10 kHz.

PINCHED—Narrowband. Midrange or upper midrange peak in the frequency response. Pinched dynamics are overly compressed.

PRESENT, PRESENCE—Adequate or emphasized response around 5 kHz for most instruments, or around 2 to 5 kHz for kick drum and bass. Having some edge, punch, detail, closeness, and clarity.

PUFFY—Bump in the response around 400 to 700 Hz.

PUNCHY—Good reproduction of dynamics. Good transient response. Or conversely, referring to highly compressed transients (especially snare drum and kick drum) that sound like hitting a punching bag. Sometimes a bump around 5 kHz or 200 Hz.

RASPY—Harsh, like a rasp. Peaks in the response around 6 kHz which make vocals sound too sibilant or piercing.

RICH—See *Full*. Also, having euphonic distortion made of even-order harmonics.

ROUND—High-frequency rolloff or dip. Not edgy.

SHARP—See *Crisp*, *Strident*, and *Tight*.

SIBILANT, ESSY—Exaggerated "s" and "sh" sounds in vocals, too much output around 5 to 10 kHz.

SIZZLY—See *Sibilant*. Also, too much highs on cymbals.

SMEARED—Lacking detail. Poor transient response. This may be a desirable effect in large-diameter mics. Also, poorly focused images.

SMOOTH—Easy on the ears, not harsh. Flat frequency response, especially in the midrange. Lack of peaks and dips in the response. Low distortion.

SPACIOUS—Conveying a sense of space, ambience, or room around the instruments. To get this effect, mike farther back, mix in an ambience mic, add reverb, or record in stereo. Components that have opposite-polarity or out-of-phase crosstalk between channels may add false spaciousness.

SQUASHED—Overly compressed.

STEELY—Emphasized upper mids around 3 to 6 kHz. Peaky, non-flat high-frequency response. See *Glassy, Harsh, Edgy*.

STRAINED—The component sounds like it's working too hard. Distorted. Inadequate headroom or insufficient power. Opposite of effortless.

STRIDENT—See *Harsh* and *Edgy*.

SWEET—Not strident or piercing. Flat high-frequency response, low distortion. Lack of peaks in the response. Highs are extended to 15 or 20 kHz, but they aren't bumped up. Often used when referring to cymbals, percussion, strings, and sibilant sounds.

THIN—Fundamentals are weak relative to harmonics. Note that the fundamental frequencies of many instruments aren't very low. For example, violin fundamentals are around 200 to 1000 Hz. So if the 300 Hz area is weak, the violin may sound thin—even if the mic's response goes down to 40 Hz.

TIGHT—Good low-frequency transient response. Absence of ringing or resonance when reproducing the kick drum or bass. Good low-frequency detail. Absence of leakage.

TINNY, TELEPHONE-LIKE—Narrowband, weak lows, peaky mids. The music sounds like it's coming through a telephone or tin can.

TRANSPARENT—Easy to hear into the music, detailed, clear, not muddy. Wide, flat frequency response, sharp time response, very low distortion and noise.

TUBBY—See *Bloated*. Having low-frequency resonances as if you're singing in a bathtub.

VEILED—The music sounds like you put a silk veil over the speakers. Slight noise or distortion, or slightly weak high frequencies.

WARM—Good bass, adequate low frequencies, adequate fundamentals relative to harmonics. Not thin. Or, excessive bass or midbass. Or, pleasantly spacious, with adequate reverberation at low frequencies. Or, gentle highs, like from a tube amplifier. See *Rich*.

WASHED OUT—Phase interference from multiple mics picking up the same source. Too much leakage or reverberation.

WOOLY, BLANKETED—The music sounds like there's a wool blanket over the speakers. Weak high frequencies or boomy low frequencies. Sometimes, an emphasis around 250 to 600 Hz.

CHAPTER 11

Mixers and Mixing Consoles

The heart of your recording studio is the mixer. It's a control center where you plug in all sorts of signals; mix or blend them; add effects, EQ, and stereo positioning; and route the signals to recorders and monitor speakers. A mixing console (also called board or desk) is a large mixer with many controls.

A recorder–mixer combines a mixer and a multitrack recorder in a single portable chassis. This convenient unit is also called a ministudio, portable studio, or digital multitracker. Low-end recorder–mixers record 4 tracks, and high-end units record 8 to 32 tracks.

STAGES OF RECORDING

First let's review the three stages of using a mixer during a session: recording, overdubbing, and mixdown.

1. Recording (tracking): The mixer accepts mic-level signals and amplifies them up to line level. You send the line-level signal from each mic to a separate track in a multitrack recorder. The multitrack recorder records several tracks on hard disk. One track might be a lead vocal, another track might be a saxophone, and so on.
2. Overdubbing: While listening to recorded tracks over headphones, the musician records new parts on open (unused) tracks. You set up the mixer to monitor both the recorded tracks and the new live part.
3. Mixdown: After all the tracks are recorded, you use the mixer to combine or mix the tracks into 2-channel stereo with panning, EQ, and effects.

MIXER FUNCTIONS AND FORMATS

Although the knobs, buttons, and meters on a mixer may appear intimidating, you can understand them if you read the manual and practice with the equipment. A mixer is complicated because it lets you control many aspects of sound:

- The loudness of each instrument (the balance among instruments in the mix)
- The tone quality of each instrument (bass, treble, midrange)
- The room that the instruments are in (reverberation)
- The left-to-right position of each instrument (panning)
- Effects (flanging, echo, chorus, etc.)
- Track assignments (which instrument goes on which track)
- Recording level (the signal voltage going to the recorder tracks)
- Monitor selection (what you want to listen to)

Mixers come in many formats:

- Analog mixer: A control device that works on analog signals, and sends them to an external multitrack recorder and 2-track recorder.
- Digital mixer: The same as the analog mixer, but works internally with digital signals. It accepts analog or digital signals.
- Software mixer: Exists only in your computer as part of digital recording software. You control it with a mouse or a control surface (described next). The recording is done on your hard drive. Whether you use a hardware or software mixer, be sure to read the section in this chapter on analog mixers because many of the same principles apply.
- Control surface: A device that looks like a mixer with faders and knobs. It plugs into your computer's USB or FireWire port and controls the software mixer.

A mixer can be specified by the number of inputs and outputs it has. For example, an 8-in, 2-out mixer (8 × 2 mixer) has 8 signal inputs that can be mixed into 2 output channels (buses) for stereo recording. Similarly, a 16-in, 8-out (16 × 8) mixing board has 16 signal inputs and 8 output channels for multitrack recording. A 16 × 4 × 2 mixing board has 16 inputs, 4 submixes or groups (explained under the section Output Section) and 2 master outputs. There also are connectors for external equipment, such as effects devices and a monitor power amplifier. The more inputs your mixer has, the more instruments you can record at the same time. If you're recording only yourself, you may need only two inputs.

Let's look at the analog mixer in more detail. Knowing how it works will help you understand the other types.

ANALOG MIXER

A mixer can be divided into three sections: input, output, and monitor. Here are the main parts of each section and what they do:

Input section

- Inputs connect to your mics, electric instruments, and recorder outputs.
- Faders are sliding volume controls that affect the loudness of each instrument. This lets you control the balance among instruments in the mix.
- EQ knobs adjust the tone quality of each instrument (bass, treble, midrange).
- Aux knobs set the amount of reverb or other effects, and also can be used to set up a monitor mix or headphone mix.
- Pan pots place the monitored sound of each track where desired between your stereo speakers—left, center, right, or anywhere between. In some mixers, the pan pot is also used with the assign switch during recording to send signals to the desired tracks.
- Channel assign buttons route each input signal to the desired recorder track.

Output section

- Master faders set the overall level of the entire stereo mix.
- Group faders set the overall level of each group or submix.
- Outputs connect to your recorder inputs and the monitor power amplifier.
- Meters help you set the correct recording level (to prevent distortion and noise).

Monitor section

- Monitor controls select what you want to listen to.
- Aux knobs or channel faders set up the monitor mix.

Let's look at each part in more detail.

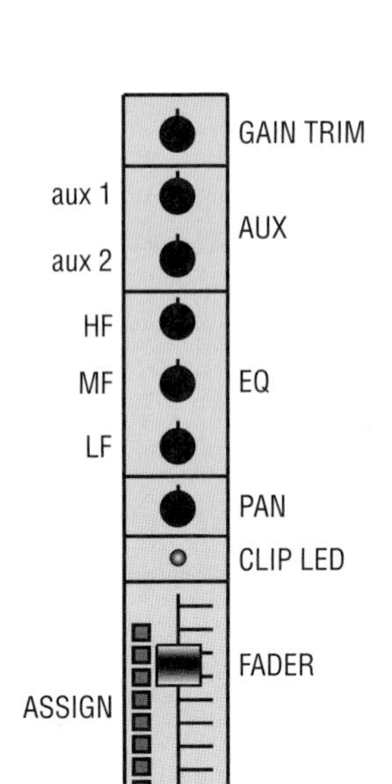

FIGURE 11.1
A typical input module.

Input Section

A mixer is made of groups of controls called modules. An input module (Figure 11.1) or channel strip affects a single input signal—from a microphone, for instance. The module is a narrow vertical strip, one per input. Several modules are lined up side by side. Each input module is the same, so if you know one, you know them all.

Let's follow the signal flow from input to output through a typical input module (Figure 11.2). Every mixer is a little different, but you're likely to find features like those described here.

INPUT CONNECTORS

On the back of each input module are input connectors with these labels:

MIC: Accepts signals from a microphone or direct box.
LINE: Accepts an electric musical instrument, or a track output of a multitrack recorder.
TRACK: Accepts a track output of a multitrack recorder; not included in all mixers.

Some units use a single jack for both mic and line inputs; others have separate jacks for each. The mic input is either an unbalanced

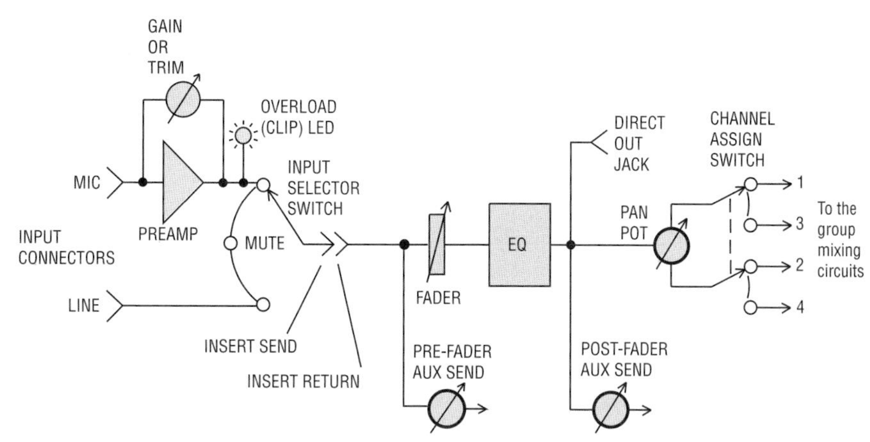

FIGURE 11.2
Signal flow through a typical input module.

1/4-inch phone jack (a 1/4-inch hole) or an XLR-type connector (with three small holes). The line input is either a 1/4-inch phone jack, an RCA phono jack (like you see on a stereo system), or an XLR-type connector.

You can plug a synthesizer directly into a phone-jack line input without using a direct box if the cable is under 10 feet; a longer cable may pick up hum. In that case, use a direct box plugged into a mic input.

In some mixers, a phone-jack input is switchable between low impedance (for microphones) or high impedance (for guitar pick-ups). Other mixers might have a separate low-impedance mic input and a high-impedance instrument input.

PHANTOM POWER (P48, +48)

This switch (not shown) turns on phantom power (48 V DC) for condenser microphones. In the mic input connector, the 48 V appears on pins 2 and 3 relative to pin 1. The microphone receives phantom power and sends audio along the same two cable leads.

MIC PREAMP

After entering the mic connector, the microphone signal goes to a mic preamplifier inside the mixer. This preamp boosts or amplifies the weak microphone signal up to a higher voltage, making it a line-level signal.

The TRIM or GAIN control adjusts the amount of amplification in the mic preamp. If the gain control is turned up full, and the incoming mic signal is very strong due to a loud instrument or vocal, this signal can overload the mic preamp. This causes distortion—a gritty sound. To get a good signal-to-noise ratio (S/N) in your mixer, set the gain control as high as possible, but not so high that the preamp distorts.

Here's how: Start with the trim turned up all the way down (counterclockwise). In most mixers, each input module has a tiny light (LED) labeled "clip," "peak," or "OL" (overload). It flashes when the mic preamp is distorting. When an instrument is playing its loudest signal through an input module, gradually turn up the trim control until the clip LED starts flashing. Then turn down the trim control just to the point where the light stays off, and turn it down another 6 dB for extra headroom.

Some low-cost mixers don't have a trim control. The input fader serves this function.

INPUT SELECTOR SWITCH

This switch lets you select the type of signal you want to work with. Some common switch labels are below:

MIC: (Mic or direct box)
LINE: (For line-level signals: a synthesizer, drum machine, electric guitar, or multitrack recorder's track output)
INPUT: (Mic or line)
TRACK: (Multitrack recorder's track output)

Using the input selector is simple. If you plugged in a microphone or direct box to record its signal, set the input selector to MIC or INPUT. If you plugged in a synth, drum machine, or electric guitar, set the selector to LINE or INPUT. When you're ready to mix, select TRACK (if available) or LINE (if the recorder track outputs are plugged into mixer line inputs).

Some mixers have no input selector. The mixer processes whatever signal is plugged in.

In some recorder–mixers, a single 1/4-inch phone jack is used both for mic-level and line-level signals. The two levels are handled either by a MIC/LINE switch or a TRIM control.

INSERT JACKS

Following the input selector switch are the insert-send jack and insert-return jack (on the back of each input module). Or the mixer might have a single insert jack, a TRS type that has the send on the tip terminal and the return on the ring terminal. Inside the mixer, the send is connected to the return so that the signal passes through to the rest of the mixer.

If you insert a plug into the insert jack(s), you break the signal path so you can insert an external device there in series with the input module's signal. You might insert a compressor into the signal path of one module for automatic volume control, or insert any other signal processor (reverb/delay, for instance). That way, if all your aux sends are tied up, you can add another signal processor. On the reverb/delay unit, set the dry/wet mix control for the desired amount of effect.

Another use for the insert jack is to send the mic-preamp output signal to a multitrack recorder track. The output of each track returns to the insert jack and continues through your mixer. In this case, you use the trim controls to set recording levels, and use the faders, EQ, and aux sends to set up a monitor mix.

Low-cost mixers omit the insert jacks. Some units have insert jacks on only two inputs.

INPUT FADER (CHANNEL FADER)

Next, the input signal that you selected goes to a fader. This is a sliding volume control for each input signal. During recording, you can use the fader in two ways. If you're recording one instrument per track, record off the insert send and use the fader to adjust the instrument's level in the monitor mix. If you're recording two or more instruments on one track, record off a group output, and use each instrument's fader to set up a mix within that group. For example, if you're recording several drum mics onto one track, set up a drum mix with the drum-mic faders.

During mixdown, you use the faders to set the loudness balance among instruments.

EQ

The signal from the input fader goes to an equalizer, which is a tone control. With EQ you can make an instrument sound more or less bassy, and more or less trebly, by boosting or cutting certain frequencies. Details are in Chapter 10 in the section Equalizer.

DIRECT OUT

The direct out is an output connector following each input fader and equalizer. The signal at the direct-out jack is an amplified, equalized version of the input signal. The fader controls the level at the direct-output jack. You can use the direct-out jack when you want to record one instrument per track (with EQ) on an external multitrack recorder. Connect the direct-out jack to a multitrack recorder track input. Because the direct output bypasses the mixing circuits farther down the chain, the result is a cleaner signal.

Suppose your mixer has 8 inputs and 2 outputs. You can use this mixer with an 8-track recorder. Just connect the direct-out jack in each input module to a separate track input. Or do the same with the insert-send jacks.

CHANNEL ASSIGN SWITCH

The equalized signal also goes through a pan pot (explained in the section below) to the channel assign switch or track selector switch. It lets you send the signal of each instrument to the recorder track you want to record that instrument on.

A mixer with four groups or buses would have an assign switch labeled 1, 2, 3, 4. If you want to record bass on track 1, for instance, find the assign switch for the input module the bass is plugged into, and push assign switch 1. If you want to record four drum mics on track 2, push assign switch 2 for all those input modules.

Some mixers assign tracks with two controls: a selector switch and a pan pot.

PAN POT

This knob sends a signal to two channels in adjustable amounts. By rotating the pan-pot knob, you control how much signal goes to each channel. Set the knob all the way left and the signal goes to one channel. Set it all the way right and the signal goes to the other channel. Set it in the middle and the signal goes to both channels equally.

Here's how you might use a pan pot to assign an instrument to a track. The channel-assign switch might have two positions labeled 1–2 and 3–4. If you turn the pan pot left, the signal goes to odd-numbered tracks (either 1 or 3, depending on how you set the assign switch). If you turn the pan pot right, the signal goes to even-numbered tracks (2 or 4). Suppose you want to assign the bass to track 1. Set the assign switch to 1–2, and turn the pan pot far left to choose the odd-numbered track (track 1).

During mixdown, the pan pot has a different function: It places images between your speakers. An image is an apparent source of sound, a point between your speakers where you hear each instrument or vocal. Set the pan pot to locate each instrument at the left speaker, right speaker, or anywhere in between. If you set the pan pot to center, the signal goes equally to both channels, and you hear an image in the center.

AUX

The aux or aux-send function (Figures 11.2 and 11.3) sends some of the input module's signal to equipment outside the mixer. A pre-fader aux send is before the fader, and usually goes to a power amp that feeds monitor speakers or headphones. You can use the pre-fader aux knobs to set up a monitor mix—a balanced blend of input signals you hear over speakers or headphones. A post-fader aux send is after the fader and EQ, and usually goes to an effects unit. You use the post-fader aux knobs to set the amount of effects (reverb, echo) heard on each instrument in a mix.

Some mixers have no aux sends, some have one aux-send control per module, and some have two or more (labeled aux 1, aux 2, etc.). The more aux sends you have, the more you can play with effects, but the greater the cost and complexity. The aux number (1 or 2) isn't

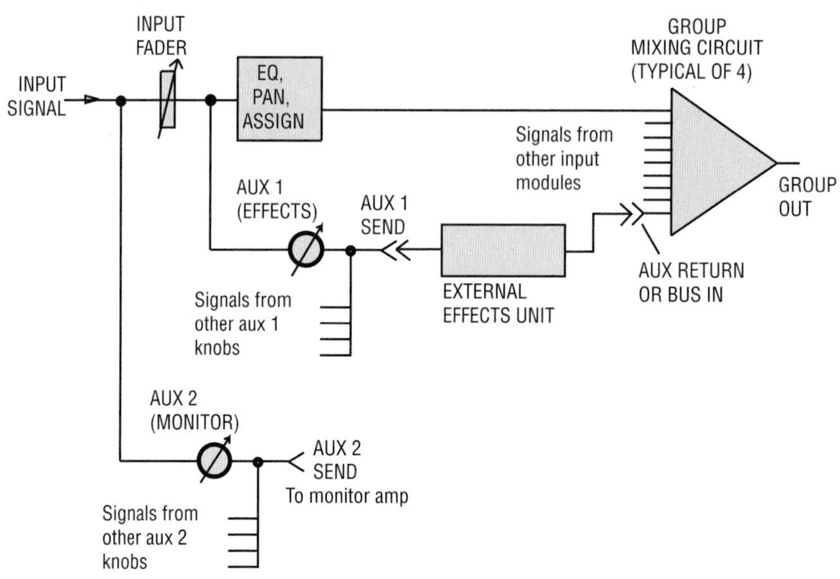

FIGURE 11.3
Aux sends.

necessarily assigned a specific function; you decide what you want aux 1 and aux 2 to do.

During recording and overdubbing, the aux knobs of all the input modules can be used to create a monitor mix. The monitor mix that you create with the aux knobs is independent of the levels going to the multitrack recorder. You use the gain-trims during recording to set recording levels, and you use the aux knobs to create an independent mix that is heard over your monitor system.

In Figure 11.3, the aux-2 send control is just before the fader. In this mixer, the signals from all the aux-2 knobs in the mixer combine at a connector jack labeled "aux-2 send." You can connect that jack to your power amplifier, which drives monitor speakers and headphones.

In Figure 11.3, the aux-1 send control is just after the fader. During mixdown, each aux-1 knob controls how much effects (reverb, echo) you hear on each track. In each input module, the aux knob adjusts how much of that input signal is sent to an external effects unit. The effected signal returns to the mixer's aux-return or bus-in jacks, where it blends with the original signal.

For example, suppose the aux-1 send is connected to a digital reverb. The more you turn up the aux-1 send knob, the more signal goes to the reverb. The output of the reverb unit returns to the mixer's bus-in jack, and blends with the original dry signal, adding a spacious effect.

A few mixers have an aux-return control (also called effects-return or bus-in) that sets the overall effects level returning to the mixer.

There might be a pre/post switch next to the aux-send knob. When an aux knob is set to pre (pre-fader), its level isn't affected by the fader setting. You use the pre setting for a headphone mix during recording or overdubbing because you don't want the fader settings to affect the monitor mix.

The post setting (post-fader) is used for effects during mixdown. In this case, the aux level follows the setting of the fader. The higher you set the track volume with the fader, the higher the effects level. But the dry/wet mix stays the same.

Output section

The output section is the final part of the signal path; the section that feeds mixed signals to the recorder tracks. It includes mixing circuits, submaster or group faders (sometimes), master faders, and meters (Figure 11.4).

MIXING CIRCUITS (ACTIVE COMBINING NETWORKS), GROUP FADERS, AND BUS OUTPUT CONNECTORS

The group mixing circuits are in the center of Figure 11.4. Recall that you use the assign switches to send each input signal to the desired channel or bus, and each bus feeds a different recorder track. A bus is a channel in a mixer containing an independent mix of signals. The bus 1 or group 1 mixing circuit accepts the signals from all the inputs you assigned to bus 1 and mixes them together to feed track 1 of the recorder. The bus 2 mixing circuit mixes all the bus 2 assignments, and so on.

Following each group mixing circuit is a group fader. If you assigned all the drum mics to group 1, the group 1 fader adjusts the overall level of the drum mix. The signal from each group fader goes to a group or bus output connector in your mixer. You can connect each bus output to a recorder track input.

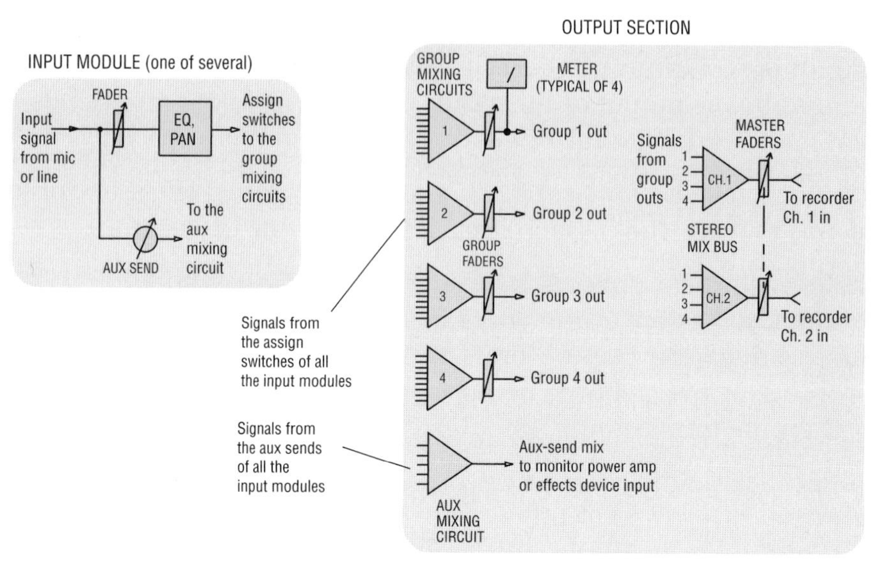

FIGURE 11.4
Input module and output section of a mixer.

STEREO MIX BUS, MASTER FADERS, MAIN OUTPUT CONNECTORS

The stereo mix bus includes two group mixing circuits: one for channel 1 and one for channel 2. Three types of signals feed into the stereo mix bus:

1. The group output signals.
2. Signals from input modules. You can assign an input module's signal directly to the stereo mix bus, bypassing the groups. This results in lower noise.
3. Effects-return signals, such as the reverberated signal from an external digital reverb.

Located on the right side of your mixer, the stereo master faders are one or two sliding volume controls that affect the overall level of the stereo mix bus. Usually, you set the master fader(s) within design center, the shaded area about three-quarters of the way up on the scale. This setting minimizes mixer noise and distortion. You can fade out the end of a mix by turning down the master faders gradually.

After the master faders, the signal goes to a pair of main output connectors, which feed a 2-track recorder of your choice.

You feed the multitrack recorder either from group outputs, direct outs, or insert sends. If you're mixing several instruments to track 5, for example, assign those instruments' signals to Group 5. Connect the Group 5 output to recorder track 5 in. If you're recording one instrument on track 5, however, connect that instrument's direct out or insert send to track 5 in. The signal is cleaner at the direct out or insert send than at the group output.

Some mixing consoles have voltage controlled amplifier (VCA) group faders. A VCA group fader acts like a remote control for groups of channel faders. For example, you could assign each drum mic's fader to a group, which is an audio path, and also assign each drum mic's fader to a VCA group, which is a fader that controls all the drum mic channels at once.

METERS

Meters are an important part of the output section. They measure the voltage level of various signals. Usually, each group or bus output has a meter to measure its signal level. If these buses feed the recorder tracks, you use the meters to set the recording level for each track.

A mixer has either VU meters or LED bar graph meters.

A VU meter (now rarely used) is a voltmeter that shows approximately the relative loudness of various audio signals. Set the record level so that the meter needle reaches +3 VU maximum for most signals, and about −6 VU maximum for drums, percussion, and piano. That's necessary because the VU meter responds too slowly to show the true level of percussive sounds.

An LED bar graph meter has a column of lights (LEDs) that shows peak recording level. Usually you set the recording level to peak near −6 dB maximum.

If your console has VU meters, and it is feeding a digital recorder with LED peak meters, set a 0 VU tone to read −20 dB on the recorder meters. This allows some headroom for peaks.

Monitor Section

The monitor section is used to control what you're listening to. It lets you select what you want to hear, and lets you create a mix over headphones or speakers to approximate the final product. The monitor mix has no effect on the levels going to your recorder.

During recording, you want to monitor a mix of the input signals. During playback or mixdown, you want to hear a mix of the recorded tracks. During overdubs, you want to hear a mix of the recorded tracks and the instrument that you're overdubbing. The monitor section lets you do this.

MONITOR SELECT BUTTONS

These buttons let you choose what signal you want to monitor or listen to. Because the configuration of these buttons varies widely among different mixers, they aren't shown in Figure 11.4.

If you want to use aux 2 as the monitor-mix bus, select the aux-2 bus as the monitor source. That lets you listen just to the aux2 mix. Some mixers have no monitor-select switches. Instead, you always monitor the stereo mix bus.

MONITOR MIX CONTROLS AND CONNECTORS

One way to set up a monitor mix is with the aux knobs. Suppose you want aux 2 to be the monitor mix. Connect the aux-2 send jack to your power amp and speakers. Or if you're using headphones, find the monitor-select switch for the headphones, and set it to aux 2. Turn up all the aux-2 knobs about halfway, then turn each knob up or down to set a good loudness balance. You do this during recording or overdubbing.

Here's another way to set up a monitor mix using the faders. Connect your multitrack recorder ins and outs to the insert jacks (Figure 11.5). Connect the insert-jack 1 tip (send) to track 1 in; connect track 1 out to the insert-jack ring (return). Make similar connections for the other tracks. Also connect the mixer's monitor-out jacks to your power amp and speakers. Monitor the stereo mix bus. With this setup, use the trim controls to set recording levels. Use the faders to set up a monitor mix, cue mix, or mixdown with EQ and effects. It's a convenient way to work.

During mixdown, monitor the stereo mix bus. This is automatic in a DAW.

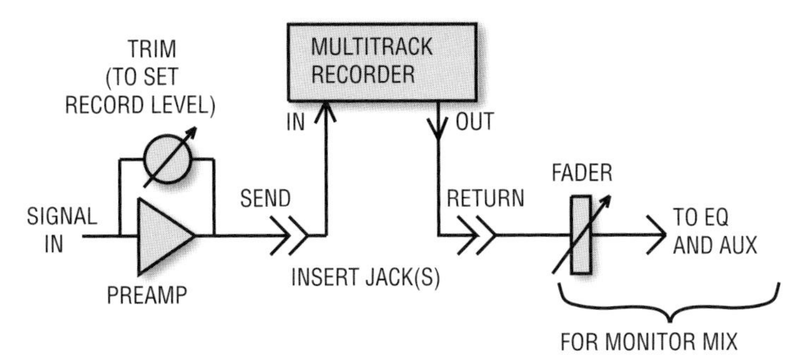

FIGURE 11.5
Using insert jacks to send each input signal to a recorder track. The track signal returns to the mixer, where you adjust level, panning, EQ, and effects.

SOLO

The SOLO button in an input module is actually a monitor function. It lets you monitor one instrument or vocal at a time so you can hear it better or so you can edit one track. By pressing two or more SOLO buttons, you can monitor more than one input signal at a time.

Suppose you hear a buzz in the audio and suspect it may be in the bass guitar signal. If you push the SOLO button in the bass guitar's input module, you'll monitor only the bass guitar. Then you can easily hear whether the buzz is in that input.

On British consoles, the SOLO function is called pre-fader listen (PFL) or after-fader listen (AFL).

MUTE

When you press the MUTE button in a channel, the signal in that channel is turned off. During mixdown, it's a good idea to mute tracks that have nothing playing at the moment to reduce noise and leakage. During overdubs, you might want to mute previous takes or certain instruments to make a cleaner headphone mix for the musician.

Additional Features in Large Mixing Consoles

Large mixing consoles have more features than small mixers. If you're working only with a small mixer or a recorder–mixer, you might want to skip this section.

FOLDBACK (FB): Another name for cue or headphone mix.

PHASE (POLARITY INVERT): Used only with balanced lines, this switch inverts the polarity of the input signal; that is, it switches XLR pins 2 and 3 to flip the phase 180 degrees at all frequencies. You might use it to correct a miswired mic cable whose polarity is reversed. If you mic a snare drum top and bottom, you need to invert the polarity of the bottom mic.

AUTOMATED MIXING CONTROLS: These controls (Read, Write, Update, Record Automation, Play Automation) set up the console for various automation functions. With automation, a memory circuit in the mixer remembers your console settings and mix moves. After leaving a mix for several days, you can recall the settings with the push of a button, then continue working on the mix. More on automated mixing is at the end of Chapter 12.

EFFECTS PANNING: This feature places the images of the effects signals wherever desired between the monitor speakers. Some consoles let you pan effects in the monitor mix as well as in the final program mix.

EFFECTS RETURN TO CUE: This is an effects-return level control that adjusts the amount of effects heard in the studio headphone mix. These monitored effects are independent of any effects being recorded.

EFFECTS RETURN TO MONITOR: This effects-return control adjusts the amount of effects heard in the monitor mix. These monitored effects are independent of any effects being recorded.

BUS/MONITOR/CUE switch for effects return: A switch that feeds the effects-return signal to your choice of three destinations: program bus (for mixdown), monitor mix (control-room speakers), cue mix (headphones), or any combination of the three.

METER SWITCHES: In many consoles, the meters can measure signal levels other than console output levels. Switches near the meters can be set so that the meters indicate bus level, aux-send level, aux-return level, monitor-mix level, etc.

Those readings help you set optimum levels for the outboard devices receiving those signals. Too low a level results in noise; too high a level causes distortion in the outboard unit. For example, if the aux-return signal sounds garbled or distorted, the cause may be an excessive aux-send level. Verify that condition by checking the meters switched to read the aux or effects bus.

DIM: A switch that reduces the monitor level by a preset amount so you can talk (as in "dim the lights").

TALKBACK: An intercom between the control room and studio. A mic built into the console lets you talk to the musicians in the studio when you push the talkback button.

SLATE: This function routes the control-room microphone signal to all the buses so you can record the name of the tune and take number.

OSCILLATOR or TONE GENERATOR: This is used to put a level-calibration tone on a tape, or to reference the recorder's meters to those on the console. You also can use it to check signal path, levels, and channel balance.

Some mixers include a FireWire or USB port. It sends the output signal of each mic preamp to a separate track in your recording software, and returns the stereo mix back to the mixer for monitoring. See Chapter 4 under the heading Computer Digital Audio Workstation.

DIGITAL MIXER

So far we've looked at the analog mixing console, which works entirely with analog signals. A digital mixing console accepts analog or digital signals. It converts the analog input signals to digital, and processes all signals internally in digital format. The signal stays in the digital domain for all mixer processing. Level changes, EQ, and so on are done by digital signal processing (DSP or computer calculations) rather than by analog circuits.

Some of the digital output signals from the mixer go through digital-to-analog (D/A) converters. The resulting analog signals feed a power amp, effects unit, and so on.

Analog and digital consoles operate differently. To use EQ in an analog console, you find the channel you want to EQ, and adjust its EQ knobs. To use EQ in a digital console, you press a button to select the channel you want to equalize, and press an EQ button. Then the EQ settings for that channel show up on an LCD screen. You press buttons and turn a knob to adjust the EQ frequency and boost/cut for that channel.

Because one knob controls the EQ for all the channels, digital consoles have fewer controls than analog consoles. One knob or switch can have several functions. This makes digital consoles a little harder to operate because you can't just reach for an EQ knob for a particular channel. You have to press a few buttons to set the EQ parameters.

Some models have an LCD touch screen that shows a bank of faders with small EQ curves, compression curves, and virtual aux knobs. To tweak EQ for one channel, find the EQ curve and tap it. The screen zooms to a larger view of the EQ settings for that channel. Rotary encoders (knobs) under the touch screen, or built into it, let you adjust parameters.

Digital consoles vary a lot in how they implement functions, so you need to read the manual.

Some digital mixers have a bank switch or layer switch. It lets you select the channels that the faders control. For example, suppose a mixer has 16 faders. Press the layer switch once to make the faders access channels 1–16. Press the switch again to make the faders access channels 17–32, and so on.

Several digital consoles have built-in effects, multitrack recording to hard disk, and a USB port for a Flash drive that saves presets and mixes.

Many digital mixers offer automated mixing. One type of automated mixing is called scene or snapshot automation. When you press the snapshot button, a memory circuit in the mixer takes a "snapshot" of all the mixer settings for later recall. Another type of automation is dynamic. Your mix moves are stored in memory for later recall.

When you recall a mix, some mixers make the faders move into the positions you set up. This feature is called "flying faders" or "motorized faders." A few mixers don't move the faders when you recall a mix. You have to set them manually by looking at a display, which is a disadvantage. Motorized faders are more expensive but easier to use.

Digital Mixer Features

Look for the following features in digital mixers when making a buying decision:

- Number of mic inputs and number of output buses
- Number and type of digital inputs and outputs: S/PDIF, AES3, MADI, TDIF, Lightpipe, USB, and FireWire. Chapter 9 describes those standards under the heading Digital Audio Signal Formats.
- Word-clock input and output
- Routing section that lets you route any input signal to any channel, or route any output signal to any output bus
- Number and type of option card slots: extra I/O, DSP, sync, effects, plug-ins
- Number of effects processors
- PC USB interface for control and/or backup
- Ease of use
- Snapshot or dynamic automation, or no automation
- Motorized or non-motorized faders
- Varispeed clock for pitch shift
- Surround-sound monitoring
- Touch screen with virtual controls

SOFTWARE MIXER

Here's another type of mixer. A software mixer (virtual mixer) is a simulated mixer you see on a computer screen (Figure 11.6). It exists only in your computer as part of digital recording software or a DAW. You control it with a mouse or with a controller surface. Recordings made with this mixer, and all the mixer settings, are stored on your hard drive.

In a software mixer, the input and output connectors are in an audio interface (such as a sound card) that connects to the computer. This interface is 2-channel or multichannel. The software mixer works with effects that are either plug-ins (software) or external (hardware). Automation is a standard feature. For more on software mixers, see Chapter 13.

FIGURE 11.6
Cakewalk Sonar Producer mix window, an example of a software mixer.

CONTROL SURFACE

A control surface or controller resembles a hardware mixer with faders and knobs (Figure 13.10). It plugs into your computer's USB or FireWire port and controls the software mixer. Some people prefer a controller because it is easier to use than a mouse. More in Chapter 13 under the heading Control Surface.

Now that you understand the typical features of mixers and mixing consoles, you're ready to learn how to use them.

CHAPTER 12

Mixer Operation

Get your hands on those knobs. You're going to operate a mixer as part of a recording session. This will be a basic run-through. Detailed session procedures are described in Chapter 15. Most of these procedures also apply to operating a virtual mixer in a Digital Audio Workstation (DAW).

First recall the stages in making a recording:

1. Session preparation
2. Recording
3. Punching in
4. Overdubbing
5. Mixdown

This chapter considers each stage in turn.

SESSION PREPARATION

If you're recording on a hard drive, make sure you have enough drive space for the project. Chapter 13 has simple equations and a table that show how much space you need.

Plan your track assignments. Write a track sheet that tells what instrument goes on which track. **Note**: If you assign multiple instruments to the same track, you can't separate their images in the stereo stage. That is, you can't pan them to different positions; all the instruments on one track sound as if they're occupying the same point in space. If you're recording 4-track, you may want to do a stereo mix of the rhythm section on tracks 1 and 2; then overdub vocals and solos on tracks 3 and 4.

Studio setup for the musicians is covered in Chapter 15.

Set Up the Mixer and Recorder

To start the process, first zero or neutralize the mixer by setting all the controls to "off," "flat," or "zero." This establishes a point of reference and avoids surprises later on. Set all faders down.

If you have a separate mixer and multitrack recorder, you need to make their meter readings match. Follow one of these procedures:

If your mixer and recorder have VU meters, play a steady tone into the mixer to get a 0 reading on the meters for all channels. Then set the multitrack recorder's record level (if any) to get 0 VU readings on all the tracks.

If your mixer has VU meters and your recorder has LED peak-reading meters, set 0VU on your mixer to equal −20dB on the recorder's peak meters.

If your mixer and recorder have LED peak-reading meters, set 0dB on the mixer to equal 0dB on the recorder's peak meters.

If you're feeding the multitrack recorder from your mixer's insert jacks or direct-out jacks, you'll need to watch the recorder meters.

Suppose you're ready to record a vocal or an acoustic instrument. Place the microphone and plug it into a mic input. If you want to record the audio output of a hardware synth or drum machine, connect a cable between the instrument's output and a line input on the mixer.

Attach a "scribble strip" of masking tape or white removable tape along the bottom of the faders, and label each fader according to what instrument you plugged into that input. Some consoles and software mixers have scribble strips that you can type on. In Figure 12.1, such a strip is on the mixer at the right.

Set the input selector (if any) to Mic or Line depending on what is plugged into each input.

Plug in headphones to hear what you're recording. Turn up the headphone volume control. Or, if you're in a control room and the musicians are in a studio, turn up the monitor level to listen over the monitor loudspeakers. On a hardware mixer, set the MONITOR

FIGURE 12.1
A studio control room with a DAW, mixer and monitor speakers.

SELECT switch to hear the signal you're recording, and turn up the musicians' cue mix.

Set the master faders about three-quarters of the way up, at 0, or within the shaded portion of fader travel. This is called design center.

If you're recording a soft synth—a software synthesizer in a DAW—there will be a short delay in the monitored signal because of the latency (signal processing delay) in the software. Set the latency or buffer setting as low as possible without getting drop-outs. A delay under 5 msec or so should be acceptable.

If you're recording a vocal or real instrument with a DAW, don't enable Echo input monitor because this causes a latency delay in the monitored signal. Instead, monitor the mic or input signal directly from your mixer or audio interface.

Assign Inputs to Tracks

If your equipment wiring is mics > mixer > insert sends > track inputs, your inputs are already assigned. Input 1 goes to track 1; input 2 goes to track 2, and so on.

If you're using a computer DAW, select the input signal source for each track. For example, you might set track 5's input to channel 5 in your audio interface. Then, whatever is plugged into interface channel 5 will go to track 5. If your sound card has just 2 channels, set track 5's input to channel 1 or 2 in your audio interface, and likewise for the other tracks. Some DAWs record stereo tracks, so you'd assign each track's input to stereo channels 1 and 2 in the audio interface.

If your equipment wiring is mics > mixer > buses > track inputs, assign each input signal to the desired output channel (bus) as specified on your track sheet. Each bus is connected to the corresponding numbered track on the multitrack recorder. **Note**: If only one instrument is assigned to a track, you can eliminate the noise of the console's combining amplifier by patching the instrument's signal directly to the recorder track. To do that, locate the direct output jack (or the insert jack) of the input module for that instrument and patch it to the desired track. Some mixing boards also require pressing a Direct button on the input module.

Set Recording Levels

Now you're ready to "get a level." Have each instrument play the loudest part of the music, one at a time or all at once. For each input signal, set the TRIM control so the recording level is as high as possible without causing distortion. While setting levels for a digital multitracker, peak each track around −6 dBFS maximum in peak meter mode, not rms. This allows some headroom for surprises. Also, musicians generally play louder during a performance than during a level check. If you exceed 0 dBFS you'll hear digital clipping which makes a loud click.

Some audio interfaces use on-screen volume controls to set recording levels. Other interfaces have level knobs for this purpose.

If you are mixing several instruments to one or two groups (as in a drum submix), follow this procedure:

1. Monitor the group(s).
2. Set its submaster fader (group fader) to design center.

3. Set the submix balances, panning, and recording level with the input faders.
4. Fine-tune each submix level with the submaster fader.

Set EQ

Although it's common to record flat (without EQ), you may want to apply equalization at this point to each instrument heard individually. Filter out frequencies above and below the range of the instrument. However, don't spend too much time on EQ until all the instruments are mixed together. The EQ that sounds right on a soloed instrument may not sound right when all the instruments are heard together. In creating the desired tonal balance, use EQ as a last resort after trying different mics and mic placements. You also can apply EQ during playback or mixdown. This may be preferable because EQ applied when recording can't always be undone if you're unhappy with it.

RECORDING

Next, set the track(s) you want to record to "record ready" mode. Now start recording. Write down the counter time for this take. "Slate" the recording: record the name of the tune and the take number. Then record a two-measure count-off. This is done to set the tempo for overdubs. For example, if the time signature is 4/4, you say, "1, 2, 3, 4, 1, 2, (rest) (rest)." The rests are silent beats. You need some silence before the song starts to make editing easier later on. A DAW has a built-in metronome (click track) that can be used for count-offs.

PLAYBACK

After recording the track, go to the beginning of the song using the return-to-zero or locate function. If necessary, set the monitor selector to "track," or "mix," and play back the recording to check the performance and sound quality. You can set a rough mix with the monitor mix (aux) knobs. If you're recording on a computer, or if your multitrack is patched to the insert jacks, use the faders, pan pots, EQ, and aux knobs to set a rough mix.

Get or write a sheet of the song's arrangement. It shows the lyrics, verses, choruses, bridge, and so on. Play the song and set a locate point (marker) at the start of each verse and chorus. That way, when the musician says "Let's fix that flat note at the end of the second chorus," you can go instantly to that part of the song.

OVERDUBBING

After your first track is recorded with a good performance, you might want to add more musical parts. This procedure is called overdubbing. When you overdub, the musician listens to tracks already recorded, and records a new part on an unused track.

Ready to overdub? Here is one way:

1. Turn off the control room speakers and listen on headphones.
2. Set up a headphone monitor mix of the recorded tracks. Some mixers use the aux knobs for this. If you're recording on a computer, or if your multitrack recorder is connected to the mixer's insert jacks, monitor the stereo mix, and set up a mix with the channel faders.
3. Plug in the mic or direct box for the instrument or vocal you want to record. Assign it to an unused track. Turn up its fader to design center (the 0 point about three-quarters of the way up).
4. Set up your multitrack recorder, mixer, or audio interface to monitor the playback of recorded tracks and the input signal of the instrument or vocal you want to record.
5. Have the musician play or sing. Can you hear the signal in the phones? Set the recording level using the TRIM (gain) control for that mic's channel.
6. Play the recording. As the musician plays or sings along, set up a good mix of the recorded tracks and the live instrument you're going to record. Ask the musician if he is hearing what he needs to hear. Change the mix if needed. Some musicians want to hear effects when they overdub; some want it dry. If you're overdubbing backup vocals one at a time, often it helps to remove (mute) certain other vocals from the headphone mix.
7. When you're ready to record the new part, go to a point about 10 seconds before the part of the song where the musician plays.

Set the recorded tracks to SAFE and set the track you're recording on to RECORD READY.

8. Before you hit that RECORD button, stop! Are you recording on the correct track(s)? Are you going to accidentally erase any tracks? Double-check your track sheet, and make sure the record-enable buttons are on only for the tracks that are safe to record over.
9. Start recording and have the musician play along with the tracks. If the musician makes a mistake, you can re-record or punch-in the new part without affecting the parts that were already recorded.

PUNCHING-IN

Punching-in is used to fix mistakes in a recorded performance, or to record a musical part in segments. A punch is also called an insert. You enable record mode on a track, play the multitrack recording, then "punch" or press the record button at the right spot, record a new part, then punch-out of record mode.

Some musicians record the same musical part over and over until the part is perfect. This process is tedious, but you have to pay attention. You need to be aware of where you are in the song at all times, and not erase anything you want to keep.

Some musicians like to record a performance a phrase at a time, perfecting each phrase as they go. Others record a complete take, then go back and fix the weak parts.

To do a punch-in, grab your song-arrangement sheet and follow these steps:

1. Go to a point about 10 seconds before the point where you want to start recording. Note the counter time and write it down. Also enter a memory location point there if your recorder can do that.
2. Play the song to the musician over headphones. The musician plays along to practice the part. Write down the counter times where you want to punch-in and punch-out. Otherwise you might erase a good take.
3. Finally you're ready to record. Before you hit that RECORD button, stop! Tell the musician what you're going to do so

there's no chance of a mistake. Be very clear. For example, "I'm recording your keyboard part on two new tracks." Or, "I'm punching-in over your old performance—is that what you wanted?"

4. Okay, ready to go. Play the recording. During a rest or pause in the music just before the part needing correction, punch-in the record button (or use a footswitch). Have the musician record the new part, and punch-out right away. You don't want to erase the rest of the track! If you're using a DAW, though, you can undo the recording.
5. Press LOCATE or GO TO MARKER to go to the location point you set, about 10 seconds before the punch. Play the recording to see if the punch was okay. If necessary, you can re-record the punch. After you go to the locate point, notice whether the musician wants to practice the part. Don't redo the punch until he is ready.

Some multitrack recorders have an autopunch function. You set the punch-in and punch-out points into the machine's memory. As the recording plays, the recorder automatically goes into and out of record mode at those points. Some recorders can loop repeatedly between those two points.

COMPOSITE TRACKS

If several open tracks are available, you can record a solo performance in several takes, each on a separate track or virtual track. Then combine the best parts of each track into a single track. Use only that track in the final mix, and you'll hear the best parts of all the takes in succession. This is called "recording composite tracks" or "comping." Here's an example of comping vocal tracks with a multitrack recorder and mixer:

1. Suppose you have four takes of a vocal recorded on tracks 10, 11, 12, and 13. Solo each track, and mark on the lyric sheet which tracks have the best performance of each section. For example, you might prefer track 13 on Verse 1, track 10 on Chorus 2, and so on.
2. Assign all the vocal tracks to an open track, which we'll call the comp track. Match their levels.
3. Start recording on the comp track.

4. As the song plays, mute and unmute the tracks to copy the best performances to the comp track. To do that, you can stop the recorder between sections, change the mute settings, and record a section at a time.
5. Once you're happy with the comp track, you can erase or archive the original tracks.

Comping with a DAW is even easier. Pick the best overall track, then copy and paste good sections from other tracks into the best track. Or divide the tracks into short sections, delete sections you don't like, and play the mix of all the tracks.

Cakewalk Sonar Producer software lets you record several takes on one track, view all the take waveforms as "lanes" in that track, select the best parts of each take, and create a composite track of those parts.

Some recorders have virtual tracks (explained in Chapter 9 in the section Recorder–Mixer (Digital Personal Studio). They let you comp a performance with virtual rather than real tracks.

GETTING MORE TRACKS

Most DAWS can record nearly an unlimited number of tracks. But what if you want to overdub more parts on a recorder–mixer, and all the tracks are full? With care, you can punch-in more instruments by recording them in the pauses on recorded tracks. For example, suppose all the tracks are full but you want to add a cymbal crash at the beginning of the chorus. Find a track that has a pause at that moment, and punch in the cymbal crash there.

As an alternative, you can bounce tracks—mix several tracks to one or two open tracks, and record the mix on that track. Then you can erase the original tracks, freeing them for more overdubs. Bouncing procedures are in the instruction manual for your recorder or software.

DRUM REPLACEMENT

Suppose you've recorded the drum tracks, but you don't like the sound of them. Maybe the kick drum is too flabby, and no amount of EQ or gating seems to help. You might try a technique called drum replacement. In a MIDI/audio recording program, you can replace an audio drum track with a MIDI track that plays drum samples. First, set up a MIDI track with a drum sample ready to play (for details, see Chapter 16). Then use one of these methods:

Method 1: In the audio drum track, insert a drum-replacement plug-in such as Drumagog or Replacer. Set it to trigger the MIDI track.

Method 2: Start recording on the MIDI drum track, and tap your MIDI controller key in sync with the playback of the audio drum track.

Method 3: Select the audio drum track, then enable the Extract Rhythm or Extract Beat feature, if any. Copy and paste the extracted MIDI beats to the MIDI drum-sample track.

Mute the original audio drum track and press PLAY. You should hear the replacement drum sample playing.

MIXDOWN

After all your tracks are recorded (maybe with some bouncing), it's time to mix or combine them to 2-track stereo. You will use the mixer faders to control the relative volumes of the instruments, use panning to set their stereo position, use EQ to adjust their tone quality, and use the aux knobs to control effects. Figure 12.2 shows the controls you will be using.

Set Up the Mixer and Recorders

Set all the mixer controls to "off," "zero," or "flat." You should start from ground zero in building a mix. Some engineers like to set all the channel faders to −12 dB and go from there.

Tape a strip of paper or masking tape along the front of the mixer to write which instrument(s) each fader affects. Keep this strip for use each time you play the recording. Or type in this information in your DAW or mixing console.

If necessary, set the input-selector switches on the mixer to "track" because you'll be mixing down the multitrack recording. Monitor the 2-track stereo mix bus. These steps are unnecessary if you're using a DAW.

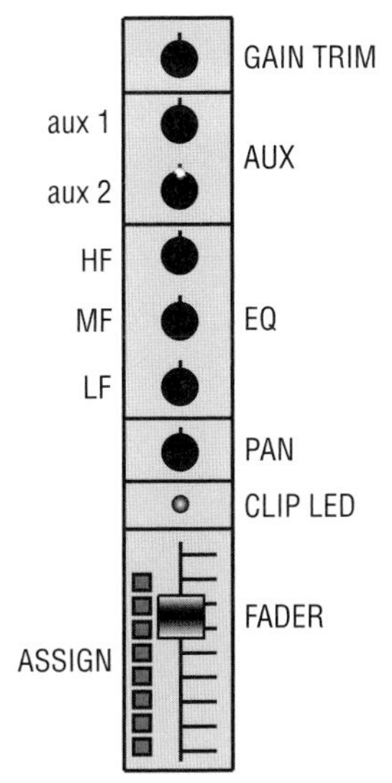

FIGURE 12.2
One input channel in a mixer, showing the fader, pan pot, EQ controls, and aux knobs.

For starters, put the master faders at design center (about three-quarters of the way up, at the shaded portion of fader travel). This sets the mixer gain structure for the best compromise between noise and headroom.

Erase or Delete Unwanted Material

Mixing will be a lot easier if you first erase noises and leakage before and after each song, and within each track.

Play the multitrack recording and listen to each track alone. Erase or delete unwanted sounds, silent (unplayed) sections, outtakes, and entire segments that don't add to the song. In a DAW you can do this visually by looking for parts of each track where there is no waveform. Zoom into a single track, highlight a silent portion, and select Edit > Cut (or a similar command). To avoid mistakes, it's best to do this while the musicians are around. If you aren't sure whether it's okay to delete part of a track, set that part's volume envelope to zero or mute the track during that part.

You can use editing to improve a song's arrangement. Typically, you'd put instrumental solos and fills only in the "holes" where the vocal is silent. Add instruments to the mix gradually so that it builds in excitement. You might add vocal harmonies only in the choruses. Reserve "all-out improvs" (where everybody plays) for the end.

Panning

You need to pan the tracks before doing the mix, because the loudness of a track depends on where it's panned. Assign each track to buses 1 and 2 (or the stereo mix bus), and use the pan pots to place each track where desired between your stereo speakers. Typically the bass, snare, kick drum, and vocals go to center; guitars can be panned left or right, and stereo keyboards and drum overheads go partly left and right (see Figure 14.2). Sometimes a mono drum mix can be punchier than a stereo one.

You can give an acoustic guitar a spacious effect by recording it twice. Pan the first take hard left and pan the second hard right. The second take could be capoed higher to add interest.

Pan tracks to many points between the monitors: left, half-left, center, half-right, right. Try to achieve a stereo stage that is well balanced either side of center. For clarity, pan to opposite sides any instruments that cover the same frequency range.

You may want some tracks to be unlocalized. Harmony singers and strings should be spread out rather than appearing as point sources. Stereo keyboard sounds can wander between speakers. You could fatten a lead-guitar solo by panning it left, and panning the solo delayed to the right. (This delay might come from a distant room mic you used while recording, or from copying and sliding a track a few milliseconds in a DAW.) Pan doubled vocals left and right for a spacious effect.

Consider creating some front-to-back depth. Leave some instruments dry so they sound close; add reverb to others so they sound farther away.

If you want the stereo imaging to be realistic (say, for a jazz combo), then pan the instruments to simulate a band as viewed from the audience. If you're sitting in an audience listening to a jazz quartet, you might hear drums on the left, piano on the right, bass in the middle, and sax slightly right. The drums and piano aren't point sources, but are somewhat spread out. If spatial realism is the goal, you should hear the same ensemble layout between your speakers. In some rock recordings, the piano and drums are spread all the way between speakers—interesting but unrealistic.

Pan-potted mono tracks often sound artificial; each instrument sounds isolated in its own little space. It helps to add some stereo reverb. It surrounds the instruments and "glues" them together.

When you monitor the mix in mono, you'll likely hear center channel buildup. Instruments in the center of the stereo stage will sound louder in mono than they did in stereo, so the mix balance will change in mono. To prevent this, note which tracks are panned hard left or right, and bring them a little toward the center: 9 and 3 o'clock on the pan knobs.

Highpass Filter Each Track

To reduce muddy buildup, highpass all tracks by ear. Set up a high-pass (low-cut) filter in each track. Adjust the filter frequency: start at a low frequency, then gradually raise it until the sound starts to thin out. Then back off a little (lower the filter frequency slightly).

Compression

Sometimes the lead vocal track might be too loud or too quiet relative to the instruments because vocals have a wider dynamic range than instruments. You can control this by running the vocal track through a compressor. It will keep the loudness of the vocal more constant, making it easier to hear throughout the mix. If you're using a hardware mixer, patch the compressor's inputs and outputs into the insert jack(s) of the vocal input module. Or connect the compressor between the recorder's vocal-track output and the mixer input. Set the desired amount of compression ratio, attack time, release time, and gain reduction. It's also common to compress the kick drum and bass. (For more information, see Chapter 10 under the heading Compressor.)

Set a Balance

Now comes the fun part. The mixdown is one of the most creative parts of recording. Here are some tips to help your mixes sound terrific.

Before doing a mix, tune up your ears. Play over your monitors some CDs whose sound you admire. This helps you get used to a commercial balance of the highs, mids, and lows.

Choose a CD with tunes like those you're recording. Check out the production. How is the balance set? How about EQ, effects, and sonic surprises? Try to figure out what techniques were used to create those sounds, and duplicate them. Of course, you might prefer to break new ground.

Using the input faders, adjust the volume of each track for a pleasing balance among instruments and vocals. You should be able to hear each instrument clearly. Some mixing consoles have trim knobs that set the playback gain of the multitrack recorder tracks. In that case, set all faders in use to design center, and adjust the trims to get a rough mix.

Recording engineer Roger Nichols offers this suggestion: Change the level of one track by +1 dB, then by −1 dB. Is it better? Is it worse? If you can hear less than 1 dB change, the fader is close to where it should be.

Here's one way to build the mix. Make all the instruments and vocals equally loud. Then turn up the most important tracks and turn down background instruments. Or, bring up one track at a time and blend it with the other tracks. For example, first bring up the kick drum to about −10dB, then add bass and balance the two together. Next add drums and set a balance. Then add guitars, keyboards, and finally vocals.

In a ballad, the lead vocal is usually on top. You might set the soloed lead-vocal level to peak at −6 dB. Bring up the monitor level so that the vocal is as loud as you like to hear it, then leave the monitor level alone. Bring in the other tracks one at a time and mix them relative to the vocal track.

When the mix is right, everything can be heard clearly, yet nothing sticks out too much. The most important instruments or voices are loudest; less important parts are in the background. In a typical rock mix, the snare is loudest, and the kick is nearly as loud. The lead vocal is next in level. Note that there's a wide latitude for musical interpretation and personal taste in making a mix.

Sometimes you don't want everything to be clearly heard. Once in a while, you might mix in certain tracks very subtly for a subconscious effect.

Recording engineer Michael Cooper offers this advice: To check the vocal level, turn the monitors way down. The lead vocal should disappear just after the instruments do. To avoid washy sound, mike acoustic guitars in mono, keep most tracks dry, use delay instead of reverb, or use pre-delay with short reverbs.

Engineer Michael Stavrou (who wrote *Mixing With Your Mind*) suggests: Turn up the most inspiring player. Mix other tracks with it, one at a time, to see which tracks fight it and which don't. Mute or delete the tracks that fight it.

If you recorded some MIDI tracks, consider replacing a MIDI instrument with another one that sounds better. You might replace a grand piano with a honky-tonk piano, replace a fretted bass with a fretless bass, and so on. Swap out the snare sound with a new one during choruses.

It's a good idea to monitor around 85 dBSPL. If you monitor louder, the bass and treble will be weak when the mix is played softly.

To test your mix, occasionally play the monitors very quietly and see if you can hear everything. Switch from large monitors to small, and make sure nothing is missing.

Set EQ

Next, set EQ for the tonal balance you want on each track. If a track sounds too dull, turn up the highs or add an enhancer. If a track sounds too bassy, turn down the lows, and so on. Cymbals should sound crisp and distinct, but not sizzly or harsh; kick drum and bass should sound deep, but not overwhelming or muddy. Be sure the bass is recorded with enough edge or harmonics to be audible on small speakers.

For suggestions on EQ settings, see Chapter 10 under the heading How to Use EQ.

You'll need to readjust the mix balances after adding EQ. The EQ that sounds right on a soloed track seldom sounds right when all the tracks are mixed together. So make EQ decisions when you have the complete mix happening. Switch EQ in and out to make sure it's an improvement.

In pop-music recordings, the tone quality or timbre of instruments doesn't have to be natural. Still, many listeners want to hear a realistic timbre from acoustic instruments, such as the guitar, flute, sax, or piano.

The overall tonal balance of the mix shouldn't be bassy or trebly; that is, the perceived spectrum shouldn't emphasize lows or highs. You should hear the low bass, mid-bass, midrange, upper midrange, and highs roughly in equal proportions. Frequency bands that are too loud can tire your ears.

When your mix is almost done, switch between your mix and a commercial CD to see whether you're competitive. If the tonal balance of your mix matches a commercial CD, you know your mix will translate to the real world. This works regardless of what monitors you use. An effective tool for this purpose is Harmonic Balancer (www.har-bal.com).

Add Effects

With the balances and EQ roughed in, it's time to add effects. If you're using a DAW, use the plug-ins. If you're using a hardware mixer, you might want to plug in a reverb or delay to add spaciousness to the sound (see Chapter 10). This device connects between your mixer's aux-send and aux-return jacks (or aux-send and bus-in jacks).

1. On the effects unit, set the dry/wet mix control all the way to wet or effect.
2. Find the AUX RETURN or BUS IN controls (if any), set them half of the way up, and pan them hard left and right.
3. Turn up the aux-send knob for each input, according to how much effect you want to hear on that input signal. Suppose you're using reverb as an effect. You might turn it up by different amounts for the vocals, drums, and lead guitar, and leave it turned down for the bass and kick drum.
4. As you're setting the aux levels, check the overload indicator on the effects unit. If it's flashing, turn down the input level on the effects unit just to the point where the overload light

stops flashing. Then turn up the output level on the effects unit (or turn up the aux return on the mixer) to achieve the same amount of effect you heard before.

Too much effects and reverb can muddy the mix. You might turn up the reverb only on a few instruments or vocals. Once you have the reverb set, try turning it down gradually and see how little you can get by with. Typically the bass gets no reverb so it retains its clarity.

Usually, short reverb decay times (under about 1 second) are best for up-tempo songs. Longer reverb times are better for slow ballads.

You might use 45 to 120 msec of pre-delay on vocals or certain instruments to separate the dry sound from the reverb. This makes the voice sound up-front even with reverb added.

The producer of a recording is the musical director and decides how the mix should sound. The producer might be the band members or yourself. Ask to hear recordings having the kind of sounds the producer desires. Try to figure out what techniques were used to create those sounds.

Also try to translate the producer's sound-quality descriptions into control settings. If the producer asks for a "warmer" sound on a particular instrument, turn up the low frequencies. If the lead guitar needs to be "fatter," try a stereo chorus on the guitar track. If the producer wants the vocal to be more "spacious," try adding reverb, and so on. *Audio clip 39 at www.esevierdirect.com/companions/9780240811444 demonstrates a mixdown.*

Set Levels

Set the overall recording level as you're mixing. To maintain the correct gain staging, keep the master faders at design center. Then adjust all the input faders by the same amount so your stereo output level peaks around −6 dB. You can touch up the master faders a few decibels if necessary. Don't exceed a 0 dB recording level if you're recording to a digital medium.

Judging the Mix

When you mix, your attention scans the inputs. Listen briefly to each instrument in turn and to the mix as a whole. If you hear something you don't like, fix it. Is the vocal too tubby? Roll off the bass on the vocal track. Is the kick drum too quiet? Turn it up. Is the lead-guitar solo too dead? Turn up its effects send.

Check the mix while listening from another room, where the lows and highs are weakened. Is the balance still good?

The mix must be appropriate for the style of music. For example, a mix that's right for rock music usually won't work for folk music or acoustic jazz. Rock mixes typically have lots of production EQ, compression, and effects; and the drums are way up front. In contrast, folk or acoustic jazz is usually mixed with no effects other than slight reverb, and the instruments and vocals sound natural. A rock guitar typically sounds bright and distorted; a straight-ahead jazz guitar usually sounds mellow and clean.

Suppose you are mixing a pop song, and you're aiming for a realistic, natural sound. Listen to the reproduced instruments and try to make them sound as if they're really playing in front of you. That is, instead of trying to make a pleasant mix or a sonically interesting recording, try to control the sound you hear to simulate real instruments—to make them believable. To do this you must be familiar with the sound of real instruments.

It's like an artist trying to draw a still-life as realistically as possible. The artist compares the drawing to the real object, notes the difference, and then modifies the drawing to reduce the difference.

When you're striving for a natural sound, compare the recorded instrument with your memory of the real thing. How does it sound different? Turn the appropriate knob on the console that reduces the difference. You're creating an illusion of accuracy. To make a recorded instrument sound real—like it's really playing in front of you—you often need to exaggerate things and add some processing.

Alternatively, when you're mixing, imagine that you're creating a sonic experience between the monitor speakers, rather than just reproducing instruments. Sometimes you don't want a recording to sound too realistic. If a recording is very accurate, it sounds like musical instruments, rather than just music itself.

This approach contradicts the basic edict of high fidelity—to reproduce the original performance as it sounded in the original environment. Some songs seem to require unreal sounds. That way, you don't connect the sounds you hear with physical instruments, but with the music behind the instruments—the composer's dream or vision.

Here's one way to reproduce pure music rather than reproducing instruments playing in a room: Mike closely or record direct to avoid picking up studio ambience. Then add reverb. Also add EQ, double-tracking, and effects to make the instrument or voice slightly unreal. The idea is to make a production, rather than a documentation—a record, rather than a recording.

Try to convey the musician's intentions through the recorded sound quality. If the musician has a loving, soft message, translate that into a warm, smooth tone quality. Add a little mid-bass or slightly reduce the highs. If the musical composition suggests grandeur or space, add reverberation with a long decay time. Ask the musicians what they are trying to express through the music, and try to express that through the sound production as well. Listen to the lyrics to see if the song is intimate (suggesting a warm, dry sound) or a proclamation (maybe with lots of reverb).

Try to keep the mix clean and clear. A clean mix is uncluttered; not too many parts play at once. It helps to arrange the music so that similar parts don't overlap. Usually, the fewer the instruments, the clearer the sound. Mix selectively, so that not too many instruments are heard at the same time. Have guitar licks fill in the holes between vocal phrases, rather than playing on top of the vocals.

In a clear-sounding recording, instruments don't "crowd" or mask each other's sound. They are separate and distinct. Clarity arises when instruments occupy different areas of the frequency spectrum. For example, the bass provides lows; keyboards might emphasize mid-bass; lead guitar may provide upper mids, and cymbals fill in the highs.

Often the rhythm guitar occupies the same frequency range as the piano, so they tend to mask each other's sound. You can aid clarity by equalizing them differently. Boost the guitar at, say, 3 kHz, and boost the piano around 10 kHz. Or pan them to opposite sides.

More on judging sound quality is in Chapter 14.

Changes During the Mix

It's rare to do a mix in which you set the faders and leave them there. Often you need to change fader levels, EQ, or effects send levels during a mix.

Level changes during the mix should be subtle, or else instruments will "jump out" for a solo and "fall back in" afterwards. Set faders to preset positions during pauses in the music. Nothing sounds more amateurish than a solo that starts too quietly then comes up as it plays. You can hear the engineer working the fader. If you need to reduce the level of a loud passage, do so at the end of the preceding soft passage before the loud one begins.

If you need to change fader levels during the mix, and you don't have automation, you might mark these levels next to each fader on a thin piece of tape. Make a cue sheet that notes the mixer changes at various counter times. For example:

0:15 Unmute vocal
1:10 Lead solo −5
1:49 Lead −10
2:42 Synth EQ +6 at 12 K
3:05 Start fade, out by 3:15.

Most recording software lets you automate the mix. Details are later in this chapter under the heading Recording Software with Automated Mixing.

Record or Export the Mix

When you're happy with the mix and recording levels, record the mix on your hard drive with recording software. If you're mixing with a DAW, export or save the mix as a 24-bit WAVE file.

If you don't have automation, and the mix is very difficult, you can record it a section at a time and then edit the sections together.

To fade out the end of the tune with a hardware mixer, pull down the mixer master faders slowly and smoothly. In a DAW, either (1) select all the tracks and insert a fade-out at the end of all the tracks or (2) export the mix to a stereo file, import the stereo mix file, and insert a fade-out at the end of the mix.

Use a fade-out that starts quickly and ends slowly. Try to have the music faded out by the end of a musical phrase. The slower the song, the slower the fade should be. The musical meaning of a fade is something like, "This song is continuing to groove, but the band is leaving on a slow train." You might want to postpone fades until mastering several songs to make an album.

The mixdown is complete. You might record several different mixes of one song, then choose the best mix. It's common to record a mix with the lead vocal up 1 dB, and another with the lead vocal down 1 dB.

Repeat these mixdown procedures for the rest of the good takes. Give your ears a rest every few hours! Otherwise, your hearing loses highs and you can't make correct judgments.

After a few days, listen to the mix on a variety of systems—car speakers, a boom box, a home system. The time lapse between mixdown and listening will allow you to hear with fresh ears. Do you want to change anything? If so, make it right. You'll end up with a mix to be proud of.

If you lack good multitrack recordings with which to practice mixing, go to www.raw-tracks.com. There you can download individual tracks in WAVE or MP3 format, or purchase a CD of raw tracks.

SUMMARY

The following are summaries of the procedures for recording, overdubbing, and mixdown. Use these steps for easy reference.

Recording

1. Turn up the headphone or monitor volume control. Monitor the aux bus that you're using for the monitor mix. If your multitrack is wired to the insert jacks, monitor the stereo bus instead. If you're using a DAW, monitoring is automatic.
2. Assign instruments to tracks. To record one instrument per track, connect its direct-out (or insert send) to a track input. If you're using a DAW, select each track and assign it an input signal from your audio interface.

3. Turn up the input faders, submaster, and master faders to design center (the shaded portion of fader travel, about three-quarters of the way up).
4. Adjust the input attenuators (trim) to set submixes and recording levels. If you're recording with a DAW, use the volume controls of the audio interface (they might be in a software application).
5. Set the monitor mix.
6. Record onto the multitrack recorder.

Overdubbing

1. Assign the instruments or vocals to be recorded to open tracks. An open track is blank or has already been bounced.
2. Turn up the monitor/cue system.
3. Turn up the submasters and master to design center.
4. Play the multitrack recording and set up a cue mix of the already-recorded tracks.
5. While a musician is playing, adjust the input attenuation and recording level.
6. Set the monitor/cue mix to include the sound of the instrument or vocal being added.
7. Record the new parts on open tracks.
8. Punch in and comp tracks as needed.

Mixdown

1. If necessary, set the input selectors to accept the multitrack recorder output signals.
2. Monitor busses 1 and 2 (or the stereo mix bus). Monitoring is automatic with a DAW.
3. Assign tracks to busses 1 and 2 (or the stereo mix bus, or the stereo output of your audio interface).
4. Turn up the master fader to design center. In some mixers, the submasters should also be up.
5. Set preliminary panning.
6. Set a rough mix with the input faders. Maybe start with all faders at −12 dB, then adjust from there.
7. Set EQ and effects.

8. Perfect the mix and set recording levels. Set up automation if your system has it (described next).
9. Record the mix onto a hard drive using recording software. If you have a DAW, export or save the mix as a WAVE file.

AUTOMATED MIXING

A multitrack mixdown is often a complicated procedure. It can be difficult to change the mixer settings correctly at all the right times. So you might want to use automated mixing—have the recorded session file store and reset the changes for you.

As you mix a song, you might adjust the mixer controls several times. For example, raise the piano's volume during a solo, then drop it back down. Mute a track to reduce noise during pauses in the performance. An automated mixing system can remember your mix moves, and later recall and reset them accordingly each time you play back the mix. You can even overdub mix moves; for example, do the vocal-fader moves on the first pass, drum moves on the second pass, and so on. You also can punch-in fader moves to correct them. Effects changes can be automated as well in some units. Automated mixing is a feature in digital mixers, software DAWs, and many recorder–mixers.

Automated mixing has many advantages. With it you can:

- Perform complicated mixes without errors
- Fine-tune the mix moves
- Recall mixes weeks or months after storing them, without having to reset the mixer manually each time
- Listen to the mix without the distraction of having to adjust faders

Types of Automation Systems

Three types of automation systems are

1. Automated mixer with non-motorized faders
2. Automated mixer with motorized faders (Flying Faders)
3. Recording software with automated mixing

Each of these is worth a closer look.

AUTOMATED MIXER WITH NON-MOTORIZED FADERS

Suppose you have mixer with non-motorized faders. The mix is playing and you are adjusting the faders. As you set the position of each fader, this action produces a MIDI signal that is recorded by a sequencer. This sequencer is either built into the mixer or is external.

When the mix plays back, so does the sequence of mix moves. MIDI signals from the sequencer set the volume level for each channel. This is done by varying the gain of a voltage controlled amplifier (VCA) or digitally controlled amplifier (DCA) in each channel. Because the faders do not move as the mix plays back, the fader positions do not represent the mix you are hearing.

AUTOMATED MIXER WITH MOTORIZED FADERS

This system works as follows. First, you adjust the faders to set up a mix. This generates MIDI signals that are recorded by a sequencer. When you play back the mix, the sequencer makes the motorized faders move up and down as if controlled by a ghost, matching your mix moves. The position of the mixer faders show the mix levels. VCAs aren't needed, which is a benefit because VCAs can degrade the audio slightly.

RECORDING SOFTWARE WITH AUTOMATED MIXING

Suppose you use a computer DAW to record tracks and edit them. The monitor screen shows virtual faders that you adjust with a mouse or control surface. As you set up a mix, the automation feature in the DAW remembers your fader moves, EQ, panning and plug-in settings. All those moves and settings are reproduced each time you open a project and play it back.

Automation is easy in DAWs that let you insert a volume envelope on a track. This envelope is a graph of a track's fader setting versus time (Figure 12.3). You might prefer to automate the mix by tweaking a track's volume envelope on screen instead of moving the fader. As the song plays, the fader for that track will move up and down, following the envelope. Many DAWs let you automate other parameters as well, such as reverb-send level, panning, and so on.

What if a track needs a major change in EQ or reverb at a certain point? Find the section in the track that needs different EQ or reverb,

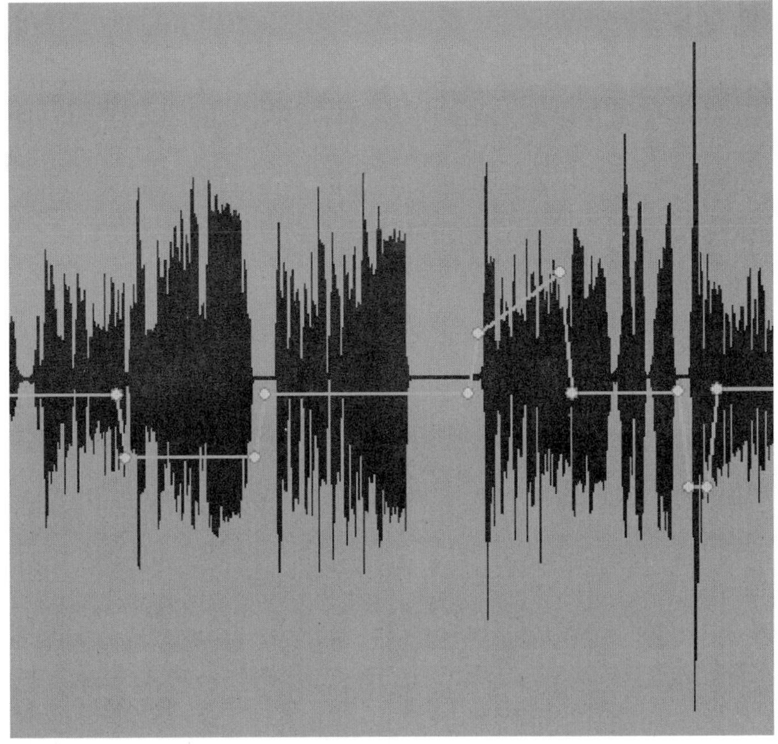

FIGURE 12.3
In a track, a volume envelope controls the track's volume in the mix over time.

create a clip of that section, and move it to another track with those plug-in settings. Or simply insert a new plug-in into the clip (if your DAW has that feature.)

As we said earlier, a MIDI/audio recording program includes both digital audio tracks and sequencer tracks. You automate the audio track levels by adjusting their gain (fader levels or volume envelopes). You automate the MIDI tracks by adjusting their MIDI volume or key-velocity scaling.

Snapshot versus Continuous Automation

Two types of mixer automation are snapshot and continuous (dynamic). With snapshot automation, you push a button to take a "snapshot" or "scene" of the mixer settings. The sequencer stores the snapshot as MIDI "program changes." To reset the mixer to any of these stored settings, you punch-up the appropriate number (MIDI program change) as the song plays. Recording software also can record and play snapshots of the mixer settings.

With continuous or dynamic automation, the sequencer records the motion of the mixer controls—not just their position. Continuous automation costs more than snapshot and consumes more memory, but permits finer resolution of mix moves. Some digital mixers do both snapshot and continuous automation with their internal sequencer. Other mixers require an external MIDI sequencer to record continuous automation moves.

Some automated mixers let you specify a fade time between snapshots. When the sequencer changes from one snapshot to the next, the sequencer fades or adjusts the control settings gradually between the two snapshots. This acts like continuous or dynamic automation.

Automated Mixing Procedure

Here is a suggested order of steps in doing an automated mix:

1. Set up a rough mix manually. You might prefer to start with a song section where all the instruments are playing.
2. Record the mutes. Mute tracks that aren't being used, or that don't add to the arrangement. For example, a chorus might sound better without the horns. Unmute each muted track just before you want it to be heard. If you're using a DAW, you have already deleted unwanted portions of each track.
3. Record the fader moves or, in a DAW, draw a volume envelope for each track. For example, turn up the guitar during the solo, turn up the toms during the fills, turn down the background vocals when they get too loud. Save your changes often.
4. Record the effects changes. You might bring up the chorusing on the acoustic guitar during the bridge, and so on.
5. Sit back and listen to the mix. Is there anything you want to change?

Another way to do an automated mix is with snapshots. At several points in the song where you need mutes or fader changes, set the mute or fader level on the appropriate track and take a snapshot.

LO-FI RECORDING: HOW TO TRASH YOUR TRACKS

It's an exciting time for home recording. We have low-cost digital recorders and mixers, soft synths, and good-sounding cheap mics. All these tools make it affordable to record with sonic purity and accuracy. That's a wonderful development.

But now that anyone can record high-quality sounds, it isn't such a big deal anymore. Sure, there's a place for clean, accurate recordings. But in many of today's records, you'll hear instances of lo-fi sound: fuzzy vocals, tinny drums, and humming guitar amps.

Lo-fi is the opposite of hi-fi. High fidelity recordings sound like real musical instruments. Technically, hi-fi sound implies a flat frequency response, no noise or distortion, and low wow and flutter. In contrast, lo-fi sounds might have a narrow frequency response (a thin, cheap sound), and might include noise such as hiss or record scratches. They could be distorted or wobbly in pitch.

Also, the lo-fi attitude snubs its nose at "clean" recordings. Clean can mean a lot of things: free of noise, free of leakage, free of excess room reverb. So a lo-fi sound might have tape hiss, some leakage, and some room acoustics as part of the mix.

Lo-fi really took off with rap music, in which the drum sound was the opposite of the usual polished studio sound. Instead of a tight kick, we heard a boomy kick. Instead of a wide-range snare with a full thump and crisp attack, rap gave us a tinny, trashy snare that was all midrange.

Want to put some lo-fi sonic interest in your mixes? Let's look at a few ways to trash those clean sounds that are so easy to record.

Lo-Fi Frequency Response

A flat frequency response in a mic or mixer tends to give an accurate, natural sound. If the response is flat, all the musical fundamentals and harmonics are reproduced in their original proportions. So the recording sounds accurate or hi-fi.

You can easily make a lo-fi effect by messing up the response. Cut the highs and lows, boost the mids. Or create a raggedy response with lots of bumps and dips. Some ways to do this are with EQ, mic choice, and mic placement.

Play a snare-drum track through your mixer, and turn down the LF EQ and HF EQ. Boost around 1 kHz or nearby frequencies. Your snare sound will change from high-budget to bargain-basement. Beck's "Soul-Suckin' Jerk" from the EP *Loser* is a good example of a lo-fi drum set.

Or try recording a child's drum set with its small heavy cymbals and boomy kick drum. You might loop a hi-hat beat made from this set, and mix it with a full-range recording of a quality drum set.

Find a toilet-paper tube, or a flexible plastic tube that extends gutter downspouts. Put the tube in front of a mic and sing through the tube. The resonances in the tube will color the sound in a wild way.

You might track down some cheap old mics at a garage sale, on eBay, or from vintage mic collectors. Record a few tracks using those mics. Their frequency response tends to be a complex series of peaks and valleys that you can't duplicate with EQ.

Unusual mic placements are fun. Record a guitar amp or vocal with the mic placed in a wastebasket (Figure 12.4). Hit a cymbal with a cheap mic while recording its signal. Mike a snare drum from underneath for a thin, zippy effect. If you pick up a crash cymbal off its edge, the sound will waver as the cymbal tilts when it is struck.

Distortion

Distortion is the addition of harmonics that didn't exist in the original sound. Here's an obvious way to create distortion: drive a piece of recording gear at very high levels—beyond what it can handle. For example, record drums on a cassette recorder with the meters pinning. Or yell into a "bullet" type harmonica mic so that the mic distorts. In a DAW, use a distortion plug-in such as iZotope Trash shown in Figure 12.5 (www.izotope.com).

Another method to add distortion to a track: feed the signal through an effects box with a distortion setting. Run a drum track through a guitar stomp box, or through a broken vintage compressor. Feed a vocal through an Amp Farm or Line 6 Pod processor (Figure 12.6). You might record some instruments on a cheap cassette recorder

FIGURE 12.4
The old mic-in-a-wastebasket trick.

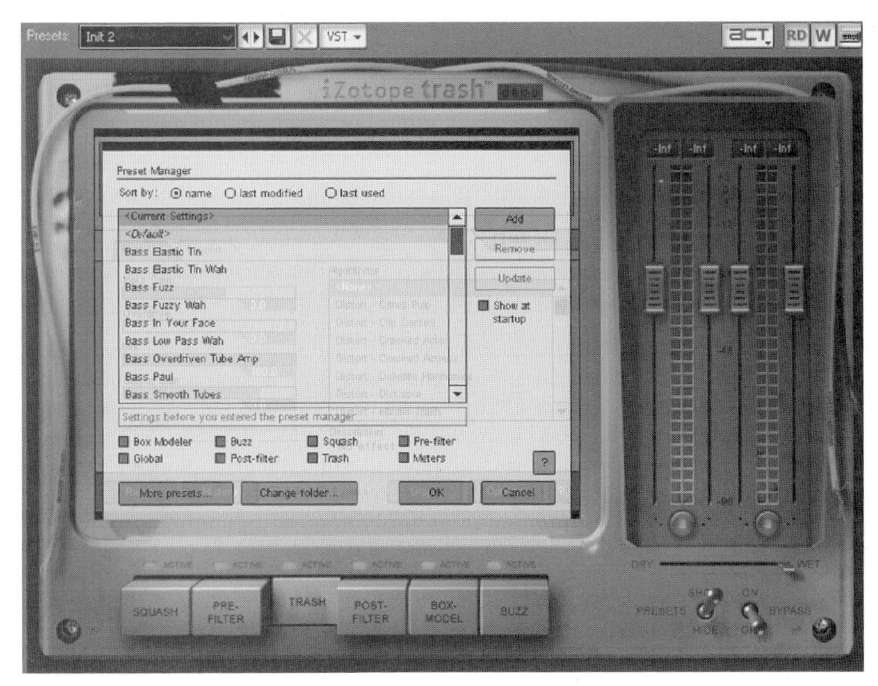

FIGURE 12.5
iZotope Trash.

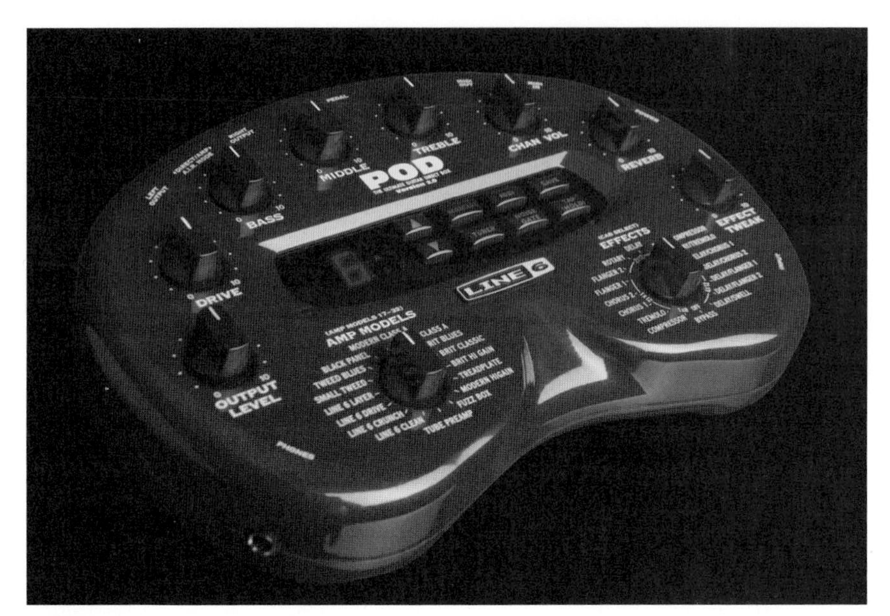

FIGURE 12.6
Line 6 Pod.

(Figure 12.7). The Rolling Stones did that to create the beginning of "Jumping Jack Flash."

Noise

Try iZotope's Vinyl simulation plug-in to add record scratches and other noises. Another way to have noises in your mix is to record noisy instruments! You might find an old, creaky upright piano and do the keyboard tracks with it. Or use an old guitar amp with blown tubes to record a sound full of hum and hiss.

Leakage

If your mixes are too sterile or studio-clean, you might want to record some leakage. Leakage (or bleed or spill) is the pickup of an instrument by another instrument's microphone. For example, if a guitar mic picks up the drums from across the room, that pickup of the drums is called leakage. Since the guitar mic picks up the drums at a distance, the leakage changes the recorded sound of the drums from tight to muddy.

FIGURE 12.7
Recording with a cheap cassette recorder.

Normally we try to isolate instruments or mike close to prevent leakage. But if you want to add some creative leakage to your mixes, mike a little farther away than you normally do, and record all the instruments at once.

Room Sound

In the quest for clean, tight recordings, it's standard practice to partly cover the walls of the studio with sound-dampening material. This reduces early reflections, which are less than about 20 msec after the direct sound from the instrument being recorded. Those early reflections tell the ear that the instrument was recorded in a small room. Normally we get rid of those early reflections and replace them with artificial reverb. But a lo-fi recording often includes the sound of the room as part of the sound of the recorded instrument.

To pick up room reflections, mike farther away than usual and leave the walls uncovered. Use the room for its coloration, rather than rejecting the room.

Want to make a synth track more organic and spacious? Run the synth track through a guitar amp, and mike the amp in stereo from several feet away. The guitar amp rolls off the highs and adds distortion as well. You could even record a vocal or harmonica through the amp this way.

For a spacious effect, consider recording several instruments in stereo with two mics. Pick up instruments or vocals in a hallway, a bathroom, a box, or even outdoors.

Lo-Fi Aesthetics

It's common to include hi-fi sounds along with lo-fi sounds in the same mix. This makes a statement to your listeners: "I can record hi-fi sounds, but I'm choosing not to. The trashy sounds are due to a conscious choice rather than a lack of recording chops." If you have nothing but lo-fi sounds in your mixes, it might sound like you don't know what you're doing—you have no command over the recording process.

So, your mix might be mostly pristine, full and sparkling, but tainted with a vocal that sounds like it came through a telephone. The contrast of clean and dirty sounds, modern and vintage, can add a lot of sonic interest.

The ear delights in complexity. We can add some of that to our recordings with intriguing, colorful lo-fi sounds.

I recommend the album *Mule Variations* by Tom Waits. It is a brilliantly creative lo-fi masterpiece. So is Beck's *Timebomb*. Others are "digital hardcore" genre albums by Ronin, Technology Scum, and Cheap Czad.

CHAPTER 13

Computer Recording

With recording software and a sound card, you can turn your computer into a powerful digital recording studio. This Digital Audio Workstation (DAW) lets you record dozens of audio tracks, edit them, add effects, do a mixdown with automation, and burn a professional-quality CD—all in your computer. The cost is only a few hundred dollars.

In addition to recording audio, most DAW software can act as a sequencer by recording MIDI data. You can record, edit, and play both audio and MIDI tracks in the same program.

Another function of a DAW is to edit tracks that were originally recorded on a hardware multitrack recorder. You can transfer eight or more tracks at once to your computer by using a sound card with Alesis Lightpipe or Tascam TDIF connectors, or with an Alesis FirePort that works with an Alesis HD24 hard-disk recorder.

A DAW has three parts (Figure 13.1):

1. A fast computer with lots of memory and a large hard drive.
2. An audio interface to get audio and MIDI into and out of your computer.
3. Recording software.

You also need some powered monitor speakers and at least one mic. If you're using a sound card that has a low-quality mic preamp, you'll also need a good mic preamp or small mixer. Some optional extras are a control surface and DSP cards. These are explained later in this chapter.

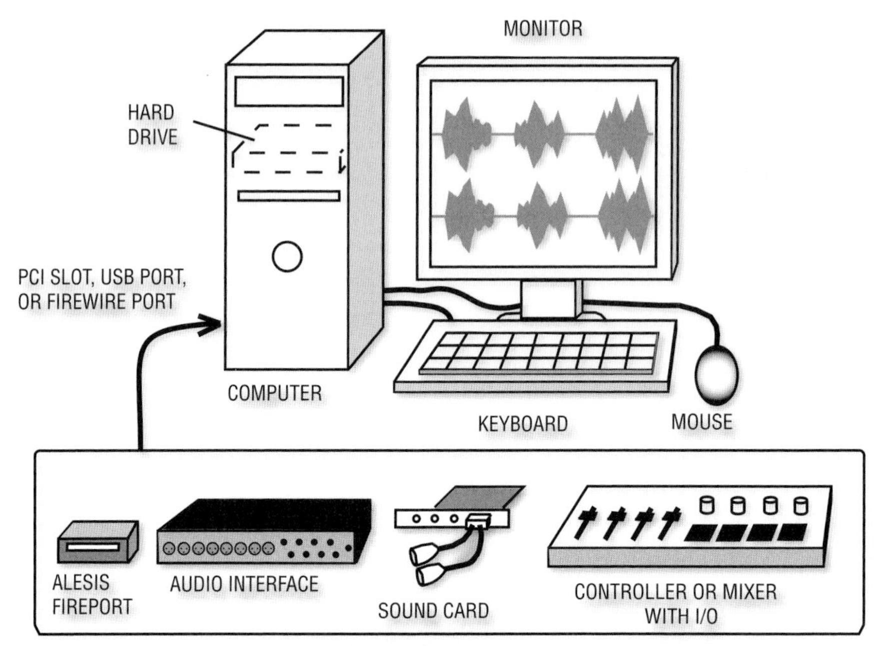

FIGURE 13.1
A computer DAW.

BASIC OPERATION

When you launch the recording software, you see simulated tracks and recorder transport buttons such as RECORD, STOP, and PLAY. You also see a mixer with virtual controls: simulated faders, knobs, buttons, and meters.

Most DAW software has a few windows or views (Figure 13.2). You view and manipulate the tracks in the Track window, do edits in the Track or Edit window, and adjust mixer controls in the Mixer window. Each window can be opened to fill the entire screen. Often you can do all the work in the Track window.

Recording and Playback

Here's how the system operates. Audio from your mic, instrument, mixer, or mic preamp goes to the inputs on the audio interface, which converts the audio to computer data and sends it to the computer.

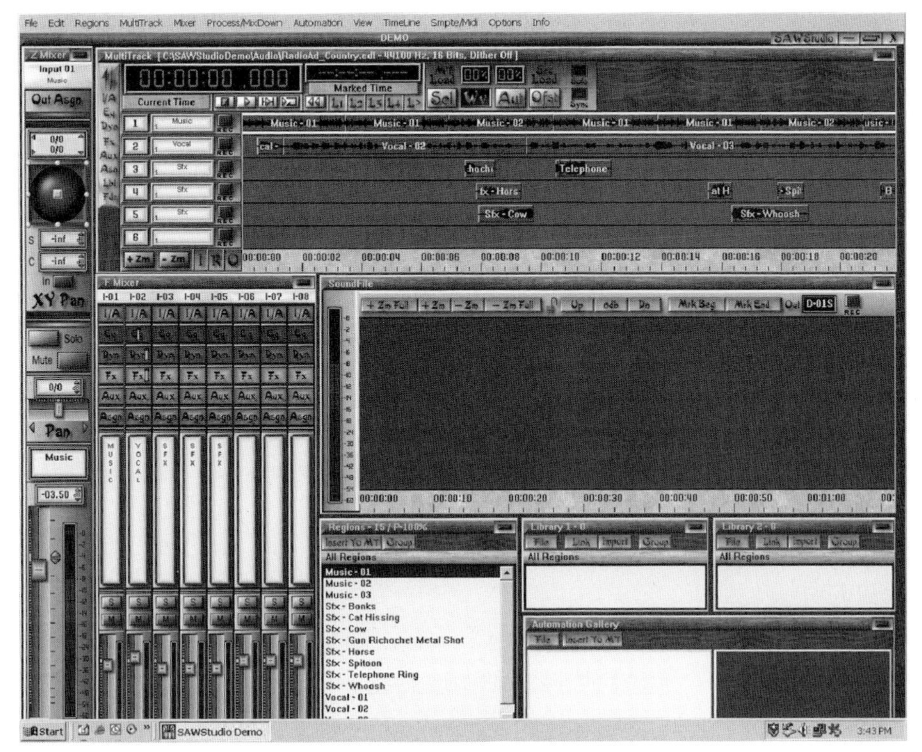

FIGURE 13.2
An example of DAW software windows.

The software lets you record this data on the computer's hard drive. During playback, the recorded data streams from the hard drive into the interface, which converts the data back into audio at the interface outputs. Monitor speakers connected to those outputs play the audio.

Audio data on the hard drive is read by an electromagnetic head. Because the head can be controlled to jump to any location on disk, it has random access: it can instantly locate any part of the audio program. As the head is jumping around, the data pieces it picks up are read into a buffer memory, then read out at a constant rate.

The waveform of the recorded audio appears on your monitor screen (Figure 13.3). You can zoom out to see the entire program, or zoom in to see individual samples.

Like a multitrack tape recorder or hard-disk recorder, a DAW has several tracks to record on. You might record a MIDI drum pattern on track 1, a bass audio signal on track 2, guitar on 3, keys on 4 and 5,

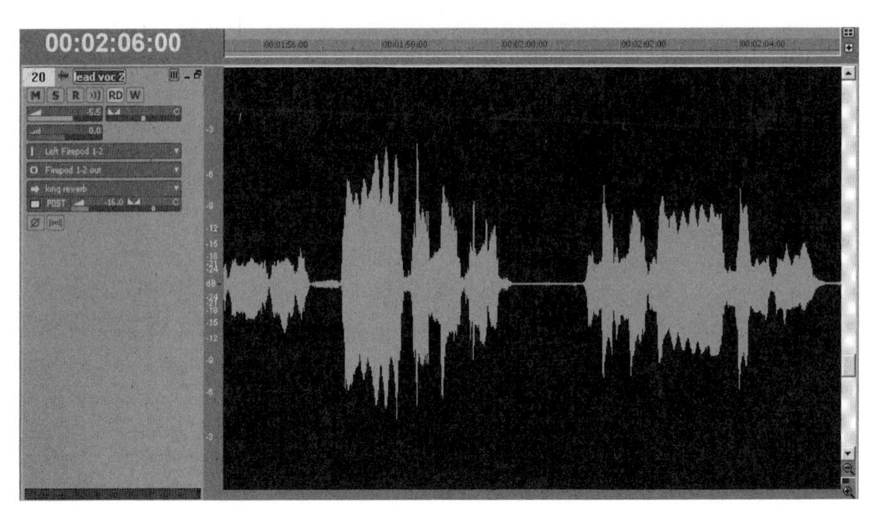

FIGURE 13.3
A waveform editing screen (SONAR Producer 7.0.2).

vocals on 6, and so on. During playback those tracks mix or combine into a stereo signal that plays through your audio interface.

Editing

Editing is a wonderful feature of all computer recording programs. When you edit audio, you select a part of the audio waveform on a track (or tracks). Then you cut, copy, or paste that audio, or slide it in time. It's like a word processor for audio.

Editing lets you clean up noises, replace bad notes with good ones, fix bad timing, or create a stuttering effect. With editing you can rearrange song sections or make choruses repeat without having to record each one. You might fade out the end of a song or crossfade between two songs. We cover editing in detail later in this chapter under the heading Editing Tips.

Mixdown

Once all your tracks are recorded and edited, it's time for mixdown. Here is the general procedure:

1. Adjust each on-screen fader with your mouse to set the level of each track until you create a good balance among tracks.

2. Make selections with the mouse to add panning, EQ, compression, and effects to various tracks.
3. Set up automation so that the computer remembers your mix moves and reproduces them each time you play the mix.
4. Once your mix is perfected, export it to a stereo WAVE file or AIFF file on your hard drive.
5. Repeat steps 1 to 4 for all the songs in a demo or album.
6. Burn a CD of the finished mixes or convert them to MP3 files.
7. Alternatively, import the mixes one after the other into a stereo track. Put the mixes in order with some silence between them. Burn a CD of the ordered mixes.

With this basic understanding, let's take a closer look at DAW components: the computer, audio interface, and recording software.

THE COMPUTER

Either Mac or PC will give great results with audio software. Just be sure that the software you want to use is compatible with your computer platform.

To play a lot of tracks and effects in real time, you need a computer with a fast central processing unit (CPU); lots of RAM; and a large, fast hard drive. A minimum system would be a 2 GHz CPU, 512 MB RAM, and an 80 GB hard drive. Two hard drives are faster: one for system files and programs, and another for audio data. The drive should be capable of high-sustained transfer rate (thruput). Current ATA-100 or ATA-133 drives can sustain up to about 40 MB per second, which is fast enough for multitrack recording.

Multitrack audio consumes a lot of disk storage space. Table 13.1 shows the amount of hard-drive space needed for a 1-hour recording with various recording formats.

If you plan to record a 1-hour album with about four takes per song, multiply the Storage Needed by four. In addition to the disk storage for each song's tracks, you will need about 30 to 200 MB per song mix, and up to 700 MB for an 80-minute CD album of the song mixes.

Table 13.1 Hard-Drive Storage Required for a 1-Hour Recording

No. of tracks	Bit depth	Sampling rate	Storage needed
2	16	44.1 kHz	606 MB
2	24	44.1 kHz	909 MB
2	24	96 kHz	1.9 GB
8	16	44.1 kHz	2.4 GB
8	24	44.1 kHz	3.6 GB
8	24	96 kHz	7.7 GB
16	16	44.1 kHz	4.8 GB
16	24	44.1 kHz	7.1 GB
16	24	96 kHz	15.4 GB
24	16	44.1 kHz	7.1 GB
24	24	44.1 kHz	10.7 GB
24	24	96 kHz	23.3 GB

AUDIO INTERFACE

Once you have a fast computer with a large hard drive, you need a way to get audio signals into and out of the computer. An audio interface does the job. Four types of interface are listed below, and we'll look at each one (see Figure 13.1).

1. Sound card (also called a PCI audio interface)
2. FireWire or USB audio interface
3. Control surface with I/O (Input/Output connectors)
4. Alesis FirePort. This converts WAVE files from an Alesis HD24 FST-formatted hard drive to FireWire, and sends the data to your computer. FireWire (IEEE-1394) is a high-speed data link between digital devices.

Sound Card (PCI Audio Interface)

The simplest form of interface is a 2-channel sound card, which plugs into a PCI user slot in your computer's motherboard. Low-cost sound cards have unbalanced 1/8-inch (mini) phone jack connectors, which include a mic input, stereo line input, and stereo line output. Generally, the sound quality and connectors of low-cost cards aren't up to professional standards.

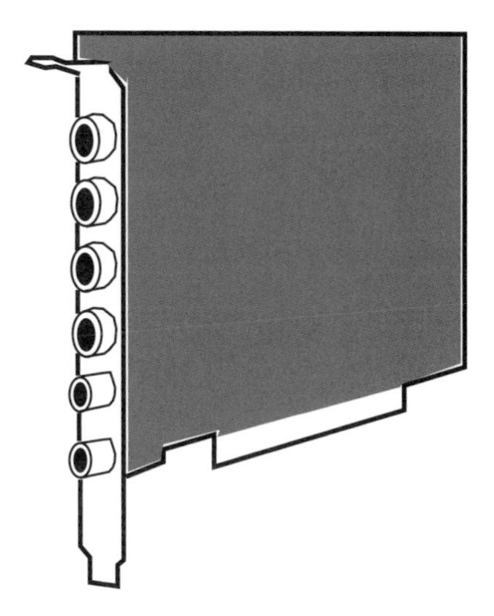

FIGURE 13.4
A sound card.

The next step up is a sound card with 1/4-inch TRS connectors (Figure 13.4) or XLR connectors on cables. It offers 2 to 8 balanced inputs. Current high-quality sound cards can record with 24-bit resolution. Many sound cards have MIDI connectors and an onboard synthesizer. Some have a FireWire port. Examples are sound cards by RME, Frontier Design, M-Audio, Lynx Studio Technology, E-MU, and Echo Digital Audio. A comparison of some sound cards is at www.pcavtech.com/soundcards/.

Choose either a 2-channel or multichannel card. A 2-channel sound card is adequate if you're recording one instrument at a time (such as a vocal, sax, keyboard, guitar, or a mix of several drum mics). You can overdub more parts using the same stereo input on different tracks. Also, a 2-channel sound card is sufficient if you're using an Alesis FirePort, which copies several Alesis HD24 recorder tracks to your hard drive, bypassing the sound card. In that case, you'd use the sound card just for monitoring.

You'll need a multichannel sound card (or a multitrack hard-disk recorder) if you want to record several instruments at once, such as a band or individual mics on a drum set. This type of card has several connectors on short cables, one per channel. Each instrument's signal goes to a separate channel in the sound card, and each channel's signal is recorded on a separate track. You can install two or more sound cards side by side to get more channels.

Check the sound card's Web site to make sure that it's compatible with your computer and recording software.

Here are some desirable features to look for when shopping for a sound card:

- 24-bit, 44.1-kHz minimum, full duplex recording. A full duplex card can record and play back simultaneously. The card works on two DMA channels.
- 85 dB or greater S/N ratio, 20 Hz to 20 kHz frequency response (±0.5 dB or less).
- XLR connectors, 1/4-inch TRS phone jacks, RCA phono jacks, MIDI connectors.
- Operates at +4 dBu as well as −10 dBV.
- Comes with drivers that work with your operating system and your recording software.

If the sound card includes an onboard synthesizer chip, look for these features:

- General MIDI (GM) compatible
- MPU-401 MIDI interface compatible
- Wavetable synthesis (better sound than FM synthesis)
- Programmable synth patches
- 24-note polyphony minimum
- 2 MB of wavetable ROM or RAM minimum

Some PCI cards have an I/O box, in which all the connectors are in a common chassis (see Figure 13.5). Because the interface's analog

FIGURE 13.5
E-MU 1616M, an example of a PCI audio interface I./O box.

circuits are outside of the computer, they tend to pick up less computer electrical noise than analog sound cards. This interface accepts analog audio or MIDI signals, converts them into computer data, and sends this data to the computer via a card in a PCI slot.

FireWire or USB Audio Interface

This is a chassis with 2 to 16 mic preamps, A/D converters, and a USB or FireWire port that sends digital audio on a single cable to your computer (Figure 13.6). Some units have MIDI ins and out, and a high-impedance instrument input for an electric guitar, electric bass, acoustic guitar pickup, or synth.

A USB or FireWire interface connects to a mating connector on the outside of your computer. To install a PCI interface you must open the computer.

Some mixers could be called audio interfaces if they include a FireWire or USB port. It sends the output signal of each mic preamp to a separate track in your recording software, and returns the stereo mix back to the mixer for monitoring. This can be a good system for recording a band because it offers easy setup of monitor mixes with effects. Chapter 4 lists some models of audio interfaces and mixers under the heading Computer Digital Audio Workstation.

Figure 13.7 shows a USB port and cable connector.

FireWire comes in two speeds: FireWire 400 (IEEE 1394) which runs at 400 Mbps (megabits per second), and FireWire 800 (IEEE 1394b) which runs at 800 Mbps (100 MB per second). In contrast, USB 2.0 high-speed is 480 Mbps.

FIGURE 13.6
PreSonus FireStudio Project, an example of a FireWire audio interface.

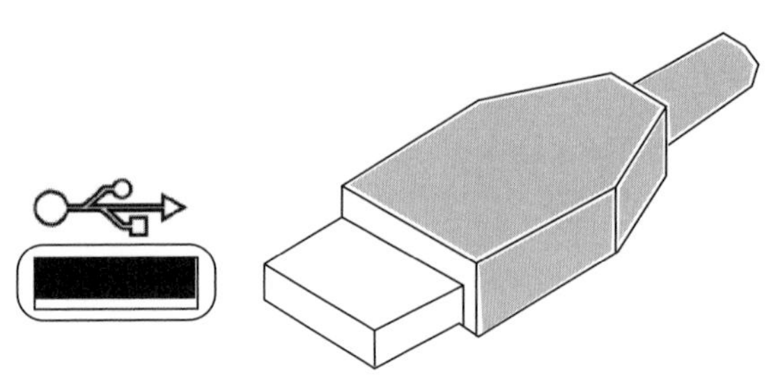

FIGURE 13.7
USB port and cable connector.

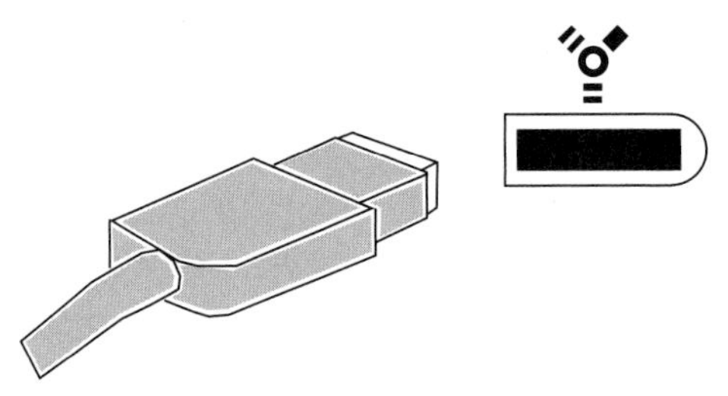

FIGURE 13.8
FireWire port and cable connector.

You can add a FireWire port to your computer with a $35 FireWire card, and some computers even have FireWire built in. To prevent data errors, get a FireWire card with a Texas Instruments (TI) chipset. Figure 13.8 shows a FireWire port and cable connector.

USB and FireWire devices are hot-swappable: you can insert or remove the connector while the computer is on. Both formats are compatible with Mac or PC.

Both ports are also available in PCMCIA cards and CardBus cards, which fit into laptops. A PCMCIA card is a credit-card-size memory card or I/O device that connects to a slot in a computer. CardBus is an advanced PCMCIA card with faster speed due to its direct memory access (DMA) and 32-bit data transfer.

Audio Interface Features

When deciding which interface to purchase, consider the features below.

DIGITAL I/O

Some interfaces have digital inputs and outputs as well as analog. Digital I/O comes in various formats (AES3, S/PDIF, Lightpipe), so make sure that the interface's format matches that of your digital recorder, digital mixer, or outboard A/D converter. See Chapter 9 under the heading Digital Audio Signal Formats.

ANALOG I/O

Check whether the analog I/O on the interface is balanced or unbalanced. Balanced connections reduce electrical interference picked up by cables. However, unbalanced connections cost less, and usually are adequate for cable runs under 10 feet. Balanced connectors are XLR or 1/4-inch TRS phone jacks. Unbalanced connectors are RCA phono jacks or 1/4-inch TS phone jacks.

SAMPLING RATE AND BIT DEPTH

Available sampling rates in audio interfaces are 44.1 to 192 kHz. Recordings made with higher sampling rates may give better sound quality but consume more hard-drive space. If you're making a CD or MP3 file, a sampling rate of 44.1 kHz does the job.

Most interfaces can handle 24-bit as well as 16-bit signals. Recordings made with 24-bit resolution sound a little smoother and cleaner (less distorted) than 16-bit recordings. Even if your final product is a 16-bit CD or MP3 file, it will sound better with a 24-bit recording that is dithered down to 16 bits during mastering. See Chapter 9 under the heading Dither.

MIDI PORTS

Many interfaces have MIDI ports. A MIDI IN port accepts MIDI signals from a MIDI controller and sends them to your computer's sequencing software. A MIDI OUT port sends MIDI signals from your computer's sequencing software to a synthesizer or sound generator. If your audio interface has MIDI ports, you don't need a separate MIDI card in your computer.

WORD CLOCK

Some interfaces offer word clock connectors, which send and receive timing signals for digital audio. Word clock is used to synchronize the digital devices in a large studio for real-time transfer of digital audio. See Chapter 9 under the headings The Clock and Jitter.

DRIVER SUPPORT

An audio driver is a program that allows recording software to transfer audio to and from an audio interface. Most interfaces are sold with several types of drivers. Be sure that your interface comes with drivers that work with your operating system and your recording software.

A good driver has a minimum latency spec under about 4 msec. Latency is the signal delay through the driver and interface to the monitor output. This can be a problem in recording a soft synth, because you hear the playback of the synth a few msec after striking the keys on the MIDI controller. See Appendix B, Minimizing Latency.

The most popular audio driver formats are ASIO, WDM, DAE, Direct I/O, GSIF, MAS, SoundManager, WAVE, and Core Audio. Here's a brief description of each format:

- ASIO (Audio Streaming Input and Output; Mac OS 9+, PC): A very popular driver developed by Steinberg. Allows multiple channels of simultaneous input and output, and low latency with software synthesizers.
- WDM (PC): Windows Driver Model with Kernel Streaming, a multichannel driver. Allows low latency with WDM-compatible audio hardware and DXi software instruments. DXi stands for DirectX Instruments, Cakewalk's virtual instrument integration standard. DirectX audio effects can be used live on input signals, not just during playback. This lets you monitor and record effects in real-time as you're recording.
- DAE (Mac, PC): Used only with Digidesign audio interfaces. It's a multichannel driver that runs with a compatible host such as Pro Tools, Logic, and Digital Performer. DAE lets you use RTAS and/or TDM plug-ins (explained later under the heading Digidesign Pro Tools).
- Direct I/O (Mac, PC): Works with Digidesign interfaces as a multichannel driver only. Doesn't let you run RTAS or TDM effects.
- GSIF (PC): Permits very low latency when playing samples from hard disk with TASCAM's GigaStudio software sampler.
- MAS (Mac): Developed by Mark of the Unicorn (MOTU). Offers resolutions up to 24/96 and multiple simultaneous input and

output channels. It's also a format for plug-ins (software audio effects).

- SoundManager (Mac): MacIntosh's standard audio driver. It lets you record and play mono and stereo files up to 16-bit and 44.1 kHz. Has a moderate amount of latency.
- WAVE (PC): The PC standard audio driver. WAVE can be used with a variety of audio interfaces (like Sound Blaster-type sound cards) to record and play mono or stereo audio. Has a moderate amount of latency.
- Core Audio. Apple's low-latency driver used in Mac OS X.
- MME is an older driver that offers lower performance than newer ones.

ASIO or WDM are the main choices for PC. Cubase seems to work best with ASIO; Cakewalk Sonar works with WDM or ASIO. Choose drivers recommended by (or supplied by) the audio interface manufacturer.

Caution: If you have multiple drivers installed, they may conflict. Then the computer might crash or the recording software might not access the audio interface. Delete unused drivers.

OTHER OPTIONS

Listed below are several features that some interfaces offer.

- Zero-latency monitoring: The input signal to the interface is fed to its output, so you hear the input signal without any signal-processing delay (latency).
- Pro Tools compatible: The interface works with Pro Tools recording software and hardware.
- Surround sound: Provides 5.1 or 7.1 surround sound monitoring.
- Powered by FireWire bus or USB bus: The FireWire or USB connection powers the interface; you don't need another power supply.
- Battery powering: This makes the interface portable for on-location recording with a laptop.
- Supplied recording software: The interface is packaged with recording software, so you might not need to buy other software.
- SMPTE sync: The interface will synchronize with a SMPTE timecode signal (explained later in this chapter under the heading Audio for Video.).

Control Surface

So far we've covered the computer and audio interface. Let's look at other hardware for a computer studio.

Using a mouse to adjust the controls on-screen is slow and can lead to repetitive stress syndrome. An alternative is a DAW control surface or controller (Figure 13.9). It's a chassis with physical controls for software functions. Resembling a mixer with real faders, knobs, and transport buttons, it lets you adjust the software's virtual controls that you see on the monitor screen. The controller attaches to the computer through a MIDI connector, Ethernet connector, USB, or FireWire port.

Many control surfaces also act as audio interfaces: they include mic/line inputs and outputs (I/O) and MIDI connectors.

Some control surfaces are dedicated to one DAW recording program, such as Pro Tools Mbox or 003 Series. Others are universal and

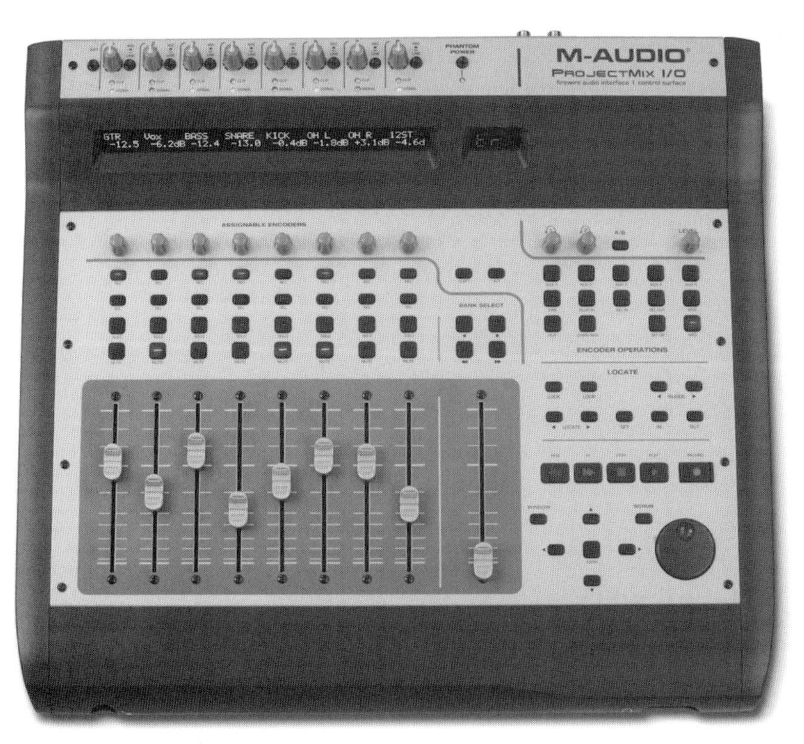

FIGURE 13.9
M-Audio Project Mix, an example of a control surface.

they work with several different DAW programs. Be sure to check whether the controller you want to buy will work with your recording software.

What if your controller has 8 faders, but your project has 16 virtual faders on screen? You can make the controller access groups (banks) of virtual faders, 8 at a time, by pressing a bank switch. Normally, your controller faders will affect virtual faders 1 through 8. Press the bank switch to make the controller faders affect virtual faders 9 through 16. Press it again to access virtual faders 17 through 24, and so on. Controllers that have 4 faders can access banks of virtual faders 4 at a time.

Advanced controllers offer these features:

- Motorized faders: The faders in the control surface are motorized, so they move like the virtual faders on screen.
- Stand-alone mixer mode: You can use the control surface as a regular mixer.
- Footswitch jack: Accepts a footswitch for punch-in/out. The footswitch works only if your recording software supports this function.
- Meters.
- Monitoring section.
- Aux send/return: Allows you to connect an external analog signal processor.
- Insert jacks: An insert jack in series with each channel's signal allows you to plug in an analog compressor.
- Expandability: You can add more controllers, side by side, to control more virtual faders at once.

Note that a MIDI controller or a MIDI fader box can control a sequencer's MIDI tracks, but not audio virtual controls. A control surface can control both.

The Frontier Design Group TranzPort is a wireless remote control for a DAW, letting you control your DAW from the studio. It includes transport controls, metering, a footswitch jack, and more. See the Web site www.frontierdesign.com for more information.

The Frontier Design Group AlphaTrack (Figure 13.10) and the PreSonus Faderport are controllers with several buttons and one long-throw fader. They are easier to use than a mouse and have a small footprint.

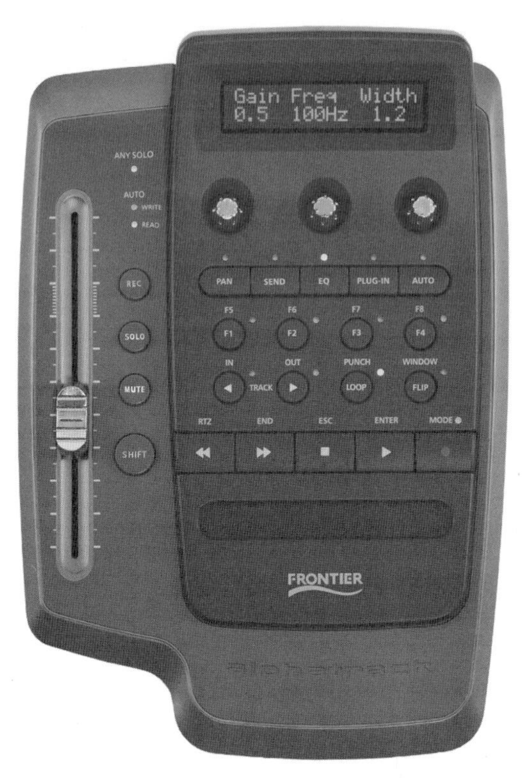

FIGURE 13.10
Frontier Design AlphaTrack, an example of a single-fader controller.

Alesis FirePort

This is a special type of audio interface that transfers audio tracks from an Alesis HD24 multitrack recorder to a computer. The HD24 records tracks as WAVE files on an Alesis FST-formatted hard drive that is plugged into the recorder. After recording tracks, you can remove the hard drive and plug it into an Alesis FirePort (Figure 13.1), a small device that converts the WAVE files of those tracks to FireWire format and sends them to your computer hard drive for editing and mixing.

As we've seen, there are a wide range of features and connectors among different interface models. A good interface is well worth the price because of its top-quality sound, convenient connections, and easy control of audio signals.

DSP CARD

Another DAW option is a digital signal processing (DSP) card that you plug into a user slot in your computer's motherboard. Why is it helpful? The more software effects (plug-ins) you use, the fewer tracks you can mix, because effects put a big load on the CPU. Normally this isn't a problem except in mixes with more than 24 tracks and loads of plug-ins in use. You can resolve this issue by installing a DSP card, which handles the processing for the plug-ins. The plug-ins access the card rather than the CPU. You can install multiple cards to expand the system. Examples include: TC Works Powercore (www.tcelectronic.com), Digidesign TDM-based Pro Tools cards (www.digidesign.com), Mackie UAD-1 (www.mackie.com), and Sonic Core (www.soniccore.com) PCI cards.

ANALOG SUMMING AMPLIFIER

This device is a mixer without faders, used to mix tracks or submixes from a DAW through analog circuitry. Two examples are the Dangerous 2BUS 16 × 2 mixer and the Manley Labs 16 × 2. By using a summing amplifier, you bypass the DAW's internal mix bus which might create slight distortion due to rounding errors, numeric overload, and bit truncation. (Some DAW software claims not to produce those errors, such as Pro Tools TDM systems.) You connect the analog output of each DAW track to a summing amplifier input, and use the DAW's faders and automation to set mix levels. Some engineers claim they avoid the distortion of mixing entirely inside the computer by recording and mixing at lower levels.

RECORDING SOFTWARE

Now let's turn from hardware to software. Some popular programs for recording, editing, and mixing are listed in Chapter 4 under the heading Computer Digital Audio Workstation. They share a lot of the same features but implement them differently. Also, each product has unique features and its own GUI (graphical user interface; Figure 13.11).

Check out the products online to see how you might like working with them. Some Web sites have interactive demos; others have free trial versions that you can download. Many sites have tutorials.

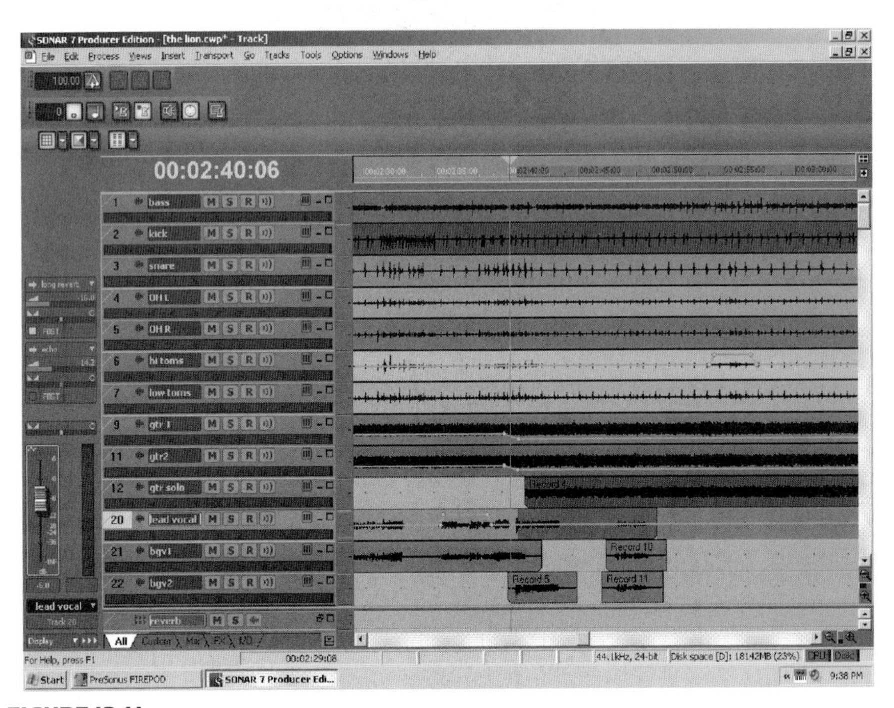

FIGURE 13.11
Cakewalk Sonar Producer track view.

When you first install recording software or hardware, expect some frustration. Your software may not be compatible with your hardware, or may not perform as expected. You might need to modify certain settings in your computer system or recording software. Read the manual and any readme files that came with your product. Also, check out the products' Web sites for knowledge bases, FAQs, etc. Each DAW program has online discussion groups or forums that can be very helpful. If you're still having problems, call or e-mail the tech support for your product.

For each recording program, books are available that provide operating details and power-user tips. To find them, search www.amazon.com or the references in Appendix D. Type the name of the program and manufacturer in the search field.

This chapter isn't meant to duplicate the information in those specialized books or in manufacturers' Web sites. Instead, it provides an overview of recording software.

Features

Here are some features to look for in recording programs:

- Multitrack digital recording from 2 tracks on up: Some programs can record and play back an unlimited number of tracks depending on the speed of your computer, RAM, and hard drive.
- MIDI sequencer: This lets you import or record several tracks of MIDI performance data, edit it, and play it. MIDI sequencer tracks can be mixed with audio tracks. Some programs include MIDI effects, MIDI plug-ins, and input quantizing.
- Freeze function mixes down effects and edits on a track to another track with embedded effects. This frees up CPU resources. Freeze also converts MIDI soft-synth tracks to audio tracks, which also reduces CPU loading. Unfreeze returns to the original tracks for editing.
- Peak and rms metering.
- Compatibility with control surfaces.
- Surround mixing (in high-end programs).
- Customizable GUI: Change the GUI to suit the way you work. Create configurations to use as templates for similar projects.
- Keyboard shortcuts: Instead of constantly using a mouse, tap certain keys on your computer keyboard. This speeds operation and gives your mouse hand a rest.
- Virtual tracks: Extra takes of a musical performance that are recorded on the hard drive. During mixdown you can choose the best takes to use, or the best parts of each take.
- Automated mixing: All settings for a song project can be saved and automated. The computer remembers your mix moves, and sets the mixer faders and pan pots accordingly during subsequent mixdowns. Some programs let you automate effects settings as well. For more information, see the section on Automated Mixing in Chapter 12.
- Loop-based composition, construction, and editing tools, ACID-loop and MIDI Groove Clip support. All these are covered later in this chapter. Loops and samples are included in high-end programs.
- Locate and marker points: Mark several points in a song (intro, verse, chorus, solos) so that you can go to them instantly.
- Routing or virtual patchbay: Assign any input to any track.

- Automatic pitch correction of audio (by Autotune, Melodyne, or V-Vocal).
- Automatic timing correction of notes by Beat Detective (Pro Tools) or Audiosnap (Cakewalk Sonar Producer) let you lock the tempo of audio tracks, such as drums, to a tempo map. You can also make one track follow another track's tempo, or extract timing from audio tracks and apply them to MIDI tracks.
- Audio-to-MIDI conversion: For example, have a singer's notes play the same pitches on a synthesized cello.
- Video display: An on-screen window that shows video clips. Found in high-end software, this feature lets you synchronize music and sound effects to a video program.
- CD recording: Record your mixes on a CD burner.
- PQ editing: Set up a list of CD song-start times before burning a CD of your mixes.
- Import and export various file types (such as WAV, MP3, WMA, OMF), sampling rates and bit depths.
- 32- or 64-bit floating-point resolution in the DSP.
- Support for 64-bit operating systems and dual-core CPUs.
- ASIO and WDM compatibility.
- Ability to learn the function of each controller encoder or button, mapping the controller to the on-screen controls.
- Spectral analysis: A display of level versus frequency of the audio program as it progresses in time.
- Notation application: Converts a MIDI file to musical notes on a treble and bass clef.
- Sync: Synchronization to SMPTE timecode or MIDI timecode (described later in this chapter under the heading Audio for Video).
- Support for DirectX and VST audio effects. (DirectX is a standard audio driver.)
- Support for DXi and VSTi soft synths. (VSTi is an acronym for Virtual Studio Technology for Instruments, Steinberg's virtual instrument integration standard.)
- Included soft synths and samplers.
- Convolution reverb, vintage compressors, audio restoration, mastering tools.
- Plug-ins with delay compensation: Described in the next section.

Plug-Ins

As we described in Chapter 10, a plug-in is a software module that adds digital signal processing or effects to a DAW recording program. Recording software comes with several plug-ins already installed. You also can purchase and download plug-ins from the Web. Unlike hardware effects, plug-ins are instantly upgradeable—just download the latest version.

When you click on the name of a plug-in, a screen pops up that looks like a hardware processor with knobs, faders, lights, and meters. Parameters (such as reverb time, chorus depth, or compression ratio) can be adjusted, and most parameters can be saved and automated.

A number of plug-ins can simulate or model the sound of specific devices. For example, Amp Farm by Line 6 is a guitar-amp modeling plug-in, while Antares Microphone Modeler can simulate the sound of popular studio mics. Vintage compressors, tube mic preamps, concert halls, and analog tape recorders have been modeled as well.

Some plug-ins run in real-time as the audio is playing. Others process a sound file and create a new file as with normalization or time compression/expansion.

Plug-in effects can be applied in three ways:

1. Master effect: The effect is on the master output bus, and processes the entire mix. Multiband compression is a typical master effect.
2. Aux-send or loop effect: The effect is on an aux bus. Set the plug-in's dry/wet mix control all the way to wet or 100% mix. Adjust the amount of effect on each track by turning up the track's aux send control. This option reduces the number of plug-ins needed for a mix, and so reduces the load on your CPU. Delay effects—reverb, echo, chorus, and flanging—are aux effects.
3. Insert effect: This effect is on a specific track and affects only that track. EQ and compression are insert effects. If you insert a delay effect into a track, you adjust the dry/wet mix control in the effect to control the amount of effect that you hear.

Plug-ins come in several formats; the most popular for Windows are Steinberg's Virtual Studio Technology (VST) and DirectX. DirectX is a group of application-program interfaces that enhance multimedia (video and audio) on Windows systems.

You can find plug-ins on their manufacturers' Web sites or on the Web sites of recording-software companies. For example, Digidesign's Web site lists dozens of plug-in partners. Another resource is www.kvraudio.com.

Digidesign Pro Tools

Pro Tools is the industry standard DAW system. However, other excellent programs are available. People often use the term "Pro Tools" generically to mean "digital audio workstation." Because Pro Tools is so popular, and has a number of configurations that can be confusing, I'll describe them here. There are three basic categories: Pro Tools LE, M-Powered, and HD.

Pro Tools LE and M-Powered are 32-track recording systems that use your computer's CPU for all processing. Your CPU speed, hard-drive speed, and amount of RAM determine how many plug-ins and soft synths you can use simultaneously.

Pro Tools LE software has nearly the same user interface as the professional Pro Tools HD systems (Figure 13.12). It's intended for singer/songwriters, project and personal studios, remote recording engineers, and bands. Pros can use it at home because it is compatible with the high-end Pro Tools HD system. Cost is $500 to $3000.

LE components include Pro Tools LE software and a choice of audio interface: Mbox 2 Mini, Mbox, Mbox 2 Pro, 003 Rack Factory, 003 Factory control surface, or Command|8 USB control surface.

- The Mbox series are USB or FireWire audio interfaces with one or two Focusrite mic preamps and various I/O options.
- 003 Factory is a control surface and audio interface that communicates with your computer via FireWire. It includes 8 analog ins and outs, 4 mic preamps, 8 channels of ADAT I/O, stereo S/PDIF and MIDI I/O, 8 touch-sensitive faders, 8 motion-sensitive rotary encoders, transport buttons, and automation. Any changes you make in your Pro Tools software are duplicated on the control surface and vice versa.

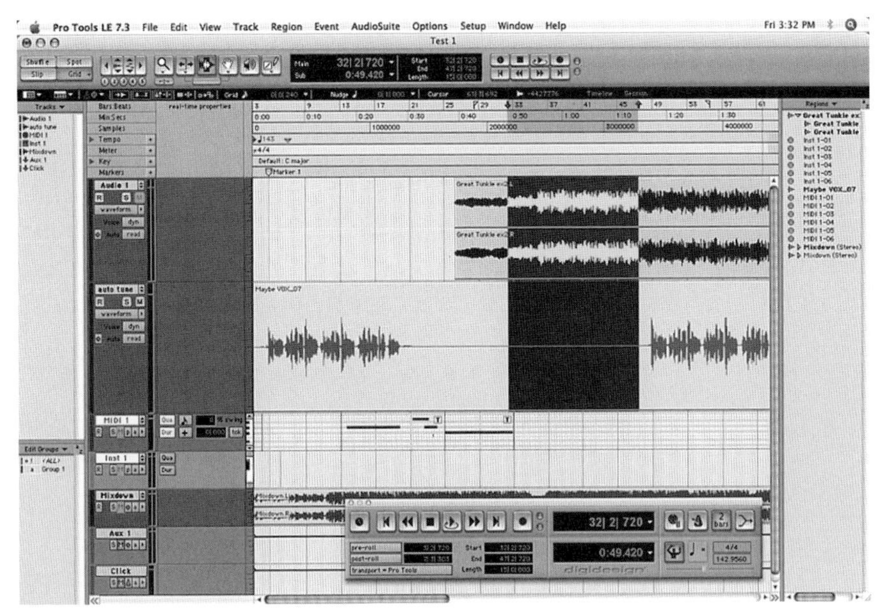

FIGURE 13.12
Pro Tools LE screen view. Source: www.en.wikipedia.org.

- 003 Rack Factory is 003 Factory in a 2U rack-mount chassis, without the control surface.
- Command|8 is a USB control surface with 8 bankable motorized faders, rotary encoders, transport controls, and displays.

Pro Tools M-Powered™ software is designed to work with certain M-Audio® audio interfaces and control surfaces. You can also use other Pro Tools products such as the Command|8® control surface. M-Powered software includes many of the same features as the professional LE and HD systems, and M-Powered sessions will open in those other systems. Easy to use, the software works with Windows XP and Vista, and Max OS-X computers. The software costs $299 while the interfaces range from $100 to $500. The M-Audio ProjectMix I/O control surface sells for $1249.

Pro Tools HD is a high-resolution professional system with DSP cards, up to 192 kHz sample rate, expandable I/O, and many options. It comes in three Core systems (each with more DSP power and more I/O) and one or more audio interfaces. Each system requires the Pro Tools 192 I/O or the 96 I/O audio interface, and supports legacy Pro Tools audio interfaces for additional inputs and outputs. Cost is $10,000 to $125,000 or more.

Pro Tools HD Systems offer dedicated DSP cards that free up your CPU for other tasks. The result is you can run more tracks, more plug-ins, and more soft synths at the same time. HD offers a wide range of sample rates, hi-res audio, and up to 192 tracks. An HD system can have 8 to 96 ins and outs. The three basic Core systems have one, two, or three DSP cards, and a system can use up to seven DSP cards.

Listed below are the three Core systems. Each can be expanded by adding more HD Accel cards.

- Pro Tools HD 1 includes the HD Core DSP card, which handles up to 32 channels of I/O and up to 96 tracks.
- Pro Tools HD 2 Accel includes the HD Core card and an HD Accel card, which provide more than four times the mixing and processing power of HD 1 systems and 64 channels of I/O and 192 tracks.
- Pro Tools HD 3 Accel, the most powerful system, includes the HD Core card and two HD Accel cards, which provide up to 96 channels of I/O and 192 tracks.

Digidesign Pro Tools HD systems also include these interface options.

- The 96 I/O Audio Interface converts audio at up to 24-bit/96 kHz. It includes 8 channels of analog I/O, 8 channels of ADAT optical I/O, 2 channels of AES/EBU and S/PDIF I/O, and word clock. It can link to other 96 I/O interfaces and to Digidesign 888/24, 882/20, 1622, or 24-bit ADAT Bridge.
- The 192 I/O Audio Interface converts audio at up to 24-bit/192 kHz. It includes 8 channels of analog I/O, 8 channels of AES/EBU, 8 channels of TDIF, 16 channels of ADAT, and 2 additional channels of ADAT or S/PDIF digital I/O. A 192 Digital expansion card provides 8 more channels of AES/EBU, TDIF, and ADAT I/O connections and a Legacy port to link to older Digidesign interfaces.
- The ICON D-Control worksurface is a large mixing console/controller surface designed to work with Pro Tools software. C|24 is another powerful control surface.

New features of the latest Pro Tools software are on the www.protools.com Web site.

Plug-ins for Pro Tools come in four formats: TDM, HTDM, RTAS (Real Time Audio Suite), and AudioSuite. TDM plug-ins work only with Pro Tools' professional TDM systems and use dedicated cards that plug into your computer. HTDM is the same for the Pro Tools HD (high definition) system. Many RTAS plug-ins are the same as TDM plug-ins, except that they use your computer's processor instead of specialized TDM DSP cards. All the above plug-ins work in real time.

AudioSuite plug-ins don't work in real time. Instead, they process a track's signal and create a new track in which the effect is part of the recorded sound; that is, the effects are embedded in the track. Playing a track with embedded effects is less work for the CPU than playing a track with real-time effects. Less load on the CPU means you can play more tracks without drop-outs.

Loads of plug-ins are available for Pro Tools. Many are included, and some can be downloaded from third-party vendors. Use the plug-in finder on the Pro Tools Web site (www.protools.com).

Included with every Pro Tools system is a Music Production Enhancement Suite, which includes soft synths, looping, guitar-amp modeling, a sample player and tube EQ.

OPTIMIZING YOUR COMPUTER FOR MULTITRACK RECORDING

Once you've chosen some recording software and installed it, you'll want to make your computer run as fast and glitch-free as possible. You need to optimize its settings for best results. Appendix B suggests some ways to speed up your hard drive, reduce software interruptions, and reduce the CPU usage. If you follow those tips, you will have a faster system that handles more plug-ins and more tracks at once. Also, when you play tracks or burn CD-Rs, clicks and drop-outs in the audio will be less likely.

USING A DAW

Let's say you purchased some DAW hardware and software, and tweaked your computer to make it fast and reliable. It's ready to rock. Here are some tips on setting up and using your DAW.

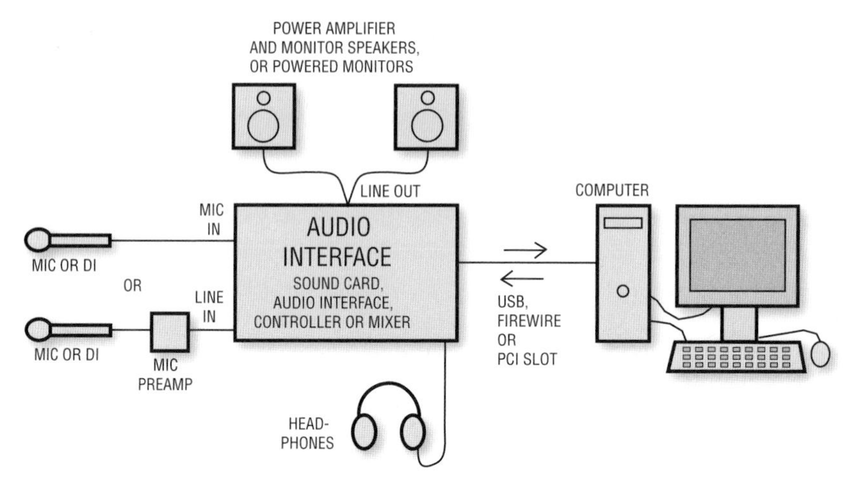

FIGURE 13.13
Connecting audio equipment to a DAW.

Connections

First, connect your audio equipment to the DAW as shown in Figure 13.13. You may need some adapter cables between devices. The following adapter cable connects a sound card's unbalanced I/O to a power amp, powered speakers, or mixer: Stereo 1/8-inch phone plug to two unbalanced phone plugs, or two RCA phono plugs (whatever matches your equipment).

Software Settings

Starting from your computer's desktop view, select Start > Settings > Control Panel. Double-click Sound and Audio Devices > Audio tab. Set the Playback and the Recording default devices to the sound card or interface you're using for audio editing.

Before you record, set a bit depth and sampling rate for the project, such as 24-bit/44.1 kHz.

If you want to record from a mic or instrument, you'll record their amplified signal on an audio track. If you want to record from a MIDI controller, you will record its MIDI signal on a MIDI track. Details on recording MIDI sequences are in Chapter 16.

Set up the audio drivers in your recording software. This setting might be called "Record timing master" and "playback timing master." Specify the input and output devices for each audio track, such as "sound-card left channel input" and "sound-card stereo outputs." Also, specify the input and output devices for each MIDI track, such as "sound-card MIDI IN" and "sound-card stereo outputs."

If you're recording many tracks simultaneously, or at a high sample rate, this may generate more data than your CPU and buffers can handle—causing drop-outs. A drop-out is a short silence or a noise burst (glitch) in the recorded audio. To prevent drop-outs while recording, turn off the track playback metering, effects, and waveform preview. Increase buffer size. Turn off or reduce video acceleration. Don't zoom or scroll while recording. Also, follow the tips in Appendix B on optimizing your computer for multitrack recording.

The buffer setting is critical. A large buffer lets you use many tracks and plug-ins without drop-outs, but it creates a lot of latency (monitoring delay). If latency is high, and you press a key on a MIDI keyboard, a soft synth will respond after a noticeable delay. A small buffer minimizes latency so soft synths respond faster, but can cause drop-outs. Generally it's best to set the buffer/latency small (under 4 msec) while tracking soft synths and set it large otherwise (maybe 25 msec). See Appendix B under the heading Minimizing Latency.

To monitor a live mic signal without latency while overdubbing, listen to a mix of the live mic signal and the previously recorded tracks. Mute the mic input signal in the sound card's software and disable Echo Input Monitoring in the mic's track.

Editing Tips

Now that your DAW is set up, you can record and edit audio. When you select a portion of the audio program, you create a clip or region. Examples of clips are an entire song, the chorus of a song, a guitar track, a drum riff, or a single note. Using a mouse, you can create a clip by marking its beginning and end points in the waveform. In Figure 13.14, a clip is selected and highlighted.

A clip is actually a pair of pointers to part of an audio file on your hard drive. One pointer is the data address for the beginning of the clip, and the other pointer is the address for the end of the clip. Rather than containing audio data, clips tell the software which section(s) of the audio file to play.

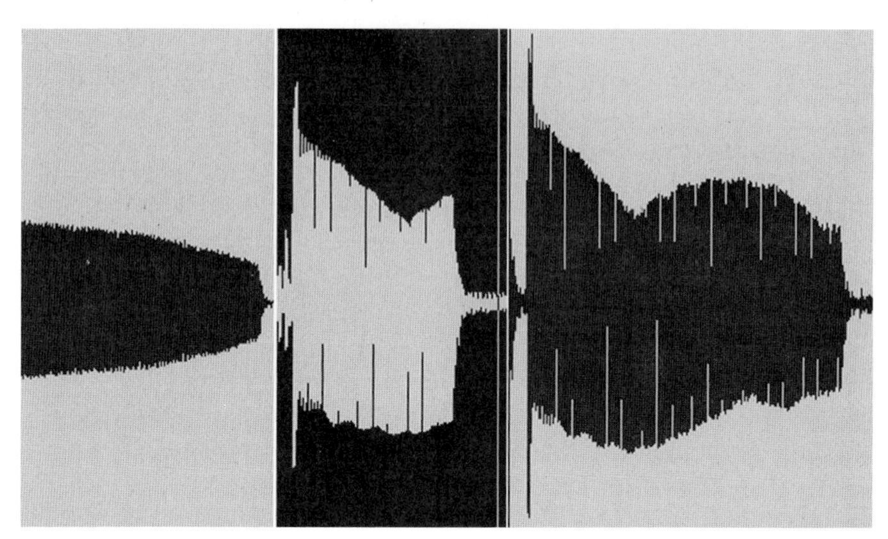

FIGURE 13.14
A clip is selected and highlighted.

When you set up the sequence of clips in a mix, you're telling the hard-drive head which pointers to play in what order. Or when you delete a clip, you're telling the hard-drive head to skip the clip's pointers during playback. If you copy an audio clip, the software doesn't make copies of the audio file it points to—instead, it plays the same audio file each time it sees a clip pointing to that file. Or when you split a clip into parts and put them in a different order, the audio file itself isn't split. Instead, the software plays the sections of the audio file that the clips point to, in the desired order.

These types of edits are called nondestructive edits. Only the pointers change; the data on disk are not changed or destroyed. Nondestructive edits aren't permanent. If you don't like an edit, you can undo it and try it again.

Some types of edits or processing are destructive: they write over the data on the disk. However, some recording programs save the data before you edit it. Then you can undo the change by reverting to the saved data.

I'll give examples of several things you can do with editing.

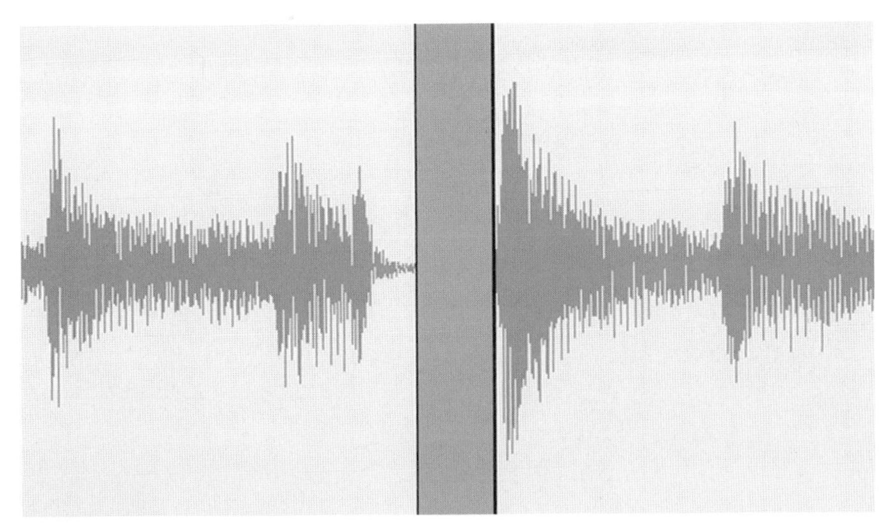

FIGURE 13.15
A guitar track with a noise deleted.

DELETE NOISES

Suppose you recorded an acoustic guitar track, but it has a chair squeak between the verse and chorus. Zoom in on that noise, play it, highlight it, and delete it (type Ctrl-X with most DAWs). You might have the option to close up the space you created—don't do that in this case. Figure 13.15 shows a guitar track with a noise deleted between notes.

CLEAN UP TRACKS

Before the beginning of a recorded song, you'll typically hear noises and a count-off. To remove them, define the group of noises as a clip and delete it (Ctrl-X). Some software lets you slip-edit or drag the beginning of each track to the right to remove noises there.

You can cut out silent areas in each track to reduce leakage and background noise. Say you've recorded a real drum set. Look at the tom-tom tracks. The tom mics pick up a lot of unwanted leakage from the snare drum and cymbals. In the tom-tom tracks, you can cut out the space between tom hits. This prevents the leakage and tightens up the sound. Figure 13.16 shows some edited tom-tom tracks with

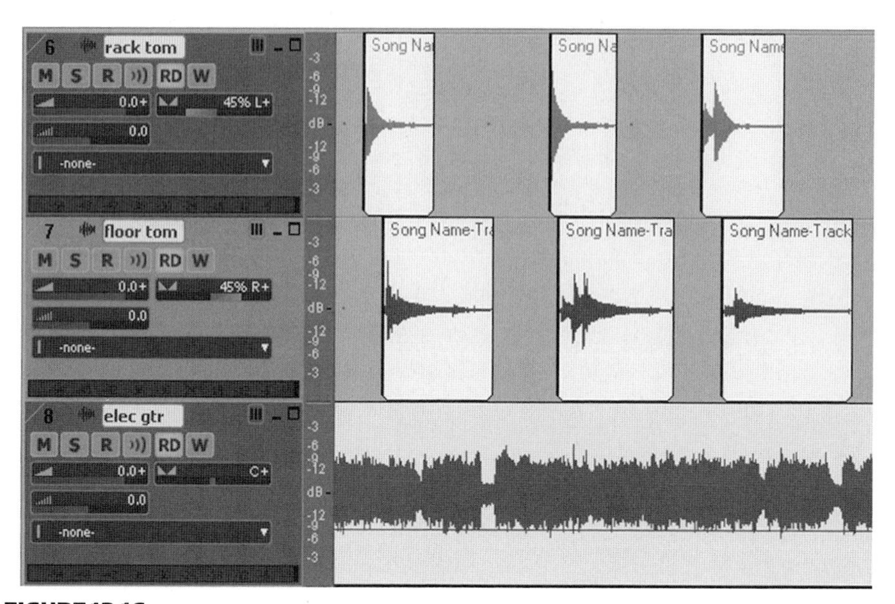

FIGURE 13.16
Edited tom-tom tracks with only the hits remaining.

only the hits remaining. You could use a noise gate instead of doing this editing.

DUPLICATE MUSICAL PARTS

Suppose you have a perfect take of the background vocals in a chorus. Rather than recording more parts for the other choruses, you could copy the vocals from one chorus and paste them at the measures where the chorus is repeated. Take care to align the vocal clips in time with the other tracks.

REPLACE WRONG NOTES

Does the bass track have a few fluffed notes? Simply delete a wrong note, copy an in-tune note, and paste it where the wrong note was. Zoom-in to adjust the timing with precision.

MAKE SPECIAL EFFECTS

Editing can create unusual effects. For instance, copy a syllable in a vocal track and paste it several times in a row to create stuttering. Suppose you want to double a guitar that is in the left channel to make it stereo. Copy the guitar track, paste it to another track, slide

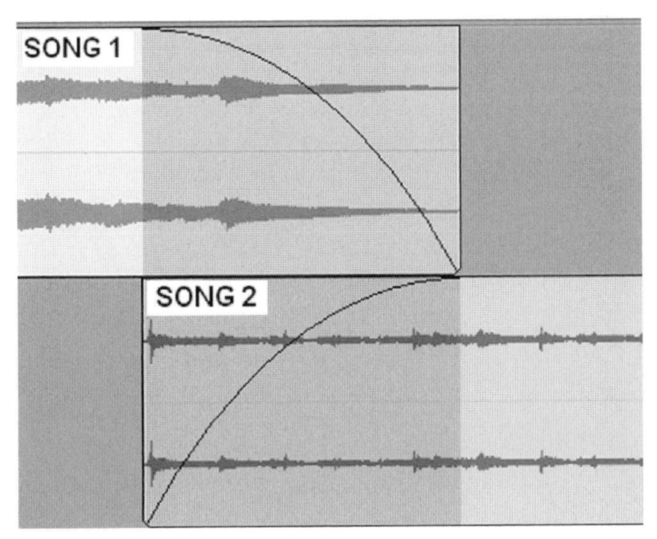

FIGURE 13.17
A crossfade between two songs.

the pasted guitar track to the right about 20 to 30 msec (which delays the guitar signal), and pan the delayed guitar track to the right. Adjust the track faders so that you hear the guitar spread evenly between your monitor speakers.

FADE-OUT, FADE-IN

A fade-out at the end of a song is a slow drop in volume to silence. DAW programs let you insert a fade-out at the end of a finished mix (or on individual tracks). If you want to fade-in a synth-pad intro, you can do that too. Fade-out and fade-in curves come in various shapes, so use whatever sounds best.

CROSSFADE

This means, "fade-out of one song while fading into another." Or crossfade across an edit point to make it sound smoother. Figure 13.17 shows two songs overlapping. The first song fades out as the other song fades in, and the two songs blend where they overlap.

FIX TIMING ERRORS

With most DAW programs you can slip each track forward or backward in time (left or right on the screen) independent of other

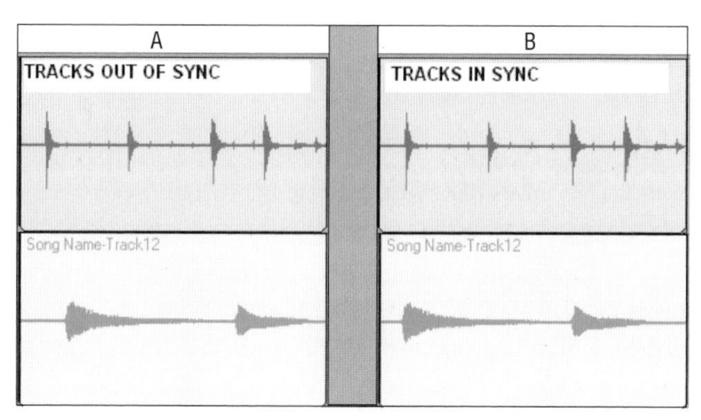

FIGURE 13.18
(A) Two musical parts out of sync. (B) Two parts in sync.

tracks. You also can move single notes in time. If a note in a bass track comes in too late, highlight that note as a separate clip, and slide it or nudge it to the left until it is on the beat. By visually aligning the start times (attacks) of notes in various tracks, you can make a recorded band sound very tight.

Figure 13.18 shows two tracks that I wanted to synchronize. I set the DAW to display only those two tracks. In Figure 13.18A, the notes in the bottom track play later than the notes in the top track. In Figure 13.18B, the notes in the bottom track have been aligned with the notes in the top track—they play in sync.

REMOVE OR REARRANGE SONG SECTIONS

You do this type of edit on a finished stereo mix, not necessarily on individual tracks.

Suppose you want to shorten a song by removing a section. Highlight the section you want to delete, cut it (Ctrl-X) and close up the space. Zoom into the edit point and slide the beginning or end of the clips there to make a smooth transition. Try to create a zero-crossing at the edit point, so that the waveform crosses the 0 dB line there (as in Figure 16.8). This method also works for editing mistakes out of spoken-word recordings. Just be careful not to leave double breaths during pauses.

Now suppose you want to remove a section of a song and put it somewhere else in the song. If you think the bridge should come earlier, you can remove it from its current location and put it where you want it.

1. Highlight the bridge waveform.
2. Cut it (Ctrl-X) and close up the space.
3. Split the song where you want the bridge to go.
4. Slide the right-side part of the song to the right to create a space.
5. Paste the bridge waveform in that space.
6. Slide the waveform segments to create a seamless edit.

CHANGE THE SONG ORDER WHILE MASTERING

When mastering a program for CD or MP3 format, you can select entire songs and move them around to change the order they will play.

Those are just a few ways that editing can improve your mixes and song arrangements.

Maintaining Audio Quality

Do your computer mixes sound harsh, grainy, or "digital"? Here are some tips to make your in-the-box mixes sound as smooth as analog.

Suppose you're recording and mixing with a DAW set to 24-bit resolution. You think you're doing everything right, but the final product still sounds vaguely distorted. You record tracks near 0 dBFS but no higher, and your meters don't show any clipping. Still, your mixes lack that smooth, effortless sound you used to get when recording analog. The sound is harsh or fatiguing.

What's going on? Chances are the signal is clipping but the meters don't indicate it. We'll explain this phenomenon, and will suggest ways to solve that problem. We'll also give other tips on helping your audio to sound less processed and more musical.

BACKGROUND

First we need to go over some digital audio basics. The A-to-D converter in your audio interface measures or samples the incoming analog signal many thousand times a second and converts those samples to binary numbers, which you record on your hard drive.

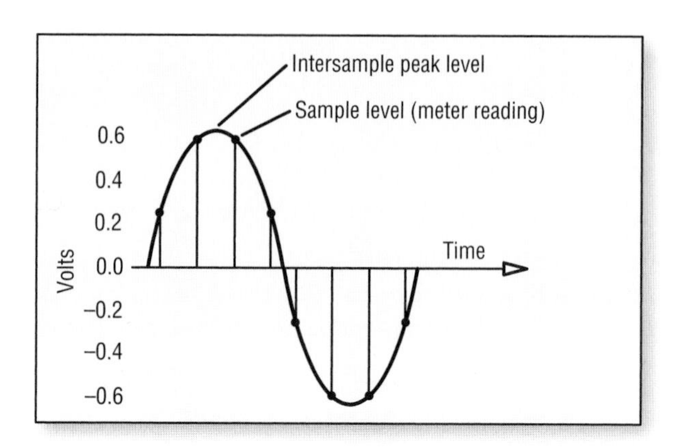

FIGURE 13.19
Analog signal levels can exceed digital sample levels.

Figure 13.19 shows an analog sine wave that is sampled periodically. The digital meters in your DAW show those sample levels, but not necessarily the peak signal levels which appear between samples. In your D-to-A converter, a lowpass filter (reconstruction filter) creates occasional peaks between the samples that can be up to 3 dB higher than the measured level at certain high frequencies. If your recorded levels are near 0 dBFS, this can cause clipping (overs or overloads) that the meters don't display. An "over" is a segment of audio in which three or more consecutive samples are at 0 dBFS (all bits on).

Similarly, some plug-ins can create larger peaks than those that went in—without any increase in volume or audible change in the program. Even an EQ cut or rolloff (subtractive EQ) can increase the signal level by producing intersample peaks due to ringing at the cut frequency.

With these facts in mind, let's explore some ways to prevent harsh digital sound.

USE METERS CORRECTLY

First, let's review the concept of average levels and peak levels. As shown in Figure 13.20, a musical signal changes in level (voltage) continuously as it plays. Imagine a musical passage with a low-level synth pad, but with high-level drum hits. The average level or volume of the passage is low, but the transient peak levels are high.

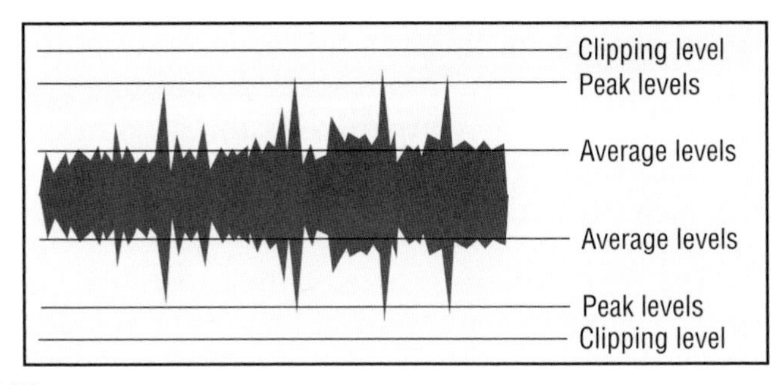

FIGURE 13.20
Average levels and peak levels in a musical signal.

Peak levels may be up to 24 dB above average levels depending on the type of signal. Percussive sounds have much higher peaks than continuous sounds do (synth pads, organ, flute)—even if the two signals have similar average levels.

The meters in your DAW can show signal levels in two modes: rms and peak. The rms (or root-mean-square) readings correspond to the average levels, while peak readings show the level of peaks or short transients. The average or rms level indicates approximately how loud the sound is, while the peak level shows how close the signal is to clipping.

Since you don't want to clip or distort the signal while recording or mixing, use peak metering, not rms.

REDUCE RECORDING LEVELS

To prevent intersample clipping, reduce your recording levels by turning down the gain in your mic preamps. Record at about −6 dBFS maximum in peak meter mode. Make sure that the recording meters are set to pre-fader.

Some DAWs, such as Cakewalk Sonar Producer, color the top 6 dB of the meters red as an aid to setting levels correctly.

One benefit of lower recording levels is that you won't overdrive your mic preamp. The distortion of most analog gear increases as you approach maximum gain. The mic preamp in your audio interface is less likely to distort its analog signal at 6 dB below maximum than at 3 to 0 dB below maximum.

Another benefit of reduced levels is that it creates some headroom for your plug-ins; 6 dB of headroom should eliminate any invisible overs. Then you can set up your mix balances without having to adjust levels every time you insert a plug-in. With lower track levels, you won't hear distortion from plug-ins and your D/A converter while you're mixing.

Signals can clip during any gain boost, such as by a fader, compressor, or EQ. So keep recorded levels low (−6 dBFS max) so you don't have to change plug-in levels during a mix. Each track's waveform may appear small, but most DAWs let you zoom in vertically.

Remember, the recorded level on each track drives the plug-ins—not the track fader, which comes after the plugs. If a track was recorded at 0 dBFS maximum, use your DAW's trim control (or insert a −6 dB trim plug-in) to reduce the signal level going into your plugs. Do this before setting up a compressor or gate, because trim affects their gain reduction.

Doesn't a low recording level increase noise? Yes, but in practice it's not a problem. A 24-bit recording has a theoretical S/N ratio of 144 dB. So even if you record at −6 dB max, you're still far above the noise floor and you won't hear any increase in hiss.

Low noise is important because the more tracks in use, the higher the noise. For instance adding 16 tracks together at equal fader settings will increase noise by about 12 dB. Still, the S/N is more than you need to have a clean recording. So you can use that extra S/N to create headroom by recording and mixing at lower levels. If your peak recording level is −6 dBFS, you create 6 dB of headroom. If your peaks are −10 dBFS, you create 10 dB of headroom.

An alternative is to get an oversampling meter plug-in which shows the actual intersample signal levels. By modeling the D/A conversion process, this type of meter reconstructs intersample peaks that regular meters miss. Examples include: Elemental Audio Inspector XL and TL MasterMeter for ProTools, Sonoris Meter for SAW Studio, Sony Oxford Limiter, TCE 6000 MD3 algorithm, TCE MD3 PowerCore plug-in, and Voxengo Elephant.

If you just keep your levels down, you don't have to worry about intersample peaks.

REDUCE LEVELS AT OTHER POINTS

Listed below are other spots in the signal path to set your level to −6 dB maximum:

- On any stem (group or submix).
- At the output of each plug-in. Check for distortion (clipping) in each plug-in, especially in an equalizer set to boost. Find the plug's output gain control and turn it down until clipping stops. You might have several plugs in series, and you don't want the output of one plug to overload the input of another.
- At the output of any upsampling plug-ins. These DSP processes upsample the signal (increase its sample rate), do the effect, then downsample the signal at the output. The downsampler acts like a partial reconstruction filter. So even with the effect turned off, the up and down sampling can create overs.

Level setting is important during mixdown, too. Set the maximum peak level of the master output bus meters to −3 to −6 dBFS to allow for signal peaks that don't display on sample-reading meters. You can make up the gain later during mastering, where you can boost the overall level or peak-limit/normalize to make a hot CD or MP3 file.

Try to keep the master faders at or near 0 dB. If the master faders are set low, you will turn up the channel faders to get a good mix level. That results in high-level signals that can overload the mix bus. Still, if a master-fader meter clips, it's okay to lower the master fader a few dB until the clipping stops.

Note that mix-bus overload isn't a serious problem if the DAW uses 48- or 64-bit floating-point math in the mix bus, because float processing can pass signals above 0 dBFS without clipping. Still, it's good practice to control the levels hitting the mix bus. Summing calculations become less accurate when signals exceed full scale.

Consider not using makeup gain in compressors. Makeup gain can cause the initial uncompressed transient to clip by boosting its level. Raise the track fader instead.

If your mix is going to a mastering engineer, omit any stereo bus processing. Record the file at −6 to −3 dBFS max so that the mix doesn't overload the mastering engineer's D/A converter. If you're mastering your own mixes (say, for customer approval), use an oversampling meter that displays the actual reconstructed signal, or set the limiter ceiling at −3 dB to −0.3 dB, not 0 dB or higher. This avoids making files that sound distorted on your customer's playback systems. Many CD or MP3 players distort with samples above the clip level.

So far we looked at level reduction as a path to smoother sound. Now let's consider other methods to make digital audio easier to listen to.

WORK ON THE SOUND SOURCE

Sometimes a musical instrument or vocal sounds edgy, even when recorded at low levels. The cause might be the source itself, the microphone frequency response or mic placement. Here are some ways to make the source sound better:

- With EQ, consider using less high-frequency boost. Consider cutting just a few dB around 3 to 7 kHz, which tends to reduce harshness. Try a 4 kHz lowpass (high-cut) filter on distorted guitar amps to take the edge off.
- Try multiband compression on voices that get raspy at high volume. Set the compressor so it kicks in above 3 kHz or so during loud passages.
- Many condenser mics have high-frequency peaks in their response which can sound brittle or sibilant. Prevent that effect by applying a high-frequency rolloff or shelf in your EQ. Better yet, use ribbon mics or flat-response condenser mics. Many ribbon microphones have reduced output at high frequencies which can give a rounder tone.
- Move the mic to a spot that sounds more mellow. For example, mike a vocal partly off to the side if it sounds too edgy when miked in front. Mike a trumpet off-axis. Mike a guitar amp near the edge of the speaker cone.
- Try using a tube mic preamp to warm up the sound. Not all tube preamplifiers do that, but some have euphonic even-order distortion that some people call added "warmth."

OTHER TIPS

- Dirty contacts can cause signal rectification and distortion. Clean connectors with agents such as Caig Labs DeoxIT (www.caig.com). Consider using high-quality cables.

- Use better converters.
- Use audio interfaces designed for low jitter.
- During mastering, add dither when truncating from 24- to 16-bit. This eliminates quantization distortion at low signal levels.
- Minimize the number of calculations that your DAW has to do. Each calculation creates some error or distortion by increasing the word length beyond 24 bits. Rather than boosting the gain of a musical section and dropping it later, undo the changes and apply only the correct amount of gain change.
- Consider using volume envelopes (fader moves) rather than compression. Compression adds distortion because it changes the waveform. Too-fast release times cause distortion in low-frequency notes and also cause pumping. A fader-setting change on a series of notes is less audible than a compressor working on each note.
- Limiting adds distortion. When you peak-limit and normalize a stereo mix to make a hot CD or MP3 file, try not to exceed 6 to 7 dB of peak reduction. Use even less limiting on mixes that don't have loud transients. Rather than normalizing the limited signal to 0 dBFS, normalize to −3 dB to −1 dB to allow for intersample peaks.
- Experiment with an analog tape plug-in to warm up a track (e.g., Crane Song Phoenix). Or record to analog multitrack tape and dump the tracks to your DAW for editing/mixing.
- To reduce graininess in a reverb plug-in, use a high-density setting or use a convolution reverb.
- For best quality, record at 24-bits and stay there throughout the project. Then turn on dither and export to a 16-bit file that you will master for a CD or MP3 file. If you're sending 24-bit data files to a mastering engineer, don't turn on dither—let the engineer do it.
- Normalization is a DAW process that raises the gain of the entire program so that the highest peak reaches 0 dBFS (in peak meter mode, not rms). Don't normalize the program until final mastering.
- Use cables designed for digital audio, and use short cables. Use internal sync on your A/D converter to reduce jitter.
- Avoid A/D and D/A conversions. Once your signal is digital, try to keep it that way. After converting the analog mic signal to digital in your audio interface, do all your processing in the computer if possible, then burn a CD or MP3 file of the final product. The only D/A conversion will be when the CD or MP3 file plays back.

Give all these suggestions a try—your music will sound better.

As we've seen, there is a whole world of computer music applications. Download the free trial versions to see what you like. Set up your DAW system, read the software manual, and have fun creating music.

AUDIO FOR VIDEO

You can use your DAW to create a sound track for a video or film production.

First we need to define the Society of Motion Picture and Television Engineers (SMPTE) timecode, which is a standardized timecode signal for use in video productions. SMPTE timecode is something like a digital tape counter, where the counter time is recorded as a signal on tape or hard disk.

Pictures on a video screen are updated approximately 30 frames per second, where a frame is a still picture. SMPTE timecode assigns a unique number (address) to each video frame—8 digits that specify hours:minutes:seconds:frames.

Suppose you want create an audio sound track for a video program. First you copy the video onto your computer's hard drive. When you open the video file in your audio editing program, it shows up in a window on your computer screen (Figure 13.21, top). While watching the video, you create the sound track by editing audio and music clips.

Here is the process:

- First, you're handed a DVD or sent a file which is a video recording of the program you're working on, plus location audio. Copy the file(s) to your hard drive.
- Launch your audio software, import the video file, and watch the picture. Using soft synths, compose and record musical parts related to the video scenes. Slide the audio clips so they sync with the video.
- Missing or poorly recorded dialog is replaced during a process called looping or automatic dialog replacement (ADR). Have actors watch the video and lip-sync their lines using the DAW.

FIGURE 13.21
A video window and sound clips in a DAW screen.

- Audition several sound effects from CD libraries, pick the ones you like, and import them into tracks in the DAW.
- Using slow motion or freeze frame, go through the video and note where each sound effect should occur. Nudge audio clips in time to align with video events.

Below is a more detailed procedure. I'm using these terms in the steps below:

AUDIO EDITOR: Performs the audio edits and maybe composes the music.

AUDIO MIXER: Mixes the end product for a feature film. There might be several mixers: one for dialog, one for sound effects and one for music.

1. During a video shoot, the audio from the location audio mixer is recorded directly to camera.
2. The audio editor receives a video file with the location sound track embedded (but extractable). In addition the audio editor

gets either a WAVE, OMF, or an AAF file of the audio to mix. The audio track that is embedded in the video file is usually a scratch mix to make sure the audio from the WAVE, OMF, or AAF file is in sync with the video. You line them up by ear (and eye).

3. In Hollywood film productions, the audio editor delivers to the audio mixer either an OMF or AAF file that contains all the audio edits in sync with the video. The audio in this file is in multitrack format so the audio mixer has maximum flexibility in mixing.
4. The audio mixer lines up the video and audio file at SMPTE time 01:00:00:00, the traditional project start time, and starts mixing. The video and audio in the computer are synchronized by the audio interface at the resolution of the sample rate of the mixer's system (typically every 48,000th of a second).
5. After the audio mixer finishes the mix using the audio elements from the audio editor, he or she sends the final mixed file (WAV or AIFF) to the video editor. The video editor marries the mixed track to the video on a video editing system.

In smaller productions, the audio editor/mixer follows this procedure:

1. Receive a WAVE, OMF, or AAF file from the video editor.
2. Import location audio, narration, and sound effects. Compose and record music. Sync these elements to picture and do a mix.
3. Export the mix (usually as a Broadcast WAV or AIFF file). Put a timecode stamp (01:00:00:00) at the beginning of the file.
4. Send the mix to the video editor.

Thanks to Matthew Gellert of bonesteelfilms.com and David Schmidt of acapellaaudio.com for their expert help with this section.

CHAPTER 14

Judging Sound Quality

Seat an engineer behind a mixing console and ask him to do a mix. It sounds great. Then seat another engineer behind the same console and again ask for a mix. It sounds terrible. What happened?

The difference lies mainly in their ears—their critical listening ability. Some engineers have a clear idea of what they want to hear and how to get it. Others haven't acquired the essential ability to recognize good sound. By knowing what to listen for, you can improve your artistic judgments during recording and mixdown. You're able to hear errors in microphone placement, equalization, and so on, and correct them.

To train your hearing, try to analyze recorded sound into its components—such as frequency response, noise, reverberation—and concentrate on each one in turn. It's easier to hear sonic flaws if you focus on a single aspect of sound reproduction at a time. This chapter is a guide to help you do this.

CLASSICAL VERSUS POPULAR RECORDING

Classical and popular music have different standards of "good sound." One goal in recording classical music (and often folk music or jazz) is to accurately reproduce the live performance. This is a worthy aim because the sound of an orchestra in a good hall can be quite beautiful. The music was composed and the instruments were designed to sound best when heard live in a concert hall. The recording engineer, out of respect for the music, should always try to translate that sound to disc with as little technical intrusion as possible.

By contrast, the accurate translation of sound to disc isn't always the goal in recording popular music. Although the aim may be to reproduce the original sound, the producer or engineer may also want to play with that sound to create a new sonic experience, or to do some of both.

In fact, the artistic manipulation of sounds through studio techniques has become an end in itself. Creating an interesting new sound is as valid a goal as re-creating the original sound. There are two games to play, each with its own measures of success.

If the aim of a recording is realism or accurate reproduction, the recording is successful when it matches the live performance heard in the best seat in the concert hall. The sound of musical instruments is the standard by which such recordings are judged.

When the goal is to enhance the sound or produce special effects (as in most pop-music recordings), the desired sonic effect is less defined. The live sound of a pop group could be a reference, but pop-music recordings generally sound better than live performances—recorded vocals are clearer and less harsh, the bass is cleaner and tighter, and so on. The sound of pop music reproduced over speakers has developed its own standards of quality apart from accurate reproduction.

GOOD SOUND IN A POP-MUSIC RECORDING

Currently, a good-sounding pop recording might be described as follows (there are always exceptions):

- Well-mixed
- Wide-range
- Tonally balanced
- Clean
- Clear
- Smooth
- Spacious

It also has:

- Presence
- Sharp transients
- Tight bass and drums
- Wide and detailed stereo imaging

- Wide but controlled dynamic range
- Interesting sounds
- Suitable production

The next sections explore each one of these qualities in detail so that you know what to listen for. Assume that the monitor system is accurate, so that any colorations heard are in the recording and not in the monitors.

A Good Mix

In a good mix, the loudness of instruments and vocals is in a pleasing balance. Everything can be clearly heard, yet nothing is obtrusive. The most important instruments or voices are loudest; less important parts are in the background.

A successful mix goes unnoticed. When all the tracks are balanced correctly, nothing sticks out and nothing is hidden. Of course, there's a wide latitude for musical interpretation and personal taste in making a mix. Dance mixes, for example, can be very severe sonically.

Sometimes you don't want everything to be clearly heard. On rare occasions you may want to mix in certain tracks very subtly for a subconscious effect.

The mix must be appropriate for the style of music. For example, a mix that's right for loud rock music usually won't work for a pop ballad. A rock mix typically has the drums way up front and the vocals only slightly louder than the accompaniment. In contrast, a pop ballad has the vocals loudest, with the drums used just as "seasoning" in the background.

Level changes during the mix should be subtle, or should make sense. Otherwise, instruments jump out for a solo and fall back in afterwards. Move faders slowly, or set them to preset positions during pauses in the music. Nothing sounds more amateurish than a solo that starts too quietly and then comes up as it plays—you can hear the engineer working the fader.

Wide Range

Wide range means extended low- and high-frequency response. Cymbals should sound crisp and distinct, but not sizzly or harsh;

kick drum and bass should sound deep, but not overwhelming or muddy. Wide-range sound results from using high-quality microphones and adequate EQ.

You might want to combine "hi-fi" and "low-fi" sounds in a single mix. The low-fi sounds generally cover a narrow frequency range and might be distorted.

Good Tonal Balance

The overall tonal balance of a recording should be neither bassy nor trebly; that is, the perceived spectrum shouldn't emphasize low or high frequencies. Low bass, mid-bass, midrange, upper midrange, and highs should be heard in equal proportions (Figure 14.1). Emphasis of any one frequency band over the other eventually causes listening fatigue. Dance club mixes, however, are heavy on the bass end to get the crowd moving.

Recorded tonal balance is inversely related to the frequency response of the studio's monitor system. If the monitors have a high-frequency rolloff, the engineer will compensate by boosting highs in the recording to make the monitors sound correct. The result is a bright recording.

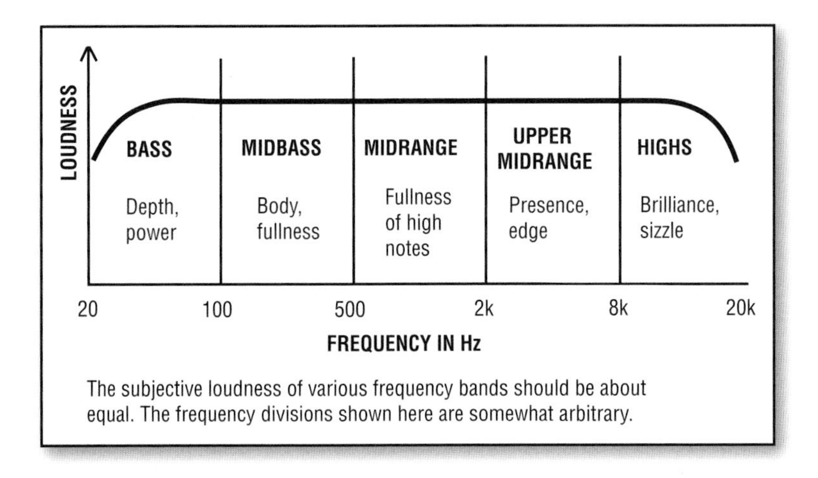

FIGURE 14.1
Loudness versus frequency of a pop recording with good sound.

Before doing a mix, play over the monitors some commercial recordings whose sound you admire, to become accustomed to a commercial spectral balance. After your mix is recorded, play it back and alternately switch between your mix and a commercial recording. This comparison indicates how well you matched a commercial spectral balance. An effective tool for this purpose is Harmonic Balancer (www.har-bal.com). Of course, you may not care to duplicate what others are doing.

In pop-music recordings, the tonal balance or timbre of instruments doesn't necessarily have to be natural. Still, many listeners want to hear a realistic timbre from acoustic instruments, such as the guitar, flute, sax, or piano. The reproduced timbre depends on microphone frequency response, microphone placement, the musical instruments themselves, and equalization.

Clean Sound

Clean means free of noise and distortion. Hiss, hum, and distortion are inaudible in a good recording. Distortion in this case means distortion added by the recording process, not distortion already present in the sound of electric-guitar amps or Leslie speakers. There are exceptions to this guideline; some popular recordings have noise or distortion added intentionally.

Clean also means "not muddy" or free of low-frequency ringing and leakage. A clean mix is one that is uncluttered or free of excess instrumentation. This is achieved by arranging the music so that similar parts don't overlap, and not too many instruments play at once in the same frequency range. Usually, the fewer the instruments, the cleaner the sound. Too many overdubs can muddy the mix.

Clarity

In a clear-sounding recording, instruments don't crowd or mask each other. They are separate and distinct. As with a clean sound, clarity arises when instrumentation is sparse, or when instruments occupy different areas of the frequency spectrum. For example, the bass provides low frequencies, keyboards might emphasize mid-bass, lead guitar provides upper midrange, and cymbals fill in the highs.

In addition, a clear recording has adequate reproduction of each instrument's harmonics; that is, the high-frequency response is not rolled off.

Smoothness

Smooth means easy on the ears, not harsh, uncolored. Sibilant sounds are clear but not piercing. A smooth, effortless sound allows relaxation; a strained or irritating sound causes muscle tension in the ears or body. Smoothness is a lack of sharp peaks or dips in the frequency response, as well as a lack of excessive boost in the midrange or upper midrange. It is also low distortion, such as provided by a 24-bit recording.

Presence

Presence is the apparent sense of closeness of the instruments—a feeling that they're present in the listening room. Synonyms are clarity, detail, and punch.

Presence is achieved by close miking, overdubbing, and using microphones with a presence peak or emphasis around 5 kHz. Using less reverb and effects can help. Upper midrange boost helps, too. Most instruments have a frequency range that, if boosted, makes the instrument stand out more clearly or become better defined. Presence sometimes conflicts with smoothness because presence often involves an upper midrange boost, while a smooth sound is free of such emphasis. You have to find a tasteful compromise between the two.

Spaciousness

When the sound is spacious or airy, there is a sense of air around the instruments. Without air or ambience, instruments sound as if they are isolated in stuffed closets. (Sometimes, though, this is the desired effect.) You achieve spaciousness by adding reverb, recording instruments in stereo, using room mics, or miking farther away.

Sharp Transients

The attack of cymbals and drums generally should be sharp and clear. A bass guitar and piano may or may not require sharp attacks, depending on the song.

Tight Bass and Drums

The kick drum and bass guitar should "lock" together so that they sound like a single instrument—a bass with a percussive attack. The drummer and bassist should work out their parts together so they hit accents simultaneously, if this is desired.

To further tighten the sound, damp the kick drum and record the bass direct. Rap music, however, has its own sound—the kick drum usually is undamped and boomy, sometimes with short reverb added. Equalize the kick and bass in complementary ways so that they don't mask each other; for example:

Kick: Boost 60 to 80 Hz, cut 600 Hz, boost 3 kHz.
Bass: Cut 60 to 80 Hz, boost 600 Hz, cut above 3 kHz.

Wide and Detailed Stereo Imaging

Stereo means more than just left and right. Usually, tracks should be panned to many points across the stereo stage between the monitor speakers. Some instruments should be hard-left or hard-right, some should be in the center, others should be half-left or half-right (panned to 10 o'clock and 2 o'clock). Try to achieve a stereo stage that is well balanced between left and right (Figure 14.2). Instruments that occupy the same frequency range can be made more distinct by panning them to opposite sides of center.

You may want some tracks to be unlocalized. Backup choruses and strings should be spread out rather than appearing as point sources. Stereo keyboard sounds can wander between speakers. A lead-guitar solo can have a fat, spacious sound.

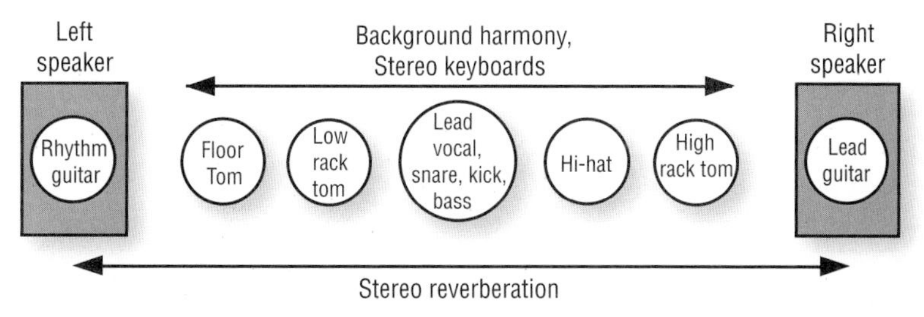

FIGURE 14.2
An example of image placement between speakers.

There should also be some front-to-back depth. Some instruments should sound close or up front; others should sound farther away. Use different miking distances or different amounts of reverb on various tracks.

If you want the stereo imaging to be realistic (for a jazz combo, for example), the reproduced ensemble should simulate the spatial layout of the live ensemble. If you're sitting in an audience listening to a jazz quartet, you might hear drums on the left, piano on the right, bass in the middle, and sax slightly right. The drums and piano aren't point sources, but are somewhat spread out. If spatial realism is the goal, you should hear the same ensemble layout between your speakers. On some commercial CDs, the piano and drums are spread all the way between speakers—an interesting effect, but unrealistic.

Pan-potted mono tracks often sound artificial in that each instrument sounds isolated in its own little space. It helps to add some stereo reverberation around the instruments to "glue" them together.

Often, TV mixes are heard in mono. Hard-panned signals sound weak in mono relative to center-panned signals. So pan sound sources to 3 and 9 o'clock, not hard-right and hard-left.

Wide but Controlled Dynamic Range

Dynamic range is the range of volume levels from softest to loudest. A recording with a wide dynamic range becomes noticeably louder and softer, adding excitement to the music. To achieve this, don't add too much compression (automatic volume control). An overly compressed recording sounds squashed—crescendos and quiet interludes lose their impact, and the sound becomes fatiguing.

Vocals often need some compression or gain-riding because they have more dynamic range than the instrumental backup. A vocalist may sing too loudly and blast the listener, or sing too softly and become buried in the mix. A compressor can even out these extreme level variations, keeping the vocals at a constant loudness. Bass guitar also can benefit from compression.

Interesting Sounds

The recorded sound may be too flat or neutral, lacking character or color. In contrast, a recording with creative production has unique musical instrument sounds, and typically uses effects. Some of these are equalization, echo, reverberation, doubling, chorus, flanging, compression, distortion, and stereo effects.

Making sounds interesting or colorful can conflict with accuracy or fidelity. That's okay, but you should know the trade-off.

Suitable Production

The way a recording sounds should imply the same message as the musical style or lyrics. In other words, the sound should be appropriate for the particular tune being recorded.

For example, some rock music is rough and raw. The sound should be, too. A clean, polished production doesn't always work for high-energy rock 'n' roll. There might even be a lot of leakage or ambience to suggest a garage studio or nightclub environment. The role of the drums is important, so they should be loud in the mix. The toms should ring, if that's desired.

New Age, disco, rhythm and blues, contemporary Christian, or pop music is slickly produced. The sound is usually tight, smooth, and spacious. Folk music and acoustic jazz typically sound natural and have little or no reverb. Music in the digital hardcore genre has lots of distortion.

Actually, each style of music is not locked into a particular style of production. You tailor the sound to complement the music of each individual tune. Doing this may break some of the guidelines of good sound, but that's usually okay as long as the song is enhanced by its sonic presentation.

GOOD SOUND IN A CLASSICAL-MUSIC RECORDING

As with pop music, classical music should sound clean, wide-range, and tonally balanced. But because classical recordings are meant to sound realistic—like a live performance—they also require good acoustics, a natural balance, tonal accuracy, suitable perspective, and accurate stereo imaging (see Chapter 18).

Good Acoustics

The acoustics of the concert hall or recital hall should be appropriate for the style of music to be performed. Specifically, the reverberation time should be neither too short (dry) nor too long (cavernous). Too short a reverberation time results in a recording without spaciousness or grandeur. Too long a reverberation time blurs notes together, giving a muddy, washed-out effect. Ideal reverberation times are around 1.2 seconds for chamber music or soloists, 1.5 seconds for symphonic works, and 2 seconds for organ recitals. To get a rough idea of the reverb time of a room, clap your hands once, loudly, and count the seconds it takes for the reverb to fade to silence.

A Natural Balance

When a recording is well-balanced, the relative loudness of instruments is similar to that heard in an ideal seat in the audience area. For example, violins aren't too loud or soft compared to the rest of the orchestra; harmonizing or contrapuntal melody lines are in proportion.

Generally, the conductor, composer, and musicians balance the music acoustically, and you capture that balance with your stereo mic pair. But sometimes you need to mike certain instruments or sections to enhance definition or balance. Then you mix all the mics. In either case, consult the conductor for proper balances.

Tonal Accuracy

The reproduced timbre or tone quality should match that of live instruments. Fundamentals and harmonics should be reproduced in their original proportion.

Suitable Perspective

Perspective is the sense of distance of the performers from the listener—how far away the stage sounds. Do the performers sound like they're eight rows in front of you, in your lap, or in another room?

The style of music suggests a suitable perspective. Incisive, rhythmically motivated works (such as Stravinsky's *Rite of Spring*) sound best with closer miking; lush, romantic pieces (a Bruckner symphony) are best served by more distant miking. The chosen perspective depends on the taste of the producer.

Closely related to perspective is the amount of recorded ambience or reverberation. A good miking distance yields a pleasing balance of direct sound from the orchestra and ambience from the concert hall.

Accurate Imaging

Reproduced instruments should appear in the same relative locations as they were in the live performance. Instruments in the center of the ensemble should be heard in the center between the speakers; instruments at the left or right side of the ensemble should be heard from the left or right speaker. Instruments halfway to one side should be heard halfway off center, and so on. A large ensemble should spread from speaker to speaker, while a quartet or soloist can have a narrower spread.

It's important to sit equidistant from the speakers when judging stereo imaging, otherwise the images shift toward the side on which you're sitting. Sit as far from the speakers as they are spaced apart. Then the speakers appear to be 60 degrees apart, which is about the same angle an orchestra fills when viewed from the typical ideal seat in the audience (tenth row center, for example).

The reproduced size of an instrument or instrumental section should match its size in real life. A guitar should be a point source; a piano or string section should have some stereo spread. Each instrument's location should be as clearly defined as it was heard from the ideal seat in the concert hall.

Reproduced reverberation (concert-hall ambience) should surround the listener, or at least it should spread evenly between the speakers. Surround-sound technology is needed to make the recorded ambience surround the listener, although spaced-microphone recordings have some of this effect. Accurate imaging is illustrated in Figure 14.3.

There should be a sense of stage depth: front-row instruments sound closer than back-row instruments.

TRAINING YOUR HEARING

The critical process is easier if you focus on one aspect of sound reproduction at a time. You might concentrate first on the tonal

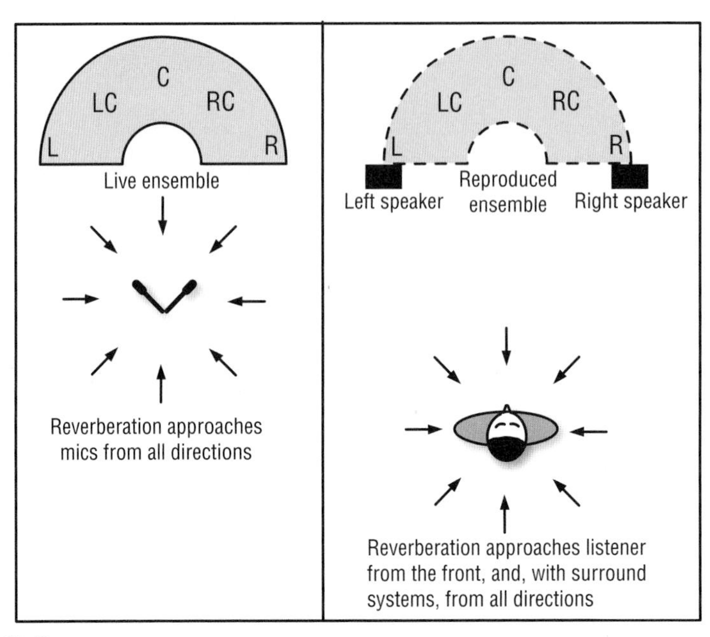

FIGURE 14.3
With accurate imaging, the sound-source location and size, and the reverberant field, are reproduced during playback.

balance—try to pinpoint what frequency ranges are being emphasized or slighted. Next listen to the mix, the clarity, and so on. Soon you have a lengthy description of the sound quality of your recording.

Developing an analytical ear is a continuing learning process. Train your hearing by listening carefully to recordings—both good and bad. Make a checklist of all the qualities mentioned in this chapter. Compare your own recordings to live instruments and to commercial recordings. Check out the Golden-Ear ear training Web site at www.moultonlabs.com/gold.htm and click on "Product."

A pop-music record that excels in all the attributes of good sound is The Sheffield Track Record (Sheffield Labs, Lab 20), engineered and produced by Bill Schnee. In effect, it's a course in state-of-the-art sound—required listening for any recording engineer or producer.

Another record with brilliant production is *The Nightfly* by Donald Fagen (Warner Brothers 23696-2), engineered by Roger Nichols, Daniel Lazerus, and Elliot Scheiner, produced by Gary Katz, and mastered by Bob Ludwig. This recording, and Steely Dan recordings by Roger Nichols, sound razor sharp, very tight and clear, elegant, and tasteful; and the music just pops out of the speakers.

The following listings are more examples of outstanding rock production, and set high standards:

Songs:

"I Need Somebody," by Bryan Adams; producer, Bob Clearmountain
"The Power of Love," by Huey Lewis & The News; producer, Huey Lewis & The News

Albums:

90125, by Yes; producer, Trevor Horn
Synchronicity, by The Police; producer, Hugh Padgham and The Police
Dark Side of the Moon, by Pink Floyd; producer, Alan Parsons
Thriller, by Michael Jackson; engineer, Bruce Swedien; producer, Quincy Jones
Avalon, by Roxy Music; engineer, Bob Clearmountain; producer, Roxy Music
Nevermind, by Nirvana; producer, Butch Vig
Come Away With Me, by Norah Jones; engineer, Jay Newland
Genius Loves Company, by Ray Charles; several engineers
Give, by The Bad Plus; engineer, Tchad Blake
Live in Paris and *The Look of Love*, by Diana Krall; engineer, Al Schmitt
Smile, by Brian Wilson; engineer, Mark Linett.

Some other recordings of note are those by Supertramp, Pat Metheny Group, Alison Krauss, The Wailin' Jennys, Heart, and Radiohead; and by producers George Martin and Andy Wallace.

Then there are the incredibly clean recordings of Tom Jung (with DMP records) and George Massenberg. Some classical-music recordings with outstanding sound are on the Telarc, Delos, and Chesky labels. You can learn a lot by emulating these superb recordings and many others.

Once you're making recordings that are competent technically—clean, natural, and well-mixed—the next stage is to produce imaginative sounds. You're in command; you can tailor the mix to sound any way that pleases you or the band you're recording. The supreme achievement is to produce a recording that is a sonic knockout, beautiful, or thrilling.

TROUBLESHOOTING BAD SOUND

Now you know how to recognize good sound, but can you recognize bad sound? Suppose you're monitoring a recording in progress, or listening to a recording you've already made. Something doesn't sound right. How can you pinpoint what's wrong, and how can you fix it?

The rest of this chapter includes step-by-step procedures to solve audio-related problems. Read down the list of "bad sound" descriptions until you find one matching what you hear. Then try the solutions until your problem disappears. Only the most common symptoms and cures are mentioned; console maintenance isn't covered.

This troubleshooting guide is divided into four main sections:

1. Bad sound on all recordings (including those from other studios)
2. Bad sound on playback only (the mixer output sounds all right)
3. Bad sound in a pop-music recording
4. Bad sound in a classical-music recording

Before you start, check for faulty cables and connectors. Also check all control positions; rotate knobs, and flip switches to clean the contacts, and clean connectors with DeoxIT from Caig Labs.

Bad Sound on All Recordings

If you have bad sound on all your recordings, including those from other studios, follow this checklist to find the problem:

- Upgrade your monitor system.
- Adjust tweeter and woofer controls on speakers.
- Adjust the relative gains of tweeter and woofer amplifiers in a bi-amped system.
- Relocate speakers.
- Improve room acoustics.
- Equalize the monitor system.
- Try different speakers.
- Upgrade the power amp and speaker cables.

- Monitor at a moderate listening volume, such as 85 dBSPL. We hear less bass and treble in a program if it's monitored at a low volume, and vice versa. If we hear too little bass due to monitoring at a low level, we might mix in too much bass.

Bad Sound on Playback Only

You might have bad sound on your playback only, but your mixer output sounds okay. If DAT tape playback has glitches or drop-outs, try these steps:

- Clean the recorder with a dry cleaning tape.
- Before recording, fast-forward the tape to the end and rewind it to the top.
- Use better tape.
- Format videocassettes nonstop from start to finish.

If a hard-drive recording has glitches or drop-outs on playback, try this:

- Increase the latency setting in your recording software.
- Follow the tips in Appendix B.

If a digital recording sounds distorted, these suggestions might help:

- Keep the recording level as high as possible, but try not to exceed −6dBFS in peak meter mode, not rms mode.
- Avoid clipping in effects plug-ins.
- Record at a higher sampling rate or higher bit depth.
- Avoid sampling-rate conversion.
- Apply dither when going from a high bit depth to a lower bit depth.
- Try a higher quality audio interface.

Bad Sound in a Pop-Music Recording Session

Sometimes you have bad sound in a pop-music recording session.

MUDDINESS (LEAKAGE)

If the sound is muddy from excessive leakage, try the following:

- Place the microphones closer to their sound sources.
- Spread the instruments farther apart to reduce the level of the leakage.

- Place the instruments closer together to reduce the delay of the leakage.
- Use directional microphones (such as cardioids).
- Overdub the instruments.
- Record the electric instruments direct.
- Use baffles (goboes) between instruments.
- Deaden the room acoustics (add absorptive material or flexible panels).
- Filter out frequencies above and below the spectral range of each instrument. Be careful or you'll change the sound of the instrument.
- Turn down the bass amp in the studio, or monitor the bass with headphones instead.

MUDDINESS (EXCESSIVE REVERBERATION)

If the sound is muddy due to excessive reverberation, try these steps:

- Reduce the effects-send levels or effects-return levels. Or don't use effects until you figure out what the real problem is.
- Place the microphones closer to their sound sources.
- Use directional microphones (such as cardioids).
- Deaden the room acoustics.
- Filter out frequencies below the fundamental frequency of each instrument.

MUDDINESS (LACKS HIGHS)

If your sound is muddy and lacks highs, or has a dull or muffled sound, try the following:

- Use microphones with better high-frequency response, or use condenser mics instead of dynamics.
- Change the mic placement. Put the mic in a spot where there are sufficient high frequencies. Keep the high-frequency sources (such as cymbals) on-axis to the microphones.
- Use small-diameter microphones, which generally have a flatter response off-axis.
- Boost the high-frequency equalization or cut slightly around 300 Hz.
- Change musical instruments; replace guitar strings; replace drum heads. (Ask the musicians first!)

- Use an enhancer signal processor, but watch out for noise.
- Use a direct box on the electric bass. Have the bassist play percussively or use a pick if the music requires it. When compressing the bass, use a long attack time to allow the note's attack to come through. (Some songs don't require sharp bass attacks—do whatever's right for the song.)
- Damp the kick drum with a pillow, folded towel, or blanket, and mike it next to the center of the head near the beater. Use a wooden or plastic beater if the song and the drummer allow it.
- Don't plug an electric guitar directly into a mic input. Use a direct box or a high-impedance input.
- Apply high-frequency boost after compression, not before.

MUDDINESS (LACKS CLARITY)

If your sound is muddy because it lacks clarity, try these steps:

- Consider using fewer instruments in the musical arrangement. Maybe turn down synth pads in the mix.
- Equalize instruments differently so that their spectra don't overlap.
- Try less reverberation.
- Using equalizers, boost the presence range of instruments that lack clarity. Or cut 1 to 2 dB around 300 Hz.
- In a reverb unit, add about 45 to 100 msec of pre-delay.
- Pan similar-sounding instruments to opposite sides.

DISTORTION

If you hear distortion when monitoring the mics in a pop-music recording, try the following:

- Increase input attenuation (reduce input gain), or plug in a pad between the microphone and mic input.
- Readjust gain-staging: Set faders and pots to their design centers (shaded areas).
- If you still hear distortion, switch in the pad built into the microphone (if any).
- Check connectors for stray wires and bad solder joints.
- Unplug and plug-in connectors. Clean them with Caig Labs DeoxIT or Pro Gold.

TONAL IMBALANCE

If you have bad tonal balance—the sound is boomy, dull, or shrill, for example—follow these steps:

- Change musical instruments; change guitar strings; change reeds, etc.
- Change mic placement. If the sound is too bassy with a directional microphone, you may be getting proximity effect. Mike farther away or roll off the excess bass.
- Use the 3:1 rule of mic placement to avoid phase cancellations. When you mix two or more mics to the same channel, the distance between mics should be at least three times the mic-to-source distance.
- Try another microphone. If the proximity effect of a cardioid mic is causing a bass boost, try an omnidirectional mic instead.
- If you must place a microphone near a hard, reflective surface, try a boundary microphone on the surface to prevent phase cancellations.
- If you're recording a singer/guitarist, delay the vocal mic signal by about 1 msec.
- Change the equalization. Avoid excessive boost. Maybe cut slightly around 300 Hz if the sound is muddy, or cut around 3 kHz if the sound is harsh.
- Use equalizers with a broad bandwidth, rather than a narrow, peaked response.
- During mastering, try a spectrum analyzer/equalizer such as Harmonic Balancer (www.har-bal.com).

LIFELESSNESS

If your pop-music recording has a lifeless sound and is unexciting, these steps might help you solve it:

- Work on the live sound of the instruments in the studio to come up with unique effects.
- Add effects: reverberation, echo, exciter, doubling, equalization, etc.
- Use and combine recording equipment in unusual ways.
- Try overdubbing little vocal licks or synthesized sound effects.

If your sound seems lifeless due to dry or dead acoustics, try these:

- If leakage isn't a problem, put microphones far enough from instruments to pick up wall reflections. If you don't like the sound this produces, try the next suggestion.
- Add reverb or echo to dry tracks. (Not all tracks require reverberation. Also, some songs may need very little reverberation so that they sound intimate.)
- Use omnidirectional microphones.
- Add hard, reflective surfaces in the studio, or record in a hard-walled room.
- Allow a little leakage between microphones. Put mics far enough from instruments to pick up off-mic sounds from other instruments. Don't overdo it, though, or the sound becomes muddy and track separation becomes poor.

NOISE (HISS)

Sometimes your pop-music recording has extra noise on it. If your sound has hiss, try these:

- Check for noisy guitar amps or keyboards.
- Switch out the pad built into the microphone (if any).
- Reduce mixer input attenuation (increase input gain).
- Use a more sensitive microphone.
- Use an impedance-matching adapter (a low- to high-Z step-up transformer) between microphones and phone-jack mic inputs.
- Use a quieter microphone (one with low self-noise).
- Increase the sound pressure level at the microphone by miking closer. If you're using PZMs, mount them on a large surface or in a corner.
- Apply any high-frequency boost during recording, rather than during mixdown.
- If possible, feed recorder tracks from mixer direct outs or insert sends instead of group or bus outputs.
- Use a lowpass filter (high-cut filter).
- As a last resort, use a noise gate.

NOISE (RUMBLE)

If the noise is a low-frequency rumble, follow these steps:

- Reduce air-conditioning noise or shut off the air conditioning temporarily.

- Use a highpass filter (low-cut filter) that is set around 40 to 100 Hz.
- Use microphones with limited low-frequency response.
- See the following section Noise (Thumps).

NOISE (THUMPS)

- Change the microphone position.
- Change the musical instrument.
- Use a highpass filter set around 40 to 100 Hz.
- If the cause is mechanical vibration traveling up the mic stand, put the mic in a shock-mount stand adapter, or place the mic stand on some carpet padding. Try to use a microphone that is less susceptible to mechanical vibration, such as an omnidirectional mic or a unidirectional mic with a good internal shock mount.
- Use a microphone with a limited low-frequency response.
- If the cause is piano pedal thumps, also try working on the pedal mechanism.

HUM

Hum is a subject in itself. See the tips on hum prevention near the end of Chapter 4.

POP

Pops are explosive breath sounds in a vocalist's microphone. If your pop-music recording has pops, try these solutions:

- Place the microphone above or to the side of the mouth.
- Place a foam windscreen (pop filter) on the microphone.
- Stretch a nylon stocking over a crochet hoop, and mount it on a mic stand a few inches from the microphone (or use an equivalent commercial product).
- Place the microphone farther from the vocalist.
- Use a microphone with a built-in pop filter (ball grille).
- Use an omnidirectional microphone, because it is likely to pop less than a directional (cardioid) microphone.
- Switch in a highpass filter (low-cut filter) set around 80 to 100 Hz.

SIBILANCE

Sibilance is an overemphasis of "s" and "sh" sounds. If you're getting sibilance on your pop-music recording, try these steps:

- Use a de-esser signal processor or plug-in. Or use a multiband compressor, and compress only the range from 3 to 10 kHz.
- Place the microphone farther from the vocalist.
- Place the microphone toward one side of the vocalist, or at nose height, rather than directly in front.
- Cut equalization in the range from 5 to 10 kHz.
- Change to a duller sounding microphone such as a ribbon.

BAD MIX

Some instruments or voices are too loud or too quiet. To improve a bad mix, try the following:

- Change the mix. (Maybe change the mix engineer!)
- Compress vocals or instruments that occasionally get buried.
- Change the EQ on certain instruments to help them stand out.
- During mixdown, continuously change the mix to highlight certain instruments according to the demands of the music.
- Change the musical arrangement so that different musical parts don't play at the same time. That is, consider having a call-and-response arrangement (fill-in-the-holes) instead of everything playing at once, all the time.

UNNATURAL DYNAMICS

When your pop-music recording has unnatural dynamics, loud sounds don't get loud enough. If this happens, try these steps:

- Use less compression or limiting.
- Avoid overall compression.
- Use multiband compression on the stereo mix instead of wide-band (full-range) compression.

ISOLATED SOUND

If some of the instruments on your recording sound too isolated, as if they aren't in the same room as the others, follow these steps:

- In general, allow a little crosstalk between the left and right channels. If tracks are totally isolated, it's hard to achieve the illusion that all the instruments are playing in the same room at the same time. You need some crosstalk or correlation between channels. Some right-channel information should leak into the left channel, and vice versa.

- Place microphones farther from their sound sources to increase leakage.
- Use omnidirectional mics to increase leakage.
- Use stereo reverberation or echo.
- Pan effects returns to the channel opposite the channel of the dry sound source.
- Pan extreme left-and-right tracks slightly toward center.
- Make the effects-send levels more similar for various tracks.
- To give a lead-guitar solo a fat, spacious sound, use a stereo chorus. Or send its signal through a delay unit, pan the direct sound hard left, and pan the delayed sound hard right.

LACK OF DEPTH

If the mix lacks depth, try these steps:

- Achieve depth by miking instruments at different distances.
- Use varied amounts of reverberation on each instrument. The higher the ratio of reverberant sound to direct sound, the more distant the track sounds.
- To make instruments sound closer, use a longer pre-delay setting in your reverb (maybe 40 to 100 msec). Use a shorter predelay (under 30 msec) to make instruments sound farther away.

Bad Sound in a Classical-Music Recording

Check the following procedures if you have problems recording classical music:

TOO DEAD

If the sound in your classical recording is too dead—there isn't enough ambience or reverberation—try these measures to solve the problem:

- Place the microphones farther from the performers.
- Use omnidirectional microphones.
- Record in a concert hall with better acoustics (longer reverberation time).

- Turn up the hall mics (if used).
- Add artificial reverberation.

TOO CLOSE

If the sound is too detailed, too close, or too edgy, follow these steps:

- Place the microphones farther from the performers.
- Place the microphones lower or on the floor (as with a boundary microphone).
- Roll off the high frequencies.
- Use mellow-sounding microphones (many ribbon mics have this quality).
- Turn up the hall mics (if used).
- Increase the reverb-send level.

TOO DISTANT

If the sound is distant and there is too much reverberation, these steps might help:

- Place the microphones closer to the performers.
- Use directional microphones (such as cardioids).
- Record with a spaced pair of directional mics aiming straight ahead.
- Record in a concert hall that is less live (reverberant).
- Turn down the hall mics (if used).
- Decrease the reverb-send level.

STEREO SPREAD TOO NARROW OR TOO WIDE

If your classical-music recording has a narrow stereo spread, try these steps:

- Angle or space the main microphone pair farther apart.
- If you're doing mid-side stereo recording, turn up the side output of the stereo microphone.
- Place the main microphone pair closer to the ensemble.

If the sound has excessive stereo spread (or "hole-in-the-middle"), try the following:

- Angle or space the main microphone pair closer together.
- If you're doing mid-side stereo recording, turn down the side output of the stereo microphone.

- In spaced-pair recording, add a microphone midway between the outer pair, and pan its signal to the center.
- Place the microphones farther from the performers.

LACK OF DEPTH

Try the following to bring more depth into your classical-music recording:

- Use only a single pair of microphones out front. Avoid multi-miking.
- If you must use spot mics, keep their level low in the mix.
- Add more artificial reverberation to the distant instruments than to the close instruments.

BAD BALANCE

If your classical-music recording has bad balance, try the following:

- Place the microphones higher or farther from the performer.
- Ask the conductor or performers to change the instruments' written dynamics. Be tactful!
- Add spot microphones close to instruments or sections needing reinforcement. Mix them in subtly with the main microphones' signals.

MUDDY BASS

If your recording has a muddy bass sound, follow these steps:

- Aim the bass-drum head at the microphones.
- Put the microphone stands and bass-drum stand on resilient isolation mounts (such as a carpet pad), or place the microphones in shock-mount stand adapters.
- Roll off the low frequencies or use a highpass filter set around 40 to 80 Hz.
- Use artificial reverb with a shorter decay time at low frequencies.
- Record in a concert hall with less low-frequency reverberation.

RUMBLE

Sometimes your classical-music recording picks up rumble from air conditioning, trucks, and other sources. Try the following to clear this up:

- Check the hall for background rumble problems.
- Temporarily shut off the air conditioning.
- Record in a quieter location.
- Use a highpass filter set around 40 to 80 Hz.
- Use microphones with limited low-frequency response.
- Mike closer and add artificial reverb.

DISTORTION

If your classical-music recording has distortion, try the following:

- Switch in the pads built into the microphones (if any).
- Increase the mixer input attenuation (turn down the input trim).
- Check connectors for stray wires or bad solder joints.
- Avoid sample-rate conversion.
- Apply dithering when going from 24 to 16 bits.

BAD TONAL BALANCE

Bad tonal balance expresses itself in a sound that is too dull, too bright, or colored. If your recording has this problem, follow these steps:

- Change the microphones. Generally, use flat-response microphones with minimal off-axis coloration.
- Follow the 3:1 rule mentioned in Chapter 7.
- If a microphone must be placed near a hard, reflective surface, use a boundary microphone on the surface to prevent phase cancellations between direct and reflected sounds.
- Adjust EQ. Try a spectrum analyzer/equalizer such as Harmonic Balancer (www.har-bal.com).
- Place the mics at a reasonable distance from the ensemble (too-close miking sounds shrill).
- Avoid mic positions that pick up standing waves or room modes. Experiment with small changes in mic position.

This chapter described a set of standards for good sound quality in both popular and classical music recordings. These standards are somewhat arbitrary, but the engineer and producer need guidelines to judge the effectiveness of the recording. The next time you hear something you don't like in a recording, the lists in this chapter will help you define the problem and find a solution.

CHAPTER 15

Session Procedures, Mastering, and CD Burning

"We're rolling. Take One." These words begin the recording session. It can be an exhilarating or an exasperating experience, depending on how smoothly you run it.

The musicians need an engineer who works quickly yet carefully. Otherwise, they may lose their creative inspiration while waiting for the engineer to get it together. And the client, paying by the hour, wastes money unless the engineer has prepared for the session in advance.

This chapter describes how to conduct a multitrack recording session. These procedures should help you keep track of things and run the session efficiently.

There are some spontaneous sessions, especially in home studios, that just "grow organically" without advance planning. The instrumentation isn't known until the song is done! You just try out different musical ideas and instruments until you find a pleasing combination.

In this way, a band that has its own recording gear can afford to take the time to find out what works musically before going into a professional studio. In addition, if the band is recording itself where it practices, the microphone setup and some of the console settings can be more or less permanent. This chapter, however, describes procedures usually followed at professional studios, where time is money.

PREPRODUCTION

Long before the session starts, you're involved in preproduction—planning what you're going to do at the session, in terms of overdubbing, track assignments, instrument layout, and mic selection.

Instrumentation

The first step is to find out from the producer or the band what the instrumentation will be and how many tracks will be needed. Make a list of the instruments and vocals that will be used in each song. Include such details as the number of tom toms, whether acoustic or electric guitars will be used, and so on.

Recording Order

Next, decide which of these instruments will be recorded at the same time and which will be overdubbed one at a time. It's common to record the instruments in the following order, but there are always exceptions:

1. Loud rhythm instruments—bass, drums, electric guitar, electric keyboards
2. Quiet rhythm instruments—acoustic guitar, piano
3. Lead vocal and doubled lead vocal (if desired)
4. Backup vocals (in stereo)
5. Overdubs—solos, percussion, synthesizer, sound effects
6. Sweetening—horns, strings

The lead vocalist usually sings a guide vocal or scratch vocal along with the rhythm section so that the musicians can get a feel for the tune and keep track of where they are in the song. You record the vocalist's performance but probably you will re-record it later. That eliminates leakage and lets you focus on the lead vocal.

In a MIDI studio, a typical order might be:

- Drum machine or soft-synth drum set (playing programmed patterns)
- Synthesizer bass sound
- Synthesizer chords
- Synth melody
- Synth solos, extra parts
- Vocals and miked solos

Track Assignments

Now you can plan your track assignments. Decide what instruments will go on which tracks of the multitrack recorder. The producer may have a fixed plan already.

What if you have more instruments than tracks? Decide what groups of instruments to put on each track. In a 4-track recording, for example, you might put guitars on track 1, bass and drums on track 2, vocals on track 3, and keyboards on track 4.

Remember that when several instruments are assigned to the same track, you can't separate their images in the stereo stage. That is, you can't pan them to different positions—all the instruments on one track sound as if they're occupying the same point in space. For this reason, you may want to do a stereo mix of the rhythm section on tracks 1 and 2, for instance, and then overdub vocals and solos on tracks 3 and 4.

It's possible to overdub more than four parts on a 4-track recorder. To do this, bounce (mix) several tracks onto one, then record new parts over the original tracks. Your recorder manual describes this procedure.

If you have many tracks available, leave several tracks open for experimentation. For example, you can record several takes of a vocal part using a separate track for each take, so that no take is lost. Then combine the best parts of each take into a single final performance on one track. Most recorder-mixers let you do these extra takes on virtual tracks.

It's also a good idea to record the monitor mix on one or two unused tracks. The recorded monitor mix can be used as a cue mix for overdubs, or to make a recording for the client to take home and evaluate.

Session Sheet

Once you know what you're going to record and when, you can fill out a session sheet (Figure 15.1). This simple document is adequate for home studios. "OD" indicates an overdub. Note the recorder-counter time for each take, and circle the best take.

Production Schedule

In a professional recording studio, the planned sequence of recording basic tracks and overdubs is listed on a production schedule (Figure 15.2).

SONG: Escape to Air Island

TRACK	INSTRUMENT	MICROPHONE
1	BASS	DIRECT
2	KICK	AKG D-112
3	DRUMS	CROWN GLM-100
4	LEAD VOC OD	STUDIO PROJECTS B1
5	HARM. VOC OD	STUDIO PROJECTS B1
6	LEAD GUIT OD	SHURE SM57
7	KEYS L	DIRECT
8	KEYS R	DIRECT

TAKE 1 03:21 - 06:18 FS
2 06:25 - 09:24 INC
3 10:01 - 13:02

FIGURE 15.1
A session sheet for a home studio.

WEST WIND STUDIOS
PRODUCTION SCHEDULE

Artist: Steve Mills
Album: Long Distance Music
Producer: B. Brauning
Engineer: D. Scriven
Date: 1-17-09

Project files: c:\cakewalk projects\mills\
mr_potato_head.cwp and sambatina.cwp

Audio data folders: d:\mills\mr_potato_head\ and \sambatina\

1. Record Mr. Potato Head
Instrumentation: Electric bass, drums, electric rhythm guitar, electric lead guitar, acoustic piano, sax, lead vocal.
Comments: Record rhythm section with scratch vocal. OD sax, piano, lead vocal. Double lead guitar in stereo.

2. Record Sambatina
Instrumentation: Electric bass, drums, acoustic guitar, percussion, synth
Comments: Record rhythm section with scratch acoustic guitar. Record synth MIDI. OD acoustic guitar, percussion and synth.

3. Mix Mr. Potato Head
Comments: Add 40msec echo to toms. Increase reverb on sax solo.

4. Mix Sambatina
Comments: Add flanger to bass on intro only. Do automated flange on percussion.

FIGURE 15.2
A production schedule.

WEST WIND STUDIOS
TRACK SHEET

Artist: S. Mills — Album: Long Distance Music
Producer: B. Brauning — Engineer: D. Scriven
Date: 1-17-09 — Song title: Dig Up Nebraska

Take	Start	Stop
Take # 1	Start 00:31	Stop 03:31
Take # 2 NC	Start 03:54	Stop 06:49
Take # 3 LFS	Start 07:21	Stop 08:38
Take # 4 FS	Start 09:01	Stop 09:12
Take #(5)	Start 09:25	Stop 12:27

Track 1: elec bass (Fender Precision DI)
Track 2: kick
Track 3: snare
Track 4: rack tom
Track 5: floor tom
Track 6: overhead L
Track 7: overhead R
Track 8: rhythm gtr L (Martin HD-28)
Track 9: rhythm gtr R (Taylor 814ce)
Track 10: lead gtr (Parker PM-20PRO, Roland Cube 60)
Track 11: keys L (Prophet '08 DI)
Track 12: keys R (Prophet '08 DI)
Track 13: scratch lead vocal
Track 14: lead vocal
Track 15: background vocals 1
Track 16: background vocals 2

FIGURE 15.3
A track sheet (multitrack log).

Track Sheet

Another document used in a pro studio is the track sheet or multitrack log (Figure 15.3). Write down which instrument or vocal goes on which track. The track sheet also has blanks for other information such as take numbers. If you're using a DAW, you can enter this information by typing on-screen.

Microphone Input List

Make up a microphone input list similar to that seen in Table 15.1. Later you will place this list by the mic snake box and by the mixing console.

Table 15.1 A Microphone Input List

Input	Instrument	Microphone
1	Bass	Direct
2	Kick	EV N/D868
3	Snare	AKG C451
4	Overhead L	Shure SM81
5	Overhead R	Shure SM81
6	High toms	Sennheiser MD421-II
7	Floor tom	Sennheiser MD421
8	Electric lead guitar	Shure SM57
9	Electric lead guitar	Shure SM57
10	Piano treble	Crown PZM-6D
11	Piano bass	Crown PZM-6D
12	Scratch vocal	Beyer M88

Be flexible in your microphone choices—you may need to experiment with various mics during the session to find one giving the best sound with the least console equalization. During lead-guitar overdubs, for example, you can set up a direct box, three close-up microphones, and one distant microphone—then find a combination that sounds best.

Find out what sound the producer wants—a "tight" sound; a "loose, live" sound; an accurate, realistic sound. Ask to hear recordings having the kind of sound the producer desires. Try to figure out what techniques were used to create those sounds, and plan your mic techniques and effects accordingly. Tips on choosing a microphone are given in Chapter 6.

Instrument Layout Chart

Work out an instrument layout chart, indicating where each instrument will be located in the studio, and where baffles and isolation booths will be used (if any). In planning the layout, make sure that all the musicians can see each other and are close enough together to play as an ensemble. Often a circular arrangement works well.

Keep all these documents in a single folder or notebook labeled with the band's name and recording date. Also include contact information, time sheets, invoices, and all the correspondence about the project.

SETTING UP THE STUDIO

About an hour before the session starts, clean up the studio to promote a professional atmosphere. Lay down rugs and place AC power boxes according to your layout chart.

Now position the baffles (if any). Put out chairs, stools, and music stands according to the layout. Run a headphone extension cable from each artist's location to the headphone junction box in the studio.

Place mic stands approximately where they will be used. Wrap one end of a microphone cable around each microphone-stand boom, leaving a few extra coils of cable near the mic-stand base to allow slack for moving. Run the rest of the cable back to the mic input panel or snake box. Plug each cable into the appropriate wall-panel or snake-box input according to your mic input list.

Some engineers prefer to run cables in reverse order, connecting to the input panel first and running the cable out to the microphone stand. That procedure leaves less of a confusing tangle at the input panel where connections might be changed.

Now set up the microphones. Check each mic to make sure its switches are in the desired positions. Put the mics in their stand adapters, connect the cables, and balance the weight of each boom against the microphone.

Finally, connect the musicians' headphones for cueing. Set up a spare cue line and microphone for last-minute changes.

SETTING UP THE CONTROL ROOM

Having prepared the studio, run through the Session Preparation section of Chapter 12 to make sure the control room is ready for the session. Then turn up the monitor system. Carefully bring up each fader one at a time and listen to each microphone. You should hear normal studio noise. If you hear any problems such as dead or noisy microphones, hum, bad cables, or faulty power supplies, correct them before the session.

Verify the mic input list. Have an assistant scratch each mic grille with a fingernail and identify the instrument the microphone is intended to pick up. If you have no assistant, listen on headphones as you scratch the grilles.

Check all the cue headphones by playing a tone or music through them and listening while wiggling each cable.

SESSION OVERVIEW

This is the typical sequence of events:

1. For efficiency, record the basic rhythm tracks for several songs on the first session.
2. Do the overdubs for all the songs in a dubbing session.
3. Mix all the tunes in a mixdown session.
4. Edit the tunes and master the album.

After the musicians arrive, allow them 1/2 hour to 1 hour free set up time for seating, tuning, and mic placement. Show them where to sit, and work out new seating arrangements if necessary to make them more comfortable.

Once the instruments are set up, listen to their live sound in the studio and do what you can to improve it. A dull-sounding guitar may need new strings, a noisy guitar amp may need new tubes, the drums may need tuning, and so on. Adjust the studio lighting for the desired mood.

RECORDING

Before you start recording, you might want to make connections to record the monitor mix. This recording is for the producer to take home to evaluate the performance.

Many DAW programs have session templates: a group of tracks with EQ, compression and effects plug-ins already in place. You simply load in the template and you're ready to go. You can create your own session templates, such as "16 tracks with aux 1 and 2" or "drum set."

Some DAWs let you create a track template, which is a format for a single track. You might have a vocal track template with the input channel, reverb, EQ, and compression already set up. Each time you're about to record another vocal, import the vocal track template.

Follow the mixer recording procedures described in Chapter 12. Set recording levels, then set cue mixes for the musicians' headphones. The monitor mix affects only what is heard, not what is recorded.

When you're ready to record the tune, briefly play a metronome to the group at the desired tempo, or play a click track (an electronic metronome) through the cue system. Or just let the drummer set the tempo with stick clicks.

Start recording. Note the recorder counter time. Hit the slate button (if any) and announce the name of the tune and the take number. Then the group leader or the drummer counts off the beat, and the group starts playing.

The producer listens to the musical performance while the engineer watches levels and listens for audio problems. As the song progresses, you may need to make small level adjustments.

The assistant engineer (if any) runs the multitrack recorder and keeps track of the takes on the track sheet, noting the name of the tune, the take number, and whether the take was complete (Figure 15.3). Use a code to indicate whether the take was a false start, long false start, nearly completed, a "keeper," and so on.

While the song is in progress, don't use the solo function, because the abrupt monitoring change may disturb the producer. The producer should stop the performance if a major flub (mistake) occurs but should let the minor ones pass.

At the end of the song, the musicians should be silent for several seconds after the last note. Or, if the song ends in a fade-out, the musicians should continue playing for about 30 seconds so there is enough material for a fade-out during mixdown.

After the tune is done, you can either play it back or go on to a second take. If you connected your multitrack recorder to the insert jacks, use the faders to set a rough mix with EQ and effects. The musicians will catch their flubbed notes during playback; you just listen for audio quality.

Now record other takes or tunes. Pick the best takes. Punch-in to correct errors on each best take, or fix errors with editing.

To protect your hearing and to prevent fatigue, try to limit tracking sessions to four hours or less. Take breaks to give your ears and body a rest.

Relating to the Musicians

During a session, the engineer needs not only technical skills, but also people skills. It's important to respect artistic personalities and to keep the creative energy flowing.

First, learn the names of the band members and refer to them by name during the session.

Musicians are often nervous at the beginning of a session. Having a sense of humor and exuding confidence helps to put the artists at ease. Talk about the band's music, origin, or instruments. Be sure that the artists are comfortable—adjust the layout and lighting, offer beverages and snacks. Tell new clients that mistakes are normal and are easy to fix. Ask band members, "Is the mix okay in your headphones? Is the volume all right?" "Does your recorded instrument sound okay?"

Don't badmouth other musicians, and respect the privacy of your clients. Each musician you record should feel confident that you're not telling others about their mistakes.

When you want to try something that could enhance a song, don't just do it. Ask first. Say, "What would you think about doubling the guitar in stereo?" or "I want your input on this."

During a playback, try not to point out the errors. The musicians will hear them and will try to do better on the next take. Don't say, "That sucked." Say "It's almost there—do you want to punch in a few spots?" Or say "That was good, but how about another take?" The goal is to make the best record you can, not to make clients feel bad about themselves. Give a musician time to practice during overdubs, too.

If the band members are getting tired or are having trouble getting through a particular song, you might say "Let's take a break" or "Let's try another song and come back to this one later."

Finally, if a piece of gear breaks or your software develops a glitch, try to work around it quietly without making it obvious. Musicians need an engineer who handles technical problems in a professional way.

OVERDUBBING

After recording the basic or rhythm tracks for all the tunes, add overdubs. A musician listens to recorded tracks over headphones and records a new part on an open track. See the Chapter 12 sections Overdubbing and Composite Tracks.

You might have the musician play in the control room while overdubbing. You can patch a synth or electric guitar into the console through a direct box, and feed the direct signal to a guitar amp in the studio via a cue line. Pick up the amp with a microphone, and record and monitor the mic signal.

Do any drum overdubs right after the rhythm session because the microphones are already set up, and the overdubbed sound will match the sound of the original drum track.

BREAKING DOWN

When the session is over, tear down the microphones, mic stands, and cables. Put the microphones back in their protective boxes and bags. Wind the mic cables onto a power-cable spool, connecting one cable to the next. Wipe off the cables with a damp rag if necessary. Some engineers hang each cable in big loops on each mic stand. Others wrap the cable "lasso style" with every other loop reversed. You learn this on the job.

Put the labeled recording and session documents in a labeled box or file folder. Normally the studio keeps the multitrack recording for possible future work unless the group wants to buy or rent it. Backup the multitrack recordings (audio files) and session files (track setups for each song) to another medium, such as an external hard drive, CD-R, or DVD-R.

If you're using a hardware mixing console, log the console and effects settings by writing them in the track sheet or reading them slowly into a portable recorder. At a future session you can play back the recording and reset the console the way it was for the original session. Automated consoles can store and recall the console's control adjustments. DAWs remember both the mixer settings and the plug-in settings.

MIXDOWN

After all the parts are recorded, you're ready for mixdown. Follow the mixdown procedures in Chapter 12 and repeat for all the song mixes. Again, be sure to backup your session files (the mix and edit settings) and audio files (track audio recordings and mixes.)

MASTERING

After recording the mixes, you can burn a CD of the mixes as they are, or you can master an album or demo of those mixes. Mastering is the last creative step before burning the final CD used for duplication or replication. In mastering you edit out noises or false starts before the beginning of each song, put the songs in the desired order, insert a few seconds of silence between songs, and make each song the same loudness and the same tonal balance. The goal is a consistent sound from track to track, so that everything flows better and the album sounds unified. You also might try to make the CD as loud or "hot" as possible (without destroying its sound quality).

At this point, you have three choices:

1. If the CD will be just a reference, not a demo or album, you can burn a CD of the unedited mixes and stop there.
2. If the CD will be a demo or album, burn a CD of the unedited mixes, then send it to a mastering engineer for mastering.
3. Or you can edit and master the mixes yourself, then burn a CD. It will be the CD pre-master for duplication or replication. "Duplication" is making copies with a CD-R burner, while "replication" is stamping CDs from a glass master (the professional standard).

Let's look at each option above.

Burning a Reference CD

You can use the CD-burning software that came with your computer CD burner, or use other software. Some examples of CD-burning programs are Roxio's Easy Media Creator and Record Now Media Lab, Nero, and Stomp's Click 'N Burn. Some DAW software includes a CD-burning application.

Before you start, note that most CD burning software puts 2 seconds of silence between songs. If you want to have extra seconds of silence at the end of a song, record the silence as part of the song's WAVE file. You might want to normalize each song's WAVE file. This raises the level of the song so that the highest peak in the song reaches maximum level: 0 dBFS or whatever level you specify. Normalizing doesn't make the tracks the same loudness, because loudness depends on the average signal level, not the peak level.

Here's a CD-burning procedure:

1. Start the CD-R burning software.
2. Select which WAVE files you want to put on the CD-R, drag them to the playlist, and arrange them in the desired order. The total playing time must be less than the CD-R length (74 or 80 minutes).
3. Choose "On the fly" or "Disc image." When recording on-the-fly, the computer grabs the sound files from random locations on hard disk and puts them in order as the CD-R does a burn. When recording a disk image, the computer rewrites the sound files to a single, contiguous space on your hard disk and puts them in order. Then you copy the disk image from hard disk to CD-R. Disc image is less likely to produce CD errors than On-the-fly.
4. Set the burn speed. Some CD recorders and blank CDs can record up to 52 times normal speed. Normal speed is 172 kBps (kilobytes per second; Bytes is capital B), double speed is 344 kBps, etc. A 52× recorder can burn a 74-minute CD in less than 2 minutes. Some CD burners automatically select the optimum speed based on what the blank CD-R can handle. High speeds don't degrade sound quality seriously, but they tend to increase errors—which the CD player may or may not correct accurately. The recommended speed to

prevent errors is 2× to 4× when making a final master CD to be sent out for replication. But some CD recorders and blank CD-Rs have fewer errors at higher speeds.

5. Set the software to Disc-at-Once Mode. Start recording the CD-R. The WAVE files will transfer in order to the CD-R disc. To prevent glitches, don't multitask while the CD is recording. CD-R discs can't be erased and used over again, so try to make everything right before you burn a disc. You could do a practice burn on a CD-RW.
6. As soon as the recording is done, the display will indicate that the table of contents is being written. Eventually, the system will beep and eject the disc. To prevent error-causing fingerprints, be sure to handle the disc only by the edges. Pop the disc in an audio CD player, press Play, and check that all the tracks play correctly.

Sending Out Your CD for Mastering

You might prefer to send your CD of mixes to a good mastering engineer. This person can listen to your program with fresh ears, then suggest processing for your album that will make it sound more commercial. She is likely to have a better monitor system and better equipment than yours. They have heard hundreds of recording projects done by others, and know how to make your CD sonically competitive.

If you plan to have your program mastered outside, don't apply any signal processing to your finished mixes such as editing, level changes, compression, normalization, fades, or EQ. Let the mastering house do it with their better equipment and software. Also, leave some headroom by recording the finished mixes at about −3 dBFS maximum in peak meter mode.

Deliver your mixes in the highest possible resolution that the mastering engineer can accept, such as 96 to 192 kHz and 24 to 32 bits on a data CD (ISO 9660 format). However, a 24-bit/44.1 kHz format is usually adequate. Copy the mixes' WAVE files to CD, rather than creating CD audio tracks (.cda files). The resulting data CD can be read by a CD-ROM but can't be played on a CD player. If your CD-burning program doesn't support data CDs, either get one that does, or create an audio or music CD at 16-bits/44.1 kHz.

Also send the mastering engineer the sequence of songs and any suggestions on how you would like certain songs to be improved or edited. The mastering engineer might suggest that you need to re-do some mixes.

Mastering Your Own Album

You can master an album with your multitrack recording software, or with a mastering program such as Steinberg Nuendo, Sony Sound Forge, BIAS Peak Pro, Sonalksis Mastering Suite, 1K Multimedia T-Racks 24, or Magix Samplitude.

First, discuss the order of the songs on the recording. The first track on the album should be a strong, accessible, up-tempo piece. Change keys from song to song. The last track should be as good as or better than the first to leave a good final impression.

Mastering engineer Bob Katz offers this advice for song sequencing: Create the album in sets of one to three songs of the same tempo. Make sure the songs flow from one to the next. Here's a suggested song order:

1. An up-tempo, exciting song that hooks the listener.
2. After a short space, an up-tempo or mid-tempo song.
3. After three or four songs, slow down the tempo.
4. Reach a climax near the end of the album.
5. The last song should sound relaxed and intimate, perhaps using fewer members of the band.

You can drag the WAVE files of your song mixes into iTunes and experiment with various sequences.

To get your mastered songs onto a CD, you need CD-burning software, either consumer or pro. With a consumer CD-burning program—such as Roxio's Easy Media Creator or Record Now Media Lab, Stomp's Click 'N Burn, or Nero—you drag the WAVE files of each of your song mixes into a playlist, then burn the CD.

With a professional program, you create one long WAVE file of all the song mixes. You can adjust the timing between each pair of songs. You specify when each song starts, and the software creates a table of contents on the CD made of song-start IDs. This feature is called

PQ subcode editing. Some programs that do this are Sony Creative Software CD Architect, Adaptec Jam, and Goldenhawk Technology CDRWIN. Some DAW software includes a CD-burning application that lets you set start IDs.

Here's a suggested procedure to master an album:

1. If you already recorded your mixes directly onto hard disk as stereo WAVE files, skip to Step 4.
2. If your song mixes are WAVE files on a CD-R, put the CD-R in your CD-ROM drive and copy each song to your hard drive. Then go to Step 4. If your song mixes are audio tracks (.cda files) on a CD-R, convert each track to a WAVE file on your hard drive using ripper software. Skip to Step 4.
3. If your mixes are on a stand-alone 2-track recorder without a USB or FireWire port, plug the recorder's digital output into the digital input of your audio interface. If your interface lacks a digital input, use analog out to analog in. Launch the recording software. Set levels if necessary, start recording on your hard drive, and play the 2-track recording containing your mixes. All the mixes will copy in real time to your hard drive to create a single long WAVE file.

   If your 2-track recorder has a USB or FireWire port, plug it into your computer's matching port. The recorder will appear as a hard drive to your computer. Copy the WAVE file(s) of the mixes to your computer hard drive.
4. Now that all the song mixes are on your hard drive, launch your DAW software. Import the first song's WAVE file to a track at time zero. The waveform of the song appears as a clip or segment of audio. Slip-edit or trim the beginning and end of the song clip to remove extra space and noises. Repeat for the other songs.

Click on the trimmed Song 1 and slide it 10 frames to the right in your mastering software. That way the CD player won't miss the beginning of Song 1.

It's convenient to put each song clip on a different track, one after another (Figure 15.4). That way you can easily adjust the spacing between songs, and apply different fader settings and processing to each song as needed. Assign all the tracks to the same stereo bus so you can apply bus processing to them later.

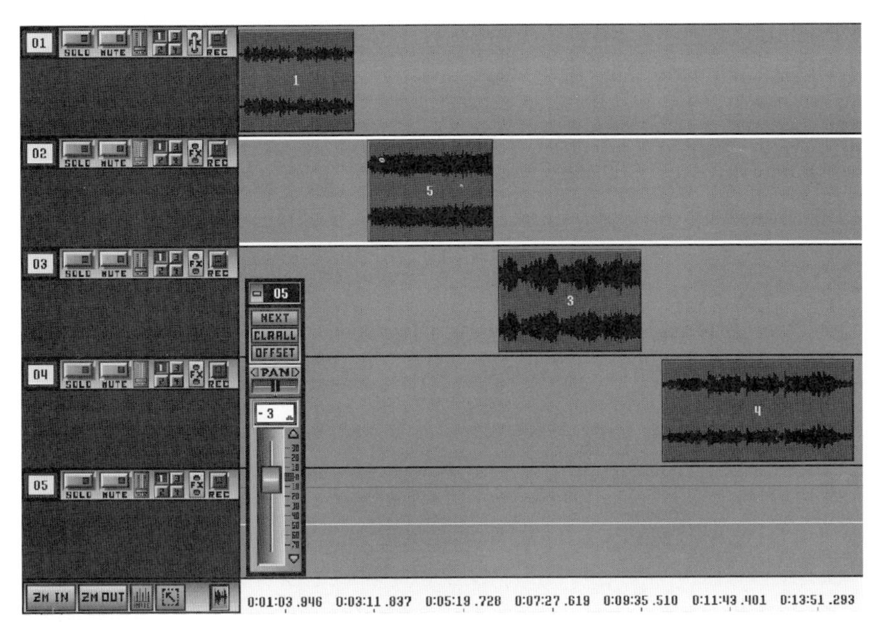

FIGURE 15.4
Placing song clips on successive tracks makes mastering easier.

5. Now that all the songs are in place and trimmed, add a fade-out at the end of certain songs if desired. If your CD-burning program can write PQ subcodes (start IDs), you can crossfade between two songs: overlap their clips near the transition point, and use the crossfade function in your DAW. Or fade out the end of one song and fade in the beginning of the next (as shown in Figure 13.17).
6. If your CD-burning program can write PQ subcodes (start IDs), you can adjust the spacing or gap between songs. Two to three seconds of silence between songs is typical, but go by ear. Use a longer space if you want to change moods between songs. Use a shorter space to make similar songs flow together. A short space also works well after a long fade-out because the fade-out acts like a long pause between songs.
7. Click on and play part of each song's waveform to check for loudness. Make all the songs equally loud by adjusting the track's fader for each song. If you put all the songs on the same stereo track, set up an automation volume envelope.

Here's one method to match song levels. If most songs are equally loud, set the fader to zero before each of those songs in order to avoid processing. Then turn down any louder songs and turn up quieter songs to match the rest of the songs. Do this by ear. You might use post-fader rms metering on the stereo mix bus as a rough indicator of average levels.

Caution: If you increase a song's level, make sure the peaks in the song don't clip.

Another way to match song levels is to find the song with the highest peaks, and leave its level alone Adjust the levels of other songs so they match the loudness of the song with the highest peaks. Again, do this by ear. You may need to turn up the intro of a song so it works in context with the previous song. Consider using Harmonic Balancer (www.har-bal.com), a tool to match song levels and tonal balances.

8. Apply EQ to songs that need it. Put songs with different EQs on different tracks. If necessary, touch up levels after adding EQ.
9. If desired, apply multiband compression, limiting, and normalization to the mix bus in order to get a "hot" or loud CD. Don't overdo it, or your project will be distorted and fatiguing to listen to. Squashed dynamics can suck the life out of music. If you apply only peak limiting and normalization, you'll get a hot CD without altering the musical dynamics. *Play audio clip 40 at www.elsevierdirect.com/companions/9780240811444*.

The idea is to knock down the peaks in the waveform because they don't contribute to perceived loudness—the average level does. Once you limit the peaks about 6 dB, you can normalize (raise the overall level) and thus create a louder program. Normalizing to 100% of full scale can create errors with some D/A converters, so you might normalize to 96% or −0.3 dB.

THE LOUDNESS WARS

Musicians and record labels, more than recording engineers or record producers, are requesting hotter levels on CDs. One reason is that a low-level CD played with others in a CD changer tends to sound less exciting. Musicians also want their songs to stand out on the radio.

Loudness depends mostly on average (rms) level rather than peak level. So in an attempt to raise average levels, engineers limit or clip the peaks. Overall compression of the mix is common too. Unfortunately, the song loses its dynamic range and its musical excitement.

When I master a pop-music CD, I often compromise by limiting peaks by no more than 7 dB, then normalizing without overall compression or clipping. That retains the dynamic range but loses a little transient clarity.

If playback equipment had switchable automatic volume control (compression), then we wouldn't need to maximize levels in the recording.

According to mastering engineer Bob Katz, hot CDs and quiet CDs end up at the same level when processed with radio-station compressor/limiters. Over-compressed CDs don't sound louder on the air; they just sound more distorted. So ignore the "loudness wars" and avoid excessive compression. Bob recommends that pop music be mastered at an average or rms level of −14 dBFS while listening at 83 dBSPL C-weighted, slow meter reading. He suggests an average level of −20 dBFS for classical music.

10. After your musical program is edited, you have two options: (1) If you have professional CD-burning software that can write PQ subcodes, export (save) the mastered song mixes to a single 16-bit/44.1 kHz stereo WAVE file. If your recording was 24-bit, first turn on dithering. That retains much of the 24-bit sound quality when converting to 16-bit format. (2) If you have consumer software that burns a CD from a list of song WAVE files, click on one song's clip in your DAW to highlight it, then export the mix to a single 16-bit/44.1 kHz stereo WAVE file. If your recording was 24-bit, first turn on dithering. Repeat for all the song clips.

TRANSFERRING THE MASTERED PROGRAM TO CD-R

At this point your hard drive contains either one long WAVE file of all the mastered songs, or several WAVE files, each one a mastered song. If the latter, launch your CD-burning application and drag the

WAVE files in order to the playlist. To prevent errors on the CD, burn it at 4 × speed and don't multitask. Check the finished CD in a CD player.

Suppose you have pro CD-burning software. It creates CD tracks from a single WAVE file by following a cue sheet: a text file that lists the start time of each song. When the CD is recorded, the software creates a Start ID for each song based on the cue sheet. With this method you can create CD song-start IDs that occur even during a continuous program, such as a live concert recording.

CDRWIN at www.goldenhawk.com is such a program. I'll describe how to use it as an example.

1. Note the start time of each song in the DAW editing software's Edit Decision List or playlist. Or check the start time of each song clip by right-clicking on it or by looking at its beginning on the project time line. The start time of each song is specified in your DAW in minutes:seconds:frames, and there are 30 frames per second. Write down the start time of each song.
2. In the cue sheet, make each song's start time about one-third of a second (10 frames) before the actual start time of each song. That way the CD player's laser will have time to find the track before playing audio. For example, if a song actually starts at 12:47:28, make its start time in your cue sheet 12:47:18. List Song 1's start time as 00:00:00. The cue sheet should also include the name and filepath of the mastered program's WAVE file.
3. Place a blank CD-R in your CD burner. Don't put a paper label on the CD before recording because the label can cause jitter.
4. Open your CD-burning program and load in the cue sheet text file. Set the recording speed to 2 × to 4 × speed to reduce errors. Some CD-recorders and blank CD-Rs have fewer errors at higher speeds.
5. Start recording the CD. The CD-burning program writes the cue sheet's start times in the CD-R's table of contents, and copies the mastered program's WAVE file to the CD-R. Each CD track starts at the time you specified in your cue sheet. To prevent glitches, don't multitask while the CD is burning.

6. When the CD is finished, handle it by its edges to prevent error-causing fingerprints. Play the CD from start to finish to check for glitches. Press the track-advance button to make sure each song starts at the right time. If you look at the CD's contents in your computer, you'll see a separate .cda (CD-Audio) track for each song.

You can identify the CD by writing on the label side with a felt-tip marker. To avoid damaging the CD plastic layer, use water-based Sharpie or Memorex CD marking pens, available in the CD-R section of many stores. Some software can print liner notes, including the track numbers, titles, and timing. If you're duplicating CDs yourself, you might want to use a CD printer rather than labels. The adhesive in labels can damage the top CD layer or warp the CD over time, making the CD unplayable.

Another way to create master CDs is to use the Alesis MasterLink: a CD recorder with a built-in hard drive and mastering tools. With it you can record your stereo mixes, edit and master them, and burn a finished CD. Sample rates can be up to 96 kHz and word length can be 16-, 20-, or 24-bit on Masterlink's CD24 format discs. The MasterLink includes sample-rate conversion and noise shaping to alter sample rates and bit resolution as needed.

Several editing/mastering tools are offered with MasterLink, such as 16 different playlists, gain control, editing start and end points, joining or splitting song sections, EQ, compression, normalization, and peak limiting.

CD-Text and ISRC Code

You can add non-audio data to your CD when you create the cue sheet. If you want the listener to see on-screen each song's title and performer as the CD plays, create a CD-Text file using the CD-Text editor that comes with most CD-burning software. Include the file-path of the CD-Text file in the cuesheet. Not all CD players support CD Text.

Some players get CD-Text information from an online database rather than from the CD. You upload the CD-Text information to the database so that the player can display it. For example, iTunes needs to access an online CD database called Gracenote (formerly CDDB) to get the artist and song titles. Gracenote is at www.gracenote.com.

Every database has a different procedure for uploading CD-Text. When using the iTunes database,

1. Insert your CD and open iTunes.
2. Select all tracks in the iTunes window.
3. Select Get Info and enter the album name, artist, year, genre, copyright info, and so on.
4. Select each track individually, select Get Info, and enter the track's correct name.
5. Go to Advanced > Submit CD track names. Your information will upload to the Gracenote database. When someone puts your CD into iTunes, your info will show up.

When using the Windows Media Player database, send a copy of your CD to AMG (All Media Guide) at www.allmediaguide.com/labels.html. Many players access the freedb database at www.freedb.org. Go there to see a list of supported players you can use to submit CD information. Popular applications are Exact Audio Copy and Nero. Another database is Muse at www.muze.com/html/industry.

iTunes before version 7 didn't display CD-Text. iTunes 7 can display CD-Text information, but only if it has been uploaded to the Gracenote database.

The International Standard Recording Code (ISRC) is a 12-character code that identifies each track. It's a tool for royalty collection and anti-piracy. Information on applying for ISRC codes is at www.recordlabelresource.com/isrc-registration.html and www.riaa.com. If the CD tracks have ISRC codes, include them in the cue sheet as well. Some online music providers, such as CDBaby, can create the ISRC codes for you.

MASTER LOG

Type or print a log describing the CD-R master (such as shown in Table 15.2) and include it with your master.

Table 15.2 A CD Master Log

CD Master Log

Album title: Don't Press That Button (demo)

Artist: Puff Daddy Bartlett

Mastering date: 2-26-06

Track	Start MM:SS:FF	Duration	Title
1	00:00:00	03:23	60 Gigabytes to Go
2	03:24:20	03:15	Unplugged Plug-In
3	06:41:00	02:49	Digital Dropout Blues
4	09:32:05	02:55	Late and See
5	12:29:11	03:07	Buffer the Vampire

Total running time: 15:47

Also include the CD liner notes such as song lyrics, instrumentation, composers, credits, arrangers, publishers, copyright dates, and artwork. Include contact information for the artist, engineer, and mastering engineer. Stay in contact with the duplication or replication house, especially about artwork.

Before you send out the master, be sure to make another copy that you keep in your studio. This copy can be used if the master is lost or damaged.

Note that the CD-R pre-master doesn't leave the studio until all studio time is paid for! When this is done, send the CD-R to the duplication house.

The document "Recommendations for Delivery of Recorded Music Projects" provides examples of session sheets, track sheets, and so on. It also recommends media for delivering and backing up master recordings. It can be found at www.aes.org/technical/documents. Search for AESTD1002.1.03–10_1.pdf.

It's amazing how the long hours of work with lots of complex equipment have been concentrated into that little CD-R—but it's been fun. You crafted a product you can be proud of. When played, it recreates a musical experience in the ears and mind of the listener—no small achievement.

COPYRIGHTS AND ROYALTIES

It's important to copyright original songs so that others can't claim them as their own. Go to www.copyright.gov/forms to download form SR (for Sound Recordings) and form CON (continuation of form SR to add more songs). The filing fee is $45 for submission by regular mail or $35 by electronic submission. That fee lets you copyright several songs by the same composer(s) as a single work on one CD. Just be sure to list all the song titles. Send the forms, CD, and fee in a box to the U.S. Copyright Office, address at the above Web site.

If you record and sell cover tunes (songs that have been recorded by other artists), or if you sample other artists' music in your own recording, you must obtain a "mechanical license" and pay a royalty fee to BMI or ASCAP. The Harry Fox Agency (HFA) is the premier musical industry resource for licensing. They offer a service called Songfile, a simple, fast way to obtain licenses for 25 to 2,500 copies of a recording as a CD, cassette, LP, or digital download. If the cover song is under 5 minutes long, the royalty fee is about $45 per song for every 500 CDs that you distribute. For details check out www.harryfox.com, www.bmi.com, and www.ascap.com.

CHAPTER 16

The MIDI Studio: Equipment and Recording Procedures

Welcome to the computerized world of MIDI equipment—samplers, synthesizers, sampling keyboards, drum machines, and sequencers. Because other texts explain this equipment in detail, brief definitions serve the purpose here. See Appendix D for books on MIDI.

Musical Instrument Digital Interface (MIDI) is a standard connection between electronic musical instruments and computers that allows them to communicate with each other.

Here are some things that MIDI lets you do:

- Play a piano-style keyboard, drum pads, MIDI guitar, or breath controller to produce a sound like any instrument.
- Record and play back a musical performance that is independent of the sound of that performance. You can hear your performance played by any musical instrument or synthesized sound, or change the instrument sound after your performance is recorded.
- Record a difficult part slowly, or one note at a time, and play it back at any tempo you wish.
- Enter the notes on a musical scale or on a piano-roll grid, one at a time, and play them back at any tempo.
- Edit any note of the performance—change its timing, length, or pitch.
- Create the effect of a band playing. Record a performance that is played by drum samples, overdub another performance that is played by a synthesized bass, overdub another performance played by piano samples, and so on.

- Combine or layer the sounds of two electronic musical instruments by playing them both with the same performance.
- Quantize the notes to make them rhythmically precise.

The MIDI signal is a stream of digital data—not an audio signal—running at 31,250 bits per second. It sends information about the notes you play on a MIDI controller, such as a piano-style keyboard or drum pads. Up to 16 channels of information can be sent on a single MIDI cable.

There are three types of MIDI ports on MIDI devices:

1. MIDI IN receives data going into the device.
2. MIDI OUT sends out data generated by the device.
3. MIDI THRU is like MIDI OUT, but duplicates the data that are at the MIDI IN port.

Connect MIDI OUT from the sending device to MIDI IN of the receiving device. For example, connect a keyboard controller's MIDI OUT to the MIDI IN connector of a MIDI/audio interface, a sequencer, or a sound module. Use MIDI THRU to connect two or more receiving devices in a row. For example, connect a keyboard controller's MIDI OUT to sequencer MIDI IN, and connect sequencer MIDI THRU to sound module MIDI IN.

See MIDI Clock and MIDI Timecode in the Glossary.

MIDI-STUDIO COMPONENTS

The following equipment typically is used in a MIDI studio:

- MIDI controller
- Sequencer
- Synthesizer
- Sampler and sample CDs
- Drum machine or soft synth playing drum-set samples
- Power amplifier and speakers, or powered speakers
- Personal computer
- MIDI computer interface, or audio interface with MIDI I/O
- Mixer (optional)
- Recorder–mixer (optional)
- Effects or plug-ins
- Audio cables

FIGURE 16.1
M-Audio Oxygen 8v2, and example of a MIDI keyboard controller.

- MIDI cables
- Equipment stand

You have learned about most of these in previous chapters, but a review might help at this point.

A MIDI controller is an instrument that generates MIDI data when you play on it. Examples are a piano-style keyboard (Figure 16.1), drum pads, MIDI guitar, or a MIDI breath controller. A synthesizer or drum machine can act as a controller.

Some keyboard controllers are compact and lightweight, and connect to your computer via a USB cable. Examples are Korg's 49-key Kontrol49, M-Audio's 25-key Oxygen 8v2, E-MU's Xboard Series, Edirol's 49-key PCR-500, and Novation's ReMoteslCo25. Because they have fewer keys than a standard keyboard, they use octave up/down buttons.

A sequencer is a device or program that lets you record, edit, and play back MIDI data. A recording done on a sequencer is called a sequence, which is a MIDI song. Unlike an audio recorder, a sequencer doesn't

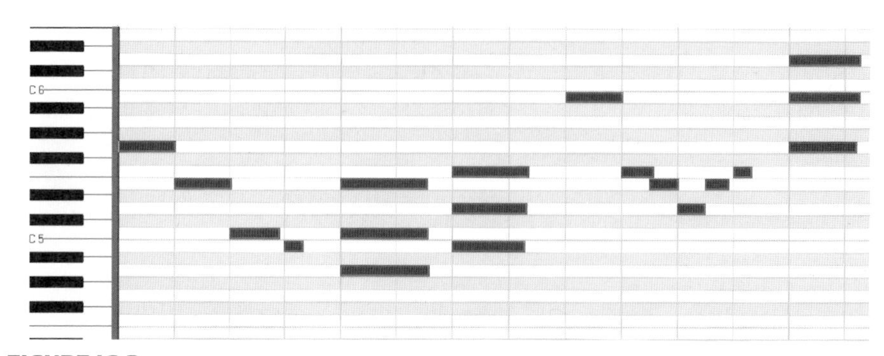

FIGURE 16.2
A screen shot of a piano-roll view, part of a MIDI sequencer program.

record audio. Instead, it records the key number of each note you play, note-on signals, note-off signals, and other parameters such as velocity, pitch-bend, and so on. A sequencer captures a performance, not the sound. Then any sound you want can be played by that performance. You can edit the recorded performance (sequence) to fix wrong notes, etc.

A sequencer records and plays MIDI files (.mid files). They are sequencer recordings of your own performances, or are MIDI files that can be downloaded from the Web.

The sequencer can be a stand-alone unit, a circuit built into a keyboard instrument, or a computer running a sequencer program (Figure 16.2). Like a multitrack audio recorder, a sequencer can record 8 or more tracks, with each track containing a performance of a different instrument.

A synthesizer is a musical instrument that creates sounds electronically. It can play MIDI data, either from a built-in keyboard, a separate MIDI controller, a MIDI sequence, or a MIDI file downloaded from the Web. Synthesizers come in four forms: piano-style, sound module, software, or a synth chip on a sound card.

- A piano-style synth has a piano keyboard and built-in sound generators. You might want to use more than one synth to expand your palette of sounds.
- A sound module or tone module is a synthesizer without a keyboard. This stand-alone device is triggered by a sequencer or a MIDI controller.

- A soft synth or virtual instrument is a synthesizer that is simulated in software. It runs in your CPU. The GUI of the software looks like a hardware synthesizer. A wavetable soft synth plays samples of real instruments, which sound more realistic than an FM soft synth.
- Another option is a synth chip, which is built into many sound cards.

Some examples of soft synths are E-MU Proteus, Arturia ARP, Cakewalk Dimension Pro and Pentagon, McDSP Synthesizer One, Native Instruments Reaktor, and TASCAM GigaStudio Orchestra.

The musical sounds that synthesizers make—such as a fretless bass, grand piano, sax, or drum set—are called patches, programs, or presets. A multitimbral synthesizer can play two or more patches at once. A polyphonic synthesizer can play several notes at once (chords) with a single patch.

A sampler is a device that records sound events, or samples, into computer memory and plays them back when activated by a sequencer MIDI file or MIDI controller. A sample is a digital recording of one note of a real sound source: a flute note, a bass pluck, a drum hit, etc. A sample also can be a digital recording of a short segment of another recording. The sampling process is described in Chapter 9 under Digital Recording.

A soft sampler is software that plays samples and lets you map them along your keyboard. In other words, you can tell the sampler which notes on the keyboard play which samples. A sampler lets you import any sound, such as a WAVE file of a single note that you recorded, or a file from a sample CD. In contrast, a soft synth generates its own sounds—either the ones supplied with it, or your own customized versions.

Some software sample players are Korg Legacy Collection, IK Sampletank, Native Instruments Kontakt 2, and Steinberg HALion. Also check out these Web sites: www.synthzone.com/softsyn.htm, www.sonicspot.com/softwaresynth.html, www.hitsquad.com/smm/, and www.kvraudio.com.

A soundfont is an audio sample (an instrument patch) in a special .sf2 format. It's like a WAVE file, but also includes a key range so that when you play a MIDI note number (keyboard key), it plays the sample pitch assigned to that note number. Soundfonts also include velocity switching, note envelope, looping, release sample, filter, and low-frequency oscillator (LFO) settings. A single soundfont can contain many WAVE files of different pitches. You can import a soundfont into a sample player. To use soundfonts, you need an E-MU or SoundBlaster sound card, a MIDI controller, MIDI interface, and MIDI sequencing software.

Often a sampler is built into a sample-playing keyboard, which resembles an electronic piano. It contains samples of several different musical instruments. When you play on the keyboard, you hear the sample notes. The higher the key you press, the higher the pitch of the reproduced sample.

You can buy CDs with samples for use in your own projects. You copy the samples to your hard drive, load them into a software sample player, then trigger them with a sequencer or MIDI controller. For example, TASCAM's GigaStudio is sample-playing software with a huge sample library. FXpansion's BFD is a drum sample library. Other libraries are available from www.MI7libraries.com and www.sonomic.com. Several DAW recording programs have samples included.

You can also download samples from the Web. Samples are included in Steinberg's HALion Player (www.steinberg.net) and Maxim Digital Audio Piano (freeware at http://mda.smartelectronix.com).

A drum machine or beat box is a device that plays built-in samples of all the sounds of a drum set and percussion (Figure 16.3). It also is a sequencer that records and plays back drum patterns played or programmed with built-in keys or drum pads. Some units can sample sounds. Most recorder–mixers have a drum machine built in. In a DAW, you might use a soft sampler that plays drum samples—a virtual drummer.

Some drum plug-ins for MIDI/audio recording programs include EastWest Storm Drum, Toontrack Drumkit, Steinberg Groove Agent, M-Audio Drum & Bass Rig, Submersible DrumCore, and Cakewalk Session Drummer.

A power amplifier and speakers (or powered speakers) let you hear what you're performing and recording. Usually these are small monitor speakers set up in a Nearfield arrangement (about 3 feet apart and 3 feet from you).

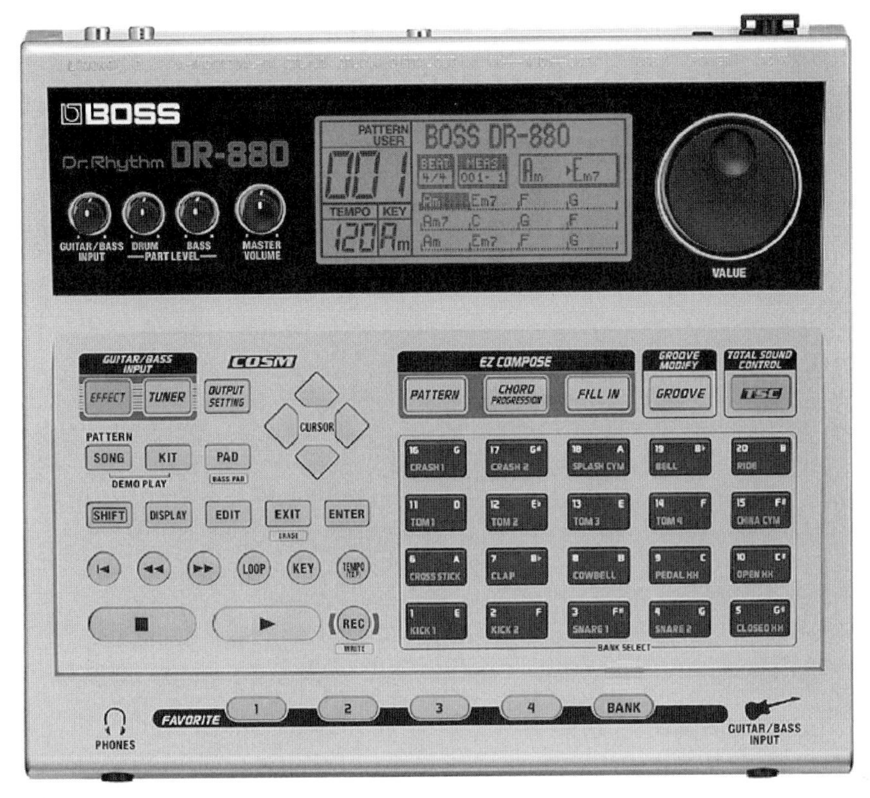

FIGURE 16.3
BOSS DR-880, an example of a beat box.

A computer is used mainly to run a sequencer program, which replaces the stand-alone sequencer and its tiny LCD screen. The computer monitor screen displays much more information at a glance, making editing easier and more intuitive.

A step up from a sequencer program is a MIDI/audio recording program. It records both MIDI sequences and digital audio tracks on your hard drive, and keeps them in sync. In other words, this program lets you add microphone signals such as a vocal, sax, etc., to MIDI sequences. This software and your computer form a Digital Audio Workstation (DAW). Details are given in Chapter 13, Computer Recording.

An alternative to a computer with MIDI software is a recorder–mixer (hardware DAW). It combines a multitrack audio recorder, MIDI sequencer, and mixer into a single chassis.

A librarian program organizes your collection of samples and synth patches.

A voice editor program lets you create your own synth patches.

A notation program converts your performance to standard musical notation. You can edit the notes, add lyrics and chords, and print out a copy.

A MIDI interface plugs into a user port in your computer and converts MIDI signals into computer data so that it can be recorded, edited, and played back with sequencer software. Many sound cards and audio interfaces include a MIDI interface.

If you have two or more hardware synthesizers, or a synth and a drum machine, you need a line mixer to blend their audio outputs into a single stereo signal.

Audio effects include compression, reverb, echo, gating, chorus, and so on. MIDI effects are effects that process MIDI signals rather than audio signals. Some MIDI effects are arpeggiate, transpose, and delay.

Audio cables carry audio signals and typically have a 1/4-inch phone plug on each end. They connect synths, sound modules, and drum machines to the line inputs in your mixer or audio interface.

MIDI cables carry MIDI signals and are used to connect synths, drum machines, and computers together so that they can communicate with each other. A MIDI cable is a 2-conductor shielded cable with a 5-pin DIN plug on each end. Pins 4 and 5 are the MIDI signal, pin 2 is shield, and pins 1 and 3 aren't connected.

An equipment stand is a system of tubes, rods, and platforms that supports all your equipment in a convenient arrangement. It allows comfort, shorter cable lengths, and more floor area for other activities.

A keyboard workstation (Figure 16.4) includes several MIDI components in one chassis: a keyboard, a sample player, a sequencer, and perhaps a synthesizer, and disk drive. That's everything you need to compose, perform, and record instrumental music. Many workstations include drum sounds so that you can get by without a separate drum machine. A few include a mic input so you can record vocals or acoustic instruments.

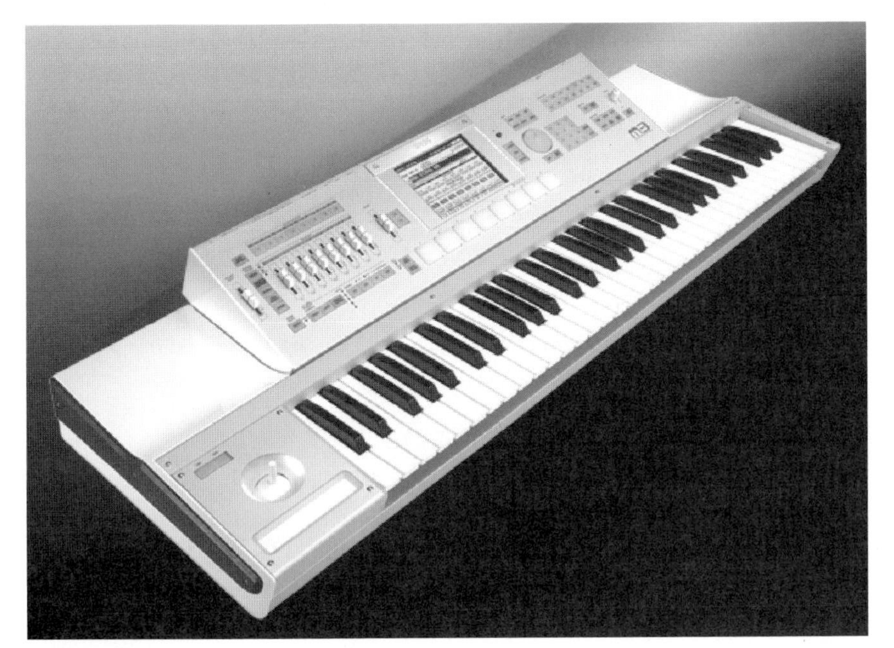

FIGURE 16.4
The Korg M3, an example of a keyboard workstation.

The rest of this chapter describes recording procedures for some MIDI studio setups. You'll also need to read your instruction manuals thoroughly and simplify them into step-by-step procedures for various operations. Note that each piece of MIDI gear has its own idiosyncrasies, and the instructions may have errors or omissions. If you have questions, call or e-mail tech support for your equipment. Also look for books and videos devoted to your specific musical instrument.

RECORDING MUSIC MADE BY SOFT SYNTHS

In this method, you will play a MIDI controller and record the performance on a MIDI track in a MIDI/audio recording program on your computer. Software synths, samples, and drums in your computer will play the sounds during recording and playback (Figure 16.5). They are included with most recording software, or you can buy them separately.

FIGURE 16.5
rgc:audio Pentagon, an example of a soft synth.

The MIDI Signal Chain

Figure 16.6 shows a MIDI signal chain that uses a soft synth. From start to finish (left to right), these are the components and their functions:

1. **MIDI controller.** A piano-style keyboard, drum pads, or breath controller that outputs MIDI code (note on, note off, key number, etc.) when you play it. The code is tagged with a channel number from 1 to 16, which you set in the controller.
2. A **MIDI cable** from your controller's MIDI OUT to your interface's MIDI IN.
3. A **MIDI interface** or MIDI/audio interface. It converts the MIDI signal into a data format that your computer can understand.
4. A **PCI, USB, or FireWire** connection from the interface to the computer.

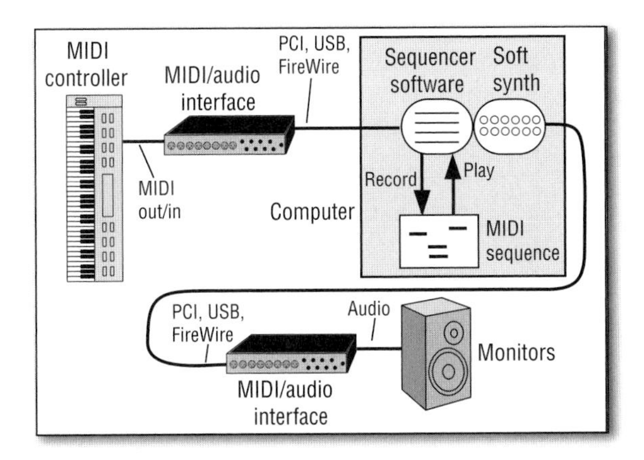

FIGURE 16.6
A MIDI signal chain that uses a soft synth.

5. A **MIDI driver** (not shown). A small program in your computer that allows your recording software to transfer data to and from an audio/MIDI interface.
6. **Sequencer** (MIDI/audio) software running in your computer. It records the MIDI signal to the computer's hard drive and plays it back.
7. **Soft synth**. A plug-in in your MIDI/audio host program. As you record your performance, you're listening to the software synthesizer play music according to your keyboard presses. So in addition to being recorded, the MIDI signal goes to a soft synth or sample player of your choice.
8. **PCI, USB, or FireWire** connection (same as #4 above). The synth's digital audio is sent back to your MIDI/audio interface through the same PCI, USB, or FireWire cable that is sending MIDI data to the computer sequencer.
9. The **interface** (same interface as #3 above) converts the synth's digital audio to analog, and sends that musical signal to its stereo output channels.
10. **Monitors**. The stereo output of your interface feeds amplified speakers or headphones so you can hear the sound of your performance.

During playback, the recorded MIDI sequence "plays" or triggers the soft-synth patch, and its audio is routed through your interface and monitors.

MIDI Recording Procedure

Now that you understand the MIDI signal chain, the steps below will make more sense. Here's how to create the audio and MIDI tracks in your sequencer (MIDI/audio) program.

1. Set Up the Audio and MIDI Tracks
   Insert a MIDI track to record your MIDI sequence, and insert an audio track that contains the soft synth as a plug-in. (Your sequencer might combine the MIDI and audio tracks into one track).
   Synth and sample patches are stored in groups called banks, and each bank contains several different patches In the MIDI track, choose a synth bank and patch (bass, drums, piano, etc).
   In the MIDI track input, specify the MIDI channel that the synth will respond to: either Omni mode (any channel) or the same channel number that your MIDI controller is set to.
   In the MIDI track output, specify the synth or sample player that you want the MIDI track to play.
   Set the audio track output to the stereo output of your interface.
2. Start Recording
   In your interface control program, or in your recording program, set the latency or buffer size low (under 4 msec if possible) while tracking so that the synth responds without a noticeable delay. See Appendix B under the heading Minimizing Latency.
   Choose a tempo in your sequencer. Set the metronome to count off two measures.
   Enable record mode on the MIDI track. Using your mouse, click on the Record button on-screen or on your computer keyboard. You'll hear the sequencer's metronome ticking at the tempo you set.
3. Play Music on Your Keyboard
   Listen to the metronome and play along with its beat. The sequencer keeps track of the measures, beats, and pulses. As you play, MIDI data from your keyboard goes from keyboard MIDI OUT to interface MIDI IN, and is recorded as a MIDI file or sequence. When the song is done, click on STOP.
   Another way to record your performance is in step-time, one note at a time. You also can set the sequencer tempo very slow, record while playing the synth at that tempo, and then play back the sequence at a faster tempo.
4. Play Back the Sequencer Recording

Click on PLAY. You'll hear the sequence playing your soft synth. If you change the patch and click on PLAY, you'll hear the same performance played by a different instrument.

5. Layer the Drum Track
   Using the recording software's sound-on-sound setting, you can record several passes of a MIDI drum track, adding another instrument each time. On the first pass tap-in the kick and snare, on the next pass add hi-hat, and so on.
6. Quantize the Track
   Quantizing automatically corrects the timing of each note to the nearest note value (quarter note, eighth note, and so on). If you wish, quantize the performance by the desired amount. **Caution**: Quantizing can de-humanize the performance, making it too rhythmically perfect. It's better to adjust the timing of certain notes only, and to the smallest note value that works. Quantizing is essential if you want to use a notation program.
7. Punch-In/Out to Correct Mistakes
   To correct a musical error, you can punch-in to record mode before the mistake, record a new performance, and then punch-out of record mode. Another way to correct errors is to edit the MIDI notes in the piano-roll view, described in Step 8, Edit the Sequence Recording. Here's one way to punch in:

   1. Go to a point in the song a few bars before the mistake.
   2. Just before you get to the mistake, punch-in to record mode and play a new, correct performance.
   3. As soon as you finish the correction, punch-out of record mode.

Alternatively, you can use autopunch. With this feature, the computer punches in and out automatically at preset measures; all you have to do is play the corrected musical part. Perform an autopunch as follows:

1. Using the computer keyboard or mouse, set the punch-out point (the measure, beat, and pulse where you want to go out of record mode).
2. Set the punch-in point (just before the part you want to correct).
3. Set the cue or pre-roll point (where you want the track to start playing before the punch).

4. Click on PLAY.
5. When the screen indicates punch-in mode, or when the appropriate measure comes up, play the corrected part.
6. The sequencer punches out automatically at the specified point in the song.

These punch-in routines were done in real time. You can also punch-in/out in step-time:

1. Go to a point in the song just before the mistake.
2. Set the sequencer to step-time mode.
3. Step through the sequence note by note, and punch-in to record mode at the proper point.
4. Record the proper note in step-time.
5. Punch-out of record mode.

8. Edit the Sequence Recording
   You might find it easier to edit the MIDI performance. Go to the MIDI edit screen or piano-roll view (Figure 16.2). It's a grid showing pitch verses time. The pitch of each note is represented by its height on the grid, and the duration of each note is represented by it length. You can grab incorrectly played notes and put them at the correct pitch and timing, delete unwanted notes, copy and paste phrases of notes, stretch or shorten note durations, and so on.
   When you're finished editing one MIDI performance, overdub other MIDI tracks, edit them, and set up a mix.
9. Arrange the Song by Combining Sequences
   At this point you can put together your composition. Many songs have repeated sections: The verse and chorus are each repeated several times. If you wish, you can record the verse and chorus once. Then copy the verse section and paste it every place it occurs in the song. Do the same for the chorus.
10. Add Audio Tracks

    Suppose you want to add vocals or live instruments to the mix.

    Here's a suggested procedure.

    1. If you're using a sound card, plug a mic into a mic preamp or mic input of a mixer. Connect the preamp or mixer line output to your sound-card line input. If you're using an audio interface, plug a mic into one of its mic connectors. Connect the interface's PCI, USB, or FireWire port to your computer.

2. In your recording software, insert an audio track, and set its input source to the audio interface or sound-card channel that the mic signal is plugged into. For example if you plugged a mic into interface input 3, set the track's input to interface input 3. Enable Record mode on that track.
3. Set the recording level by adjusting the microphone input trim or gain on your preamp, mixer, or interface.
4. Go to the beginning of the tune and hit PLAY in your DAW. The MIDI sequences that you recorded earlier should start playing. (You may need to press the PLAY key on a drum machine first if it is an outboard device, rather than a part of software.)
5. While listening to the MIDI tracks playing through headphones, record the vocal on an audio track. Then overdub more vocals and non-MIDI instruments on other open tracks.

You might record a few takes and then cut and paste selected portions to create a perfect take. For example, record one good chorus and copy it in each chorus section in your song. You also can edit audio notes to correct their pitch or timing.

When you play back what you recorded, the recorded MIDI tracks play the soft synths, and the digital audio tracks play their audio signals. A mix of the soft synths and audio tracks plays from the audio interface line output, which is connected to your monitors.

11. Mix, Export the Mix, and Burn a CD

    1. After all your tracks are recorded, use the on-screen mixer to set up a mix of the audio tracks and MIDI instruments. Adjust levels, panning, and effects (plug-ins).
    2. Play the song several times to perfect the mix and to set up automation. See Chapter 12 for mixing procedures and automation.
    3. Export the mix to a stereo WAVE file on your hard drive. Then you can burn a CD of the song or convert it to an MP3 file.

RECORDING A HARDWARE SYNTH

Suppose you like the sound of a hardware synth. You want to record its sound and also record a MIDI sequence of your performance so you can edit it. Figure 16.7 shows the connections.

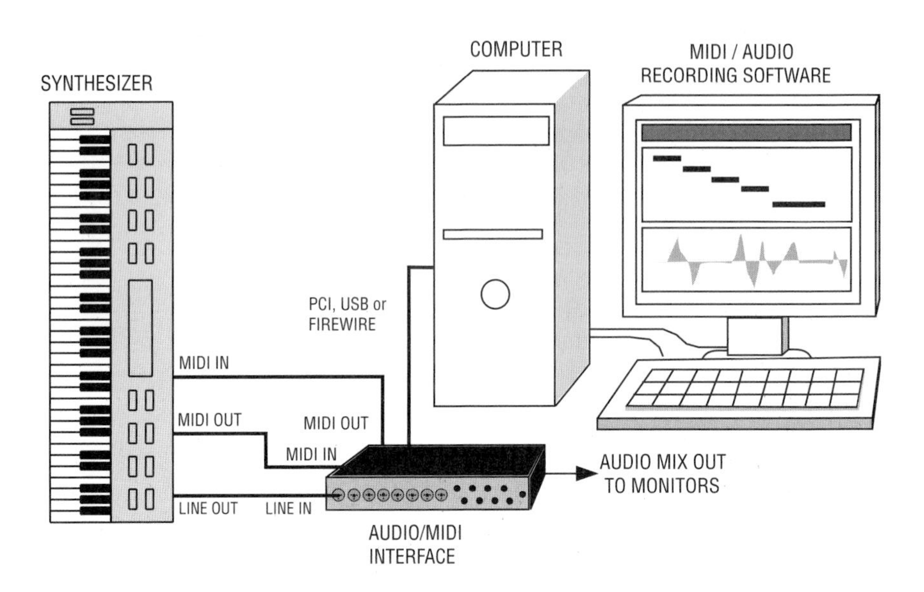

FIGURE 16.7
Setup for recording a hardware synth on a computer.

1. Monitor a mix of the synth audio and sequencer playback. In other words, monitor a mix of the interface input signal and output signal.
2. Insert a MIDI track in your sequencer program. Set its input to your MIDI/audio interface in Omni channel mode. Set the output of the MIDI track to your MIDI/audio interface. That way, during playback the recorded sequence will pass through the MIDI/audio interface to your hardware synth and will trigger it to play music.
3. Enable Record mode in the MIDI track. Record your performance on the MIDI track. (The sequencer will also record whatever MIDI channel number is set in the synth.)
4. If necessary, set your synthesizer to External Clock (MIDI clock) so it will follow the timing of the recorded sequence.
5. Play the sequence and edit it in the piano-roll view.
6. Insert an audio track. Set its input to the interface line input that is connected to the synth audio output. Disable Record mode on the MIDI track, and enable record mode on the audio track.
7. Make sure that the synth MIDI channel is unchanged. Set levels and start recording. In other words, play the MIDI track

while recording the synth audio on the audio track. The MIDI sequence will "play" or trigger the synthesizer as you record its output on an audio track.

8. When the song is done, you have an audio track produced by the hardware synth. If you wish, you can disable the MIDI track that drove the synth.

"NO SOUND" MIDI TROUBLESHOOTING

Suppose you load a MIDI file and hit PLAY on your sequencer, but you hear nothing. Or you play notes on your MIDI controller, but there's no sound. Not a peep.

That's not too surprising, considering that the MIDI signal chain is a long one. Every link in the chain has to work, whether it's hardware, software, or their settings. I'll give some MIDI troubleshooting advice that should coax your system to speak when you ask it to.

If any part of the chain isn't turned on, or has the wrong settings, you won't hear a sound when you monitor the synth—no matter how hard you pound the keys! Listed below are some possible reasons why your synth can't sing, and what to do about it.

- MIDI OUT or MIDI THRU isn't connected to MIDI IN somewhere in your system. Trace the cable connections from beginning to end, and see whether a MIDI cable isn't plugged in where it should be. The cable itself might be broken.
- Your MIDI interface isn't communicating with your computer. A missing or flaky PCI, USB, or FireWire connection can disrupt the data flow between interface and computer. Re-plug or replace the cable. You might need to restart your sequencer program too. Some recording software has an on-screen indicator that flashes when the software is receiving MIDI data, so check it out.
- When you're tracking, the MIDI sequencer track isn't selected. Some sequencer software doesn't monitor the soft synth unless you select its MIDI track. In some sequencers, you can live-monitor only one synth at a time, but you can play back multiple synth tracks at once.
- The MIDI sequencer track has no MIDI channel assigned, or is set to a different channel than the MIDI controller. Just as

you need to set your TV to a certain channel to see a station's programming, you need to set your MIDI sequencer's track to the same channel your controller is sending data on. It might be easiest just to set the MIDI track input to your interface in Omni mode, so the track will hear any channel that the controller is sending. You can set multiple MIDI tracks to Omni; just record one track at a time.

- The MIDI sequencer track output isn't pointing at your soft synth. You have to tell the MIDI track which synth to play.
- The wrong sound bank was selected in a synth. You might have specified a bank that contains no sounds. Or the MIDI mapping on your controller is set so the notes aren't playing the patches you want them to play. Check the bank setting, patch settings, and MIDI mapping.
- In the sequencer's options menu, under MIDI devices, your MIDI interface isn't selected. The sequencer doesn't know where to find the driver that communicates with your interface, so you don't hear anything. If your sequencer has a MIDI data indicator that is NOT flashing in response to your playing, maybe you forgot to select the MIDI driver.
- The synth's audio-track volume, or the MIDI track's key velocity, is turned down. Disable any mute buttons and turn up the faders for those tracks. Check that no other tracks are soloed.
- You started playing the MIDI track in the middle of a long note, rather than at its beginning. A synth needs to receive a Note-On command to play a note. If you start playing the sequence from a point *after* the beginning of a note, the synth doesn't receive the note-on code, so it remains silent.
- Your hardware synthesizer lost the program-change command. Although you told the synth which patch to play, sometimes this information is lost. Change the patch on the synth and then set it back as it was. Or record the program change at the beginning of the song.
- The input to your sequencer track lost the setting for its MIDI input device. Sometimes when you close a sequencer program and reopen it, the MIDI track inputs might forget their settings. Temporarily set the MIDI track input to another device, then set it back as it was.
- The MIDI driver is buggy. Download the latest update from the interface manufacturer.

- The synth audio output isn't assigned to your interface's stereo output channels. In other words, the synth is making music but not sending it anywhere.
- Can you hear any audio playing through your monitors? If not, the monitor power amp, or your mixer's monitor controls, might be off or turned down. Maybe the cables between your interface and monitors aren't plugged in. Seems obvious, but it's happened to me.

I hope these suggestions make your speakers play notes when you press keys. The better you understand the MIDI/audio signal path, the better you can find errors and fix them.

RECORDING WITH A KEYBOARD WORKSTATION

This is another way to record in a MIDI studio. Basically you follow the steps for multitrack sequencer recording above, but do them in the keyboard workstation rather than in a computer. The workstation's manual tells how to do it.

You can access the keyboard's effects menus to set up the effects that are heard with a patch: hall reverb, chorus, flanging, echo, distortion, and so on. Press the correct number on the numeric keypad to get to the effects menus. (Note that these are built-in keyboard effects, not outboard studio effects.)

Save the completed song (MIDI sequence) in multitrack form to the keyboard's internal memory or to a plug-in RAM card. If you're satisfied with the final results, use a DAW to record the stereo output signals of the workstation. Then you can burn a CD or create an MP3 file from the WAVE file you just recorded on your hard drive.

In addition to these basic operations, you can:

- Bounce tracks: copy one track's performance to another track so it will play another patch
- Edit each note event
- Create and copy patterns—for a drum or bass part, for example
- Modify track and song parameters
- Insert/delete/erase measures
- Modify sounds and effects
- Change the instrument (patch) that each track plays

RECORDING WITH A DRUM MACHINE AND SYNTH

Suppose you want to record a drum pattern on a hardware drum machine, and then you want to add a synth part using your computer.

First record the drum pattern. Here's one suggested method:

1. On the drum machine, set the tempo, time signature, and pattern length in measures. In this example, the pattern will be 2 bars long.
2. Add a count-off (a few measures of clicks) at the beginning so that overdubs made later can start at the correct time.
3. Start recording, and play the hi-hat key in time with the metronome beat.
4. At the end of 2 bars, the hi-hat pattern you tapped repeats over and over (loops).
5. While this is happening, you can add a kick drum beat.
6. While the hi-hat and kick drum are looping, add a snare drum beat, and so on.
7. Mix the recording by adjusting the faders or keys on the drum machine for each instrument.

Next, you repeat the process for a different rhythmic pattern—say, a drum fill—and store this as Pattern 2. Then develop other patterns. Finally, you make a song by repeating patterns and chaining them together as described in the drum machine's instruction manual. A song is a list of patterns in order.

Some musicians like to compose by programming a simple repeating drum groove first. While listening to this, they improvise a synth part. After recording the synth part, they redo the drum part in detail, adding hand claps, tom-tom fills, accents, and so on.

Finally, copy the audio signal of the drum pattern to a stereo track in your computer recording program. Play back that track while overdubbing other parts.

USING EFFECTS

No matter how you record with MIDI, effects are an important part of the mix. To keep the sound lively, try to vary the effects throughout the song, or use several types of effects in the mix.

For example, suppose you have a multitimbral synth, and you want to add a different effect to each patch. Whether or not you can do this depends on your synth. If it has a separate output for each patch, you can use a different effect on each patch. But if your synth has only a single output (mono or stereo) and you run it through an effects device, the same effect is on all the patches.

If your song includes program changes (patch changes), you can have the effects change when the patch changes. Set up a MIDI-driven multi-effects processor so that each synth program change corresponds to the desired effect. When the synth program changes, the effect changes also.

What if you want the effect, but not the synth patch, to change during a mix? Reserve a track and channel just for effects program changes. You don't hear these program changes in your synth, but you do hear the effects change. During a mixdown, it's usually easier to change effects automatically with your sequencer, rather than manually.

In a sample player, some samples might have reverberation or some other effect already on them: the effect is part of the sampled sound. Note that the sampled reverberation cuts off every time you play a new note. Although this sounds unnatural, you can use it for special effect.

Because effects are audio signals, audio recorders can record effects but MIDI sequencers can't. MIDI sequencers, however, can play synths or patches that have effects built in.

Some recording software lets you convert a MIDI track to an audio track, then apply audio effects to that track. Other software uses a separate MIDI track and audio track for the same instrument patch. You apply effects to the audio track.

MIDI effects (MFX) are non-audio processes applied to MIDI signals, such as an arpeggiator, echo/delay, chord analyzer, quantize, transpose MIDI event filter, or velocity change. They are plug-ins in MIDI tracks.

LOOP-BASED RECORDING

Let's turn to a different aspect of computer MIDI recording. It's possible to compose, record, and perform music entirely in software.

You might start by creating a variety of loops or grooves, which are constantly repeated rhythmic or musical patterns.

You can buy loops or make your own. You might create a loop by recording a 4-measure sample of a drum beat or bass riff that you play to a metronome or click track. Or use a handheld recorder to sample some interesting sounds in the field. The sounds might create a rhythm when repeated. If not, you can edit the timing of individual sounds until you get a rhythm you like. Soft synths are another source for loops, and some are bundled with ready-to-use patterns. Many recording programs include loop patterns. You can also create loops from parts of tracks that you recorded.

Making Your Own Loops

Let's make a 4-bar loop in 4/4 time using the editing function in a DAW. Start with a musical pattern longer than 4 measures that was made to a metronome or click track. Ideally the pattern has little or no reverb so that the reverb won't be cut off when you trim the loop.

Trim the start of the loop waveform just before beat 1, and trim the end of the loop after measure 4-beat 4, and just before the next beat 1. To avoid a click in the audio when the loop repeats, both trim points should be at zero crossings where the waveform crosses the 0-volt line (Figure 16.8)

Copy the trimmed 4-measure clip and paste it on the beat several times to repeat it. Or right-click the clip and select Loop Clip. If you hear no clicks and the rhythm is steady, export the 4-measure loop

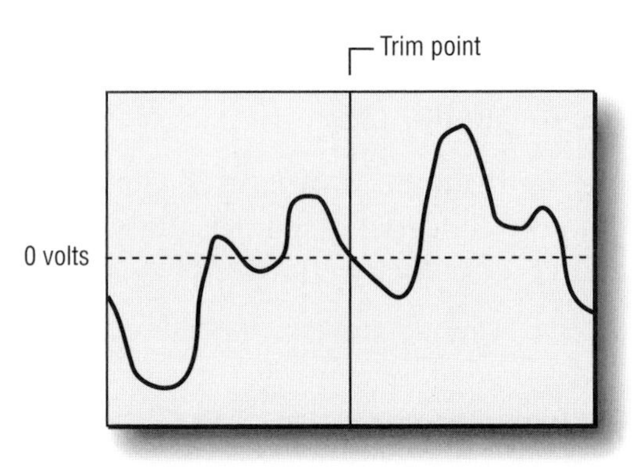

FIGURE 16.8
Trim loops at zero crossings in the waveform.

to your loop library. Store your loop collection by tempo, style, and type; for example, 120 bpm, jazz, drum set with brushes.

Once you have created several loops, you can drag and drop them (or import them) into your DAW software. There, you can copy and paste loops on the beat to repeat them. A SONAR groove clip (explained next) can be repeated by click-dragging the end of the clip to the right.

Types of Loops

There are five types of loop audio files based on their ability to have their tempo changed:

- A loop made from a standard digital audio file. It has a fixed tempo, so you must build your song around that tempo.
- A processed audio file. The tempo or pitch of the loop can be changed by a time-stretch or pitch-shift algorithm in your audio editing program. But if you need to change the tempo of a song, you need to adjust all the loops you stretched.
- A file with REX-based time-stretching. The transients in the audio file are cut into slices, whose spacing depends on the tempo. REX files follow tempo changes in your composition.
- An acidized (RIFF) WAV file. Pitch and tempo information are in the file header, and the audio is sliced at transients as with REX files. Acidized files follow tempo and key changes. Slowing down the loop by more than 15 or 20 bpm can add artifacts, so it helps to start with a slow tempo of less than 120 bpm.
- Cakewalk SONAR recording software includes Groove Clips that have embedded tempo and root-note pitch. SONAR can stretch the clips to match changes in tempo, or transpose clips to match the project's key. You repeat a Groove Clip by dragging its end to the right in the track. Almost any audio clip can be converted to a SONAR Groove Clip; just select the clip and press Ctrl-L.

Working with Loops

Once you've made several repetitions of the groove, you can add effects, distortion, or pitch/tempo changes. You might put some echo or reverb on a drum loop and vary the dry/wet mix with automation as the loop plays. When the loops are complete, overdub audio and MIDI tracks, such as soft-synth parts, percussion, vocals, and acoustic instruments. Import other loops and place them where you want them in the song.

You can also create loops externally, then import them into a DAW recording program. For example, compose a repetitive beat of drums, synth, and samples in a beat box. Launch your DAW software, set up a stereo track, and record the box output to that track as it plays the beat. On other DAW tracks you can overdub vocals, doubled vocals, rap sections, and harmony.

Repetitive loops can get boring, so edit or remove various notes in the piano-roll view to add some variety. You might insert some spaces here and there in the song to create breaks for solos or record scratches.

Some loop programs offer groove quantizing, which transfers the timing and dynamics from one groove to another. It allows human-like variations in timing and key velocity.

Loop Libraries

Loop libraries are collections of sampled drum and bass beats that let you loop (repeat), change tempo, etc. The beats come as MIDI files and WAVE files. Some examples are M-Audio ProSessions, Hark Loops, Cakewalk loop libraries, Sony Creative Software Artist Integrated loop libraries, Sonic Foundry Acid loop libraries, GarageBand, Smart Loops, FL Studio, Beatboy, Keyfax Twiddly Bits, Groove Monkee, FXpansion, Discrete Drums, APO Multimedia Mix It, Multiloops Naked Drums, Pocket Fuel RADS series, ILIO.com, Wizoo VST Drum Sessions, www.bigfishaudio.com, and www.kellysmusicandcomputers.com/loop_sample_libraries.htm. Do a Google search for "Loop Libraries" or "ACID loops." Complete looped rhythm tracks are available online for purchase. Just add vocals.

Loop Creation Software

Many DAW recording programs include loop-creation software, which is also available separately. For example, Propellerheads Software (www.propellerheads.se) has a variety of programs to create and modify loops. They also offer soft synths, drum machines, and sequencers. Here are some of their products:

ReCycle is a stand-alone tool that starts with a loop and lets you change its tempo and pitch, and replace and process sounds within the loop. By detecting peaks in the waveform, ReCycle automatically breaks a loop into parts or slices. A slice might be a snare-drum hit, a kick-drum hit, or a kick-and-snare hit. When a loop's tempo is changed, the start point of each slice moves in time so that the beats occur at the right time. (You might need to touch this up manually.)

If the tempo is slowed down, ReCycle creates a decay after each drum hit to fill in the gaps. You can delete slices, change their length, attack, decay, and pitch; and add compression or EQ. Then you can import the improved loop as an REX2 file into an audio track in your sampler or sequencer program. There you can control all aspects of the loop.

Reason is a group of synthesizers and samplers, a drum machine, mixer, effects, pattern sequencer, and more (Figure 16.9). It's all-in-one and easy to learn. Reason can import and play ReCycle's REX files.

Reason Adapted is a lite version that is distributed as part of some software bundles.

ReBirth is a software emulation of two analog bass-line synths and two classic drum machines. Also included are delay, distortion, compression, and filtering. It integrates into sequencer software and allows real-time audio streaming.

Reload is a utility that converts AKAI S1000 and S3000 formatted media into formats (such as WAVE files) that can be used with Reason, ReCycle, and other audio applications.

ReWire is a useful feature found in some loop programs. It transfers or streams audio data between two computer applications in real time, almost like a cable. This allows programs to communicate with each other and synchronize together.

Other loop-based tools to compose and record music are listed below.

Ableton Live lets you compose with soft synths, record hi-resolution multitrack audio, and play loops in a live performance. With Live you can make stuttering drums, create arrangements, do MIDI loops, and modify grooves, timing, pitch, volume, and effects.

Sony Creative Software's **ACID** products let you select audio loops from Windows Explorer, drag them into a recording program's track

FIGURE 16.9
The Thor synthesizer screen in Reason software.

view, and arrange them into multitrack projects. The tempo and key of each loop are automatically matched to your project's music in real time, using the slicing technique described before. ACID uses time-stretching algorithms to lengthen sounds when the tempo is slowed down.

Other loop-intensive DAWs include Image Line Software FL Studio (PC) and Apple GarageBand (with Apple Loops included). Some excellent articles on loops were in the July 2004 and August 2002 issues of *EQ Magazine* and in the January 2007 issue of *Recording* magazine.

CHAPTER 17

On-Location Recording of Popular Music

Sooner or later you'll want to record a band—maybe your own—playing in a club or concert hall. Many bands want to be recorded in concert because they feel that's when they play best. Your job is to capture that performance on a recorder and bring it back alive.

You can record pop-music concerts in several ways. Listed below are a range of techniques from simple to elaborate. In general, sound quality improves as the recording setup becomes more complex.

- Record off the board (PA mixer) into a portable stereo recorder.
- Record with two mics and a portable stereo recorder.
- Record with a 4-tracker: Record with a stereo mic on tracks 1 and 2, while recording the PA mixer output on tracks 3 and 4. Mix the tracks later.
- Feed the PA mixer's insert-send signals to a multitrack recorder. Edit and mix the recording back in your studio.
- Use a mic splitter on stage to feed the PA mixer, monitor mixer, and recording mixer. Record to multitrack. Edit and mix the recording back in your studio.
- Do the multitrack recording in a van or truck.

We'll describe each method so you can decide what's best for you and try it out.

RECORD OFF THE BOARD

Let's start with the easiest technique: connect the PA mixing console (board) to a portable digital recorder.

Using cables with the appropriate mating connectors, connect the PA mixer's tape-out or 2-track-out connectors to your stereo recorder's line input(s). If your recorder's line input is a stereo mini phone jack, use an adapter cable: either two 1/4-inch phone plugs to a stereo mini phone, or two RCA connectors to a stereo mini phone. They are available at Radio Shack and other electronic suppliers. Outside the U.S., if your recorder's line input is a stereo mini socket, use an adapter cable: either two 6.35 mm jack plugs to a stereo mini jack, or two phono connectors to a stereo mini jack.

The recorded mix might be poor with this method. Here's why: The sound mixer hears a combination of the band's instruments, the stage monitors, and the house speakers. So the board mix is intended to *augment* the sound of the instruments and monitors on stage—not to sound good by itself. For example, if the bass-guitar amp is very loud on stage, it will be turned down in the mix that you record off the board. Vocals will be turned up high in the mix because they can't be heard otherwise. Board mixes can sound good if there isn't much sound coming off the stage (as with acoustic groups), and the venue is large or outdoors. It helps to monitor the board mix with headphones or isolating earphones to hear what you're recording.

RECORD WITH MICS AND A PORTABLE DIGITAL RECORDER

Here's another simple method. Try a stereo mic and a portable stereo recorder. The process is easy and the required gear costs as low as $200. A stereo recording may not offer the sound quality of a professional multitrack recording. But if you record how the band sounds from a seat in the audience, that may be good enough—especially if the recording is just a demo, or is for the band members.

You can use a recorder with built-in microphones, or you can plug two mics (or a stereo mic) into the recorder. You'll pick up the group as a whole from several feet away. The mics will capture not only the musicians, but also the room reverb and background noise. You could call it a "documentary" or "audio snapshot" recording.

As you record, what you hear with headphones or earphones is what you get. There's no mixing back in the studio.

If you place the mics a few feet from a folk group or jazz group without a PA system, the sound can be quite good. But most bands use a PA. When you record the band you're also recording the sound of the PA speakers, so the mix or balance you get depends on the PA engineer's skill. The sound will be more distant and muddy than you get with several close-up mics.

Gear

You'll need a portable digital recorder, described in Chapter 4. A mic stand is optional.

Headphones or earphones let you know whether the mics are working correctly, and let you hear what the mics are picking up. If the band and PA are loud, it's hard to hear what's being recorded unless you use isolating headphones (Remote Audio HN-7506) or isolating earphones (Etymotic ER-4S, ER-4P, and ER-6i; Shure E3 and E4.)

Preparing for the Session

Before going on the road, install fresh batteries and clean the connectors with isopropyl alcohol or DeoxIT from Caig Labs (www.caig.com). Do a trial recording to make sure everything works.

If possible, record the gig in a room where the audience is attentive and the background noise is low. You might visit some potential venues to check out the noise and acoustics. Avoid very live rooms because they can make the recording muddy.

Be sure you have enough free space on your Flash memory card before going on location. Listed below are approximate recording times for a 1 GB card. Double these times for a 2 GB card.

24-bit/ 44.1 kHz stereo WAV—1.0 hour
16-bit/ 44.1 kHz stereo WAV—1.5 hours
256 kbps stereo MP3—9.0 hours
128 kbps stereo MP3—16.5 hours

The recorder might have a record-level switch labeled "manual" and "auto." Set it to manual to retain the dynamics of the performance. If the switch is labeled "AGC" (Automatic Gain Control), set it to "off."

At the Gig

Plug in some headphones and turn on Record Monitor mode. You'll hear the room acoustics and any background noise (audience, air

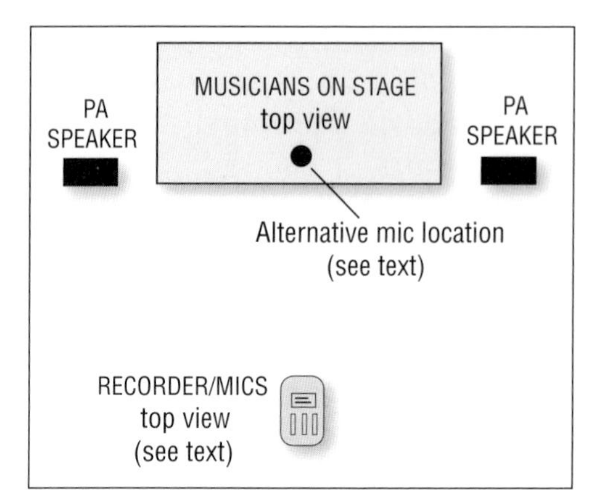

FIGURE 17.1
Typical mic placement for a stereo recording of a pop group.

conditioning, traffic). Room noises that you wouldn't otherwise notice become obvious when you listen on headphones. Also listen for buzzes, distortion, and crackles from bad cables or connections.

Where do you place the mics? The closer the mics are to the group, the clearer and cleaner the sound is. In other words, close placement captures more of the music and less of the room sound and background noise. Try to place the mics as close as possible to the stage where you still pick up the house PA speakers well: about a stage-width away from the stage (Figure 17.1). Keep the mics away from a bar or other noise sources.

If there are dancers near the stage and the ceiling is low, you might try boundary mics (such as two PZM®s) gaffer-taped to the ceiling or mini mics hung from the ceiling.

To eliminate muddy room sound, you can record acoustic groups off the PA mixer into a line input. Or try placing the recorder/mics on stage (on a stool or mic stand) and record the sound of the monitor speakers plus the band. Some bluegrass or old-time bands sing into one microphone, so mount the recorder/mics next to that mic.

If the group plays in a circle (as in an outdoor jam or rehearsal), try placing omnidirectional mics in the center. Or just walk around with the mics while monitoring the mics over headphones. Find a spot where you hear a good balance and put the mics there. Ask the musicians' permission to record them during a break in the music.

Set the recording level so the meter reads about −6 dB maximum. That allows some headroom for surprises. Peak levels above 0 dB result in distortion. Some portable recorders include a limiter which prevents recording levels above 0 dB. Others have a clip or peak LED that flashes when the recording level is too high.

Recorders have a mic gain switch or pad (attenuator) to prevent mic-preamp distortion. If you have to set the recording-level control less than 1/3 up to achieve a 0 dB recording level, use the low-gain setting or switch in the pad.

If you're using large-diaphragm condenser mics plugged into a phantom power supply, you might need to plug the output of the supply into the recorder line input—rather than the mic input—to prevent distortion.

A Recording Session with No PA

You might be able to record a small folk group or acoustic jazz group on location without a PA system or audience. This gives you the freedom to improve the sound.

1. If the room is too live (reverberant), put up some packing blankets, comforters, rugs, acoustic foam, or cushions.
2. Often a good-sounding spot for the musicians is near the center of a large room. Place the musicians around the stereo mic pair where you want them to appear in the recording. For example, you might place two singing guitarists on the left and right, with bass in the center.
3. Experiment with microphone height to vary the vocal/guitar balance. Try different miking distances to vary the amount of ambience or room sound. A typical distance is three to six feet (Figure 17.2).
4. As the musicians are playing and during playback, monitor the mic signals with headphones. If some instruments or vocalists are too quiet, move them closer to the mics, and vice versa, until the balance sounds right.
5. While the band is rehearsing a loud song, make a trial recording. Set the recording level and mic gain as described earlier.
6. Record a tune. If someone makes a mistake, either record another take of the entire tune, or record starting from a few bars before the mistake, and edit the takes together later.

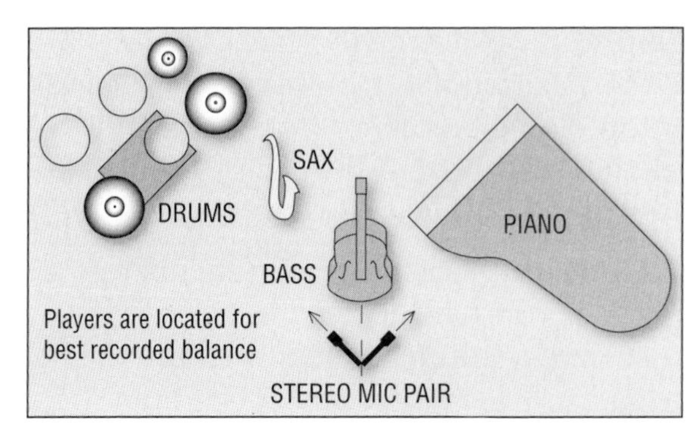

FIGURE 17.2
A method of stereo miking a jazz group.

After the Gig

Back in the studio, connect the USB port in the recorder to the USB port in your computer. The recorder shows up as a storage device on your computer screen. Drag-and-drop the recorded sound files to the computer's hard drive for editing and CD burning. The files transfer in a few minutes. Then the Flash memory card is empty and you're free to make more recordings.

Now you can edit the recording and adjust its tonal balance (equalize it) with DAW recording software or with Harmonic Balancer software (www.har-bal.com). You might cut a few dB around 300 Hz to reduce boomy reverb and get a clearer recording. If your mics are weak in the bass, compensate by boosting a few dB at 50 to 100 Hz.

Try out these tips and enjoy your recordings!

RECORD WITH A 4-TRACKER

This method is fairly simple and provides good sound. Put a stereo mic or mic pair at the front-of-house (FOH) PA mixer position. Plug the mic connectors into mic inputs 1 and 2 of a 4-track recorder–mixer (or any size recorder–mixer). Connect the PA mixer's tape outputs or 2-track outputs to line inputs 3 and 4 (Figure 17.3). Mix the recording to stereo back in the studio.

The FOH mics pick up the band as the audience hears it: lots of room reverb, lots of bass, but rather muddy or distant. The PA mixer output sounds tight and clear, but typically is thin in the bass.

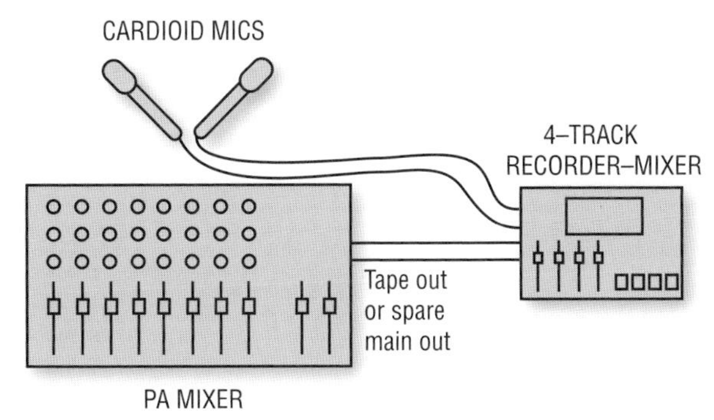

FIGURE 17.3
Recording two mics and a PA mix on a four-track recorder–mixer.

Luckily, a mix of all four tracks can sound surprisingly good. Tracks 1 and 2 provide ambience and bass; tracks 3 and 4 provide definition and clarity.

When you mix the four tracks, you might hear an echo because the FOH mics pick up the band with a delay (caused by the sound travel time from stage to mics). To remove the echo, import all the tracks into digital recording software, and delay the PA mixer tracks by sliding them a little to the right. Align the waveforms of the mic tracks and mixer tracks at big peaks.

CONNECT THE PA-MIXER INSERT SENDS TO A MULTITRACK RECORDER

Now we get into professional techniques. This is an easy way to record, and it offers very good sound quality with minimal equipment (Figure 17.4). Connect the mixer insert sends to the inputs of a multitrack hard-drive recorder. Use the mixer gain trims to set recording levels during the sound check. You don't have to mix while recording—instead, mix and monitor back in your studio. As for drawbacks, you might have to ask the PA operator to adjust the gain trims during the show to prevent recorder clipping.

Connections

This section describes how to connect a multitrack hard-drive recorder to a mixer's insert-send connectors.

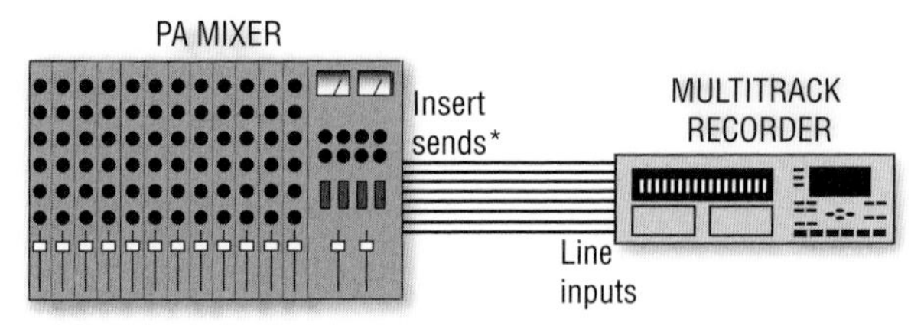

FIGURE 17.4
Connecting insert sends to a multitrack recorder provides great sound and easy setup.

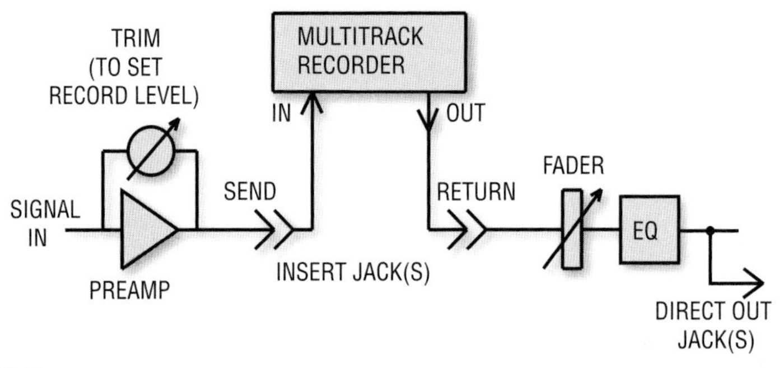

FIGURE 17.5
Simplified signal flow through part of a mixing console, showing insert and direct out.

In the console are several mic preamplifiers, one per mic, which amplify the mic-level signal up to line level. For each mic channel, this line-level signal typically appears at two connectors on the back of the mixer: *direct out* and *insert send*. That's where to connect to the recorder inputs.

Usually the *insert send* is the best connector to use. Here's why. Typically the direct-out signal is post-fader (Figure 17.5). This means the signal at the direct-out connector comes after the fader, so the signal is affected by the fader (volume) settings. Any fader movements will show up on your recording, which is undesirable. It's better to connect recorder tracks to insert sends which are usually pre-fader, pre-EQ. However, any changes the PA operator makes in the trim settings during the show will affect your recording levels.

In some mixing consoles, the direct outs can be set to pre-fader. If so, use those, because insert sends are often in use to connect compressors or gates.

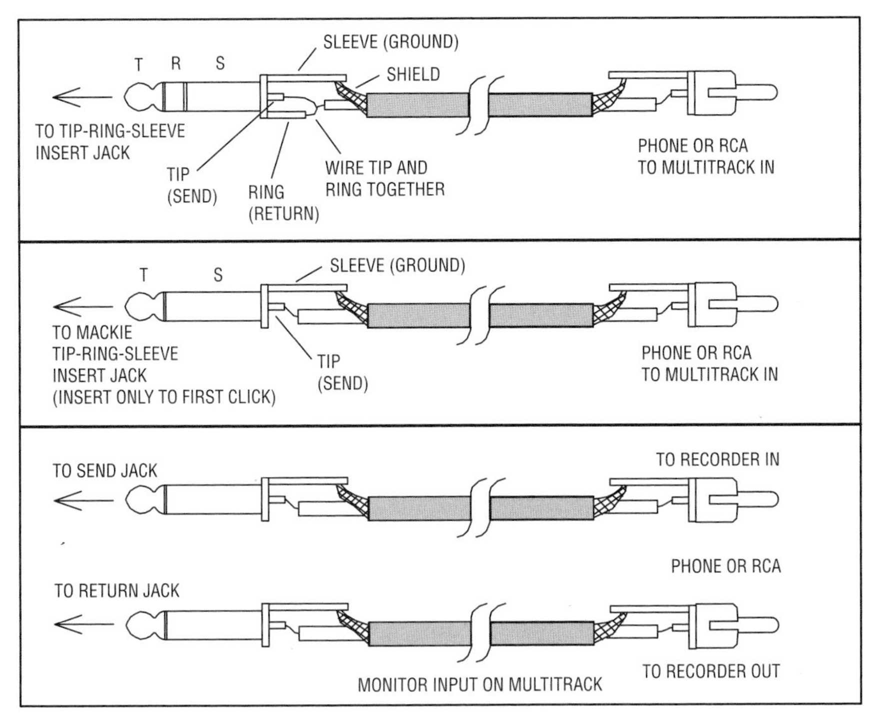

FIGURE 17.6
Three ways to wire cables based on the type of insert connector.

First, find out what kind of insert connectors the PA mixer has, and what kind of input connectors your multitrack recorder has. Buy or make some shielded cables (or a snake) that mate with those connectors. Figure 17.6 shows three ways to wire cables based on the type of insert connector.

Some boards have a single TRS insert connector per channel, instead of separate insert-send and insert-return connectors. Usually the tip is send and the ring is return. In the TRS connector you plug into the TRS insert, wire tip, and ring together, and also to the cable hot conductor (Figure 17.6, top). That way, the insert-send goes directly to the insert-return. If nothing is connected to the insert return, no mic signal goes through the mixer.

On some PA mixers with a TRS insert connector, you can use a TS connector (Figure 17.6, middle). Plug it in halfway to the first click so you don't break the signal path—the mic signal still goes through the PA mixer. If you plug in all the way to the second click, the signal doesn't go through the PA mixer, just to the recorder. Cover the connections with a mixer case cover or board because someone could bump into the mixer and dislodge a cable.

If the PA mixer has separate send and return insert connectors, connect the send to the recorder track input, and connect the return to the recorder track output (Figure 17.6, bottom). If necessary, set your multitrack recorder to monitor the input analog signal so that the PA mixer will receive a signal. Another option: Use an insert snake with TRS connectors at the mixer end. Carry some TRS-to-dual TS adapters to handle mixers that have separate connectors for insert send and return.

The insert sends are balanced or unbalanced, and the same is true of the recorder inputs. To connect balanced and unbalanced equipment correctly, see the article "Sound System Interconnection" on the Rane Web site, www.rane.com/note110.html.

You often encounter PA consoles where some insert sends are tied up with signal processors. You must use the direct-out connectors of those channels instead, which usually are post-fader (unless they can be switched to pre-fader). Another option is to "Y" the insert send to your recorder and to the processor input. Or, assign those mic channels to unused groups (busses) and get your recording signals from there.

Some rack-mounted mic preamps have insert connectors. You could plug the mic-snake XLRs into these preamps, then connect the preamp's inserts to the PA mixer line inputs. That way, any gain changes made on the PA console won't affect your recording. **Caution**: Any gain changes you make on the mic preamps will affect the PA levels.

If the PA mixer has a FireWire or USB port, simply connect the port to a laptop that is running recording software. Set up the software to recognize the mixer as its input/output device. The signal from each mic channel goes to a separate track in the software.

What if you want to record several instruments on one track, such as a drum mix? Assign all the drum mics to one or two output buses in the PA mixer. Plug the bus-out insert connector to the recorder track input. Use two buses for stereo. Adjust the faders to get a good drum mix.

Monitoring

To monitor the quality of the signals you're recording, you generally let the PA system be your monitor system. But you may want to set up a monitor mix over isolating headphones or earphones so you can hear more clearly.

Here is a suggested procedure. Connect all the recorder outputs to unused line inputs in the PA mixer, or to a separate mixer. Use those faders to set up a monitor mix. Assign them to an unused bus, and monitor that bus with headphones/earphones. If you can spare only a few inputs, plug in just one track at a time to check its sound quality. Listen closely for any hum, noise, or distortion.

Setting Levels

Set recording levels with the PA mixer's gain-trim or input-attenuator knobs. This affects the levels in the PA mix, so be sure to discuss your trim adjustment in advance with the PA mixer operator. If you turn down an input trim, the PA operator must compensate by turning up that channel's fader and monitor send.

As we said, if the PA operator changes the input trim during the show, these changes will show up on your recording.

Set recording levels before the concert during the sound check (if any!). It is better to set the levels a little too low than too high because during mixdown you can reduce noise but not distortion. A suggested starting level is −10 dB, which allows for surprises. Don't exceed 0 dB because the signal will distort. Also, if you set the recording level conservatively, you're less likely to change the gain trims during the performance. You don't want to hassle the PA operator.

Some recording engineers run each insert-send signal through a potentiometer to set the recording level. By adjusting levels on a rack-mounted panel of potentiometers, you don't have to ask the PA operator to change the gain trims for you. Of course, if you're operating the PA console, you can set the gain trims yourself.

If you have a spare recorder, record a safety copy on it at the same time. This provides a backup in case one recorder fails.

SPLITTING THE MIC SIGNALS

Now let's consider a different way to make a multitrack recording. Plug each mic into a mic splitter, which sends the mic signal to two destinations: the PA mixer and recording mixer. This gives you independent control of each microphone. The splitter has one XLR input and two or three XLR outputs per mic. The third output on some splitters goes to a monitor mixer. You can use transformer-isolated splitters or Y-cable splitters.

Splitters are made by such companies as Pro Co, Radial, Whirlwind, Rapco, Klark Technik, and Wireworks.

As shown in Figure 17.7, connect the outputs from all the splitter channels to the PA snake and to your recording snake. Connect the recording snake to your recording mixer (or audio interface) mic inputs. Connect the recording mixer's insert sends to a multitrack recorder, or connect the interface USB/FireWire cable to a laptop computer. When the recording is done, you can edit and mix the recording back in your studio.

Let's explain the advantages of splitting the mic signals. You use your own mic preamps, so you aren't dependent on the quality of the PA console mic preamps. Also, you aren't hassling the operator about adjusting gain trims. Each mix engineer can work without interfering with the other. The FOH engineer can change trims, level, or EQ, and it will have no effect on the signals going to the recording engineer. Another plus: A splitter provides consistent, unprocessed recordings

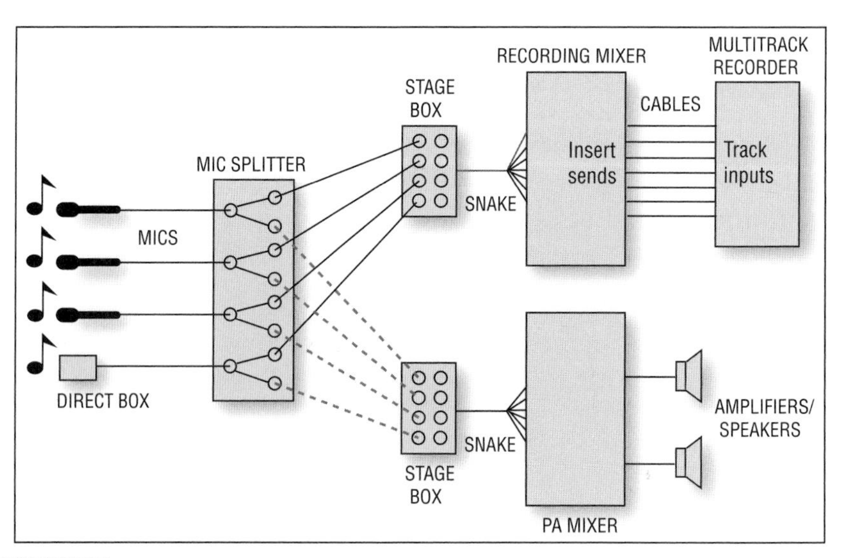

FIGURE 17.7
Splitting the mic signals to the recording mixer and PA mixer.

of the mic signals. This consistency makes it easy to edit between different performances. What's more, splitters let you use mic preamps on stage if you wish. That way, the cable from each mic to its preamp is short, which reduces hum and radio-frequency interference.

Using Splitters

To use a splitter or multichannel splitter, plug each mic into a splitter input. Decide which mixer you want to supply phantom power (usually the PA mixer). Connect the splitter's direct outputs to that mixer's snake. Connect one set of isolated outputs to the recording snake, and another set to the monitor snake (if used). Plug the snakes into the mixers. Figure 17.8 shows splitter connections to three mixers.

Splitters have a ground-lift switch on each output channel. This switch connects or disconnects (floats) the cable shield from pin 1 of the XLR connector. When the ground-lift switches are set correctly, you should get no ground loops and their resulting buzzes.

How do you set the ground-lift switches?

1. First, turn *off* phantom power in each console. Turn down all the faders.

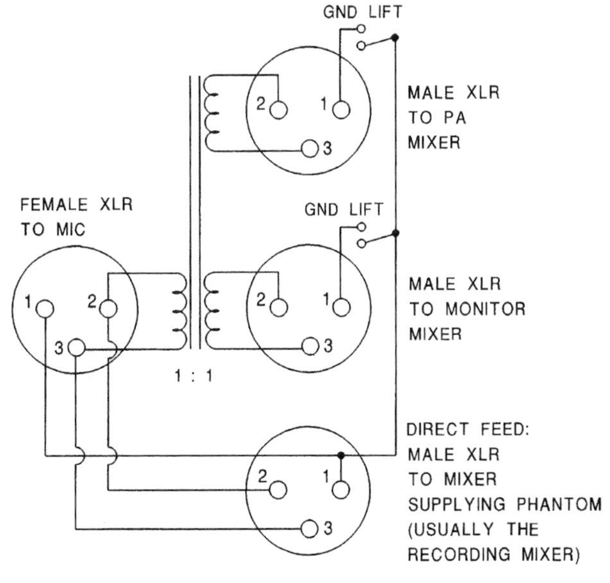

FIGURE 17.8
A transformer-isolated microphone splitter.

2. Make sure the direct feed's ground-lift switches (if any) are set to *ground*, not *lift*. Otherwise phantom power won't work.
3. Go to the mixer connected to the direct feed. Turn on the mixer, switch on phantom power, and bring up each fader to listen for a signal.
4. On the splitter, find the ground-lift switches for the recording-mixer feed. Set them to the position where you monitor the least hum and buzz at the recording mixer. In XLR Y-cable splitters, solder a ground-lift switch between the shield and pin 1 in the non-direct feeds.
5. Repeat step 4 for the monitor mixer.

MULTITRACK RECORDING IN A TRUCK

Here's the ultimate setup. Each mic is split three ways to feed the stage boxes for the recording, reinforcement, and monitor consoles. A long snake is run to a recording truck or van parked outside the concert hall or club. In the truck, the snake connects to a mixing console plugged into a multitrack recorder. The interior of the truck is acoustically treated, and powered Nearfield monitor speakers allow accurate monitoring of the signal. Designing a recording truck is a subject in itself, and is beyond the scope of this book.

The truck contains the console, monitors, and recorders, so you don't need to cart them into the venue. This saves setup time. Also, a truck provides a quiet, controlled monitoring environment.

There you have a full choice of methods for recording live. Check out all the options, and you'll find a system that works well for your style of recording.

PREPARING FOR THE SESSION

Ready to record a live gig? The recording will go a lot smoother if you plan what you're going to do. So sit down, grab a pen, and make some lists and diagrams as described here. We'll go over the steps to plan a recording.

Preproduction Meeting

Call or meet with the sound-reinforcement company and the production company putting on the event. Find out the date of the event, location, phone numbers, and e-mails of everyone involved; when the job starts; when you can get into the hall; when the second set starts; and other pertinent information. Decide who will provide the split, which system will be plugged in first, second, and so on. Draw block diagrams for the audio system and communications (comm) system. Determine who will provide the comm headphones.

If you're using a mic splitter, work out the splitter feeds. The mixer getting the direct side of the split provides phantom power for condenser mics that aren't powered on stage. If the house system has been in use for a long time, give them the direct side of the split.

Overloud stage monitors can ruin a recording, so work with the sound-reinforcement people toward a workable compromise. Ask them to start with the monitors quiet because the musicians always want them turned up louder.

Make copies of the meeting notes for all participants. Don't leave things unresolved. Know who is responsible for supplying what equipment.

Site Survey

If possible, visit the recording site in advance and go through the following checklist:

- Check the AC power to make sure the voltage is adequate, the third pin is grounded, and the waveform is clean.
- Listen for ambient noises: ice machines, coolers, 400-Hz generators, heating pipes, air conditioning, nearby discos, etc. Try to have these noise sources under control by the day of the concert.
- Sketch dimensions of all rooms related to the job. Estimate distances for cable runs.
- Turn on the sound-reinforcement system to see if it functions okay by itself (no hum, and so on). Turn the lighting on at various levels with the sound system on. Listen for buzzes. Try to correct any problem so that you don't document bad PA sound on your recording.
- Determine locations for any audience/ambience mics. Keep them away from air-conditioning ducts and noisy machinery.
- Plan your cable runs from stage to recording mixer.
- If you plan to hang mic cables, feel the supports for vibration. You may need microphone shock mounts. If there's a breeze in the room, plan on taking windscreens.

- Make a file on each recording venue including the dimensions and the location of the circuit breakers.
- Determine where the control room will be. Find out what surrounds it—any noisy machinery?
- Visit the site when a crowd is there to see where there may be traffic problems.

Mic List

Now write down all the instruments and vocals in the band. If you want to put several mics on the drum kit, list each drum that you want to mike. As for keyboards, decide whether you want to record off each keyboard's output, or off the keyboard mixer (if any).

Next, write down the mic or direct box you want to use on each instrument (see Table 17.1).

Table 17.1 Mic List (Example)

Input	Instrument	Mic
1	Bass	DI
2	Kick	AKG D112
3	Snare	Shure Beta 57
4	Hi-hat	Crown CM-700
5	Small rack tom	Shure SM57
6	Large rack tom	Shure SM57
7	Small floor tom	Senn. MD421
8	Large floor tom	Senn. MD421
9	Overhead L	AT 4051A
10	Overhead R	AT 4051A
11	Lead guitar	Shure SM57
12	Rhythm guitar	Shure SM57
13	Keyboard mixer	DI
14	Lead vocal	Beyer M88
15	Harmony vocal	Crown CM-311A

Make copies of this mic list. At the gig, you'll place one list by the stage box, and the other by each mixing console. The list will act as a guide to keep things organized.

Track Sheet

Next, decide what will go on each track of your multitrack recorder. If you have enough tracks, your job is easy: Just assign each instrument or vocal to its own track: bass to track 1, kick to track 2, and so on. If you need to assign several instruments or vocals to the same track, set up a submix as described earlier. Table 17.2 shows a sample track list.

Block Diagram

Now that your track assignments are planned, you can figure out what equipment you'll need. Draw a block diagram of your recording setup from input to output (such as Figure 17.9). Include mics, mic cables, mic stands and booms, DI boxes, insert cables, multitrack recorder(s), outlet strip and extension cord, and recording media. You might bring your own mixer and snake, or use those from the house system. On your diagram, label the cable connectors on each end so you'll know what kinds of cables to bring. It's a good idea to keep a file of system block diagrams for various recording venues.

In Figure 17.9, the block diagram shows a typical recording method: feeding FOH console insert jacks to a multitrack recorder. We'll use this example in the rest of this chapter.

Table 17.2 Track Sheet (Example)

Track	Instrument
1	Drum mix L
2	Drum mix R
3	Bass
4	Lead guitar
5	Rhythm guitar
6	Keys mix
7	Lead vocal
8	Harmony vocal

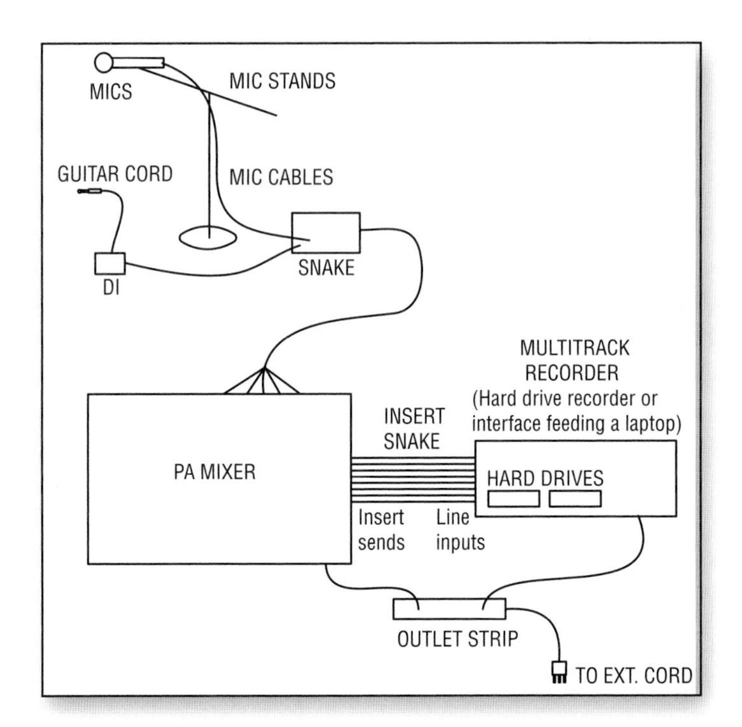

FIGURE 17.9
Sample block diagram.

Equipment List

Generate a list of recording equipment from your block diagram. Based on Figure 17.9, you'd need the following recording gear:

- Direct boxes, mics, mic cables, mic stands, and booms (unless these are part of the house system)
- Insert cables (for example, 8 stereo phone plugs to 8 RCA connectors)
- Multitrack recorder
- Outlet strip and extension cord
- Recording medium (bring enough for the duration of the gig)

Don't forget the incidentals: a cleaning tape, pen, notebook, flashlight, guitar picks, heavy-duty guitar cords, drum keys, spare tape, mic pop filters, gaffer's tape, guitar tuner, ear plugs, audio-connector adapters, audio cable ground-lift adapters, in-line pads, in-line polarity reversers, spare cables, gooseneck lights for the console, spare batteries, water, food—and aspirin!

Bring a tool kit with screwdrivers, pliers, soldering iron and solder, AC-outlet checkers, fuses, a pocket radio to listen for interference, ferrite beads of various sizes for RFI suppression, canned air to shoot out dirt, cotton swabs and pipe cleaners, and DeoxIT from Caig Labs to remove oxide from connectors.

Check off each item on the list as you pack it. After the gig, you can check the list to see whether you reclaimed all your gear.

PREPARING FOR EASIER SETUP

You want to make your setup as fast and easy as possible. Here are some tips to help this process.

Protective Cases

Mount your console and recorders in protective carrying cases. Install casters or swivel wheels under racks and carrying cases so you can roll them in. Rolling is so much easier than lifting and carrying. You might permanently install the multitrack recorders in SKB carrying cases that act as racks. When a remote job comes up, just grab them and go.

A very helpful item is a dolly, wheeled cart, or hand truck to transport heavy equipment into the venue. Consider getting some lightweight tubular carts. Being collapsible, they store easily in your car or truck. One maker of equipment carts is www.kart-a-bag.com.

Pack mics, headphones, and other small pieces in cloth bags, trunks, or milk crates.

Caution: Keep tapes and hard drives separated from magnets, such as in headphones, monitor speakers, and dynamic mics. (However, tape erasure is much less of a problem with digital media than it is with analog tapes.)

You might want to build a mic container: a big box full of foam rubber with cutouts for all the mics. Or construct a wheeled cabinet with drawers for mics, DIs, and speaker cables.

In an article in the September–October 1985 issue of *db Magazine*, remote recording engineer Ron Streicher offered these suggestions:

> "Especially for international travel, make sure your documentation is up to date and matches the equipment you're carrying. Make a list of everything you take: all the details, such as each pencil, razor blade, connector, etc. Also make sure your insurance is up to date. You need insurance for en route as well as at the destination.
>
> "I organize my cases so I know where every item is. They're ready to go anytime and make setup much faster. The cables are packed with their associated equipment, not in a cables case. I check everything coming and going, and try to have 100% redundancy, such as a small mixer to substitute for the large console."

Mic Mounts

If you'll be recording a singer/guitarist, take a short mic mount that clamps onto the singer's mic stand. Put the guitar mic in the mount. Also bring some short mounts or mini mics to clamp onto drum rims. By using these mounts, you eliminate the weight and clutter of several mic stands.

Some examples of short mounts are the Mic-Eze units by Ac-cetera (www.ac-cetera.com). They have standard 5/8–27 threads and mic clamps that either spring shut or are screw-tightened. Flex-Eze is two clamps joined by a short gooseneck and Min-Eze is two clamps joined by a swivel.

Snakes and Cables

You can store mic cables on a cable spool, available in the electrical department of a hardware store. Wrap one mic cable around the spool, plug it into the next cable and wrap it, etc.

A snake can be wrapped around a large reel or can be coiled in a trunk. Commercial snake reels are made by such companies as Whirlwind (www.whirlwindusa.com), ProCo (www.procosound.com), and Hannay (www.hannay.com).

Use wire ties to join cables that you normally run together, such as PA sends and returns.

Snake hookup is quicker if the snake has a multipin twist-lock connector (such as Whirlwind W1 or W2). This connector plugs into a mating connector that divides into several male XLRs. Those XLRs plug into the mixing console. Leave the XLRs in the console carrying case. You'll find that the snake is easier to handle without the XLR pigtails on it.

For a clean, rapid hookup of drum mics, put a small snake near the drum kit, and run it to the main stage box. Or, put the main stage box by the drum kit. Snakes are made by such companies as Whirlwind, ProCo, and Horizon (www.horizonmusic.com).

Check that all your mic cables are wired in the same polarity: pin 2 hot on both ends.

You might want to use 3-conductor shielded mic cables. Connect the shield to ground only at the male XLR end. Also use cables with 100% shielding. Those measures enhance the shielding capability of the shield and reduce pickup of lighting buzzes.

In XLR-type cable connectors, don't wire pin 1 to the shell, or you may get ground loops when the shell contacts a metallic surface.

Label all your cables on both ends according to what they plug into; for example, DSP-9 effects in, track 12 out, power amp in, snake aux 2 out. Or you might prefer to number the cables near their connectors. Cover these labels with clear heat-shrink tubing.

Label both ends of each mic cable with the cable length. Put a drop of glue on each connector screw to temporarily lock it in place.

Rack Wiring

You can speed the console wiring by using a small snake between the rack and the console, and between the multitrack recorders and the console. When packing, plug the snakes into the rack gear and multitracks, and coil the snake inside the rack and the multitrack carrying case. In other words, have all your equipment pre-wired. At the gig, pull out a bundled harness and plug it into the console jacks.

You might be feeding your multitrack from the insert jacks of the house console. If so, use a snake with TRS stereo phone plugs at the console end. Carry some TRS-to-dual-mono phone adapters to handle consoles that have separate jacks for insert send and return.

Some engineers prefer to make a clearly marked interface panel on the rear of the racks and plug into the panels. This is easier than trying to find the right connectors on each piece of equipment.

Small bands might get by with all their equipment in a single, tall rack. Mount a small mixer on top, wired to effects in the middle, with a coiled snake and power cord on the bottom.

Small snakes for rack and multitrack connections are made by Hosa (www.hosatech.com), Horizon, and ProCo, among others.

Other Tips

Here are some more helpful hints for successful on-location recordings.

- Record several gigs and pick the best performances to use in an album.
- Plan to use a talk-back mic from the board to the stage monitor speakers during sound checks. You might bring a small instrument amp for talkback so that you can always be heard.
- Hook up and use unfamiliar equipment before going on the road. Don't experiment on the job!
- Consider recording with redundant (double) systems so you have a backup if one fails.
- Walkie-talkies are okay for pre-show use, but don't use them during the performance because they cause RF interference. Use hard-wired communications headsets. Assistants can relay messages to and from the stage crew while you're mixing.
- During short set changes, use a laptop computer Local Area Network to show what set changes and mic-layout changes are coming up next; transmit this information to the monitor mixer and sound-reinforcement mixer.
- Don't put tapes or hard drives through airport X-ray machines because the transformer in these machines isn't always well shielded. Have the tapes or hard drives inspected by hand.
- Hand-carry your mics on airplanes. Arrange to load and unload your own freight containers, rather than trusting them to airline freight loaders. Expect delays here and at security checkpoints.

- Get a public-liability insurance policy to protect yourself against lawsuits.
- Call the venue and ask directions to the load-in door. Make sure that someone will be there at setup time to let you in. Ask the custodian not to lock the circuit-breaker box the day of the recording.
- A few days before the session, check out the parking situation.
- Just before you go, check out all your equipment to make sure it's working.
- Arrive several hours ahead of time for parking and setup. Expect failures—there's always something going wrong, something unexpected. Allow 50% more time for troubleshooting than you think you'll need. Have backup plans if equipment fails.
- In general, plan everything in advance so you can relax at the gig and have fun!

AT THE SESSION: SETUP

Okay! You've arrived at the venue. After parking, offload your gear to a holding area, rather than onstage, because gear on stage will likely need to be moved.

Learn the names of the house sound-crew members, and be friendly. These people can be your assets or your enemies. Think before you comment to them! Try to remain in the background and don't interfere with their normal way of doing things (for example, take the secondary side of the split).

Power Distribution System

At the job, you need to take special precautions with power distribution, interconnecting multiple sound systems, and electric guitar grounding.

Consider buying, renting, or making your own single-phase power distribution system (distro). It will greatly reduce ground loops and increase reliability. One source of AC power distribution equipment is www.furmansound.com. Figure 17.10 shows a suggested AC power distribution system.

The amp rating of the distro's main breaker box should exceed the current drain of all the equipment that will be plugged into the distro system.

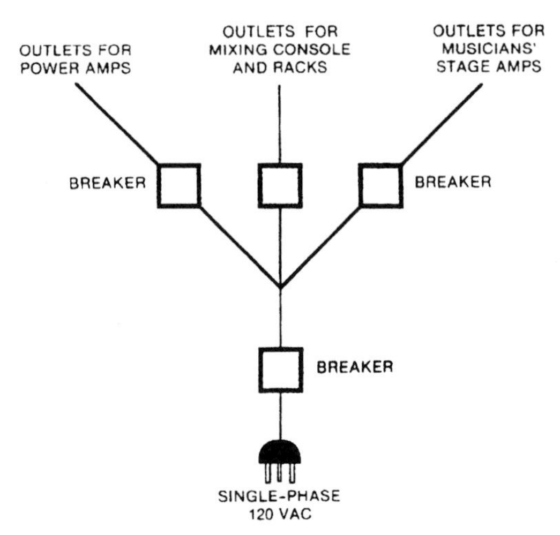

FIGURE 17.10
An AC power distribution system for a touring sound system.

Power Source

If you're using a remote truck, find a source of power that can handle the truck's power requirements, usually at a breaker panel. Some newer clubs have separate breaker boxes for sound, lights, and a remote truck. Find out whether you'll need a union electrician to make those connections. Label your breakers.

Check that your AC power source isn't shared with lighting dimmers or heavy machinery; these devices can cause noises or buzzes in the audio.

The industry-standard power connector for high-current applications is the Cam-lok, a large cylindrical connector. Male and female Cam-loks join together and lock when you twist the connector ring. Distro systems and power cables with Cam-lok connectors can be rented from rental houses for film, lighting, electrical equipment, or entertainment equipment. One such rental house is Mole-Richardson at www.mole.com.

Use an adapter from Cam-lok to bare wires. Pull the panel off the breaker box, insert the bare wires, and connect the Cam-lok to your truck's power.

Caution: Have an electrician do the wiring if you don't know what you're doing. A union electrician might be required anyway. Some breaker boxes have Cam-loks already built in.

To reduce ground-loop problems, get on the same power that the house sound system is using. From that point, run your distribution system, or at least run one or two thick (14- or 16-gauge) extension cords, to your recording system. These cords may need to be 100 to 200 feet long. Plug AC outlet strips into the extension cord; then plug all your equipment into the outlet strips.

If your recording system is one or two multitrack recorders that will connect to the FOH console, simply plug into the same outlet strip that the FOH console is using.

Measure the AC line voltage. If the AC voltage varies widely, use a line voltage regulator (power conditioner) for your recording equipment. If the AC power is noisy, you might need a power isolation transformer.

Check AC power on stage with a circuit checker. Are grounded outlets actually grounded? Is there low resistance to ground? Are the outlets the correct polarity? There should be a substantial voltage between hot and ground, and no voltage between neutral and ground.

Some recording companies have a gasoline-powered generator ready to use if the house power fails. If there are a lot of lighting and dimmer racks at the gig, you might want to put the truck on a generator to keep it isolated from the lighting power.

Interconnecting Multiple Sound Systems

If you hear hum or buzz when the systems are connected, first make sure that the signal source is clean. You might be hearing a broken snake shield or an unused bass-guitar input.

If hum persists, experiment with flipping the ground-lift switches on the splitter and on the direct boxes. On some jobs you need to lift almost every ground. On others you need to tie all the grounds. The correct ground-lift setting can change from day to day because of a change in the lighting. Expect to do some trial-and-error adjustments.

If the house system has serious hum and buzz problems, offer help. You can hear buzzes in your quiet truck that they can't hear over the main system with noise in the background.

Often, a radio station or video crew will take an audio feed from your mixing console. In this case, you can prevent a hum problem by using a console with transformer-isolated inputs and outputs. Or you can use a 1:1 audio isolation transformer between the console and the feeds. For best isolation, use a distribution amp with several transformer-isolated feeds. Lift the cable shield at the input of the system you're feeding. Some excellent isolation transformers are made by Jensen (www.jensen-transformers.com).

Connections

After unpacking, place one mic list by the stage box so you know what to plug in where. Place a duplicate list by each console. Attach a strip of white tape just below the mixer faders. Use this strip to write down the instrument that each fader affects.

Based on the mic list you wrote, you might plug the bass DI into snake input 1, plug the kick mic into snake input 2, and so on. Label fader 1 "BASS," label fader 2 "KICK," etc. Also plug in equipment cables according to your block diagram.

Have an extra microphone and cable offstage ready to use if a mic fails.

Don't unplug mics plugged into phantom power because this will make a popping noise in the sound-reinforcement system.

Running Cables

To reduce hum pickup and ground-loop problems associated with cable connectors, try to use a single mic cable between each mic and its snake-box connector.

Avoid bundling mic cables, line-level cables, and power cables together. If you must cross mic cables and power cables, do so at right angles and space them vertically.

Plug each mic cable into the stage box; then run the cable out to the each mic and plug it in. This leaves less of a mess at the stage box. Leave the excess cable at each mic stand so you can move the mics. Don't tape the mic cables down until the musicians are settled.

It's important that audience members don't trip over your cables. In high-traffic areas, cover cables with rubber floor mats or cable crossovers (metal ramps). At least tape them down with gaffer's tape pressed lengthwise onto the cables.

It helps to set up a closed-circuit TV camera and TV monitor to see what's happening on stage. You need to know when mics get moved accidentally, or when singers use the wrong mic, etc.

Recording-Console Setup

Here's a suggested procedure for setting up the recording system efficiently:

1. If the console is set up in a dressing room or locker room, add some acoustic absorption to deaden the room reflections. You might bring a carpet for the floor, plus acoustic foam or moving blankets for the walls.
2. Turn up the recording monitor system and verify that it is clean.
3. Plug in one mic at a time and monitor it to check for hums and buzzes. Troubleshooting is easier if you listen to each mic as you connect it, rather than plugging them all in, and trying to find a hum or buzz.
4. Check and clean up one system at a time: first the sound-reinforcement system, then the stage-monitor system, then the recording system. Again, this makes troubleshooting easier because you have only one system to troubleshoot.
5. Use as many designation strips as you need for complex consoles. Label the input faders bottom and top. Also label the monitor-mix knobs and the meters.
6. Monitor the reverb returns (if any) and check for a clean signal.
7. Make a short test recording and listen to the playback.
8. Verify that left and right channels are correct, and that the pan-pot action isn't reversed audibly.
9. If you're setting up a separate recording monitor mix, do a preliminary pan-pot setup. Panning similar instruments to different locations helps you identify them.

MIC TECHNIQUES

Usually the miking is left up to the sound-reinforcement company. But there are some mic-related problems you should know about, such as feedback, leakage, room acoustics, and noise. Here are some ways to control these problems:

- Use directional microphones, such as cardioid, supercardioid, or hypercardioid. These mics pick up less feedback, leakage, and noise than omnidirectional mics at the same miking distance.
- With vocal mics, aim the null of the polar pattern at the floor monitors. The null (area of least pickup) of a cardioid is at the rear of the mic—180 degrees off-axis. The null of a supercardioid is 125 degrees off-axis; hypercardioid is 110 degrees.
- To reduce breath pops with vocal mics, be sure to use foam pop filters. Allow a little spacing between the pop filter and the mic grille. It also helps to switch in a low-cut filter (100-Hz highpass filter).
- Mike close. Place each mic within a few inches of its instrument. Ask vocalists to sing with lips touching the mic's foam pop filter.
- Use direct boxes. Bass guitar and electric guitar can be recorded direct to eliminate leakage and noise in their signals. However, you might prefer the sound of a miked guitar amp. You could record the guitar direct from its effects boxes. Then use a guitar-amp emulator during mixdown. Note that sequencers and some keyboards have high-level outputs, so their DI boxes need transformers that can handle line level.
- Use contact pickups. On acoustic guitar, acoustic bass, and violin, you can avoid leakage by using a contact pickup. Such a pickup is sensitive only to the instrument's vibration, not so much to sound waves. The sound of a pickup isn't as natural as a microphone, but a pickup may be your only choice. Consider using both a pickup and a microphone on the instrument. Feed the pickup to the house and monitor speakers; feed the mic and pickup to the recording mixer.

When you're recording a band that has been on tour, should you use their PA mics or your own mics? In general, go with their mics. The artists and PA company have been using their mics for a while and may not want to change anything. Most mics currently used in PA are good quality anyway, unless they are dirty or defective.

If you're not happy with their choice, you could add your own instrument mics. Let the PA people listen to the sound in the recording truck, or in headphones. If it sounds bad because of their mic choice, ask "Would it be okay if we tried a different mic (or mic placement)?" Usually it's all right with them—it's a team effort.

Electric Guitar Grounding

While setting up mics, you need to be aware of a safety issue with the electric guitar. Electric-guitar players can receive a shock when they touch their guitar and a mic simultaneously. This occurs when the guitar amp is plugged into an electrical outlet on stage, and the mixing console (to which the mics are grounded) is plugged into a separate outlet across the room. If you're not using a power distro, these two power points may be at widely different ground voltages. So a current can flow between the grounded mic housing and the grounded guitar strings.

Caution: Electric guitar shock is especially dangerous when the guitar amp and the console are on different phases of the AC mains.

It helps to power all instrument amps and audio gear from the same AC distribution outlets. If you lack a power distro, run a heavy extension cord from a stage outlet back to the mixing console (or vice versa). Plug all the power-cord ground pins into grounded outlets. That way, you prevent shocks and hum at the same time.

If you're picking up the electric guitar direct, use a transformer-isolated direct box and set the ground-lift switch to the minimum-hum position.

Using a neon tester or voltmeter, measure the voltage between the electric-guitar strings and the metal grille of the microphones. If there is a voltage, flip the AC polarity switch on the amp (if any). Use foam windscreens for additional protection against shocks.

Audience Microphones

If you have enough mic inputs, you can use two audience mics to pick up the room acoustics and audience sounds. This helps the recording to sound "live." Without audience mics, the recording may sound too dry, as if it were done in a studio.

One easy method is to aim two cardioid mics at the audience. Put them on tall mic stands, on the stage floor, and on either side of the stage. If those mic stands must not be seen, try hanging mics (Figure 17.11).

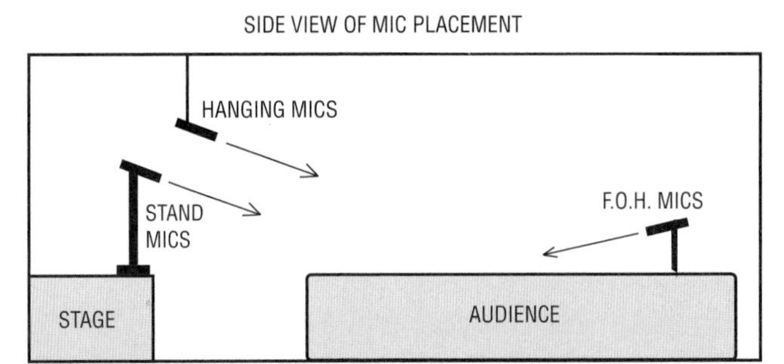

FIGURE 17.11
Some audience miking techniques.

Some engineers pick up the audience and hall with two mics placed at the FOH mix position. That's an easy, effective way to capture the crowd. But the signal from these mics is delayed relative to the on-stage mics. If the FOH mix position is far from the stage (say, 50 feet or more), this delay will cause an echo when the audience mics are mixed with the stage mics. Prevent this by placing the audience mics fairly near the stage. Or, during computer mixdown, move the FOH mic tracks to the left, earlier in time so that large peaks in the FOH tracks' waveform coincide with large peaks in on-stage tracks. If you're using the audience mics only to capture applause, the delay isn't an issue.

What if you don't have enough tracks for the audience mics? Record them on a 2-track recorder. Load this recording into your DAW along with the stage-mic mix. Align the two recordings in time as described above.

If the audience mics are run through the FOH mixer, leave the audience mics unassigned in that mixer to prevent feedback.

To get more isolation from the house speakers in the audience mics, use several mics hung close to the audience. Some engineers put up four audience mics maximum; some use eight to ten. Use directional mics and aim the rear null at the house speakers.

Another option is not miking the audience, or not using the audience tracks. Instead, during mixdown, you could simulate an audience with an audience-reaction CD. Simulate room reverb with an effects unit.

SOUND CHECK AND RECORDING

Now that the mics are set up, you might have time for a sound check. That's when you set recording levels as described earlier in the section Setting Levels.

A few minutes before the band starts playing, start recording. Keep a close eye on recording levels. If a track is going into the red, slowly turn down its input trim and note the recorder-counter time where this change occurred.

Caution: If you're recording off the FOH mixer, turning down its input trim will affect the house levels. The FOH mixer operator will need to turn up the corresponding monitor send and channel fader.

This is a touchy situation that demands cooperation. Ideally, you set enough headroom during the sound check so you won't have to change levels. But be sure the house mixer operator knows in advance that you might need to make changes. Ask the operator whether she wants to adjust the gain trims for you, so she can adjust corresponding levels at the same time. Thank the operator for helping you get a good recording.

If you're recording with a splitter and mic preamps on stage, assign someone to watch the levels and adjust them during the concert. Preamps with meters allow more precise level setting than preamps with clip LEDs.

Keep a track sheet and log as you record. For each song in the set list, note the recorder-counter time when the song starts, or press the Set Locate button on your multitrack recorder. Later, during mixdown, you can go to those counter times or locate points to find songs you want to mix. Also note where any level changes occurred so you can compensate during mixdown. It helps to note a counter time when the signal level was very high. When you mix the recording you can start at that point in setting your overall mix levels.

TEARDOWN

After the gig, pack your mics away first because they may be stolen or damaged. Refer to your equipment list as you repack everything. Note equipment failures and fix broken equipment as soon as possible.

After you haul your gear back to the studio, it's time for mixing and editing.

1. If necessary, copy all the tracks into your recording software's multitrack template.
2. Select all the tracks, and split all the tracks at the same point—several seconds before and after each song.
3. Export only the tracks in each split section to separate song folders. That way you can mix each song separately by importing the tracks from each folder into a multitrack template. Another option: Copy the tracks in each split section, then paste them into new projects.

Alternatively, use volume envelopes or automation to mix all the songs, then export one long concert mix. If you want, you can import that stereo mix and split it into individual songs, saving only the best ones. Then overlap one song's ending applause with the beginning of the next song.

Some of the information in this chapter was derived from two workshops presented at the 79th convention of the Audio Engineering Society in October 1985. These workshops were titled "On the Repeal of Murphy's Law-Interfacing Problem Solving, Planning, and General Efficiency On-Location," given by Paul Blakemore, Neil Muncy, and Skip Pizzi and "Popular Music Recording Techniques," given by Paul Blakemore, Dave Moulton, Neil Muncy, Skip Pizzi, and Curt Wittig.

CHAPTER 18

On-Location Recording of Classical Music

This chapter explains how to do an on-location stereo recording of a classical-music ensemble. We cover both equipment and procedures.

EQUIPMENT

- Microphones (low-noise condenser or ribbon type, omni or directional, free field or boundary, stereo or matched pair.)
- Multitrack recorder or stereo recorder.
- Low-noise mic preamps or low-noise mini mixer (necessary if the mic preamps in your recorder are low quality).
- Phantom-power supply (necessary if your mixer or mic preamp lacks phantom power).
- Stereo bar.
- Mic stands and booms, or fishing line to hang mics.
- Shock mount (optional).
- Mixer (necessary if you use more than two mics).
- MS matrix box (if you're recording with the Mid-Side technique).
- Headphones and/or speakers.
- Power amplifier for speakers (optional) or powered Nearfield monitors.
- Recording medium: hard drive or Flash memory.
- Power strip, extension cords.
- Notebook and pen.
- Talkback mic and powered speaker (optional).
- Tool kit.
- Fresh batteries.

First on the list are microphones. You'll need at least two or three of the same model number or one or two stereo microphones. Good microphones are essential, for the microphones—and their placement—determine the sound of your recording.

For classical-music recording, the preferred microphones are condenser or ribbon types with a wide, flat frequency response and very low self-noise (explained in Chapter 6). A self-noise spec of less than 20 dB equivalent SPL, A-weighted, is recommended.

If you want to do spaced-pair recording, you can use either omnidirectional or directional microphones. Omnis are preferred because they generally have a deeper low-frequency response. If you want to do coincident or near-coincident recording for sharper imaging, use directional microphones (cardioid, supercardioid, hypercardioid, or bidirectional). Mics with those polar patterns tend to roll off in the low frequencies, but you can compensate for this with equalization.

You need a power supply for condenser microphones: either an external phantom-power supply, a mixer or mic preamp with phantom power, or internal batteries. A low-noise stereo mic preamp (or low-noise portable mixer) lets you make recordings free of electronic hiss.

You can mount the microphones on stands or hang them from the ceiling with nylon fishing line. Make sure that the fishing line's tensile strength exceeds the weight of the mics. Check legal safety issues with hanging mics; different rules apply in different places. Stands are much easier to set up but more visually distracting at live concerts. Stands are more suitable for recording rehearsals or sessions with no audience present. Neumann makes tiltable "auditorium hangers" that suspend mic capsules of the KM 100 system from their cables.

The mic stands should have a tripod folding base and should extend at least 13 feet high. Some suitable products are telescoping stands, which are lightweight and compact. Some examples are the Shure S15A (www.shure.com), Quik Lok A85 (www.quiklok.com), AEA-13MDV (www.wesdooley.com), K&M 21411B (various vendors), and stands at www.micsupply.com. You can use baby booms or stand extenders to increase the height of regular mic stands.

A useful accessory is a stereo bar (stereo microphone adapter, stereo mic mount). This device mounts two microphones on a single stand for coincident or near-coincident stereo recording. A long stereo bar can accommodate spaced-pair miking. Some models are at www.micsupply.com.

Another needed accessory in most cases is a shock mount to prevent pickup of floor vibrations. Some models are the On-Stage MY410 (various vendors), Sabra Som SSM-1 (www.sabra-som.com), Shure A55M (www.shure.com), and AKG H85 and H100 (www.akg.com).

In difficult mounting situations, boundary microphones may come in handy. They can lie flat on the stage floor to pick up small ensembles or can be mounted on the ceiling or on the front edge of a balcony. They also can be attached to clear Plexiglas panels that are hung or mounted on mic stands. Boundary mics are made by most microphone companies.

To monitor in the same room as the musicians, you need some closed-cup, circumaural (around the ear) headphones or isolating earphones to block out the sound of the musicians. You want to hear only what is being recorded. Of course, the headphones should be wide-range and smooth for accurate monitoring. A better monitoring arrangement might be to set up powered Nearfield loudspeakers in a separate room.

If you're in the same room as the musicians, you have to sit far from them to clearly monitor what you're recording. To do that, you need a pair of 50-foot microphone extension cables. Longer extensions are needed if the mics are hung from the ceiling or if you monitor in a separate room.

A mixer is necessary when you want to record more than one source; for example, an orchestra and a choir, an orchestra with spot mics, or a band and a soloist. You might put a pair of microphones on the orchestra and another pair on the choir. Connect the insert sends from the mixer to a multitrack recorder, then mix the tracks to stereo back in the studio. If your budget doesn't allow a multitrack recorder you could mix the mic signals live to a stereo recorder.

For monitoring a mid-side recording, bring an MS matrix box that converts the MS signals to left–right signals, which you monitor.

SELECTING A VENUE

If possible, plan to record in a venue with good acoustics. It should have adequate reverberation time for the music being performed (about 2 seconds for orchestral music). This is very important, because it can make the difference between an amateur-sounding recording and a commercial-sounding one. Try to record in an auditorium, concert hall, or house of worship rather than in a band room or gymnasium. Halls with wooden surfaces and a shoebox shape tend to sound best.

You may be forced to record in a hall that is too dead: that is, the reverberation time is too short. In this case, you may want to add artificial reverberation from a digital reverb unit or plug-in, or cover the seats with plywood sheets or 4-millimeter polyethylene plastic sheeting. Strong echoes can be controlled with carpets, RPG diffusers, or drapes.

If the venue is surrounded by noisy traffic, consider recording after midnight. Turn off noisy air conditioning if possible while recording.

Call the venue manager and ask that the circuit breakers for the stage outlets be turned on the day of the session. Ask where you can load-in your equipment. Also make sure that the load-in door will be unlocked when you plan to load in.

SESSION SETUP

Be sure to test all your equipment for correct operation before going on the job. If you're using battery-powered devices, install fresh batteries in them just before the concert. Keep your equipment inside your home or studio until you're ready to leave. A recorder left outside in a cold car may become sluggish if the lubricant stiffens, and batteries may lose some voltage.

Make sure your hard drive or flash card has enough capacity for the session. Table 13.1 shows the storage required for a 1-hour recording.

Arrive at the venue a few hours early to allow for setup and for fixing problems.

First, power-up the equipment. You can use batteries or an AC extension cord plugged into an outlet near the stage. Make sure this outlet is live. Gaffer-tape the extension cord lengthwise or cover it with mats to prevent tripping. Tie your outlet strip's AC cord to the extension cord so they can't separate if someone pulls on the extension cord.

If this is a session, listen to the ensemble playing on the stage. Is the sound bad? You might want to move the musicians out onto the floor of the hall.

Mounting the Mics

Place your microphones in the desired stereo miking arrangement. As an example, suppose you're recording an orchestra rehearsal with two crossed cardioids on a stereo bar (the near-coincident method). Screw the stereo bar onto a mic stand and mount two cardioid microphones on the stereo bar. For starters, angle them 110 degrees apart and space them 7 inches apart horizontally. Aim them downward so that they'll point at the orchestra when raised. You may want to mount the microphones in shock mounts or put the stands on sponges to isolate the mics from floor vibration.

Basically, place two or three mics (or a stereo mic) several feet in front of the group, raised up high (as in Figure 18.1). The microphone placement controls the acoustic perspective or sense of distance to the ensemble, the balance among instruments, and the stereo imaging.

As a starting position, place the mic stand behind the conductor's podium, about 12 feet from the front-row musicians. Connect mic

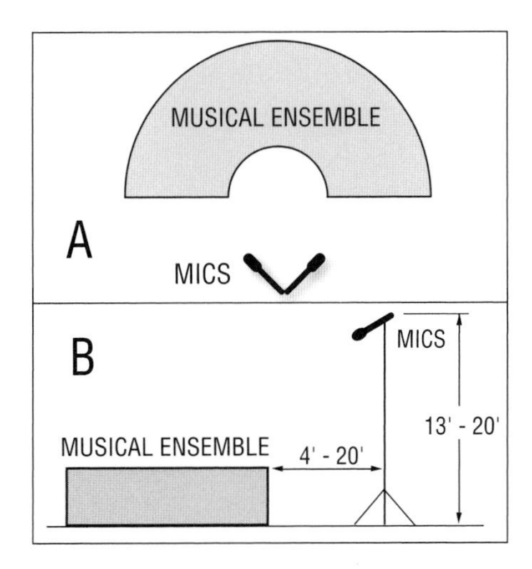

FIGURE 18.1
Typical microphone placement for on-location recording of a classical music ensemble: (a) top view, (b) side view.

cables and gaffer-tape them to the top of each mic stand for strain relief. Raise the microphones about 14 feet off the floor. This prevents overly loud pickup of the front row relative to the back row of the orchestra. It also reduces audience noise by mounting the mics farther from the audience. Gaffer-tape the mic cables to the bottom of the stand to keep it from being pulled over.

Leave some extra turns of mic cable at the base of each stand so you can re-position the stands. This slack also allows for people accidentally pulling on the cables. Try to route the mic cables where they won't be stepped on, or cover them with mats.

Live, broadcast, or filmed concerts require an inconspicuous mic placement, which may not be sonically ideal. In these cases, or for permanent installations, you probably want to hang the microphones from the ceiling rather than using stands. You can position a stereo bar with three nylon fishing lines spaced apart. Make sure that the tensile strength of the fishing line exceeds the weight of the mics and stereo bar. Hang individual mics by their cables and attach two fishing lines to the front of each mic to aim it. Another inconspicuous mic placement is on mic-stand booms projecting forward of a balcony. For drama or musicals, directional boundary mics can be placed on the stage floor near the footlights.

If you're uncertain which mic technique to use, you could set up two or more mic arrays and record them to multitrack. When you master the program, choose the best-sounding technique. You might even mix the signals of two or more mic pairs.

Connections

Now you're ready to make connections. Here are some connection methods for just two mics:

- If your stereo recorder has high-quality mic preamps and phantom power, plug the mics directly into the recorder mic inputs.
- If your recorder lacks phantom power (or the phantom voltage is lower than the mics require), plug the mics into a phantom supply, and from there into your recorder mic inputs.
- If your recorder lacks high-quality mic preamps, plug the mics into a low-noise mic preamp or mixer with phantom power. Connect cables from there into your recorder line inputs.

Here are some connection methods for more than two mics:

- Plug the mics into a stage box (described in Chapter 1) and run the snake back to your mixer. Plug the snake mic connectors into your mixer mic inputs. If you're recording to two-track, plug the mixer stereo outputs into the recorder line inputs. If you're recording to multitrack, plug the mixer insert-send jacks into the recorder line inputs.
- If you have a multichannel audio interface with mic preamps, plug the mics into the interface. Connect the interface USB or FireWire port to a laptop computer.
- If you want to feed your mic signals to several mixers—recording, broadcast, PA—plug your mic cables into a mic splitter or distribution amp (described in Chapter 17 under the heading Splitting the Mic Signals). Connect the splitter outputs to the snakes for each mixer. Supply phantom from one mixer only, on the direct side of the split. Each split will have a ground-lift switch on the splitter. Set it to *ground* for the mixer supplying phantom. Set it to *lift* for the other mixers (or to *ground* if that results in the least hum). This prevents hum caused by ground loops between the different mixers.
- If you're using directional microphones and want to make their response flat at low frequencies, you can run them through a mixer with equalization for bass boost. You might prefer to equalize the recording during mastering instead. Boost around 50 to 100 Hz until the bass sounds natural or until it matches the bass response of omni condenser mics. You won't need this EQ if the microphones have been factory-equalized for flat response at a distance.

Monitoring

Put on your headphones or listen over loudspeakers in a separate room. Sit equidistant from the speakers, as far from them as they are spaced apart. You'll probably need to use a Nearfield arrangement (speakers about 3 feet apart and 3 feet from you) to reduce coloration of the speakers' sound from the room acoustics.

Turn up the recording-level controls and monitor the signal. When the orchestra starts to tune up, set the recording levels to peak around −20 dBFS so you have a clean signal to monitor. You can set levels more carefully later on.

MICROPHONE PLACEMENT

Nothing has more effect on the production style of a classical-music recording than microphone placement. Miking distance, polar patterns, angling, spacing, and spot miking all influence the recorded sound character.

Miking Distance

The microphones must be placed closer to the musicians than a good live listening position. If you place the mics out in the audience where the live sound is good, the recording probably will sound muddy and distant when played over speakers. That is because all the recorded reverberation is reproduced up-front along a line between the monitor speakers, along with the direct sound of the orchestra. Close miking (4 to 20 ft from the front row) compensates for this effect by increasing the ratio of direct sound to reverberant sound.

The closer the mics are to the orchestra, the closer it sounds in the recording. If the instruments sound too close, too edgy, or too detailed, or if the recording lacks hall ambience, the mics are too close to the ensemble. Move the mic stand a foot or two farther from the orchestra and listen again.

If the orchestra sounds too distant, muddy, or reverberant, the mics are too far from the ensemble. Move the mic stand a little closer to the musicians and listen again.

Eventually you'll find a sweet spot, where the direct sound of the orchestra is in a pleasing balance with the ambience of the concert hall. Then the reproduced orchestra will sound neither too close nor too far.

Here's why miking distance affects the perceived closeness (perspective) of the musical ensemble: The level of reverberation is fairly constant throughout a room, but the level of the direct sound from the ensemble increases as you get closer to it. Close miking picks up a high ratio of direct-to-reverberant sound; distant miking picks up a low ratio. The higher the direct-to-reverb ratio, the closer the sound source is perceived to be.

If the recording venue is "live" because of hard surfaces — brick, glass, stone—chances are you will need to mike closely. On the other hand, if the venue is "dead" because of soft surfaces—carpet, drapes, stuffed seats—expect to mike farther away.

An alternative to finding the sweet spot is to place a stereo pair close to the ensemble (for clarity) and another stereo pair distant from the ensemble (for ambience). According to Delos recording director John Eargle, the distant pair should be no more than 30 feet from the main pair, otherwise its signal might simulate an echo. You record the two pairs to a multitrack recorder and mix them back in the studio. The advantages of this method are

- It avoids pickup of bad-sounding early reflections.
- Close miking reduces pickup of background noise.
- After the recording is finished, you can adjust the direct/reverb ratio or the perceived distance to the ensemble.
- Comb filtering due to phase cancellations between the two pairs isn't severe because the delay between them is great and their levels and spectra are different.

Similarly, Skip Pizzi recommends a "double MS" technique, which uses a close MS microphone mixed with a distant MS microphone (as shown in Figure 18.2). One MS microphone is close to the performing ensemble for clarity and sharp imaging, and the other is out in the hall for ambience and depth. The distant mic could be replaced by an XY pair for lower cost.

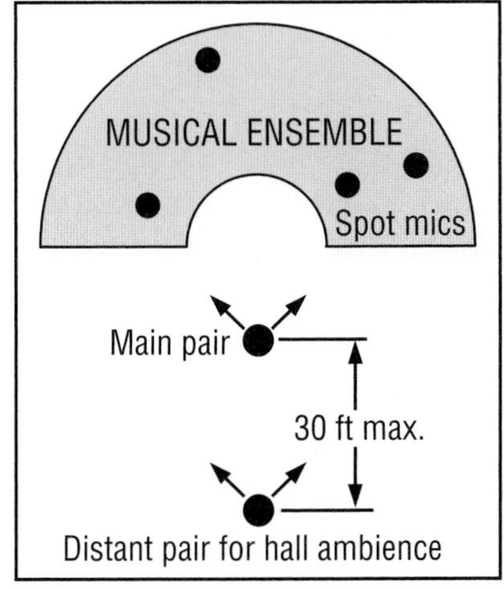

FIGURE 18.2
Double MS technique using a close main pair and a distant pair for ambience. Spot mics also are shown.

If the ensemble is amplified through a sound-reinforcement system, you might be forced to mike very close to avoid picking up amplified sound and feedback from the reinforcement speakers. In that case you'll need to add high-quality artificial reverberation or convolution reverb.

For broadcast or communications, consider miking the conductor with a wireless lavalier mic or stand-mounted mic.

Stereo-Spread Control

Concentrate on the stereo spread. If the monitored spread is too narrow, it means that the mics are angled or spaced too close together. Increase the angle or spacing between mics until localization is accurate. **Note**: increasing the angle between mics will make the instruments sound farther away; increasing the spacing won't.

If off-center instruments are heard far-left or far-right, that indicates your mics are angled or spaced too far apart. Move them closer together until localization is accurate.

If you record with a mid-side microphone, you can change the monitored stereo spread either during the recording or after.

To change the spread during the recording, connect the stereo-mic outputs to the matrix box and connect the matrix-box L–R output to the recorder. Use the stereo-spread control (M/S ratio) in the matrix box to adjust the stereo spread.

To alter the spread after the recording using a matrix box: Record the mid signal on one track and the side signal on another track. Monitor the output of the recorder with a matrix box. Back in the studio, run the mid and side tracks through the matrix box, adjust the stereo spread as desired, and record the left and right outputs.

To alter the spread after the recording using a DAW:

1. Record the mid mic on track 1; record the side mic on track 2.
2. Copy or clone track 2 to track 3. Be sure the waveforms are aligned.
3. Pan track 2 hard left; pan track 3 hard right.
4. Reverse the polarity of track 3 or use an "invert polarity" plug-in.

5. Group tracks 2 and 3 so their faders move together.
6. To change the stereo spread, vary the levels of tracks 2 and 3 relative to track 1.

If you're set up before the musicians arrive, check the localization by recording yourself speaking from various positions in front of the microphone pair while announcing your position (e.g., "left side," "mid-left," "center"). Play back the recording to judge the localization accuracy provided by your chosen stereo array. Recording this localization test is an excellent practice.

MONITORING STEREO SPREAD

Full stereo spread on speakers is a spread of images all the way between speakers, from the left speaker to the right speaker. Full stereo spread on headphones can be described as stereo spread from ear to ear. The stereo spread heard on headphones may or may not match the stereo spread heard over speakers, depending on the microphone technique used.

Due to psychoacoustic phenomena, coincident-pair recordings have less stereo spread over headphones than over loudspeakers. Take this into account when monitoring with headphones or use only loudspeakers for monitoring.

If you're monitoring your recording over headphones or anticipate headphone listening to the playback, you may want to use near-coincident miking techniques, which have similar stereo spread on headphones and loudspeakers.

Ideally, monitor speakers should be set up in a Nearfield arrangement (say, 3 feet from you and 3 feet apart) to reduce the influence of room acoustics and to improve stereo imaging. On the wall behind the monitors, attach a panel of acoustic foam that extends a few feet beyond the speaker spacing.

If you want to use large monitor speakers placed farther away from you, deaden the control-room acoustics with acoustic foam or thick fiberglass insulation (covered with muslin). Place the acoustic treatment on the walls behind and to the sides of the loudspeakers. This smoothes the frequency response and sharpens stereo imaging.

You might include a stereo/mono switch in your monitoring system to check for mono compatibility.

Soloist Pickup and Spot Microphones

Sometimes a soloist plays in front of the orchestra. You have to capture a tasteful balance between the soloist and the ensemble. That is, the main stereo pair should be placed so that the relative loudness of the soloist and the accompaniment is musically appropriate. If the soloist is too loud relative to the orchestra (as heard on headphones or loudspeakers), raise the mics. If the soloist is too quiet, lower the mics. You may want to add a spot mic (accent mic) about 3 feet from the soloist and mix it with the other microphones. Take care that the soloist appears at the proper depth relative to the orchestra.

If a sound reinforcement system is in use, place the soloist mic 8 to 12 inches away to prevent feedback. Use a foam windscreen on a vocal mic. To make the soloist mic inconspicuous, you could place a small-diaphragm mic at about chest height aiming at the mouth, and use a slender mic stand such the Schoeps Active Tube RC.

Many record companies prefer to use multiple microphones and multitrack techniques when recording classical music. Such methods provide extra control of balance and definition and are necessary in difficult situations. Often you must add spot or accent mics on various instruments or instrumental sections to improve the balance or enhance clarity (as shown in Figure 18.2). In fact, John Eargle contended that a single stereo pair of mics rarely works well.

As pkautzsch posted on Gearslutz.com:

> "Recording is not about the real picture, but about the picture we want. I like to have it similar to the real picture—that is, choir behind orchestra, soloists in front, violins left, etc.—but I'd never try to make a recording that is real. At the best, it will be credible. What spot mics do in recording for home stereo use is nothing else than what our eyes and brains do when attending a concert."

A choir that sings behind the orchestra can be miked separately with two to four cardioids. You might place the choir in the audience area facing the orchestra, and mike the choir.

If the recording mics are also used for sound reinforcement, place the choir mics about 3 feet out front and 3 feet above the head height of the back row. Add piano mics and windscreened soloist mics about 8 inches from their sound sources. Record all the mics to multitrack, and mix the tracks back in your studio. Add reverb and EQ as needed.

Pan each spot mic so that its image position coincides with that of the main microphone pair. Using the mute switches on your mixing console, alternately monitor the main pair and each spot mic to compare image positions.

When you use spot mics, mix them at a low level relative to the main pair—just loud enough to add definition but not loud enough to destroy depth. Operate the spot-mic faders subtly or leave them untouched. Otherwise the close-miked instruments may seem to jump forward when the fader is brought up, then fall back in when the fader is brought down. If you bring up a spot-mic fader for a solo, drop it only 6 dB when the solo is over, not all the way off.

Often the timbre of the instrument(s) picked up by the spot mic is too bright. You can fix it with a high-frequency rolloff or by using a mic with that characteristic. Adding artificial reverb to the spot mic can help too.

To further integrate the sound of the spots with the main pair, you might want to delay each spot's signal to coincide with those of the main pair. That way, the main and spot signals are heard at the same time. For each spot mic, the formula for the required delay is

$$T = \frac{D}{C}$$

where

T = delay time in seconds
D = distance between each spot mic and the main pair in feet
C = speed of sound, 1130 ft per second

For example, if a spot mic is 20 feet in front of the main pair, the required delay is 20/1130 or 17.7 msec. Some engineers add even more delay (10 to 15 msec) to the spot mics to make them less noticeable. As a rule of thumb, 1 foot corresponds to about 1 millisecond of delay.

A suggested track assignment for multitrack concert recordings is shown below:

Tracks 1–2: main pair
Tracks 2–3: distant pair
Tracks 4–5: "outriggers" or widely spaced pair
Tracks 6 and up: spot mics and conductor's mic (for announcing takes)

Once the microphones are positioned properly, gaffer-tape the mic stand legs to the floor (or add sandbags) so that the stands can't be knocked over.

SETTING LEVELS

Now you're ready to set recording levels. Ask the conductor to have the orchestra play the loudest part of the composition, and set the recording level for the desired meter reading. A typical recording level is −6 dB maximum on a peak-reading meter for a digital recorder. The level can go up to 0 dB maximum without distortion, but aiming for -6 dB allows for surprises. Bass-drum and tympani hits produce the highest peaks.

If you plan to record a concert with no sound check, you have to set the record-level knobs to a nearly correct position ahead of time. Do this during a pre-concert trial recording, or just go by experience: set the knobs where you did at previous sessions (assuming you're using the same microphones in the same venue). Another way to preset the recording level is to aim for a peak meter level of −20 dBFS when the orchestra tunes up.

Before going on location, you could play loud orchestral music over your studio monitors or home stereo, set up your mics and recorder, and set approximate recording levels. If you need to set the level less than one-third up to get a 0 dB meter level, you probably need to insert a pad or set the mic-gain switch to low gain.

RECORDING A CONCERT

Before the concert obtain a printed program of the musical selections. On this program, next to each piece, you will write down the recorder counter times of the beginning and end of that piece. That will help you locate and identify the pieces correctly when you edit the recording later.

Start recording when the conductor walks out (or sooner). Record the concert nonstop if your recording medium allows. Document your mic placement and recording level for future jobs.

EDITING

At this point, the recording is finished and you brought it back to your studio. The concert recording has long applause after each piece. Suppose you want to edit the applause shorter and insert a few seconds of silence between the compositions. Here is a suggested procedure for editing a stereo recording using a DAW:

1. Connect a USB cable between the computer and recorder. Then the computer recognizes the recorder as a storage device. Click-drag the recorder's WAVE file of the concert to your hard drive.
2. In the audio editing software, start a new project and set up an audio track.
3. Import the concert file from the hard drive into the audio track.
4. Play the track and locate the first piece. Refer to the counter times you wrote on your session notes or concert program.
5. Delete the part of the recording before the first piece, but don't close up the space yet.
6. Find the applause at the end of the first piece. About 10 seconds into the applause, split the clip or region. Also split the clip a few seconds before the start of the next piece. Cut out the audio between the split points, but don't close up the space.
7. Label the clip of the composition with its title.
8. Repeat Step 6 for the rest of the pieces.

Now each musical piece is in a separate, labeled clip or region on your screen. Next you will add fades and adjust the spacing between the pieces. Please refer to Figure 18.3.

1. Slide the first piece back to time zero, then slide it to the right 10 frames (1/3 second).
2. At the end of each piece, let the applause play for 3 seconds then fade it out over about 8 seconds. Use a fade that starts quickly and ends slowly.

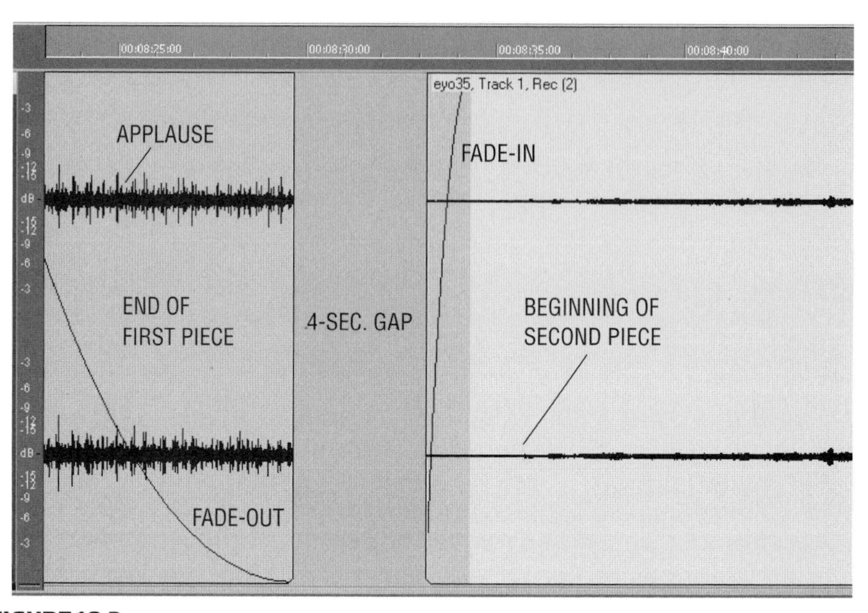

FIGURE 18.3
A DAW screen showing an applause fade-out, a 4-second gap, and a fade-in before the next piece.

3. If there is background noise such as air-conditioning rumble, insert a fade-in about 2 seconds before the beginning of each piece (Figure 18.3). Or you may want the track to start right when the music starts; i.e., with no ambience before it starts.
4. Time-slide the clips to create a 4-second gap between them (or whatever interval sounds right).

If you don't want total silence between pieces, don't fade out the applause. End each piece's clip where the applause stops, and leave about 4 seconds of recorded "room tone" between clips.

To edit a session that has no applause, set the beginning of each clip just before the first note. Set the end of each clip just after the reverberant tail fades to silence. Leave about a 4-second silent gap between clips (or whatever sounds right).

When you're finished editing, note the start time and duration of each clip. You will use this information to write a cuesheet that determines the start IDs when burning a CD. Export the edited program to a 24-bit stereo file.

Sometimes, despite your best intentions and the finest microphones, the spectral balance of the recording might be poor. The bass might be weak, the strings strident, and so on. Room acoustics play a strong part in this. Fortunately, a recording having a skewed tonal balance often can be salvaged with EQ. An effective tool for this purpose is Harmonic Balancer (www.har-bal.com), which shows the spectrum of the recording and lets you equalize it as needed.

Once the program is equalized (if necessary), import the equalized file and normalize it so that the highest peak reaches −0.3 dBFS. Finally, enable dither and export the normalized mix to a 16-bit stereo file. You will burn a CD from this file. There's your finished recording!

CHAPTER 19

Surround Sound: Techniques and Media

So far we've covered techniques that result in a 2-channel stereo recording. With stereo, you hear all the instruments and reverb in front of you, in the area between the two loudspeakers. But with surround sound, you hear audio images in every direction. For example, the musical ensemble could be up front, while the hall ambience envelops you from the sides and rear, as it does in real life.

Stereo uses two channels feeding two loudspeakers in front of you. Surround sound uses multiple channels feeding multiple speakers placed around you. A disadvantage of stereo is that you must sit in a tiny "sweet spot" to hear correct localization. In contrast, surround sound can be heard correctly in a wide area between the speakers.

Surround gives a wonderfully spacious effect. It puts you inside the concert hall with the musicians. You and the music occupy the same space—you're part of the performance. For this reason, some engineers feel that surround is more musically involving, more emotionally intense, than regular stereo. There's a sense of envelopment. Surround mixing is fast becoming a valuable new tool to offer your customers. Because many listeners have home theater systems with multiple speakers, they're already set up to play surround audio recordings.

A magazine devoted to multichannel sound production is *Surround Professional* found at www.surroundpro.com/articles/publish/article_190.shtml.

SURROUND SPEAKER ARRANGEMENT

Inherited from the film industry, surround sound uses six channels feeding six speakers placed around the listener. This forms a 5.1 surround system, where the "point 1" is the subwoofer (LFE) channel. The LFE channel is band-limited to 125 Hz and below, while the other channels are full bandwidth (20 Hz to 20 kHz). "5.1" is a channel format—a surround-sound standard that states the number of channels, their frequency response, and the speaker placement.

The six speakers are:

1. Left-front
2. Center
3. Right-front
4. Left-surround
5. Right-surround
6. Subwoofer

Figure 19.1 shows the recommended placement of monitor speakers for 5.1 surround sound. It is the standard setup proposed by the International Telecommunication Union (ITU). From the center

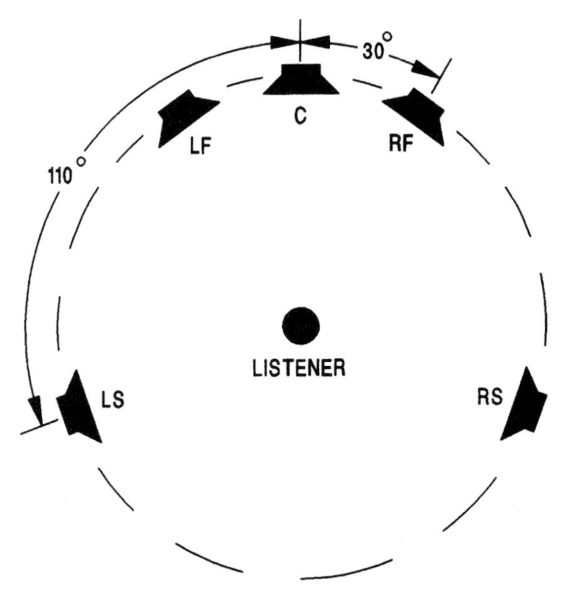

FIGURE 19.1
Recommended placement of monitor speakers for 5.1 surround sound.

speaker, the left and right speakers should be placed at ±30 degrees, and the surrounds at ±110 degrees. (Some engineers prefer 120 to 125 degrees for the surround speakers.) If you're mixing movie soundtracks, use dipole speakers for the surrounds, and place them to the sides (±90 degrees).

The left- and right-front speakers provide regular stereo. The surrounds provide a sense of envelopment due to room ambience. They also allow sound images to appear behind the listener. Deep bass is filled in by the subwoofer. Because we don't localize low frequencies below about 120 Hz, the sub can be placed almost anywhere without degrading localization.

Originally developed for theaters, the center-channel speaker is mounted directly in front of the listener. In a home-theater system, it is placed just above the TV screen, or just below and in front of the TV screen. This speaker plays center-channel information in mono, such as dialog.

Why use a center speaker, when two stereo speakers create a phantom center image? If you use only two speakers and you sit off-center, the phantom image shifts toward the side on which you're sitting. But a center-channel speaker produces a real image, which doesn't shift as you move around the listening area. The center speaker keeps the actors' dialog on-screen, regardless of where the listener sits.

Also, the phantom center image doesn't have a flat frequency response, but a center speaker does. Why is this? Remember that a center image results when you feed identical signals to both stereo speakers. The right-speaker signal reaches your right ear, but so does the left-speaker signal after a delay around your head. The same thing occurs symmetrically at your left ear. Each ear receives direct and delayed signals, which interfere and cause phase cancellations at 2 kHz and above. A center-channel speaker doesn't have this response anomaly.

With a phantom center image, the response is weak at 2 kHz because of the phase cancellation just mentioned. To compensate, recording engineers often choose mics with a presence peak in the upper midrange for vocal recording. The center-channel speaker doesn't need this compensation.

For sharpest imaging and continuity of the soundfield, all the speakers should be:

- The same distance from the listener
- The same model (except the sub)
- The same polarity
- Direct-radiator types
- Driven with identical power amps
- Matched in sound pressure level with pink noise (the sub is 10 dB louder)
- Placed in a room that is left-right symmetrical in the front and the rear.

If you're mixing movie soundtracks, the surround speakers should be dipole designs rather than direct-radiator types. The dipole speakers project sound forward and backward to create a diffuse effect.

Typically the speakers are 4 to 8 feet from the listener and 4 feet high. Use a length of string to place the monitors the same distance from your head. The sub can go along the front wall on the floor. Be sure that all the speakers sound the same so there is no change in tonal balance as you pan images around.

A DSP algorithm to reproduce surround sound on headphones has been developed by Lake Technology. It is licensed to Dolby Labs under the name Dolby Headphone. This feature is beginning to appear in consumer receivers and surround processors.

SETTING UP A SURROUND MONITORING SYSTEM

Working in surround, of course, requires more equipment than working in stereo. You'll need:

- Five "satellite" monitors and a subwoofer, and six channels of power amplification, or five powered monitors and a powered sub
- A sub/satellite crossover (a bass management system); this is built into some surround monitor packages
- A six-channel volume control (hardware or software)

Be sure to include a subwoofer in your monitor system. If you don't, you might not hear low-frequency noises that a home listener with a sub will hear. These noises include breath pops, mic-stand thumps, air-conditioning rumble, and excessively heavy deep-bass notes.

You could use a home-theater surround receiver for power amplification and bass management. It has a single volume control that simultaneously adjusts the level of all the tracks. Most home-theater receivers have five amp channels and a line output that feeds a powered subwoofer. The sub's power amp should be at least 100 watts, and the receiver should have six analog inputs for your surround mix.

A feature to look for in surround sound receivers is Dolby Digital and Digital Theater Systems (DTS) decoders (explained later under the heading Surround Encoding for CD). Receivers labeled "5.1 ready" or "Dolby Digital ready" aren't Dolby Digital compatible. The receiver must have Dolby Digital and DTS decoders to play those formats. However, you don't need those decoders to do surround mixes.

Bass Management

In the surround receiver is a "bass management" circuit. Bass management is nothing more than a subwoofer/satellite crossover filter. It sends the deep lows to the sub, and sends other frequencies to the five "satellite" speakers that surround you.

A bass management circuit routes frequencies above about 100 Hz to the five full-range speakers, and routes frequencies below about 100 Hz from all six channels to the subwoofer. In other words, the bass management circuit routes low-frequency signals from all the five channels—and the LFE channel—to your sub. By keeping the deep lows out of the full-range speakers, bass management reduces their low-frequency distortion and lets them be made relatively small for home use. Note that bass management affects only what you monitor, not what you record.

Bass management can be done by a surround-receiver circuit, a standalone box, a special circuit in a subwoofer, or a software plug-in.

You set the bass management crossover frequency as low as possible to remove the directionality of the bass, and to extend the headroom of the sub. Typically you'd set the crossover frequency to the frequency where your full-range speakers are down 3 dB on the low end. If your five main speakers extend down to 20 or 30 Hz, you don't need bass management. In some receivers, the bass management crossover frequency is adjustable among 120/100/80 Hz—whatever your system needs for flattest response.

The crossover frequency is 120 Hz with the Dolby Digital standard and 80 Hz with the DTS standard. However, the Dolby Digital, DTS, and THX standard crossover frequencies are irrelevant to the frequency of the bass-management filter for the mixing work in your studio. And the upper frequency limit of the LFE track (125 Hz) has nothing to do with the crossover frequencies of the bass management (120 to 40 Hz). As mentioned above, you set the crossover frequency according to the frequency response of your monitor speakers.

LFE Channel Filtering

The LFE channel lowpass filter used in bass management is a very steep filter (48 dB per octave), so it causes a lot of phase shift. But you can turn off the LFE filter in the encoder, and use your own gentler filter instead. Try 80 Hz, 24 dB per octave. Use any filter you like, as long as it removes all the energy at 125 Hz. Insert this filter between your console's LFE channel output and the input of the mixdown recorder's LFE track. In other words, record the LFE track pre-filtered, and turn off filtering in the encoder.

Surround Mixing Equipment

Here's what you need to mix in surround:

- Multitrack recordings in any format: analog tape, digital tape, hard disk, etc.
- If you're using a mixing console, it needs to have at least six output channels (also called busses or subgroups). Most digital consoles have a surround matrix, a section set up for mixing and monitoring in surround.
- If you're using a console, you need an 8-track recorder (MDM or hard-disk) to record the surround mix. Tracks 1 through 6 record the surround channels, while tracks 7 and 8 record a separate stereo mix.
- If you're using a DAW for surround mixing, you need an audio interface with at least 8 outputs. Set up the DAW to feed the 5.1 output channels to six of the interface outputs, then connect those outputs to your five powered satellites and subwoofer. This system replaces the surround receiver mentioned earlier.

- If you're using a DAW, you need surround mixing software. For example, MOTU's Digital Performer and Cakewalk's Sonar Producer include surround mixing. You could mix to six tracks on your hard drive, and copy the six resulting WAVE files to a CD or DVD.

Connections

Now that you know the necessary equipment for surround mixing (including bass management), you can wire the system together. Two methods are described below.

Method 1 uses an external multitrack recorder to record the surround mix tracks. Method 2 uses the DAW to record the surround mix tracks.

Method 1 using an external multitrack recorder: Basically you connect line-level signals from six busses to the associated tracks on your mixdown recorder. To monitor those six tracks, connect that recorder's outputs either to a surround receiver, or to a bass-management filter that feeds six channels of power amps. The receiver or amplifiers drive the five small speakers and the sub.

Figure 19.2 shows the connections. Patch the console's bus outputs or surround matrix outputs to the inputs of the 8-track mixdown recorder. On the back of this recorder, connect track outputs 1 through 6 to the inputs of the surround receiver or power amp inputs. On the receiver or power amps, connect the speaker outputs to the speakers. If your subwoofer is self-powered, connect the

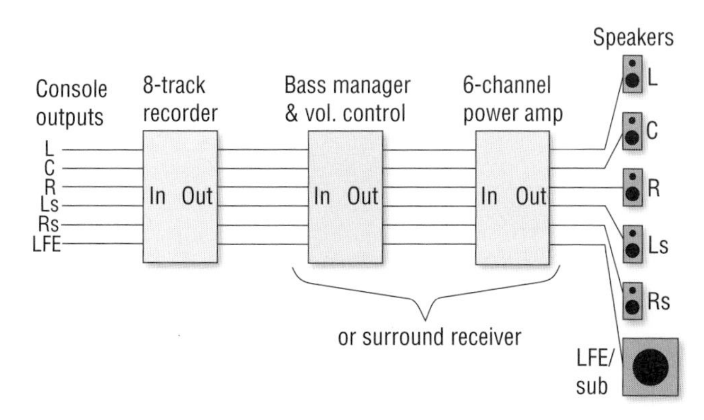

FIGURE 19.2
Mixdown and monitoring system for 5.1 surround with an external multitrack recorder.

LFE channel line output to the sub line input. If your speakers are self-powered, connect them to outputs 1 through 6 of your bass manager.

When you connect your mixer or DAW to the 8-track mixdown recorder, which signal goes on which track? The most common track assignment (the Dolby Digital, ITU, and SMPTE standard) is given below:

1. Left-front
2. Right-front
3. Center
4. LFE or subwoofer channel
5. Left-surround
6. Right-surround
7. Stereo mix left
8. Stereo mix right

Be sure to label your recording media with the track assignment.

Method 2 using the DAW to record the surround mix tracks: Connections for powered monitors are shown in Figure 19.3 (top).

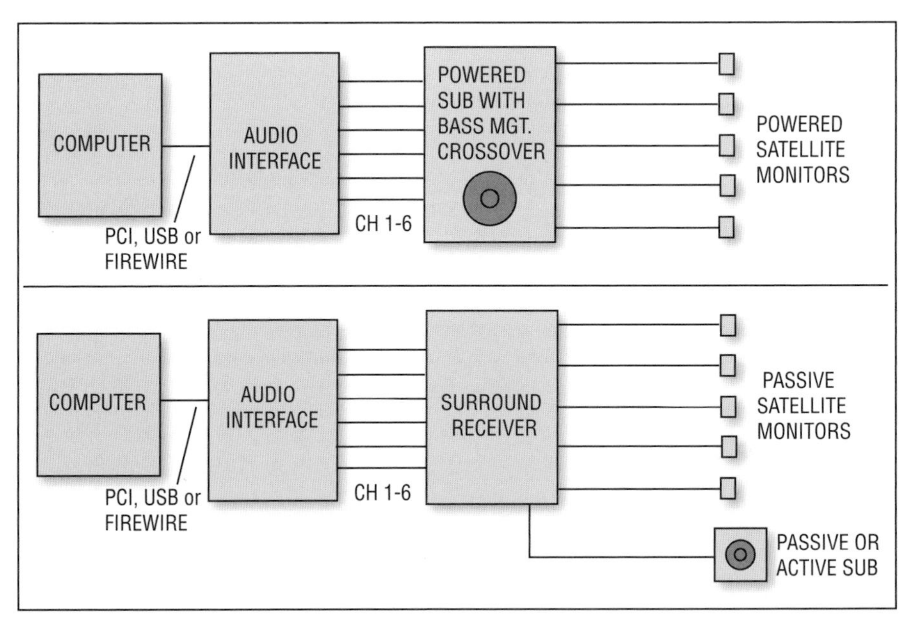

FIGURE 19.3
Mixdown and monitoring system for 5.1 surround, using a DAW as the multitrack recorder. Top: with powered monitors. Bottom: with passive monitors.

Set the overall listening level with your computer surround-monitoring software, and trim individual speaker levels at each speaker. Connections for passive monitors are shown in Figure 19.3 (bottom). Set the overall listening level with the volume control on the surround receiver. The bass-management filter is built into the powered sub or into the surround receiver.

Calibration

It's essential to calibrate your speakers so that their levels match. You need a sound level meter (a low-cost one from Radio Shack is adequate). Here's a suggested procedure:

1. Turn on the LFE channel and disable bass management.
2. Set each power amp channel to the same volume setting.
3. Get a pink noise signal from a console noise generator, receiver, or CD. Connect the signal to a console input. Other options are a surround calibration DVD, such as the 5.1 Audio Toolkit DVD from www.gold-line.com.
4. Assign that input channel to one surround channel (output bus or group fader). For example, start with the front-left channel.
5. Set the console's or DAW's bus faders and master faders to unity gain (the shaded portion of fader travel).
6. Adjust the input fader so the level on the mixer's meter is −20 dBFS, the standard level for surround mixes. This level allows for some headroom.
7. Hold the sound level meter at your mixing position and set it to C-weighting, slow response. Aim the meter's mic at the speaker. Set the level of the power-amp channel (or the powered monitor) so you read 85 dBSPL on the meter.
8. Repeat steps 4 through 7 for each channel and its speaker, one at a time, so that all speakers are putting out the same level. (In most home-theater receivers is a menu that lets you set the level of each channel.) If you're mixing for theater sound with dipole speakers, set their level to 82 dBSPL instead.
9. Assign pink noise to just the LFE channel so that you hear the noise coming from the subwoofer. The playback level of the LFE channel is set 10 dB higher than the main speaker levels because that's done in consumer surround systems. Using the

receiver's LFE gain control, or the sub's level control, set the sub's SPL to 95 dBSPL if you're using an RTA, or to 89 dBSPL if you're using a sound level meter.

10. After calibration, record your mixes at -6dBFS maximum in peak meter mode, and set your master monitor level as desired.

Why do the RTA and sound level meter require different SPL settings? Unlike an RTA, an SPL meter integrates the energy of all the octaves it is measuring to come up with a single reading. The main speakers produce 8 to 10 octaves, but the sub produces 2 octaves. The sub's 2-octave energy reads lower on the SPL meter than the main speakers' 8- to 10-octave energy. So, instead of setting the sub's SPL to 95 dB, you set it to 89 dB to allow for this energy difference.

It's important to phase-align the subwoofer's signal with the full-range speakers' signals. Improper alignment can cause a dip in the system's frequency response at the crossover frequency. If your sub has a phase-matching control, send a sine wave to all speakers at the crossover frequency. Adjust the phase control to get the loudest sound at the mix position. If the sub has a polarity switch, try it both ways, and use the position that gives the loudest bass when all the speakers are playing.

Here's another calibration method developed by Lorr Kramer of DTS and surround-sound expert Mike Sokol: Use bandwidth-limited pink noise (20 to 80 Hz) for the woofer and 500 to 2000 Hz for the main speakers. Set each speaker to the same SPL, one at a time. Using 500 Hz to 2 kHz pink noise for the main speakers keeps the signal out of the LFE channel and avoids exciting room modes. And limiting the top end to 2 kHz helps avoid directionality effects of the mic in the SPL meter.

Now that your monitors are calibrated, do the same for your mixing console or DAW recording software.

1. Feed pink noise into one channel input module.
2. Route or patch that module's signal equally to all six surround output busses.
3. Set each output bus fader to unity gain.
4. At the input module, set the trim to make the output level −20 dB on the console's meters.

5. Trim the console's ADAT or TDIF outputs (if used) so that all six tracks read −20 dBFS. (The LFE channel doesn't get 10 dB extra gain in the recording path, only in the monitoring path.)
6. On your surround receiver, set the small/large speaker switch to small, and enable bass management.

RECORDING AND MIXING POP MUSIC FOR SURROUND

Doing a multitrack recording for later mixdown in surround is almost identical to recording for mixdown to stereo. Some engineers add a few ambience mics to pick up the studio acoustics. These ambience tracks are panned to the rear speakers during mixdown.

Many producers like to monitor the original recording session in stereo, not surround, to reduce the technical issues that can slow down the session. They wait until mixdown to monitor in surround.

In Chapter 12 we showed some ways to mix multitrack recordings to 2-track stereo. Now let's look at mixing those same recordings to 5.1 surround.

Panning

To place the image of each track in space around you, some consoles include a pair of pan controls (left/right and front/rear) for each channel. Others use a trackball, mouse, or joystick. Some digital audio editing programs, or plug-ins for those programs, permit surround panning as well. Some plug-ins for Pro Tools are Neyrinck's Mix 51, Surround Zone, and Audio Research Labs Surround Stage. Minnetonka's SurCode is a program to mix 5.1 surround sound, including a "Build your own mixer" GUI and automated surround panners (www.minnetonkaaudio.com).

Suppose your console or recording software has 8 output busses, and you want to pan or move a track between front-left and rear-right. Assign the track to the front-left and rear-right busses. They should be odd-even numbered. When you turn the pan knob from left to right, the sound image should move as desired.

Where should you pan tracks in a surround mix? There are no set rules. Some producers like to put all the musicians up front, and put the hall ambience and applause in the rear. This works especially well for live concerts and recordings of classical music. This ambience can be created artificially with a digital reverb, or recorded with mics in the concert hall. Stereo reverb returns are typically sent to the rear channels. Several multichannel digital reverbs are available, both as hardware and as software.

Some producers use the rear speakers for background vocals, percussion, horn stabs, or strings, leaving the lead vocal and rhythm instruments in front. A group of five singers could be panned to place one singer in each speaker, so that the listener is in the middle of the group. You might put the lead vocal in the center speaker, and also partly in the surrounds to pull the vocal out toward the listener. Spreading instruments between left-front and left-surround, and right-front and right-surround, gives a greater sense of envelopment.

You can place instruments in fixed positions (static panning), or move them around in space (dynamic panning). Some consoles let you move a sound by drawing its path in space on your monitor screen. You might move a sound along a circle, arc, or line. Then save these pans as part of your automated mix.

Using the Center Speaker

Some engineers prefer to send center tracks equally to the left and right speakers—not the center speaker. This creates a phantom center image. However, sending a track to the center speaker gives a better tonal balance and more stable imaging than the center phantom image.

It's not recommended to send a track only to the center speaker. Some listeners don't have a center speaker. So if you send, say, a vocal to just the center speaker, those listeners without a center speaker won't hear it. Also, home listeners can solo the center channel and hear punch-ins on that channel. So many producers route center tracks mainly to the left and right speakers, and also feed a little signal to the center speaker (maybe 6 dB down). Some producers don't use the center speaker at all.

Using the LFE Channel

With music mixes you seldom need to send anything to the LFE subwoofer track—it gets deep bass from the bass management circuit in the listener's playback system, and at the proper level. Because the five main channels of 5.1 surround encoders reproduce from 20 Hz to 20 kHz, there's no need to put anything musical in the LFE channel. The consumer's bass management filter will redirect any sub-80-Hz energy to their own subwoofer anyway, whether or not you put this bass signal in the LFE track. On the other hand, suppose the end listener has a subwoofer/satellite system with six channels, but no bass management system. In that case, the sound they hear will be thin in the bass if there is no signal in the LFE channel.

With film mixes, you typically send low-frequency sound effects (explosions, tornadoes, crashes) to the LFE channel. The LFE track also helps with DJ mixes and synth mixes that have lots of lows. Even if you mix only music, you need a subwoofer to hear the very deep noises (breath pops, room rumble) that a home listener will hear with a sub.

Downmixing

Downmixing is making a stereo mix from a 5.1 surround mix. It is done in the consumer's home theater receiver. In the downmixing circuit, the left and right surround channels are blended with the left and right front channels. The center channel is blended equally with the left and right channels. The LFE channel is either mixed with the front signals or not used.

Downmixes made this way seldom create a well-balanced stereo mix, so be sure to check your 5.1 mix for stereo compatibility. Surround monitoring systems should have a downmix button so you can hear how your surround mixes will sound when downmixed to stereo by consumer receivers.

It's best to do a separate stereo mix and record it on tracks 7 and 8 of your 8-track mixdown recorder. This stereo mix can be put on DVD-Audio discs or Super Audio CDs (explained later in this chapter) along with the surround mix.

Surround Mix Delivery Format

You will deliver your final 6-track recording to a mastering facility that will encode it onto a DVD. Supply them either the multitrack mixdown recording, or a DVD or CD-R with AIFF or WAV files, one for each output channel. Be sure to label the recording with the track contents. Identify each track with a recorded slate (front-left, center, sub, etc.). Print a 30-second test tone at 1 kHz at −20 dBFS on all tracks. Also include this information: LFE channel filtered or not, LFE channel filter frequency, mixdown listening level in dBSPL, sampling rate, bit resolution, SMPTE format (if included), media formats, program length, intended final audio sampling rate and bit depth, and a note about any glitches on your master.

SURROUND-SOUND MIC TECHNIQUES

So far we've talked about mixing multitrack recordings to surround—a method intended mainly for pop music. But for classical music, you can record in surround using five microphones, which capture the spatial character of the concert hall in which the musical ensemble is playing. Each mic feeds a separate track of a multitrack recorder.

Surround mic techniques are somewhat different from stereo mic techniques. In addition to the usual front-left and front-right mics, you need two surround mics to pick up the hall ambience, and sometimes a center mic to feed to the center channel. Note that listening in surround reduces the stereo separation (stage width) because of the center speaker, but mic techniques for surround are optimized to counteract this effect.

A number of mic techniques have been developed for recording in surround. Let's take a look at them.

Soundfield 5.1 Microphone System

This system is a single, multiple-capsule microphone (Soundfield ST250 or MKV) and Soundfield Surround Decoder for recording in surround. The decoder translates the mic's B-format signals (X, Y, Z, and W) into L, C, R, LR, RR, and mono subwoofer outputs.

Delos VR^2 Surround Miking Method

The late John Eargle, Delos' director of recording, developed their VR^2 (Virtual Reality Recording) format. Recordings made with this method offer discrete surround. They also are claimed to sound good in stereo and very good with "steered" analog decoding, such as Dolby Pro Logic.

In making these recordings, Eargle typically used the mic placement shown in Figure 19.4. This method employs an ORTF pair in the center, flanked by two spaced omnis typically 12 feet apart. Two house mics (to pick up hall reverb) are placed about 23 to 52 feet behind the main pair. These house mics are cardioids aiming at the upper rear corners of the hall, spaced about 12 feet apart and about 30 feet high. Spot mics (accent mics) are placed within the orchestra to add definition to certain instruments.

The mics are assigned to various tracks of a digital multitrack recorder:

- 1 and 2: A mix of the coincident-pair mics, flanking mics, house mics, and spot mics
- 3 and 4: Coincident-pair stereo mic
- 5 and 6: Flanking mics
- 7 and 8: House mics (surround mics)

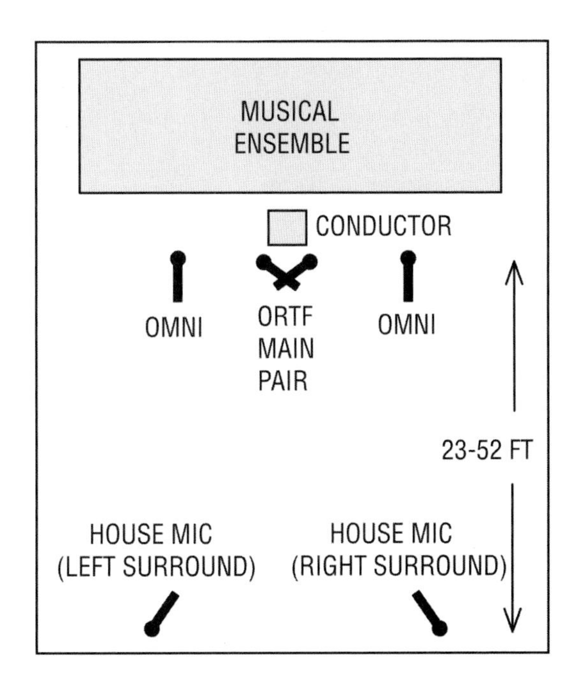

FIGURE 19.4
A Delos surround miking method.

NHK Method

The Japanese NHK Broadcast Center has worked out another surround miking method. They found that, for surround recording, cardioid mics record a more natural amount of reverb than omni mics. The mics are placed as described below:

- A center-aiming mic feeds the center channel.
- A near-coincident pair feeds front-left and front-right.
- Widely spaced flanking mics add expansiveness.
- Up to three pairs of ambience mics aim toward the rear.

Figure 19.5 shows the mic placement. The front mics are placed at the critical distance from the orchestra, where the direct-sound level matches the reverberant-sound level. Typically, this point is 12 to 15 feet from the front of the musical ensemble and 15 feet above the floor.

NHK engineers make this recommendation: When you're monitoring the surround program, the reverb volume in stereo listening should match the reverb volume in multichannel listening. That is, when you fold down or collapse the monitoring from 5.1 to stereo, the direct/reverb ratio should stay the same.

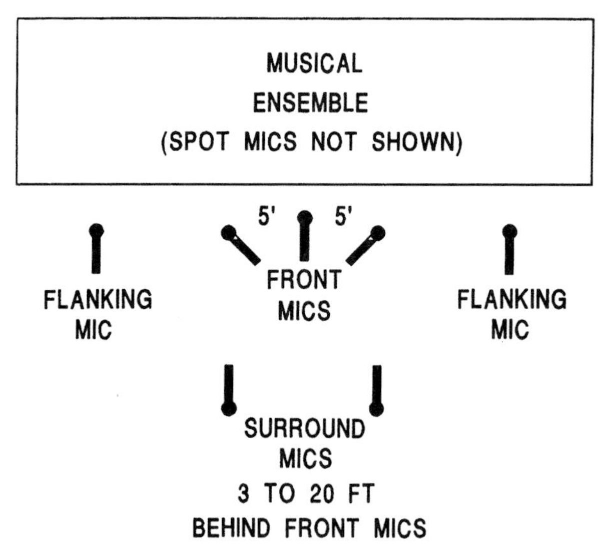

FIGURE 19.5
An NHK surround-sound miking method.

The KFM 360 Surround System

Jerry Bruck of Posthorn Recordings developed this elegant surround-miking method. It is a form of the mid-side (MS) stereo technique.

Bruck starts with a modified Schoeps KFM 6U stereo microphone, which is a pair of omni mics mounted on opposite sides of a 7-inch hard sphere. Next to those mics, nearly touching, are two figure-8 mics, one on each side of the sphere, each aiming front and back (Figure 19.6). This array creates two MS mic arrays aimed sideways in opposite directions. The mics don't seriously degrade each other's frequency response.

In the left channel, the omni and figure-8 mic signals are summed to give a front-facing cardioid pattern. They are also differenced to give a rear-facing cardioid pattern. The same thing happens symmetrically in the right channel. The sphere, acting as a boundary and a baffle, "steers" the cardioid patterns off to either side of center, and makes their polar patterns irregular.

By adjusting the relative levels of the front and back signals, the user can control the front/back separation. As a result, the mic sounds like it is moving closer to or farther from the musical ensemble.

According to Bruck:

> "The system is revelatory in its ability to recreate a live event. Perhaps most remarkable is the freedom a listener has to move around and select a favored position, as one might move around in a concert hall to select a preferred seat. The image remains stable, without a discernible sweet spot. The reproduction is unobtrusively natural and convincing in its sense of being there."

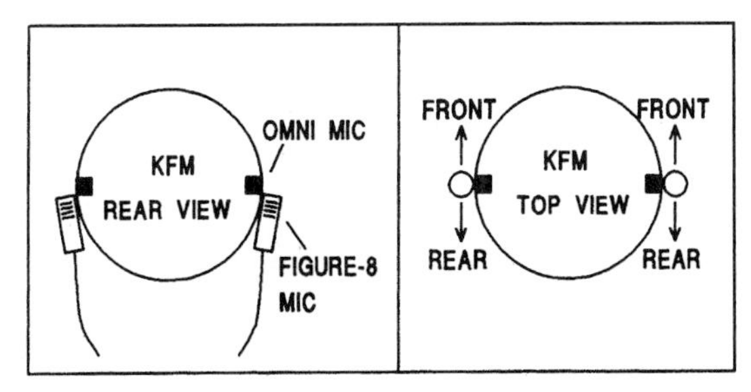

FIGURE 19.6
The KFM 360 Surround Miking System.

The four mic signals can be recorded on a 4-track recorder for later matrixing. The figure-8 mics need some EQ to compensate for their low-frequency rolloff and loss in the extreme highs. To maintain a good S/N ratio, this EQ can be applied after the signal is digitized.

Five-Channel Microphone Array with Binaural Head

This method was developed by John Klepko of McGill University. It combines an array of three directional mics with a 2-channel dummy head (Figure 19.7):

- For the front left and right channels: identical supercardioid mics
- For the center channel: a cardioid mic
- For the surround channels: a dummy head with two pressure-type omni mics fitted into the ear molds

The mics are shock mounted and have equal sensitivity and equal gains. Supercardioids are used for the front left/right pair to reduce center-channel buildup. Although the dummy head's diffraction causes peaks and dips in the response, it can be equalized to compensate.

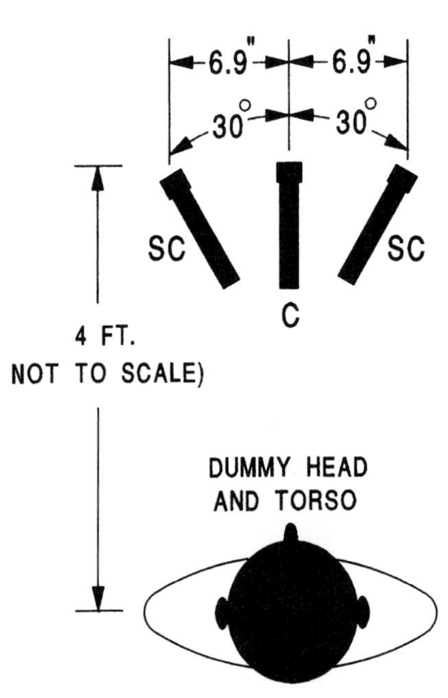

FIGURE 19.7
The Klepko surround-sound miking method.

During playback, the listener's head reduces the acoustical crosstalk that would normally occur between the surround speakers.

According to Klepko:

> "The walkaround tests form an image of a complete circle of points surrounding the listening position. Of particular interest is the imaging between ±30 degrees and ±90 degrees. The array produces continuous, clear images here where other (surround) techniques fail."
>
> "The proposed approach is downward compatible to stereo, although there will be no surround effect. However, stereo headphone reproduction will resolve a full surround effect due to the included binaural head-related signals. Downsizing to matrix multichannel (5-2-4 in this case) is feasible except that it won't properly reproduce binaural signals to the rear because of the mono surrounds. As well, some of the spatial detail recorded by the dummy-head microphone would be lost due to the usual bandpass filtering scheme (100 Hz to 7 kHz) of the surround channel in such matrix systems."

DMP Method

DMP engineer Tom Jung has recorded in surround using a Decca Tree stereo array for the band and a rear-aiming stereo pair for the surround ambience (Figure 19.8). Spot mics in the band complete the miking. The Decca Tree uses three mics spaced a few feet apart, with the center mic placed slightly closer to the performers. It feeds the center channel in the 5.1 system.

The rear-aiming mics are either a coincident stereo mic, another Decca tree, or a spaced pair whose spacing matches that of the Decca tree outer pair. Jung tries to aim the rear mics at irregular surfaces to pick up diffuse sound reflections.

Woszcyk Technique (PZM Wedge plus Opposite-Polarity, 180-Degree Coincident-Cardioid Surround Mics)

A recording instructor at McGill University, Wieslaw Woszcyk, developed an effective method for recording in surround that also works well in stereo. The orchestra is picked up by a PZM wedge made of two 18 × 29 inch hard baffle boards angled 45 degrees. A mini omni mic is mounted on or flush with each board. At least 20 feet behind

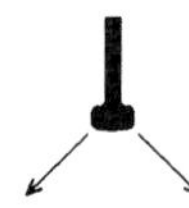

FIGURE 19.8
A DMP surround miking method.

the wedge are the surround mics: two coincident cardioids angled 180 degrees apart, aiming left and right, and in opposite polarity (Figure 19.9).

According to Woszcyk, his method has several advantages:

- Imaging is very sharp and accurate, and spaciousness is excellent due to strong pickup of lateral reflections.
- The out-of-phase impression of the surround pair disappears when a center coherent signal is added.
- The system is compatible in surround, stereo, and mono. In other words, the surround signals don't phase-interfere with the front-pair signals. That is because (1) the surround signals are delayed more than 20 msec, (2) the two mic pairs operate in separate sound fields, and (3) the surround mics form a bidirectional pattern in mono, with its null aiming at the sound source.

If a PZM wedge isn't acceptable because of its size and weight, other arrays with wide stereo separation may be substituted.

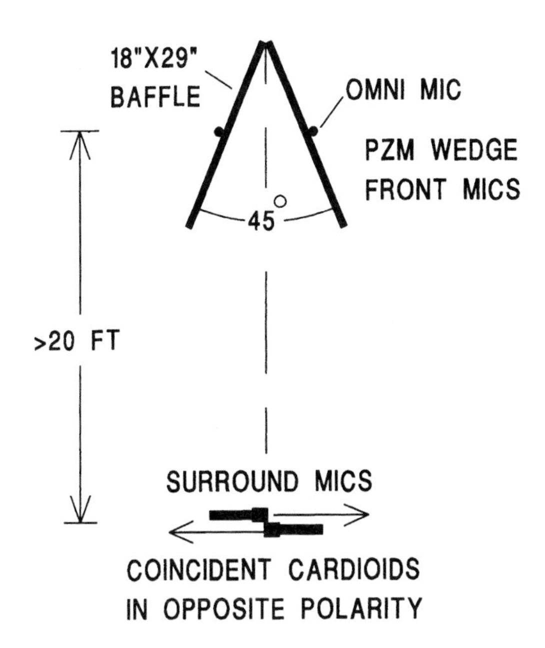

FIGURE 19.9
Woszcyk surround miking method.

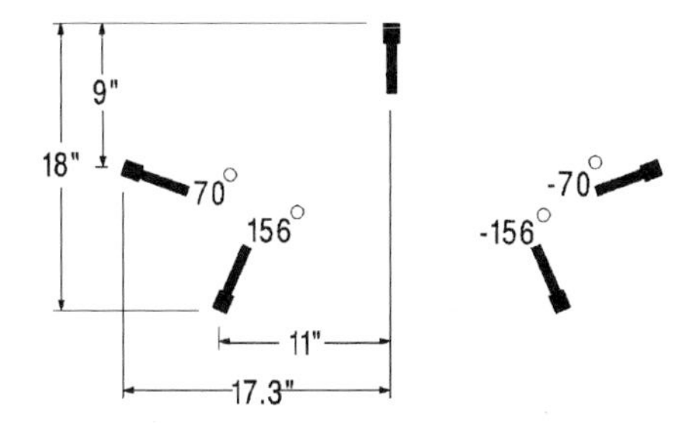

FIGURE 19.10
Williams' five cardioid mic array (multichannel microphone array; MMA).

Williams Five Cardioid Mic Array

Michael Williams, an independent audio consultant, worked out the math to determine the best cardioid microphone arrangement for realistic reproduction of surround-sound fields. His method is shown in Figure 19.10.

Double MS Technique

Developed by Curt Wittig and Neil Muncy, the double MS technique uses a front-facing mid-side mic pair for direct sound pickup and a

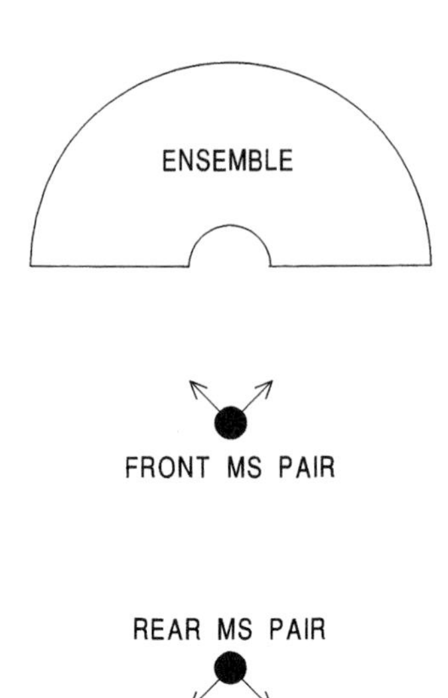

FIGURE 19.11
Double MS technique.

rear MS pair facing away from the front (Figure 19.11). The rear pair is placed at or just beyond the critical distance of the room where the reverberant sound level equals the direct sound level. The matrixed outputs feed front-left, front-right, rear-left, and rear-right speakers. No center channel mic is specified, but you could use the front-facing cardioid mic of the front MS pair for this purpose.

Surround Ambience Microphone Array

The surround ambience microphone (SAM) array was developed by Gunther Theile of the Institute für Rundfunktechnik (IRT). Four cardioid mics are placed 90 degrees to each other and 21 to 25 cm apart. No center channel is described.

Chris Burmajster Array

Based on extensive listening tests, this array was invented by Chris Burmajster of Innocent Ear Ltd. It includes an ORTF pair for left-front and right-front channels, a center mic, and two rear-facing cardioids angled 90° for the left-rear and right-rear channels. The mics are mounted on the metal bar shown in Figure 19.12 (not commercially available). According to the inventor, this arrangement provides solid central imaging. The rear ambience channels sound coherent with the front channels, rather than "disjointed" as can happen with mics

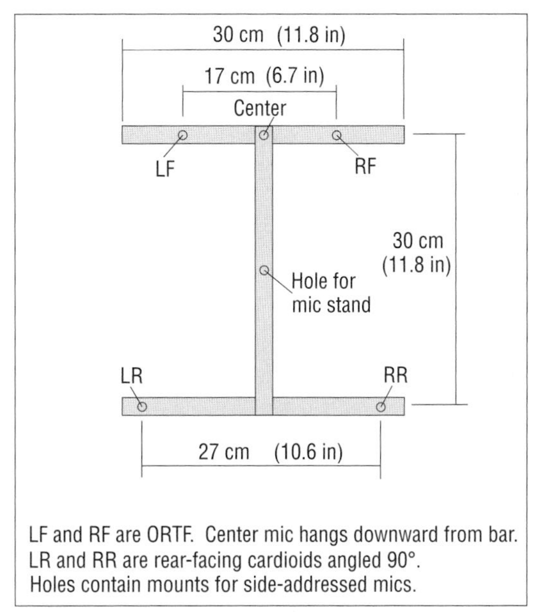

FIGURE 19.12
Mounting bar for Burmajster surround array.

far back in the hall. Details are at http://homepage.ntlworld.com/chris.burmajster (see Miking for Surround on the Web site).

Ideal Cardioid Arrangement

This system uses a special mic mount with five arms that radiate out from a center point, like a star. At the end of each arm is a condenser mic aiming outward from the center. One example includes the Microtech Gefell INA 5, which uses five M930 mics in shock mounts (www.microtechgefell.de). In the SPL Atmos 5.1 Surround Recording System (discontinued), five Brauner condenser mics feed a five-channel mixing console, which adjusts the mic polar patterns and offers panning, bass management, and surround monitoring. SPL's Web site is www.spl-usa.com. Both systems use the Ideal Cardioid Arrangement (ICA 5, ITU-775 specification, Figure 19.13) developed by Volker Henkels and Ulf Herrmann.

The Holophone H2-PRO Surround Mic

This is a surround microphone using several omni mic capsules flush-mounted in a football-shaped surface. It captures up to eight channels of discrete surround sound and has eight XLR connectors. It can be found on the Web site www.holophone.com.

Sonic Studios DSM-4CS 4-Channel Surround Dummy Head

The Sonic Studios Web site (www.sonicstudios.com) offers a four-mic array that you can put on your head, or on a dummy head, to record surround.

Slotte Method

Benedict Slotte developed a surround-miking technique intended to produce very sharp images. The front three mics (Figure 19.14) are

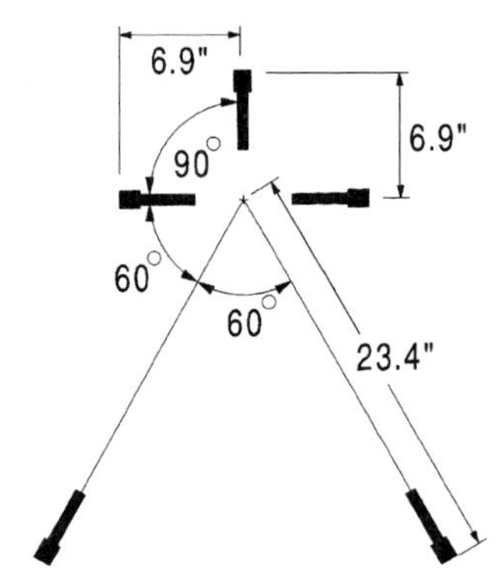

FIGURE 19.13
Ideal Cardioid Arrangement (ICA) used in the Brauner SPL Atmos 5.1 Adjustable Surround Microphone and the Microtech Gefell INA 5 Surround Microphone.

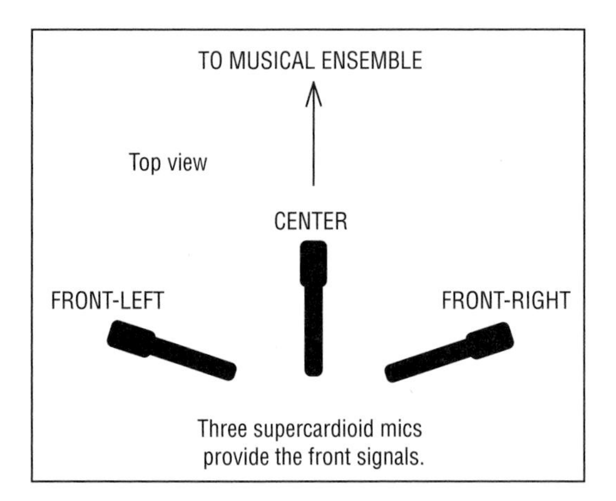

FIGURE 19.14
Slotte method.

an optimized near-coincident array using three supercardioid mics. The array was designed for minimal crosstalk and time differences between mics. The level and time differences between the left-center pair are zero for a sound source at the midpoint between them. The same is true for the right-center pair. In his AES paper, Slotte lists a range of mic angling, spacing, and relative levels that work well.

Martin Method

Geoff Martin invented a surround technique using two Blumlein pairs: one for front left-right, and one for surround left-right (Figure 19.15). The goal is high interchannel coherence for direct sounds and low interchannel coherence for reverberation. Not shown is the center-channel mic, a figure-eight which is coincident with the front pair. The center mic can aim straight down to minimize image distortion, or it can aim forward but located at least 6 cm below the main pair. Forward aiming is recommended if you want a lot of direct sound in the center speaker.

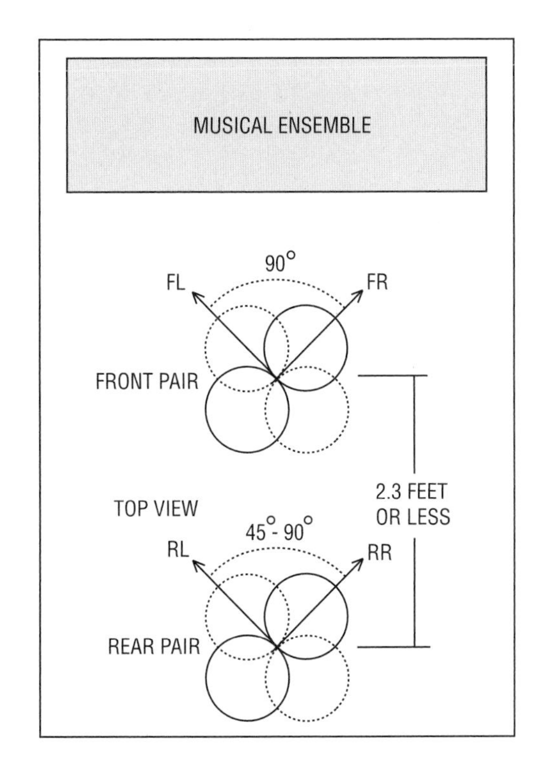

FIGURE 19.15
Martin method.

Mike Sokol's FLuRB Array

This array uses four coincident cardioid mics at 90 degrees to each other aiming to the front, left, right, and back (Figure 19.16). The four mic signals feed a matrix processor (Manley FLuRB) that delivers the correct signals for 5.1 surround and up to 8.1 surround. The array is compact, relatively low cost, and convenient to use. Plus, it will sum to stereo or mono without phase cancellations.

Stereo Pair Plus Surround Pair

In this method, the center-channel mic is omitted. You use a standard stereo pair of your choice to pick up the musical ensemble, plus another stereo pair of your choice to pick up the hall ambience. The hall mics feed the left- and right-surround channels. For example, two Crown SASS-P MKII microphones can be placed back to back, separated by several feet.

You might try a hybrid approach for a pop-music concert: Feed the front speakers a mix of multiple close-up mics on stage, and feed the rear speakers the signals from a rear-aiming stereo mic.

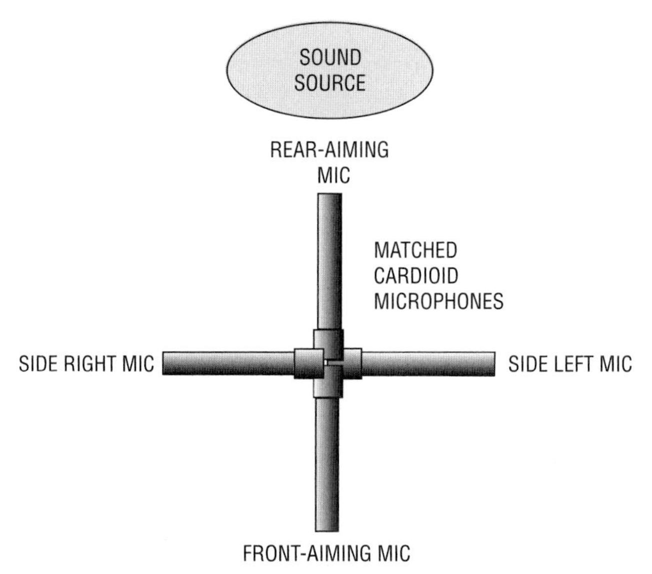

FIGURE 19.16
Mike Sokol's FLuRB Array.

SURROUND MEDIA

So far we've discussed surround recordings, and their creation either by multitrack mixdown or by surround mic techniques. Now let's turn our attention to the media that will play those surround recordings. They can be distributed to the public on CD, Super Audio CD (SACD), DVD, or Blu-ray. Two DVD formats for surround sound are DVD-Audio and DVD-Video. Let's examine all of these formats in detail.

Compact Disc

The compact disc is a 2-channel format that uses linear pulse code modulation (PCM) encoding. The CD's bit depth is fixed at 16 bits, and the sampling rate is fixed at 44.1 kHz.Storage capacity is 650 MB (74 minutes) or 700 MB (80 minutes). To fit the six channels of surround into the CD's two channels, two data-reduction encoding schemes are used: Dolby Digital or DTS. We'll explain them later under the heading Surround Encoding for CD.

DVD

Another medium for playing surround recordings is the Digital Versatile Disc or DVD. It is a high-capacity optical storage medium the size of a compact disc (4.73 inches diameter). DVD can store digital data in three formats: audio, video, and computer.

DVD COMPATIBILITY

The DVD player reads:

- CD
- CD-R (in some units)
- CD-ROM
- DVD-Video disc (video plus audio)

A DVD-Video player can play a DVD-Audio disc if the latter has a Dolby digital version of the audio in the DVD-Video zone on the disc.

DVD CAPACITY

Compared to a CD, a DVD has much greater capacity due to its smaller pits and closer tracks. The scanning laser has a shorter wavelength than in a CD player, which lets it read DVD's denser data stream. In addition, MPEG-2 data compression of the video data increases the data density on the DVD.

Some DVDs have a single layer of pits; others have a dual layer at different depths. The laser automatically focuses on the required layer. The Blu-ray Disc records and plays up to 27 GB of data on a single-sided, single-layer CD-sized disc using a 405 nanometer blue-violet laser. It uses MPEG-2 video-recording format and AC3 and MPEG-1 Layer 2 audio-recording format.

Listed below are the storage capacities of various media:

- A compact disc holds up to 700MB.
- A single-sided, single-layer DVD holds 4.7 GB (7 times CD capacity).
- A single-sided, dual-layer DVD holds 8.5 GB.
- A double-sided, single-layer DVD (DVD-RAM) holds 9.4 GB.
- A double-sided, dual-layer DVD holds 17 GB.
- A single-sided, single-layer Blu-ray Disc holds up to 27 GB.

A DVD with 4.7-GB capacity is enough to hold a 2-hour, 10-minute movie plus subtitles.

DVD PLAYERS

Most DVD players have these outputs:

- 2-channel analog outputs playing a Dolby-Surround-encoded stereo signal
- A digital output that can be connected to an external decoder for 5.1-channel surround sound (either Dolby Digital or DTS)
- 6 analog outputs fed from built-in Dolby Digital or DTS decoders (in some units)

DVD DATA RATE

The DVD data transfer rate is 10.1 megabits per second (Mbps). A 6-channel surround track, with data compression, consumes 384 or 448 kilobits per second (kbps). Linear PCM stereo, at 16 bits and 48-kHz sampling rate, takes 1.5 Mbps. A 96-kHz sample rate doubles the transfer rate to 3 Mps. A 24-bit resolution increases the rate to 4.5 Mbps.

Two forms of DVD are DVD-Video and DVD-Audio, and we'll look at them below.

DVD-VIDEO

DVD-Video is a DVD format for videos: movies with a surround-audio soundtrack. The audio tracks can be Dolby Digital (AC-3) surround; 5.1 channels, with MPEG-1 compressed audio; or they can be 2 channels, 16- to 24-bit, 48- or 96-kHz, linear PCM audio. In other words, you can have surround sound with compromised fidelity, or 2-channel sound with excellent fidelity. A DVD disc can be encoded with both formats, and listeners can choose which one they want to hear.

All players support Dolby Digital (AC-3) surround. Many software packages for DVD-Video authoring are easy to use and cost under $70. Some titles are Roxio's DVDit and MyDVD; and Cyberlink PowerDVD. Several manufacturers of DVD recorders offer bundled DVD authoring software.

DVD-AUDIO

DVD-Audio is a DVD format that features audio programs. It also can have optional still pictures (slide shows), Internet links, visual interactive menus, on-screen text and lyrics, and about 15 minutes of video clips.

DVD-Audio discs use PCM encoding. This encoding can be linear, as on CDs, or "packed" using Meridian Lossless Packing (MLP). MLP reduces the data rate up to 50%, but without any data loss: The reproduced signal is identical, bit for bit, with the original signal. While a standard CD is a 2-channel format fixed at 16-bit/44.1 kHz resolution, DVD-Audio is a multichannel surround format with higher resolutions. Compared to a standard CD, DVD-Audio permits:

- Better fidelity due to higher bit depth and higher sampling rate (up to 24-bit, 96 kHz with six channels, or 176.4 and 192 kHz with two channels).
- Longer playing time in some formats (up to 6 hours of 16-bit, 44.1-K stereo).
- More channels (up to six in the 5.1 scheme, allowing surround sound).

- Optional formats in addition to PCM, such as MPEG-2 BC, DTS, Dolby Digital, or Direct Stream Digital (DSD; the format used in the Super Audio CD).
- Option for different sampling rate and bit depth on different channels. DVD-Audio allows a wide range of sampling rates, number of channels, and quantizations. For example, a 4.7-GB, single-layer disc can hold a 75-minute program in which the left-center-right signals are 24-bit/88.2 kHz and the two surround channels are 20-bit/44.1 kHz.
- Option for different formats on different tracks. One track might be 16-bit/96 kHz surround, while another could be 24-bit/192 kHz stereo.
- Other optional media formats besides audio. For example, you can get video content from computer graphics, still photos, tracking over still photos, and mini-DV camera shots of concerts.

DVD-Audio content is stored in a separate DVD-Audio zone on the disc (the AUDIO__TS directory). The program can be copy protected by a digital signature signal and digital watermarking.

Most DVD-Audio players will also play DVD-Videos and CDs, and some will play Super Audio CDs. A DVD-Video player can play a DVD-Audio disc if the latter has a Dolby digital version of the audio in the DVD-Video zone on the disc. The digital outputs on a DVD-Audio player include PCM and Dolby Digital. Some units have DTS and DSD outputs, and all have multichannel analog outputs. Future players might have FireWire (IEEE 1394) connections.

You compile and record a DVD-Audio disc as you do a CD-R disc: with software, a hardware disc recorder, and blank discs. Specifically, you need DVD-Audio authoring software, a DVD-R recorder or DLT tape backup drive, and blank DVD discs. The write-once discs have dye on one side like a CD-R. Handling up to 4.7 GB on a single-sided disc, they can record DVD-Video, DVD-Audio, or DVD-ROM files.

Two major titles of authoring software are Avid Express Studio HD and Cirlinca DVD Audio solo. Just click the mouse to assemble a playlist of surround-sound tracks. Another major DVD-authoring package is discWelder Bronze by Minnetonka software (www.minnetonkaaudio.com or www.discwelder.com). This low-cost program lets you record PCM stereo files up to 24-bit, 192 kHz and PCM 5.1 files up to 24-bit, 48 kHz to DVD-Audio, using a computer DVD recorder.

Blu-ray Disc (BD)

Blu-ray, also known as Blu-ray Disc (BD) is an optical disc format. It records, rewrites, and plays high-definition video (HD) and stores huge amounts of data, more than five times that of the DVD, up to 25 GB on a single-layer disc and 50 GB on a dual-layer disc. Blu-ray is expected to replace DVD, but DVDs will play on Blu-ray players.

Blu-ray supports these audio codecs:

Linear PCM (LPCM)—up to 8 channels of uncompressed audio (mandatory)
Dolby Digital (DD)—format used for DVDs, 5.1-channel surround sound (mandatory)
Dolby Digital Plus (DD+)—extension of Dolby Digital, 7.1-channel surround sound (optional)
Dolby TrueHD—lossless encoding of up to 8 channels of audio (optional)
DTS Digital Surround—format used for DVDs, 5.1-channel surround sound (mandatory)
DTS-HD High Resolution Audio—extension of DTS, 7.1-channel surround sound (optional)
DTS-HD Master Audio—lossless encoding of up to 8 channels of audio (optional)

Super Audio CD

An alternative to DVD-Audio is the Super Audio CD developed by Sony and Philips. It uses the Direct Stream Digital (DSD) process, which encodes a digital signal in a 1-bit (bitstream) format at a 2.8224 MHz sampling rate. This system offers a frequency response from DC to 100 kHz with 120 dB dynamic range.

The Super Audio CD has two layers that are read from one side of the disc. On one layer is a 2-channel stereo DSD program followed by the same program in six channels for surround sound. On the other layer is a Red-Book 16-bit/44.1 K program, which makes dual inventories unnecessary. The 16-bit/44.1 kHz signal is derived from

Sony's Super Bit Mapping Direct processing, which downconverts the bitstream signal with minimal loss of DSD quality. Standard CD players can play the 16-bit/44.1 kHz layer, and future players would be able to play the 2- or 6-channel DSD layer for even better sound quality.

The TASCAM DV-RA1000HD records DSD audio or high-resolution DVD audio—up to 24-bit/192 kHz—to DVD blanks. This standalone unit also records standard CD-DA, WAV, and DSDIFF files to CD and DVD discs.

ENCODING SURROUND FOR RELEASE ON VARIOUS FORMATS

As we said, surround mixes can be released on CD, DVD-Video, DVD-Audio, or Super Audio CD (SACD). We need to consider how to get six channels of surround audio onto those formats.

In the past, DVD-Video discs and videocassettes used Dolby Surround—a matrix encoding system that combined four channels (left, center, right, surround) into two channels. The Dolby Pro Logic decoder in the consumer playback system unfolded the two channels back into four. The surround channel, which was mono and limited bandwidth, was reproduced over left and right surround speakers. Unfortunately, matrixing reduced the separation between channels because it mixed the channels together.

Recent encoding schemes let us encode six channels of surround audio onto disc, while keeping the channels discrete (not matrixed together). Let's look at how this is done for CD, DVD-Video, and DVD-Audio.

Surround Encoding for CD

A CD has only enough capacity for two channels of audio. So to get six channels of surround audio onto a CD, data reduction is needed. Two data-reduction schemes are Dolby Digital and DTS.

Dolby Digital and DTS each use a different lossy encoding process to reduce the bit rate needed to transmit the six channels via a 2-channel bitstream. "Lossy" means that the reproduced signal is missing some data that appeared in the original signal, so there is a slight loss in sound quality.

Why is this data reduction necessary? Audio at 44.1-kHz sampling frequency and 16-bit word length flows at a rate of about 700 Kbps for each channel. The rate for six channels is 4.2 Mbps. To handle this huge data flow, data reduction is needed. The data-reduction encoder in Dolby Digital is AC-3. It can be used on any number of channels.

First used in movie theaters, Dolby Digital and DTS are perceptual coding methods. They use data reduction (data compression) to remove sounds deemed inaudible due to masking. Dolby Digital's AC-3 encoder compresses the data about 12:1, while DTS compresses about 3:1. Both formats offer six discrete channels of digital surround sound. DTS resolution is 20-bit, while Dolby Digital is 16, 18, or 20 bits (perhaps 24 bits in the future). Dolby's AC-3 encoder accepts data at sample rates of 32, 44.1, or 48 kHz; DTS accepts 44.1 kHz. Compared to Dolby Digital, DTS uses less data compression and needs more data storage space and bandwidth. But some listeners say that DTS sounds more transparent—more like the original discrete tracks that were fed into the DTS encoder. Still, Dolby is the most common format.

To play a Dolby Digital or DTS recording, you need a newer DVD player or a CD player with a digital output. Plug the digital output into a decoder, which extracts the six channels of digital audio and converts them to analog. The decoder has a DSP chip that can decode both the DTS and Dolby Digital formats. Some surround receivers have DTS and Dolby Digital decoders built in.

Surround Encoding for DVD-Video

In a DVD-Video disc, much of the disc's capacity is taken up by video signals. So this format also can handle only two channels of audio. DVD-Video uses Dolby Digital's AC-3 scheme to encode (data-reduce) the six channels of surround into two channels. During playback, the two channels are decoded to recover the six channels. As we said, those six channels are separate or discrete—not matrixed.

Surround Encoding for DVD-Audio

A DVD-Audio disc has lots of capacity for audio signals. The six channels can be encoded without any data loss by using MLP.

Table 19.1 Parameters of Surround Media Formats

Format	Sample rate	Resolution	Stereo/surround systems
CD	44.1kHz	16-bit	Stereo PCM. 16-, 20-, or 24-bit DTS (6 channels)
SACD	2.82MHz	1-bit	DSD (2 or 6 channels)
DVD-V	48kHz	16- or 24-bit	Stereo PCM, Stereo MPEG-2, Dolby Digital, or DTS (2 to 6 channels)
DVD-A	48kHz	16-bit	Dolby Digital (6 channels)
	96kHz	24-bit	PCM encoded with MLP (6 channels)
	192kHz	24-bit	PCM encoded with MLP (stereo)

Summary of Media Formats

Let's summarize the surround encoding schemes used in different media:

- CD uses DTS or Dolby Digital. Both are lossy systems.
- DVD-Video uses Dolby Digital, a lossy system.
- DVD-Audio uses MLP, a lossless system.

However, a DVD-Audio disc also can include Dolby Digital mixes so that it will play on CD players and DVD-Video players.

Table 19.1 summarizes the sample rate, the resolution (bit depth), and the stereo or surround systems associated with various media formats. Figure 19.17 summarizes the types of data handled by various formats of surround media and players.

Encoding Hardware and Software for CD and DVD-Video

With the CD and DVD-Video formats, the surround encoding can be done either by you or by the mastering house that manufactures the disc. To do it yourself you need a hardware or software encoder. Some examples of software encoders are Minnetonka Audio's SurCode series, Sonic Foundry's 5.1 Surround Plugin Pack, Sonic Foundry's SoftEncode 5.1 Channel, or Universal Audio's SmartCode

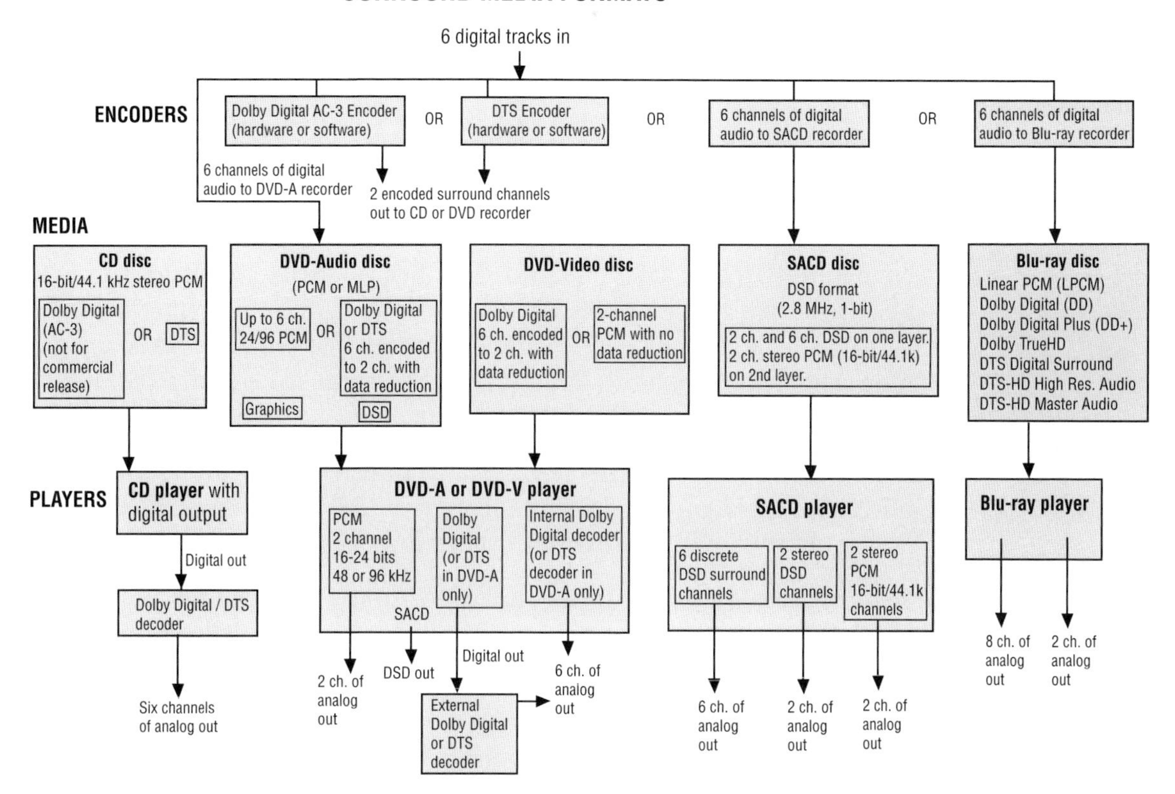

FIGURE 19.17
Surround media formats.

Pro. If you're using a hardware encoder, connect its inputs to the outputs of your 8-track mixdown deck, in parallel with the connections to the monitor system.

Surround-sound expert Mike Sokol suggested a way to put Dolby Digital or DTS encoded surround mixes on a CD-R. Load the six audio tracks as WAV files into your computer and feed them to the encoding software. Select an AC-3 WAV file output or DTS WAV file output. Encode the tracks. Finally, burn the resulting WAV file onto a CD-R. To hear the surround mix, connect the CD player's digital output to a surround decoder, such as the one in a surround receiver. Commercially released CDs must use only DTS. Details are in Mike's upcoming book titled *Surround Sound Production*.

You don't necessarily have to buy an encoder; the encoding can be done by a mastering house or DVD manufacturer. Simply send them your 8-track recording. But you might want to put an encoder in your monitor system to hear the effect of the encoder as the end listener will hear it.

DVD PRE-MASTERING FORMATS

Here are acceptable 8-track media formats to send to a DVD manufacturer. Put SMPTE on track 8, or put a stereo mix on tracks 7 and 8. For Dolby Digital, track assignments are 1, left; 2, center; 3, right; 4, LFE (low-frequency effects or subwoofer); 5, left surround; 6, right surround; 7, data channel; 8, data channel. For DTS the track assignment is 1, left; 2, right; 3, left surround; 4, right surround; 5, center; 6, LFE; 7, data; 8, data.

If you want to create the LFE or 5.1 channel, you can mix the five other channels and feed them through a 120-Hz lowpass filter. Note that most music-only surround mixes don't use the LFE channel.

You might have video slide-show data that accompanies the audio. Store the video on a removable hard drive or send the file via the Internet. TIFF and bitmap (BMP) formats are acceptable.

DOLBY UNITS FOR DVD MASTERING

This information is for those who produce the actual DVD. DVD mastering requires MPEG-2 video and Dolby Digital encoders. You can use either one for multichannel audio on DVD titles. Dolby Digital soundtracks are compatible with mono, stereo, or Dolby Pro Logic systems. Dolby encoders and decoders are described on Dolby's Web site at www.dolby.com/professional/index.html.

Special thanks to Mike Sokol for his generous technical help with this chapter.

CHAPTER 20

Web Audio and Online Collaboration

You've recorded a new song and you can't wait for people to hear it. You might want to distribute it on the Web, in addition to (or instead of) CDs.

Why put your music online? Web excerpts of your songs can entice listeners to buy your CDs. Making songs available on the Web costs less than duplicating CDs and selling them in stores. Also, distributing online can be done much more quickly than distributing CDs. Keep in mind, though, that it's hard to get your songs noticed among the millions of other titles that are online. Many books and Web sites offer suggestions on marketing your music.

STREAMING VERSUS DOWNLOADING

There are two basic ways to transmit music on the Internet: streaming files and downloading files. A streaming file plays as soon as you click on its title. A downloaded file doesn't play until you copy the entire file to your hard disk. Streaming audio is heard almost instantly (after the buffer memory fills), but often sounds muffled and can be interrupted by net congestion.

The sound quality of streaming audio depends on the speed of the modem and ranges from funky AM radio (with a 56K modem) to near-CD quality (with a cable modem). With downloaded audio you must wait a few seconds to several minutes to download the song. But when it plays, the sound is high-fidelity and continuous.

DATA COMPRESSION

Audio files that you record on your computer are usually WAV or AIFF files. They have no data reduction (data compression), so they take up lots of memory. A 3-minute song recorded at 16-bits, 44.1 kHz consumes about 32 MB. Downloading a 32-MB WAV file on the Web would take several hours using a 56K modem. So, audio WAV files intended for Web download are usually data-reduced or data-compressed by encoding them as MP3, WMA, or other formats. For example, if you compressed a 3-minute song by 10:1 it would consume about 3.2 MB and would download 10 times faster than the equivalent WAV file.

Most types of data compression tend to degrade sound quality. The sound quality of a compressed-data format depends on its bitrate, measured in kilobits per second (kbps). The higher the bitrate, the better the sound, but the greater the file size. A bitrate of 128 kbps for stereo MP3 files is considered to be a good compromise between sound quality and file size. At low bitrates (at and below 128 kbps), you can hear artifacts such as "swirly" cymbals, smeared transients such as drum hits, a general "phasey" effect, and less treble. At higher bitrates above 192 kbps, the artifacts start to disappear, and to most listeners the sound is CD quality.

In short, data compression reduces file size and download time at the expense of sound quality.

A stereo MP3 file encoded at 128 kbps is 64 kbps per channel. A mono MP3 file encoded at 64 K is equal in quality to a stereo MP3 file encoded at 128 K. Mono files consume half the file space of stereo files if both are the same bitrate.

Table 20.1 relates MP3 stereo bitrate to sound quality.

WEB-RELATED AUDIO FILES

You can put audio on the Web in several file formats. Let's look at some of the file types in current use.

WAVE (.wav): A standard PC format for audio files. It encodes sound without any data reduction by using pulse code modulation. WAVE files that are used to make audio CDs are 16-bit, 44.1 kHz. Because WAVE files consume a lot of memory (about

Table 20.1 Data Compression Ratio and Sound Quality of Various MP3 Bitrates

Bitrate (CBR)	Compression	Sound quality
64 kbps	20:1	AM radio quality
128 kbps	10:1	Cassette quality
192 kbps	7:1	Near-CD quality
256 kbps	5:1	CD quality
320 kbps	4.4:1	CD quality

32 MB for a 3-minute song in stereo), they are seldom used on the Internet for songs because they take so long to download. However, an Internet connection by fiber-optic cable (such as Verizon's Fios) is much faster than a standard copper-wire connection. Fios is claimed to offer up to 50 or 30 Mbps downloads and up to 20 or 5 Mbps uploads. With Fios, it's possible to transfer WAV files—even hi-res ones—in a short time. For example, www.itrax.com offers songs for sale online that are surround files or 24-bit/96 kHz stereo files.

AIFF (.aif or .aiff): Audio Interchange File Format, a standard Mac format for audio files. Use this format to transfer audio files between a PC and a Mac. Like WAV files, AIFF files aren't data-compressed.

MID (MIDI, or Musical Instrument Digital Interface): This is a MIDI sequence, not an audio file. A .mid file is a string of numbers that represent performance gestures, such as note on/off, which note is played, key velocity, and so on. Chapter 16 describes how to create MIDI files. Because MIDI files don't include audio, they are very compact. The user downloads the MIDI file and imports it into a MIDI sequencer or MIDI/audio DAW. The sequence generates instrumental music in a synth, sound module, or soft synth when played. To ensure that the proper instrument sounds play, most people use a General MIDI file format.

The following formats use data compression.

MP3 (MPEG Level-1 Layer-3): The most popular format. You can choose a low bitrate to make a small file with lower sound quality, or choose a high bitrate to make a larger file with higher quality.

When encoding an MP3 file, you have a choice of constant bit rate (CBR) or variable bit rate (VBR). VBR uses more bits on complex musical passages and fewer bits on simpler passages. CBR encodes every frame at the same bitrate, so it is less efficient. Use VBR to get a fixed level of quality at the lowest possible bitrate and file size. CBR is recommended only for streaming where the bitrate must be fixed. Average bit rate (ABR) is a compromise between CBR and VBR modes. Use ABR when you need to know the size of the encoded file but still allow some variation in the bit rate.

MP3PRO: An improvement over MP3. Songs encoded at 64 kbps with MP3PRO are said to sound as good as songs encoded at 128 kbps with MP3. MP3PRO offers faster downloads and nearly doubles the amount of music you can put on a Flash memory player. MP3 and MP3PRO files are compatible with each others' players, but a free MP3PRO player is needed to hear MP3PRO's improvement in sound quality. See www.mp3prozone.com.

WMA (Windows Media Audio): This popular Microsoft format provides Digital Rights Management (DSR), which is licensing technology that provides copy protection. It's useful for online music stores. For more information see www.microsoft.com/windows/windowsmedia. WMA Pro also supports multichannel and high resolution audio.

Suppose you want to download a 3-minute song. We'll assume that the download speed is 90 kbps per second (typical of DSL).

- The original WAV file of the song is 30.3 MB, CD quality or better, and downloads in about 5.6 minutes.
- An MP3 or WMA file at 128 kbps is 3.0 MB, cassette quality, and downloads in about 33 seconds.
- An MP3 or WMA file at 320 kbps is 6.9 MB, CD quality, and downloads in about 1.3 minutes.

RealAudio: This format is used for streaming audio and for music downloads from RealAudio's music store. The streaming fidelity depends on modem speed and the current Internet bandwidth. RealAudio files (.ra or. rm) are often used as short excerpts or previews of songs.

OGG (Ogg Vorbis): License-free open software. For a given file size, Vorbis sounds better than MP3. Vorbis takes up less file size than MP3 files of equal quality. For more information see www.xiph.org and www.vorbis.com.

AAC (MPEG Advanced Audio Coding): AAC offers better sound quality than MP3 at the same bitrate. Many listeners claim that AAC files made at 128 kbps sound like the original uncompressed audio source. What's more, AAC supports multichannel audio and a wide range of sample rates and bit depths. It's also used with Digital Rights Management technology by helping to control the copying and distributing of music.

Currently, WMA and MP3 are the most popular types. *Audio clip 41 at www.elsevierdirect.com/companions/978024081444.*

WHAT YOU NEED

To put your music on the Web, you need to download a few pieces of software, which are low-cost or free:

Ripper: A program that converts audio from a CD or CD-R to a WAV file. Examples are Exact Audio Copy (www.exactaudiocopy.de), CDex (http://cdexos.sourceforge.net), and FreeRip3 (www.mgshareware.com). Not all CD-ROM drives support ripping.

MP3 Encoder: A program that converts a WAV or AIFF file to an MP3 file. Examples of freeware are RazorLame (www.free-codecs.com/download/RazorLame.htm) and Lame MP3 Encoder (www.free-codecs.com/download/Lame_Encoder.htm). You need both. Get Lame version 3.97 or later. Another option is Altomp3 from www.Yuansoft.com. Some recording software includes an MP3 encoder, and so does Winamp, FreeRip3, CDex, dBpowerAmp, foobar2000, Audio Mp3 Editor, and Exact Audio Copy.

ID3 Tag Editor: This program lets you add song title, artist, genre, and other metadata information to an MP3 file. This information displays on screen when the MP3 file plays. You can add studio contact information in a comments field. Two freeware programs are Stamp ID3 Tag Editor and Mp3tag. Several MP3 playback programs include a tag editor.

WMA Encoder: A program that converts a WAV or AIFF file to a WMA file. It's a free download from Microsoft at www.microsoft.com/windows/windowsmedia/forpros/encoder/default.mspx. Windows Media Technologies has Digital Rights Management, which limits copies or format conversions.

Ogg Vorbis Encoder (optional) from www.vorbis.com. Converts CD tracks or WAV files to Vorbis format. For a given file size, Vorbis sounds better than MP3. Vorbis takes up less file size than MP3 files of equal quality.

To listen to your music that you put on the Web, you need:

MP3 Player: A program or a device that plays MP3 files. Some software player examples are Yahoo Music Jukebox from http://music.yahoo.com/jukebox/, Winamp from www.winamp.com, Windows Media Player from www.microsoft.com/windows/windowsmedia, RealPlayer from www.real.com, and QuickTime in Apple's OS or www.apple.com/quicktime/. Portable MP3 players play MP3 files downloaded from the Web.

Windows Media Player: Although intended to play mainly Windows Media files, this free program plays many file types: .avi, .asf, .asc, .rmf, .wav, .wma, .wax, .mpg, .mpeg, .m1v, .mp2, .mp3, .mpa, .mpe, .ra, .mid, .rmi, .qt, .aif, .aifc, .aiff, .mov, .au, .snd, and .vod. This player is a free download from www.microsoft.com/windows/windowsmedia/.

A source for players, rippers, and encoders is www.team-mp3.com. Some digital audio editing programs can output MP3, WMA, and RealAudio files.

HOW TO UPLOAD COMPRESSED AUDIO FILES

Got everything you need? Let's go. Basically, here's what you'll be doing:

1. Start with a WAV file of a song on your computer. Or convert your song from a CD to a WAV file by using a ripper.
2. Edit and process the WAV file to optimize it for the Internet.
3. Then use an encoder to convert the WAV file to an MP3 or WMA file.

4. Finally, send (upload) the converted file to a Web site that features music in the MP3 or WMA format.

Let's go over the procedure in more detail. You can substitute WMA for MP3 in the procedure below.

1. Start with a WAV file of the song. Edit the start and stop points of the song. You might want to edit a 30-second excerpt of a song to use as a preview of your music. If so, add a fade-in and fade-out to the preview.
2. Next, you might want to process the audio so that it will sound louder and clearer when played on the Web. To do this, reduce the bandwidth: apply low cut below 40 Hz and high cut above 15 kHz.You might want to try a higher frequency low cut and a lower frequency high cut. Apply a little compression or multiband compression, apply peak limiting, then normalize the song to maximize its level.
   Save the edited, processed song as a WAV file (PC) or AIFF file (Mac) on your hard drive.
3. Use an MP3 encoder to convert the song's WAV or AIFF file to an MP3 file on your hard drive. You might want to use a bitrate of 128 kbps, which many Web sites require. It's a good compromise between file size and sound quality.
4. If you want, add the song's title, artist, and comments using an ID3 tag editor.
5. Next you will upload your MP3 files to an MP3 server—a Web site that accepts MP3 files for distribution on the Web. Examples include: www.mp3.com, www.myspace.com, www.AWAL.com, and www.digstation.com(for independents only). The CD duplicator Disc Makers will upload your songs quickly to digstation.com for a small fee. Also see the MP3 Sites page at www.team-mp3.com. Some sites offer free downloads of their music files; others charge the listener so that you make some income.

iTunes Producer is free software that lets you prepare your music for submission to iTunes. Use it to encode your music into an AAC format; enter album, song, and artwork information; and send your files to Apple for consideration. More information at www.iTunes.com.

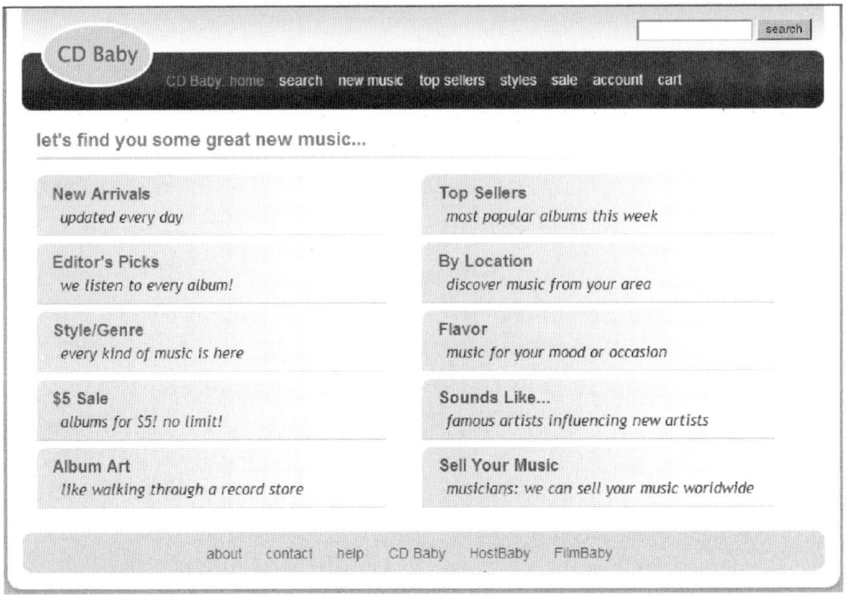

FIGURE 20.1
CD Baby Web site.

As an alternative, you can send your physical CD to CD Baby (www.cdbaby.com; Figure 20.1), and they will convert it to data-compressed files for you. They are a popular, affordable way for independent artists to sell CDs.

eMusic (www.emusic.com) doesn't work with unsigned artists, but they do accept music from independent labels. Usually this is done through a service such as The Orchard (www.theorchard.com), INgrooves (www.ingrooves.com), or TuneCore (www.tunecore.com; Figure 20.2). They can compress and deliver a label's CD catalog to all digital music stores. All income from digital sales is combined and reported to you.

Once you have chosen a site, go there and click on the button labeled "Artists Only," "Submit your music," "Sell your music," or something similar. This is where you upload your MP3 files.

Once you sign up and fill out some forms, click on "Upload" (or whatever is close to that). You can also upload scanned photos of

FIGURE 20.2
TuneCore Web site.

your band, your album cover, and text describing your band and its music. Some sites take a few days or weeks to approve your songs.

Congratulations—you're on the Web!

PUTTING YOUR MUSIC ON YOUR WEB SITE

So far I've covered how to upload your songs to an MP3 server. Now I'll explain how to put your songs on your own Web site. People visiting your site can click on a song title to hear a streaming song preview, or to download an entire song.

Let's start by creating a Web page link to each MP3 file. It's easy. You can create the link with Web page design software or with one line of HTML code.

Again, you can substitute WMA for MP3 in the procedure below.

1. Suppose you have a song on CD called "Blues Bash." Using ripper software, transfer that song to your hard drive as a WAV file (PC) or AIFF file (Mac). In this example, we'll call the file BLUES.wav. If you have a ripper/encoder, convert that song to an MP3 or WMA file and go to Step 3.

2. Use an MP3 encoder to convert the WAV file to an MP3 file. I recommend a setting of 160 kbps bitrate, VBR, stereo. This setting gives hi-fi sound with a relatively small file, so it's fairly quick to download. You now have an MP3 file called BLUES. mp3.
3. You can add song-title and artist information with an ID3 tag editor.
4. In your computer, move BLUES.mp3 to the same directory on your hard drive that your Web page is in.
5. Use Web page design software to open your band's Web page. Here we'll call it PAGE1.HTM.
6. On that page, where you want the song title to appear, type in the title of the song. In this example it's "Blues Bash." Link that title to BLUES.mp3.

When you left-click on the song title, BLUES.mp3 should load and play. When you right-click on the song title, you can select "Save Target As" to copy the MP3 file to a directory on your hard drive.

If you want to use HTML code:

1. Open your Web page with a browser such as Windows Explorer.
2. Select View > Source. You'll see the HTML code for that page in plain text format.
3. Find a spot on the page where you want the song title to appear. Type in the title of the song and the link to its MP3 file. In HTML it would look something like this:

```
<a href = "blues.mp3"> Blues Bash</a>
```

   For a WMA file it would be

```
<a href = "blues.wma">Blues Bash</a>
```

4. Save and close the text file, go to Windows Explorer, and select View > Refresh. You should see the link to the MP3 or WMA song. When you left-click on the link, you should hear the song. When you right-click on it, you can select "Save Target As" to copy the MP3 file to a directory on your hard drive.

Now that your song links are working correctly on your computer, it's time to upload them.

1. Upload (send) PAGE1.HTM and BLUES.MP3 to your Web server. Make sure that both files go to the same directory. Some Web servers have an upload page for this purpose. **Note**: Some Web servers don't allow MP3 files. Find the page on the Web server site that tells what files they permit.
2. Now start your Web browser and go to PAGE1.HTM on your site. Click "Refresh" in your browser so that you see the PAGE1. HTM that you just uploaded.
3. Left-click the song title (in this case, "Blues Bash"), and it should play after the buffer fills. Or right-click the song title and select "Save Target As." The song should download over a few minutes. Then you can play it with your MP3 player or WMA player in glorious hi-fi stereo.

You might prefer to upload your songs to SNOCAP (www.snocap .com) (Figure 20.3). They send you code to paste into your Web site or MySpace. Then viewers of your site can hear a 30-second sample of a tune and purchase it for download. SNOCAP makes it easy to sell digital downloads from your own Web site. A similar service is Micropay at www.monkrat.com.

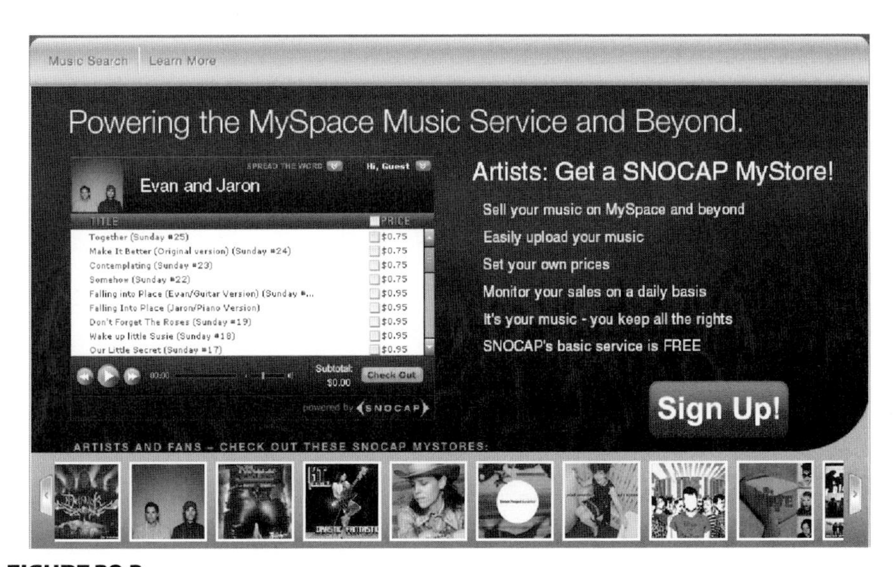

FIGURE 20.3
The SNOCAP Web site home page (copyright 2008 Snocap, Inc.).

Other good information on marketing your music online is in the article "Music's New Messiahs: You!" by Gary Mraz in the April 2008 issue of *EQ Magazine*. Also see http://newmusicstrategies.com/ebook/. CD Baby offers marketing advice after you create an account.

FLASH ANIMATION

Adobe Flash—originally called Shockwave Flash or Macromedia Flash—is a technology to create and display animation or video with an accompanying sound track in Web sites, games, and mobile phones. It uses a scripting language called ActionScript. You create Flash animations with the Adobe Flash Professional multimedia authoring program, which explains how to import your audio files into a Flash presentation.

Flash is a subject in itself and is covered in many books. For details, check out these Web sites: www.adobe.com, www.royaltyfreemusic.com/tutorials/import-flash-audio.htm, and www.digitalprosound.com/Htm/Techniques/2000/Oct/Flash.htm.

COLLABORATING BY SHARING FILES

Recording engineers and musicians in different cities can work together on a common project by sharing files—sending digital audio files to others in WAV, MP3 or WMA format.

For example, you can send a song mix to a studio musician in another city, who overdubs their part and sends it back to you as a WAV file to import back into your multitrack project. Here's the procedure:

1. Export a WAV file of the song's stereo mix.
2. Copy that WAV file to a data CD-R.
3. Mail or FedEx the CD-R to the musician.

Then the musician follows these steps:

1. Import the mix into their recording program.
2. Overdub their part.
3. Export their part as a WAV file.
4. Copy that WAV file to a data CD-R.

5. Send the CD-R back to the recording engineer.
6. The engineer imports the WAV file on that CD-R into the multitrack project for mixing and editing.

What if one studio uses MacIntosh and another uses PC? CD-Rs and DVD-Rs made on a Mac computer can't be read by a PC computer. Fortunately, we have MAC-PC conversion software, such as TransMac by acutesystems.com. It lets a PC read a Mac-burned data CD and extract the WAV files.

You could e-mail MP3 or WMA files instead. If you want to use those compressed-data formats, encode the MP3 or WMA files at 256 kbps or higher, VBR, for best sound quality.

Sending WAV, MP3, or WMA files by e-mail can be cumbersome because some e-mail programs limit attachments to 10 MB. Also, the person receiving the file attachment will have to wait a long time for the e-mail to download. A better arrangement is to upload the audio files to a file-sharing Web site such as www.yousendit.com, www.dropload.com, www.proaudiobus.com, Digidesign's Digidelivery, or Apple's iChat AV. When the recipient is ready to download the file, they can go to that Web site and do so.

Sharing Multitrack Projects

Suppose you want to send your multitrack recordings to another studio for mixing. You can send them the hard drive that the recording was made on, along with instructions for extracting the WAV files. Or export a WAV file of each track (starting at time zero), and copy those files to a data CD-R or data DVD-R. Name each file by song and instrument.

Before you send a multitrack project to another studio, follow these steps:

1. Make sure each track is named according to its instrument.
2. To reduce file size, delete unused tracks and delete unused audio files from the session. Some recording software has a tool called "Consolidate audio files" which puts all the audio files used in a project in a common folder.
3. Render (bounce or export) each track that has automation, effects, and edits to a new track that extends from the beginning

to the end of the project. Start each track at time zero and set its fader to 0 dB before bouncing. Then you will have several tracks with no edits, and with automation moves and plug-ins embedded in the audio. Give only those tracks to the other studio as 24-bit WAV files. If the other studio will do automation or effects, disable those features before bouncing.

4. Consider using plug-ins common to Windows and MacIntosh. If the other studio has the same plug-ins you do, also send them rendered tracks with the plug-ins disabled. Include notes on the plug-ins you used, along with any presets.
5. Do screen captures of the mixer view, showing the fader and pan settings.
6. Convert all files to the same bit depth.
7. Bounce any MIDI soft-synth tracks into audio tracks, or export the MIDI tracks in Standard MIDI File format.
8. Many DAWs can read OMF (Open Media Format). If you want to export your session via OMF, first follow the steps above. Then export your audio mix file using File > Export > OMF or something similar. In Pro Tools check the Include-Audio option. Recent alternatives to OMF are Advanced Authoring Format (AAF), and Material eXchange Format (MXF). See www.aafassociation.org.
9. Cakewalk SONAR saves its projects as a .bun (bundle) file, which includes all the track recordings, edits, balances, and plug-ins. Another studio with SONAR Producer can import and reproduce that project except for plug-ins that they don't have.

For more information, read the "NARAS DAW Guidelines for Music Production" at http://womb.mixerman.net/showthread.php?t=760. Read the "Pro Tools Guidelines for Music Production" at www.charlesdye.com/ptguidelines2.0.pdf. Also see www.grammy.com/Recording_Academy/Producers_And_Engineers/Guidelines/.

A revised standard has been published: AES31-3-2008 "AES standard for network and file transfer of audio—Audio-file transfer and exchange—Part 3: Simple project interchange." This standard provides a way to express audio-edit data in readable text form to enable simple, accurate computer parsing. Edit precision is sample-accurate. A TCF presentation method offers readable frame-based timecodes. All AES standards documents are available at www.aes.org/publications/standards. Enter "aes31-3" into the Quick Search window and click "Search."

When sending out your mixes for mastering for CD release, a good format is one stereo WAV file per song recorded at 24 bits/44.1 kHz. Many studios can handle higher sample rates. Ideally your stereo mixes are recorded without overall compression or limiting, and with a recorded level of −6 dBFS to −3 dBFS maximum in peak meter mode.

FINDING STUDIO MUSICIANS, PRODUCERS, AND ENGINEERS

At many Web sites you can find online session work, or find online session musicians to enhance your projects. Using their services, you can work with people all over the world from the comfort of your own studio. Some sites accept only musicians who are with major labels, or they divide their services into major-label musicians and minor-label or indie musicians.

For example, sessionplayers.com seeks major-label musicians, singers, and engineers to join their group of available studio professionals. In their Player's Network, they also offer directories of professional but more affordable session musicians for recording work and musical collaborations.

Here are some sites where session players and engineers can do online collaboration (all sites below start with www.): sessionplayers.com/playersnetwork.htm, ejamming.com, esession.com, musiciansjunction.com, bandmix.com, talentmatch.com, findmusicians.org, and shenandoahmusic.com/online-session-musicians.htm.

Also, search Google for these terms: music collaboration technology, musician finder, musician's forum, online recording session, and online session musicians.

Appendix A

dB or Not dB

In the studio, you need to know how to set and measure signal levels and match equipment levels. You also need to evaluate microphones by their sensitivity specs. To learn these skills, you must understand the decibel, the unit of measurement of audio level.

DEFINITIONS

In a recording studio, level originally meant power, and amplitude referred to voltage. Today, many audio people also define "level" in terms of voltage or sound pressure, even though this terminology is not strictly correct. You should know both definitions to communicate.

Audio level is measured in decibels (dB). One dB is the smallest change in level that most people can hear—the just-noticeable difference. Actually, the just-noticeable-difference varies from 0.1 to about 5 dB, depending on bandwidth, frequency, program material, and the individual. But 1 dB is generally accepted as the smallest change in level that most people can detect. A 6- to 10-dB increase in level is considered by most listeners to be "twice as loud." Sound pressure level, signal level, and change in signal level all are measured in dB. *Play audio clip 42 at www.elsevierdirect.com/companions/9780240811444to hear the effects of various decibel changes on loudness. Audio clip 43 wraps up the demonstration.*

SOUND PRESSURE LEVEL

Sound pressure level (SPL) is the pressure of sound vibration measured at a point. It's usually measured with a sound level meter in dB SPL (decibels of sound pressure level).

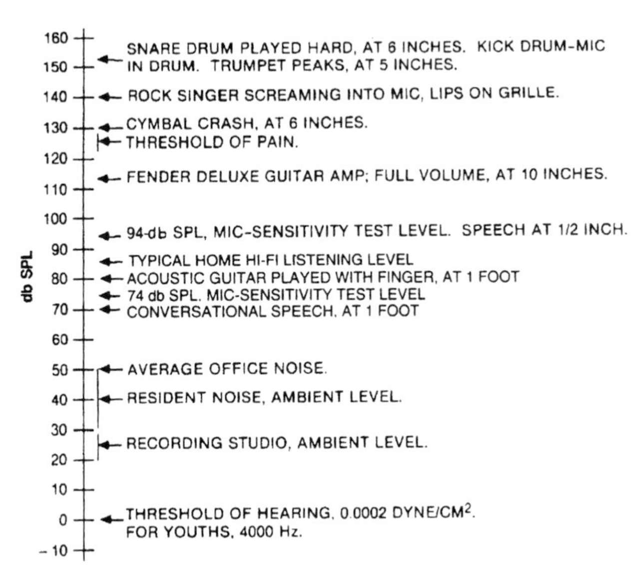

FIGURE A.1
A chart of sound pressure levels.

The higher the sound pressure level, the louder the sound (Figure A.1). The quietest sound you can hear, the threshold of hearing, is 0 dB SPL. Average conversation at 1 foot is 70 dB SPL. Average home-stereo listening level is around 85 dB SPL. The threshold of pain—so loud that the ears hurt and can be damaged—is 125 to 130 dB SPL.

Sound pressure level in decibels is 20 times the logarithm of the ratio of two sound pressures:

$$\text{dB SPL} = 20 \log P/P_{ref}$$

where P is the measured sound pressure in dynes/cm^2, and P_{ref} is a reference sound pressure: 0.0002 dyne/cm^2 (the threshold of hearing).

SIGNAL LEVEL

Signal level also is measured in dB. The level in decibels is 10 times the logarithm of the ratio of two power levels:

$$\text{dB} = 10 \log P/P_{ref}$$

where P is the measured power in watts, and Pref is a reference power in watts.

Recently it's become common to use the decibel to refer to voltage ratios as well:

$$dB = 20 \log V/V_{ref}$$

where V is the measured voltage, and V_{ref} is a reference voltage.

This expression is mathematically equivalent to the previous one, because power equals the square of the voltage divided by the circuit resistance:

$$dB = 10 \log P_1/P_2 = 10 \log ((V_1^2/R)/(V_2^2/R))$$

$$= 10 \log (V_1^2/V_2^2) = 20 \log (V_1/V_2)$$

The resistance R (or impedance) in this equation is assumed to be the same for both measurements, and thus divides out.

Signal level in decibels can be expressed in various ways, using various units of measurement:

- dBm: decibels referenced to 1 milliwatt, with no defined load impedance
- dBu: decibels referenced to 0.775 volt
- dBv: decibels referenced to 0.7746 volt
- dBV: decibels referenced to 1 volt

dBm

If you're measuring signal power, the decibel unit to use is dBm, expressed in the equation

$$dBm = 10 \log P/P_{ref}$$

where P is the measured power, and P_{ref} is the reference power (1 milliwatt).

For an example of signal power, use this equation to convert 0.01 watt to dBm:

$$dBm = 10 \log (P/P_{ref})$$

$$= 10 \log (0.01/0.001)$$

$$= 10$$

So, 0.01 watt is 10 dBm (10 decibels above 1 milliwatt).

Now convert 0.001 watt (1 milliwatt) into dBm:

$$\begin{aligned} \text{dBm} &= 10 \log (P/P_{ref}) \\ &= 10 \log (0.001/0.001) \\ &= 0 \end{aligned}$$

So, 0 dBm = 1 milliwatt. This has a bearing on voltage measurement as well. Any voltage across any resistance that results in 1 milliwatt is 0 dBm. This relationship can be expressed in the equation

$$0 \text{ dBm} = V^2/R = 1 \text{ milliwatt}$$

where V = the voltage in volts, and R is the circuit resistance in ohms.

For example, 0.7746 volt across 600 ohms is 0 dBm. One volt across 1000 ohms is 0 dBm. Each results in 1 milliwatt.

Some voltmeters are calibrated in dBm. The meter reading in dBm is accurate only when you're measuring across 600 ohms. For an accurate dBm measurement, measure the voltage and circuit resistance, then calculate:

$$\text{dBm} = 10 \log ((V^2/R)/0.001)$$

dBu

Another unit of measurement expressing the relationship of decibels to voltage is dBu. This means decibels referenced to 0.775 volt.

$$\text{dBu} = 20 \log V/V_{ref}$$

where V_{ref} is 0.775 volt.

dBv

The unit dBv with a small "v" means decibels referenced to 0.7746 volt. This figure comes from 0 dBm, which equals 0.7746 volt across 600 ohms (because 600 ohms used to be a standard impedance for audio connections):

$$\text{dBv} = 20 \log V/V_{ref}$$

where V_{ref} is 0.7746 volt.

dBV

Signal level also is measured in dBV (with a capital V), or decibels referenced to 1 volt:

$$dBV = 20 \log V/V_{ref}$$

where V_{ref} is 1 volt. For example, use this equation to convert 1 millivolt (mV; 0.001 volt) to dBV:

$$\begin{aligned} dBV &= 20 \log (V/V_{ref}) \\ &= 20 \log (0.001/1) \\ &= -60 \end{aligned}$$

So, 1 millivolt = −60 dBV (60 decibels below 1 volt). Now convert 1 volt to dBV:

$$\begin{aligned} dBV &= 20 \log (1/1) \\ &= 0 \end{aligned}$$

So, 1 volt = 0 dBV.

To convert dBV to voltage, use the formula

$$\text{Volts} = 10^{(dBV/20)}$$

Change in Signal Level

Decibels also are used to measure the change in power or voltage across a fixed resistance. The formula is

$$dB = 10 \log (P_1/P_2)$$

or

$$dB = 20 \log (V_1/V_2)$$

where P_1 is the new power level, P_2 is the old power level, V_1 is the new voltage level, and V_2 is the old voltage level.

For example, if the voltage across a resistor is 0.01 volt, and it changes to 1 volt, the change in dB is

$$
\begin{aligned}
\text{dB} &= 20 \log (V_1/V_2) \\
&= 20 \log (1/0.01) \\
&= 40 \text{ dB}
\end{aligned}
$$

Doubling the power results in an increase of 3 dB; doubling the voltage results in an increase of 6 dB.

THE VU METER, ZERO VU, AND PEAK INDICATORS

A VU meter is a voltmeter of specified transient response, calibrated in volume units or VU. It shows approximately the relative volume or loudness of the measured audio signal. VU meters are used on analog tape recorders, broadcast consoles, some live-sound mixing consoles, and older recording consoles.

The VU-meter scale is divided into volume units, which aren't necessarily the same as dB. The volume unit corresponds to the decibel only when measuring a steady sine-wave tone. In other words, a change of 1 VU is the same as a change of 1 dB only when a steady tone is applied.

Most recording engineers use 0 VU to define a convenient "zero reference level" on the VU meter. When the meter on your mixer or recorder reads "0" on a steady tone, your equipment is producing a certain level at its output. Different types of equipment produce different levels when the meter reads 0 (Figure A.2). Zero VU corresponds to:

- +8 dBm in older broadcast and telephone equipment
- +4 dBm in balanced recording equipment
- −10 dBV in unbalanced recording equipment
- −20 dBFS on a peak-reading meter in a digital recorder or recording software

A 0 VU recording level (0 on the record level meter) is the normal operating level of an analog tape recorder; it produces the desired recorded flux on tape. A "0 VU recording level" doesn't mean a "0 VU signal level."

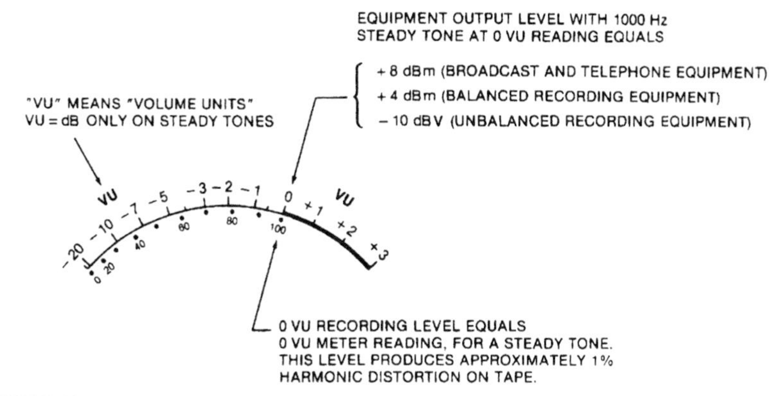

FIGURE A.2
VU-meter scale.

With a VU meter, 0 VU corresponds to a recording level 8 dB below the level that produces 3% third-harmonic distortion on tape at 400 Hz. Distortion at 0 VU typically is below 1%.

The response of a VU meter isn't fast enough to track rapid transients accurately. In addition, when a complex waveform is applied to a VU meter, the meter reads less than the peak voltage of the waveform. (This means you must allow for undisplayed peaks above 0 VU that use up headroom.)

In contrast, a peak indicator responds quickly to peak program levels, making it a more accurate indicator of recording levels. One type of peak indicator is an LED that flashes on peak overloads. Another is the LED bar graph meter commonly seen on digital recorders. Yet another is the peak program meter (PPM). It is calibrated in dB, rather than VU. Unlike the VU meter reading, the PPM reading does not correlate with perceived volume.

In a digital recorder, the meter is an LED or LCD bar graph meter that reads up to 0 dBFS (FS; full scale). In a 16-bit digital recorder, 0 dBFS means all 16 bits are ON (1) at the peak of the waveform and all 16 bits are OFF (0) at the trough of the waveform. In a 24-bit digital recorder, 0 dBFS means all 24 bits are on or off. The OVER indication means that the input level exceeded the voltage needed to produce 0 dBFS, and there is some short-duration clipping of the output analog waveform. Some manufacturers calibrate their meters so that 0 dBFS is less than 16 bits or 24 bits ON; this allows a little headroom.

BALANCED VERSUS UNBALANCED EQUIPMENT LEVELS

Generally, audio equipment with balanced (3-pin) connectors works at a higher nominal line level than equipment with unbalanced (phono) connectors. There's nothing inherent in balanced or unbalanced connections that makes them operate at different levels; they're just standardized at different levels.

These are the nominal (normal) input and output levels for the two types of equipment:

- Balanced: +4 dBu (1.23 volts)
- Unbalanced: −10 dBV (0.316 volt)

In other words, when a balanced-output recorder reads 0 VU on its meter with a steady tone, it is producing 1.23 volts at its output connector. This voltage is called +4 dBu when referenced to 1 milliwatt. When an unbalanced-output recorder reads 0 on its meter with a steady tone, it is usually producing 0.316 volt at its output connector. This voltage is called −10 dBV when referenced to 1 volt.

INTERFACING BALANCED AND UNBALANCED EQUIPMENT

There's a difference of 11.8 dB between +4 dBu and −10 dBV. To find this, convert both levels to voltages:

$$\begin{aligned} \text{dB} &= 20 \log (1.23/0.316) \\ &= 11.8 \end{aligned}$$

So, +4 dBu is 11.8 dB higher in voltage than −10 dBV (assuming the resistances are the same).

A cable carrying a nominal +4 dBu signal has a signal-to-noise ratio (S/N) 11.8 dB better than the same cable carrying a −10 dBV signal. This is an advantage in environments with strong radio-frequency or hum fields, such as in a computer. But in most studios with short cables, the difference is negligible.

Connecting a +4 dBu output to a −10 dBV input might cause distortion if the signal peaks of the +4 equipment exceed the headroom of the −10 equipment. If this happens, use a pad to attenuate the level 12 dB (Figure A.3, top). The wiring shown converts from balanced to unbalanced as well as reducing the level 12 dB. If you hear hum with this connection, add an isolation transformer as shown in Figure A-3, bottom. All the leads should be twisted.

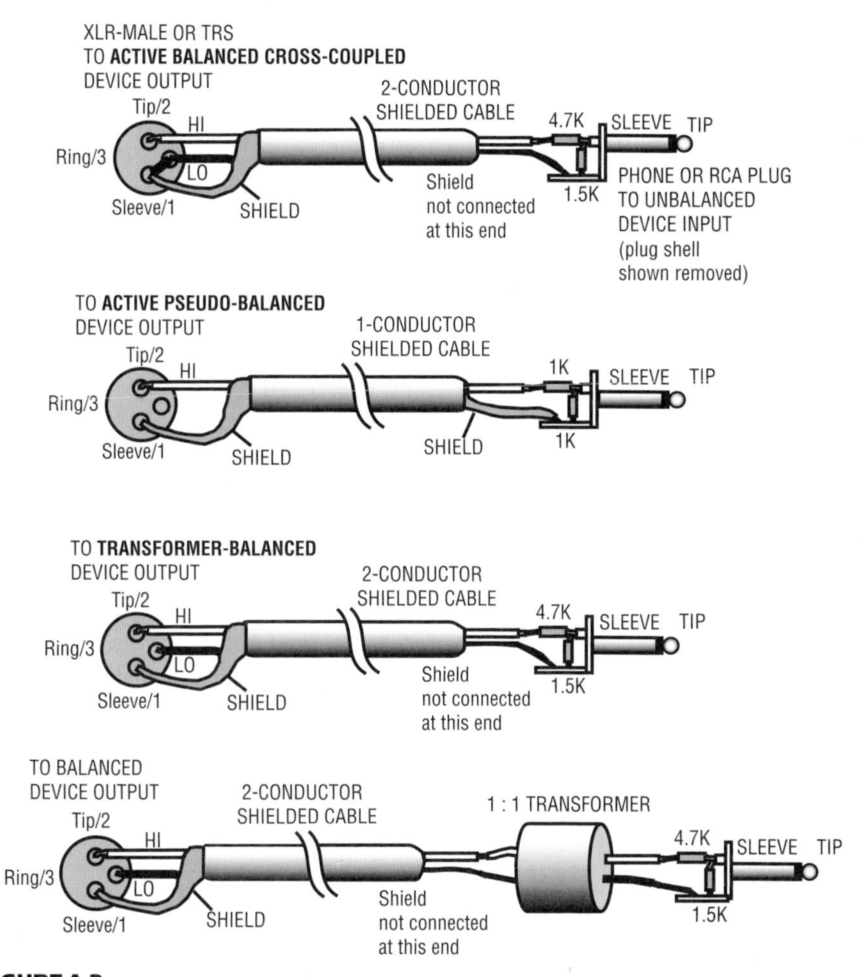

FIGURE A.3

Top: Wiring balanced out to unbalanced in. The resistors form a 12-dB pad to match the balanced +4 dBu output to the unbalanced −10 dBV input. Bottom: Same, with an isolation transformer added to reduce hum.

You don't always need that pad. Many pieces of equipment have a +4/−10 level switch. Set the switch to the nominal level of the connected equipment. If there is no such switch on either device, connect between them a +4/−10 converter box such as the Ebtech Line Level Shifter (www.ebtechaudio.com/lls-2des.html) or use the wiring shown here.

To connect an unbalanced −10 dBV output to a balanced +4 dBu input, use the wiring shown in Figure A.4. All the exposed leads should be twisted.

MICROPHONE SENSITIVITY

Decibels are an important concern in another area: microphone sensitivity. A high-sensitivity mic puts out a stronger signal (higher voltage) than a low-sensitivity mic when both are exposed to the same sound pressure level.

A microphone-sensitivity specification tells how much output (in volts) a microphone produces for a certain input (in SPL). The standard is millivolts per pascal, where one pascal (Pa) is 94 dB SPL.

A typical "open-circuit sensitivity" spec is 5.5 mV/Pa for a condenser mic and 1.8 mV/Pa for a dynamic mic. "Open-circuit" means that

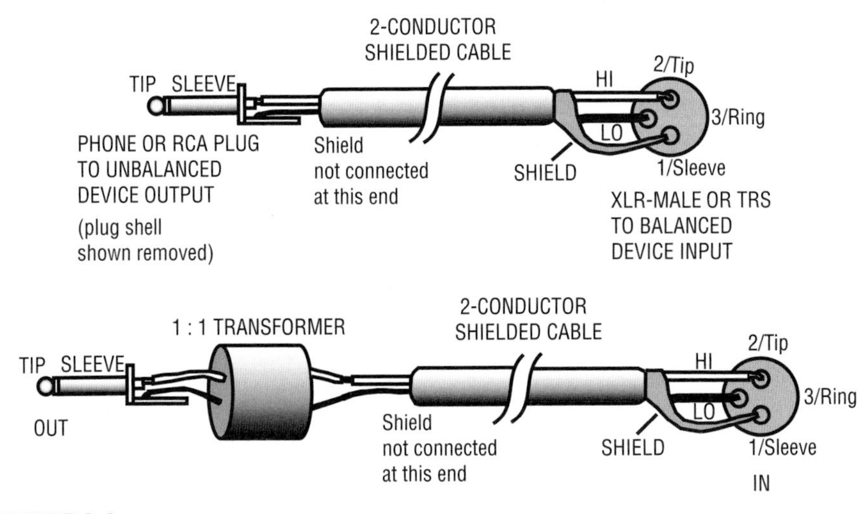

FIGURE A.4
Top: Wiring unbalanced out to balanced in. Bottom: Same, with an isolation transformer added to reduce hum.

the mic is unloaded (not connected to a load, or connected to a mic preamp with a very high input impedance). If the spec is 5.5 mV/Pa, that means the mic produces 5.5 mV when the SPL at the mic is 94 dB SPL.

If you put a microphone in a 20 dB louder soundfield, it produces 20 dB more signal voltage. For example, if 74 dB SPL in gives 0.18 mV out (−75 dBV), then 94 dB SPL in gives 1.8 mV out (−55 dBV); 150 dB SPL in gives 1.1 volt out (+1 dBV), which is approximately line level! That's why you need so much input padding when you record a kick drum or other loud source.

Appendix B

Optimizing Your Computer for Multitrack Recording

Once you've chosen some recording software and installed it, you'll want to make your computer run as fast and glitch-free as possible. Speed becomes critical as you use more tracks, more effects, high bit rates, and high sampling rates.

The data flow of multitrack digital audio places high demands on computer speed. Recording and playback of digital audio requires long, continuous periods of streaming audio data. The more tracks in use, and the higher the bit depth and sampling rate, the faster the data flow has to be. And the more soft synths and effects plug-ins that are in use, the greater the load on the CPU.

Clearly, you need a fast computer for multitrack recording. But you also need to optimize the computer's settings for best results. I will cover some ways to speed up the data flow and to reduce the CPU usage. If you follow these tips, you will have a faster system that handles more plug-ins and more tracks at once. Also, when you play tracks or burn CDs, clicks and drop-outs in the audio will be less likely.

Most of these tips apply to a PC computer. Mac advice is offered at the end of this appendix in the section Optimizing MacIntosh for Multitrack Recording.

Disclaimer: Back up your system and data files first, then proceed at your own risk. Neither I nor Focal Press will be responsible for damage to your computer system or files from making these changes. If you aren't comfortable doing a particular tweak, don't do it. However, I have done all these adjustments with no problems, and they are reversible.

SPEEDING UP YOUR HARD DRIVE

The hard drive should have a fast average access time (under 9 msec) and a high internal sustained transfer rate. Some recommended hard drives are the Seagate Barracuda, Maxtor Diamond Max, and Western Digital models.

What transfer rate is needed? The data rate of 24 tracks at 24-bits/96 kHz is 6.6 MB/sec. That is a practical minimum.

The best current hard drives can transfer about 40 MB/sec continuously with a PCI bus and an Ultra ATA/66 or Ultra ATA/100 interface (for EIDE drives), or an SCSI interface for SCSI drives. SCSI drives tend to be faster than IDE, but SCSI costs more, and many IDE drives are fast enough.

To use ATA66 or higher, your motherboard and BIOS must support it. You need an 80-pin ribbon cable that is 18 inches maximum. Plug the blue connector into the motherboard, the black one into the Master device, and the gray one into the Slave device.

Hard drives with high rotational speed or spindle rate (7200 rpm or greater) tend to have faster transfer rate. They are recommended for 24-bit multitrack productions. The typical sustained transfer rate of a 7200 rpm drive is 30 to 40 MB/sec. This provides up to 160 16-bit/44.1 kHz tracks or 50 24/96 tracks, depending on file fragmentation.

For fastest speed, use one hard drive for applications and Windows files, and another for audio data. That way the audio drive head does not waste time looking for system files. Put the Windows drive on the Primary IDE channel as a master, and put the audio drive on the Secondary IDE channel as a master. To set a drive to master or slave, change the jumpers on the back of the drives. A laptop computer can connect to an external USB or FireWire hard drive that is used just for audio files.

In your recording software under Options, indicate the drive that contains the audio data.

You can put the two hard drives on the same IDE cable (both on the Primary channel) as master and slave. Usually, but not always, they will run a little slower that way. If you have a second CD-ROM drive, put it on the secondary channel as a slave.

If you have only one hard drive, use one disk partition for audio files and another partition for programs. That way you can defragment or reformat the audio data partition frequently.

It's a good idea to have a dedicated drive just for musical instrument samples because they generate high bandwidth streams when you play them.

Defragment the audio file drive often. Defragging reorganizes files into contiguous areas on the disk so that each audio file can be accessed with minimum head movements. Before you defrag, disable programs in the System Tray. Then double-click My Computer, Right-click your audio drive, and select Properties > Tools > Defragmentation.

When you're finished using all the files on the audio data drive, reformat it to zero-out the clusters. This is necessary even if you deleted all the files.

Your system will run faster if you delete unneeded applications. Select Start > Settings > Control panel > Add/Remove programs. You must uninstall a program, rather than deleting its folders, to notice an improvement.

Install lots of RAM, at least 1 GB for audio, or at least 1.5 GB for soft synths. How does more RAM help? When your DAW application is streaming audio, it continuously feeds data into a RAM buffer for each track. If the RAM buffer is big, it needs to be filled less often, so the disk-drive heads need to search less often. A larger RAM buffer lets the disk heads work more efficiently by getting bigger continuous chunks of data into RAM. Also, having lots of RAM prevents frequent access of the swap file (described later). The goal is to have your programs running in RAM, rather than swapping data to your hard drive.

Important: Check that Direct Memory Access (DMA) is activated on hard drives and CD ROMs. DMA mode transfers data directly from the hard drive to RAM, bypassing the CPU. With DMA, your CPU overhead (while accessing the drive) will fall from about 50 to 5 percent or less. Here is the procedure:

In Windows XP, select Start > Settings > Control Panel > System > Hardware > Device Manager > IDE ATA/ATAPI Controllers (or hard disk controller) > Right-click Primary IDE channel > Properties >

Advanced Settings Tab > Transfer Mode: "DMA if available." Repeat for the Secondary IDE channel. Reboot.

Disable write-behind caching on your hard drives. Write-behind caching waits until the system is idle before writing data to the hard drive, causing a delay. Here's how to turn it off: In XP select Start > Settings > Control Panel > System > Hardware > Device Manager > Disk drives > Double-click on your audio data drive > Policies > Uncheck "Enable Write Caching" > OK.

Disable read-ahead caching on your hard drive. If it is enabled, data are read out of the cache instead of directly from the hard drive—not good for streaming audio. Your recording program might have a setting for read-ahead caching.

INCREASING PROCESSING SPEED

A computer has a central processing unit (CPU) that does most of the calculations needed to run software. For example, in some DAWs the CPU performs all the signal processing for real-time effects. Get a computer or motherboard with the fastest CPU that you can afford—a clock speed of 2 Gigahertz or greater. Consider getting a computer or motherboard with a dual-core or quad-core CPU, an × 64 processor, a 64-bit operating system, and 64-bit drivers for your hardware. The Athlon AMD64 CPU or Intel Core 2 Duo are good choices for audio work. An × 64 CPU lets the computer carry more RAM, which helpful for soft synths.

The following tips reduce the amount of data that the CPU has to handle:

Don't enable hyperthreading in BIOS unless your software requires it.

If you hear drop-outs or glitches, maybe too many real-time effects are running. Select a track that has real-time effects, and bounce (export) that track with effects to an open track. That way the effects are recorded or embedded in the bounced track, rather than running in real time. Then archive (save) the original track and delete it from the project. That reduces CPU loading.

Archive tracks that are muted, then delete those tracks from the project to further reduce the load on the processor.

It also helps to convert MIDI soft-synth tracks to audio tracks. Some DAWS have a "freeze" or "bounce MIDI to track" function for this purpose. Freezing de-activates the synth plug-in, replacing it with an audio track of the synth output. That frees up processor power for the mix.

Limit polyphony in a soft synth if you don't need it.

Don't insert the same delay effect (echo, reverb, chorus, flanging) in several tracks. Instead, set up an aux bus that has the desired effect. Use aux send controls on tracks to adjust the amount of effect on each track. This reduces the number of effects processes that are running and reduces CPU loading.

In extreme cases, consider installing a Digital Signal Processing (DSP) card, such as the TC Works PowerCore. It handles the processing for the plug-ins. The plug-ins access the card rather than the CPU. You can install multiple cards to expand the system. Pro Tools includes its own DSP cards.

Before recording, turn off the waveform drawing function of your sound clips or regions.

Reduce the number of colors: Right-click the Desktop > Properties > Settings > set Colors to 16 bit or 256 colors. A higher setting reduces audio performance when the level meters redraw in your audio application.

Reduce video acceleration: Select Start > Settings > Control panel > System > Performance > Graphics > Advanced. Reduce Hardware Acceleration as much as you can without degrading the display of your audio program.

In XP, adjust visual effects for best performance: Select Start > Right-click My Computer > Properties > Advanced > Performance > Settings > Visual Effects Tab > Adjust for best performance.

Important: Change the Performance mode for the operating system: Right-click My Computer > Properties > Advanced > Performance > Settings > Advanced > Processor scheduling > select Background Services. This setting allocates more processor time to background activities, such as streaming audio and ASIO drivers.

PREVENTING INTERRUPTIONS

Unnecessary background programs can rob CPU cycles, interrupt the audio program, and interrupt the hard-drive head from playing audio. The following tweaks to your PC can increase the number of tracks and effects you can play without drop-outs or clicks by disabling various background programs or increasing free memory.

Don't multitask (run other programs) while the audio is streaming. Each additional task slows down the system. Press Ctrl-Alt-Del to see what's running, highlight anything you don't need, and select End Task. Leave Explorer and Systray running. Don't scroll while recording or playing back.

Important: Disable system sounds such as Windows' chimes and fanfares: Select Start > Settings > Control Panel > Sounds and audio devices > Sounds > set to No Sounds.

Turn off autosave except when working just on MIDI files.

Use a PS/2 mouse instead of a USB or wireless mouse, which can cause audio instability.

Switch off your Anti-Virus program when using your programs. Do this by right clicking on its icon on the bottom right of your screen, then click "Disable." You need Anti Virus for downloading files from outside sources but not for DAW work.

In the preferences menu of your programs, disable auto update to prevent interruptions. Also select Start > Settings > Control Panel > System > Automatic Updates > Turn off automatic updates.

Don't map through sound card. Select Start > Settings > Control Panel > Sounds and Audio Devices > Hardware Tab > (highlight your sound card from the list) > Properties > Properties Tap > Audio Devices > (highlight your sound card from the list) > Properties, and check the "Do not map through this device" checkbox. The Microsoft device mapper uses unnecessary resources and can add latency. It's better to select the audio interface that you want to use from within your recording software.

Check interrupts. Select Start > Programs > Accessories > System tools > System information > Hardware resources > IRQs. You will see a complete list of all the IRQs in use. Make sure that SCSI controllers don't share interrupts with anything. Make sure the sound card has its own IRQ number. Don't share audio cards/devices IRQs with USB, SCSI, or graphics cards. Don't share Network/Modems IRQs with USB/SCSI/Video. Don't assign IRQ #9 to anything; that is the cascaded IRQ from #2. To reset IRQs, you might need to remove ALL cards except for the video card and add one card at a time in different slots.

Place your audio card in a non-IRQ-shared PCI slot. Look in the PCI table of your motherboard manual for the highest priority slot that doesn't share with another device or another slot. Consider using a shareware utility called PowerStrip, which lets you adjust PCI latency timers. Set the latencies high enough to get the needed performance, and no higher.

If you aren't using any devices on your USB, serial or parallel ports, go to the BIOS and disable them. This will reduce the number of IRQs and may prevent conflicts. Also disable or disconnect other devices that you aren't using; this frees up interrupts.

Disable background image: Right-click Desktop > Properties > Background > None.

Disable Screen Saver: Right-click Desktop > Properties > Screen Saver > None.

Important: Switch off power schemes: Start > Settings > Control Panel > Power Options > Set "Turn off hard discs" to "Never." Switch off hibernation: Start > Settings > Control Panel > Power Options > Hibernate > Uncheck "Enable hibernate support."

Remove programs and services that load on startup. Some are necessary, but many aren't. They consume memory and resources. In XP, select Start > Run. Type in MSCONFIG and press OK. Select the Startup tab and disable everything but the system tray and the speaker volume control. Keep any programs that control hardware. Select the Services tab. At the bottom of the window, check "Hide all Microsoft services." Then press the "Disable All" button. Press OK then Restart. When Windows reboots, check "Don't show this message…" and click OK.

Some experts say that you should NOT disable Multimedia Class Scheduler, Plug and Play, Superfetch (in Vista), Task Scheduler,

Windows Audio, and Windows Driver Foundation. If you use Pro Tools, re-check "MMERefresh" in the Startup tab and "Digidesign MME Refresh Service" in the Services tab.

SETTING THE BUFFER SIZE

Audio data streaming from your hard drive goes into a buffer memory, then is read out at a constant clock rate. The buffer must be filled faster than the clock empties it; otherwise there is nothing to read. This requires the CPU reading from the buffer to pause its processing while the buffer refills. Then you hear a short silence or audio drop-out. So if the hard drive is too slow, or the buffer is too small, this causes buffer underrun and results in drop-outs.

If the hard drive streams data faster than the buffer can be emptied, this causes buffer overflow or overrun. The extra data from the hard drive might be lost, causing drop-outs, clicks, or stopped playback. Or the extra data might overwrite adjacent memory locations, causing a system crash.

A small buffer forces the CPU to fill and empty the buffer much more often, giving the CPU less time for processing. In other words, the smaller the buffer, the higher the percentage of CPU usage (the harder the CPU has to work).

So, if you experience drop-outs, crashes, or high CPU percentage, increase the buffer size. You might start with a setting of 25 msec and go from there. Don't make the buffer bigger than necessary, because then playback will start with a delay.

Another buffer is the I/O buffer, used for reading and writing data to the hard drive. In your recording software, set disk I/O buffer size to 256 or 512.

MINIMIZING LATENCY

If you're using soft synths, you want them to respond as fast as possible to MIDI-controller key presses. In other words, you want to minimize latency (monitoring delay). Latency is directly related to buffer size. Latency (msec) = buffer size (samples) × 1000 divided by the sample rate. For example, if the buffer size is 256 samples and the sample rate is 44,100 Hz, the latency is 5.8 msec.

A small buffer setting reduces latency—the application responds faster—but tends to create drop-outs, or reduces the number of tracks that play without dropouts. A large buffer setting increases latency but prevents drop-outs. Which setting is best?

To solve this conflict, use two different buffer settings. In your recording program, go to the Preferences menu. Or open your audio interface's control window. Set the buffer/latency small (under 4 msec if possible) while tracking MIDI synths or while processing the input signal with effects (Echo Input Monitor setting). During playback and mixdown, set it large (maybe 25 msec). The bigger buffer will reduce the load on your CPU and let you use more plug-ins.

If you can't set the latency under 4 msec without drop-outs or crashes, freeze or bounce tracks that have effects, or temporarily disable any plug-ins. You might also convert MIDI tracks to audio, then disable soft synths.

Monitoring with headphones has less perceived latency than monitoring with speakers. That's because sound takes time to travel from the speakers to your ears. For example, if your speakers are 6 feet away, that adds 5.3 msec of perceived delay to the monitored signal.

Drivers affect latency, so download the latest drivers for your audio interface. However, note that earlier versions of drivers are sometimes faster than later versions. Also check out third-party drivers such as CEntrance Ideal Driver (www.centrance.com) and asio4all, a universal ASIO driver for WDM audio. You have to set up your audio software to use ASIO to notice any change in latency.

As we said before, set processor scheduling to "Background Services." If you use ASIO drivers, this tweak ensures low latency without drop-outs.

The round-trip latency of a MIDI keyboard playing a soft synth is less than the round-trip latency of audio in to audio out. When you trigger a soft synth with a MIDI keyboard, the round-trip latency is the sum of these latencies:

- Driver (about 4 to 6 msec depending on the driver)
- Output buffer (whatever you set it to)
- D/A conversion (about 1 to 1.5 msec)

With the buffer set to 3 msec, the calculated latency when you play a MIDI soft synth is about 8 msec. That's not bad, considering that 9 msec is just noticeable.

OTHER TIPS

- Check out www.tweakXP.com, www.Tweak3D.net, and www.extremetech.com.
- Obedia.com offers 24/7 help for computer recording: tech support, consulting and troubleshooting.
- More information on computer recording can be found at www.pcrecording.com.
- If you're using your computer professionally as part of a commercial recording studio, do NOT have a connection to the Internet on that computer. Have it be a dedicated digital audio workstation so that viruses, etc., won't destroy a customer's session. Use another computer for the Internet.
- Four vendors of audio-optimized computers are www.rainrecording.com, www.dawbox.com, www.sweetwater.com (Creation Station), and www.liquiddaw.com.
- Microsoft has a patch that fixes audio stuttering with USB devices in XP. Go to www.microsoft.com and obtain Service Pack 2.
- Go to the manufacturer Web sites of your sound card (or audio interface) and audio editing software. Check out the support sections for advice on optimizing your computer and on troubleshooting sound problems.
- Keep your sound card(s) away from the AGP graphics card to prevent hum.
- Don't install much software (it can corrupt the registry).
- Disable the computer's internal sound card when you're using your own sound card or audio interface. Click Start > settings > control panel > system > hardware > device manager > sound, video, and game controllers > Right-click the internal sound card > disable. Optional: While you're in sound controllers, remove the legacy audio drivers except the ones you're using.
- On the Web, download the latest drivers for your motherboard, video card, IDE controller, CD burner, recording software, and sound card or audio interface. Also download the Windows updates.

- Consider using an AGP video card rather than a PCI video card to reduce the load on the PCI bus.
- Free up disk space by uninstalling Windows components that you don't use. Select Start > Settings > Control Panel > Add or remove programs > Add/Remove Windows Components.
- For XP, disable Disc Indexing Service on NTFS formatted drives if you do NOT do frequent searches for filenames. Double-click My Computer > Right-click each hard drive > Properties > Uncheck "Allow Indexing Service to index this disc for fast file searching" > Apply. Choose all files and subfolders within the drive. This action takes a while as all the files are scanned. If you get a message that says Access Denied…, press "Ignore All."
- Use nLite before installing Windows XP to get a DAW-optimized installation (www.nliteos.com/nlite.html).
- If you have a utility that measures disk thruput (sustained disk transfer rate) you might want to measure your system's speed before and after each hard-drive change mentioned above. That way you will know which changes were effective. A good utility is DskBench, available free from www.prorec.com and other sources.

WINDOWS VISTA

Windows Vista offers several benefits for recording: improved stability, more processing in the graphics card and less in the CPU, WaveRT low-latency driver support, and Multimedia Class Scheduler Service (MMCSS) to allow the audio engine higher priority to the CPU.

Most of the advice for Windows XP applies to Vista. Here are some additional Vista tips:

- Vista Home Premium is recommended for audio (in 32-bit or 64-bit versions).
- Be sure to download the latest Vista drivers for your software, hardware, and plug-ins.
- Go to www.update.microsoft.com for performance-increasing patches.
- If your motherboard allows, disable HPET in BIOS.
- Use Vista Lite (www.vlite.net) before installing Vista to get a DAW-optimized installation.
- Don't change the page file (swap file or virtual memory) because Vista handles it well.

- Disable startup items that you don't need. Select Start > Run. Type Defender and press Enter. Select Tools > Software Explorer > All Users. Highlight an item you don't need and click Disable. Repeat for other items. Reboot when done.
- Turn off User Account Control (if you don't need this security feature). Select Start > Control Panel > User Accounts and Family Safety > User Accounts > Turn User Account Control on or off > Uncheck Use User Account Control (UAC) > OK > reboot.
- Disable Windows components that you don't need. Select Start > Run > appwiz.cpl. Select Turn Windows features on or off. Unselect any components that you aren't using. Click OK.
- Turn off Windows Aero. Right-click on the desktop and select Personalize. Select Windows Color and Appearance. Select "Open classic appearance properties" for more color options. In the Color Scheme list box, select Windows Vista Basic or Windows Standard for a more classic appearance. Click OK
- Turn off Windows Sidebar. Locate the Sidebar icon in your systray. Right-click on the icon and select Exit. Uncheck Start Sidebar when Windows starts. Click OK.
- Check www.extremetech.com for Vista hot fixes.
- Select Start > Settings > Control Panel. Click "Uninstall a Program" to launch Vista's "Uninstall or Change a Program Window". In the Tasks pane on the left, click "Turn Windows Features On or Off". In the list of features, uncheck any you don't need.

OPTIMIZING MACINTOSH FOR MULTITRACK RECORDING

- Many of the PC tips apply to Mac: Use a second, fast hard drive running at 7200 rpm or higher for audio files. Use a minimum of non-audio software. Disable energy-saver or sleep software. Update drivers. Reduce monitor resolution. Disable video acceleration software and cards. Check online user groups for advice.
- Check the Memory panel and turn off virtual memory.
- Check for enough application memory. Select the program's icon and press Command-I. In the info window that appears, try increasing the amount of allotted memory.

- Minimize extensions. Extensions are small programs that extend the functionality of the operating system when they are loaded. But they use up CPU clock cycles and slightly reduce stability. You might disable all extensions but these: sound card, Appearance, QuickTime, SystemAV, and video drivers. Consider loading the CD-ROM driver only when you need it. If the recording application freezes after boot-up, check the application's documentation for known extension conflicts.
- Upgrade to OS X.
- For maximum speed, use computers with PCI Express (PCIe).
- Use a Mac Pro, an 8-core system.
- Use HFS on the audio drive—not HFS (extended). HFS can make the cluster sizes too small for efficient streaming of audio.
- Buy more RAM: 2 GB or more is recommended.
- Delete outtakes and unused files to prevent a "too many files open" message.
- If you can't play enough tracks, increase the buffer size.
- If you hear pops or clicks, check your sound card's clock settings. When recording an analog source, set it to Internal Clock or Sync. If your source is digital, set it to external S/PDIF sync or the format you're using. Be sure sample rates match. Move audio and digital cables away from USB cables and power lines. Try turning off USB hubs, modems, and printers.
- If your system crashes frequently, try removing downloaded software. Delete your application's preferences file if it has become corrupt.
- If the application stops in the middle of a recording, check to see whether you have set a maximum recording time in advance.

Appendix C

Impedance

Impedance is one of audio's more confusing concepts. To clarify this topic, I'll present a few questions and answers about impedance.

WHAT IS IMPEDANCE?

Impedance (Z) is the resistance of a circuit to alternating current, such as an audio signal. Technically, impedance is the total opposition (including resistance and reactance) that a circuit has to passing alternating current.

A high-impedance circuit tends to have high voltage and low current. A low-impedance circuit tends to have relatively low voltage and high current.

I'M CONNECTING TWO AUDIO DEVICES. IS IT IMPORTANT TO MATCH THEIR IMPEDANCES? WHAT IF I DON'T?

First some definitions. When you connect two devices, one is the source and one is the load. The source is the device that puts out a signal. The load is the device you are feeding the signal into. The source has a certain output impedance, and the load has a certain input impedance.

A few decades ago in the vacuum-tube era, it was important to match the output impedance of the source to the input impedance of the load. Usually the source and load impedances were both 600 ohms. If the source impedance equals the load impedance, this is called "matching" impedances. It results in maximum power transfer from the source to the load.

In contrast, suppose the source is low Z and the load is high Z. If the load impedance is 10 times or more the source impedance, it is called a "bridging" impedance. Bridging results in maximum voltage transfer from the source to the load. Today, nearly all devices are connected bridging—low-Z out to high-Z in—because we want the most voltage transferred between components.

For best sound quality and highest voltage transfer, the input impedance of the connector you're plugging into should be at least 7 to 10 times higher than the output impedance of the source (such as an electric guitar or microphone).

For example, if a mic's impedance is 200 ohms, the input impedance of a mic input (that you plug the mic into) should be 7 to 10 times higher, or 1400 to 2000 ohms. If you look at the input impedance spec for mixer mic inputs, it's typically around 1500 ohms.

Similarly, the impedance of an electric-guitar pickup is typically 20 to 40 K ohms (20,000 to 40,000) ohms. So the ideal input impedance of a guitar amp (or direct box input, or instrument input) is at least 7 to 10 times higher, or at least 280 to 400 K ohms.

If you connect a low-Z source to a high-Z load, there is no distortion or frequency-response change caused by this connection. But if you connect a high-Z source to a low-Z load, you might get distortion or altered response. For example, suppose you connect an electric bass guitar (a high-Z device) into an XLR-type mic input (a low-Z load). The low frequencies in the signal will roll off, so the bass will sound thin. And the highs might roll off, making the sound dull.

We want the bass guitar to be loaded by a high impedance, and we want the mic input to be fed by a low-impedance signal. A direct box or impedance-matching adapter does this (Figure C.1). Such an adapter is available from Radio Shack, part no. 274-017.

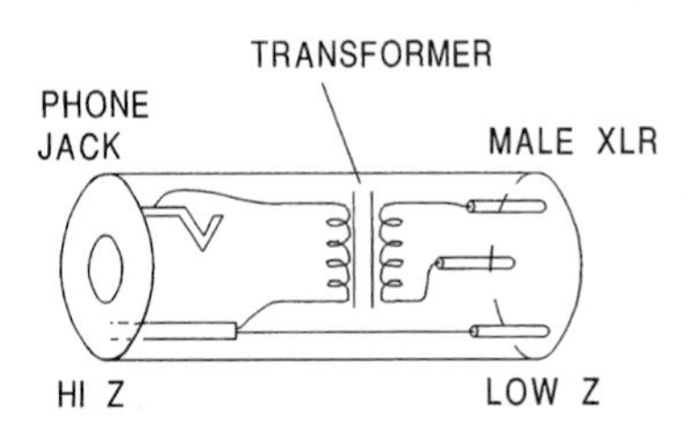

FIGURE C.1
High-to-low impedance matching adapter.

The adapter is a tube with a phone jack on one end and a male XLR connector on the other. Inside the tube is a transformer. Its primary winding is high Z, wired to the phone jack. The transformer's secondary winding is low Z, wired to the XLR. You plug the guitar cord into the phone jack, and plug the XLR into a mic input in a snake or mixer. Use it with a bass guitar, electric guitar, or synth.

This impedance-matching adapter works, but isn't ideal. The load it presents to the bass guitar might be 12 kilohms, which will slightly load down the high-Z guitar pickup, causing thin bass. An active direct box solves this problem. In place of a transformer, the active DI usually has a field effect transistor (FET). The FET has a very high input impedance that doesn't load down the bass guitar.

In general, the input impedance of a direct box (or an instrument jack) is higher than the input impedance of a line input.

WHAT ABOUT MICROPHONE IMPEDANCE?

Recording-quality mics have XLR (3-pin) connectors and are low Z (150 to 300 ohms). A low-Z mic can be used with hundreds of feet of cable without picking up hum or losing high frequencies.

I'M CONNECTING A MIC TO A MIXER. IS IMPEDANCE A CONSIDERATION?

Yes. If your mixer has phone-jack inputs, they are probably high Z. But most mics are low Z. When you plug a low-Z mic into a high-Z input you get a weak signal. That's because a high-Z mic input is designed to receive a relatively high voltage from a high-Z mic, and so the input is designed to have low gain. So you don't get much signal amplification.

If you can't get enough level when you plug a mic into a phone-jack input, here's a solution: Between the mic cable and the input jack, connect an impedance matching adapter (Figure C.2) such as Radio Shack 274-016. It steps up the voltage of the mic, giving it a stronger signal.

The adapter is a tube with a female XLR input and a phone-plug output. Inside the tube is a transformer. Its primary winding is low Z, wired to the XLR. It secondary winding is high Z, wired to the phone plug. Connect the mic to the XLR; connect the phone plug to the mixer's phone jack. Then the mixer will receive a strong signal from the mic.

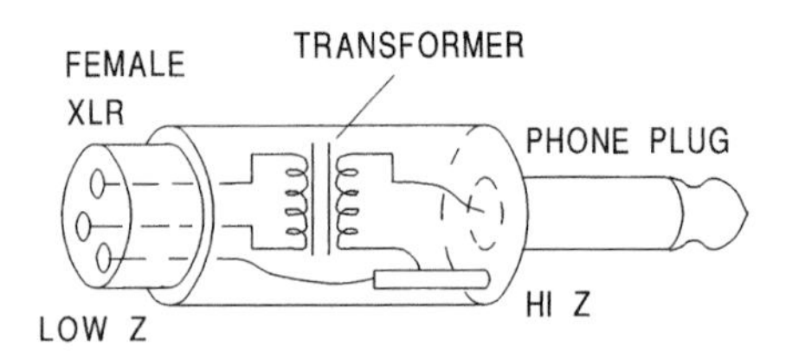

FIGURE C.2
Low-to-high impedance matching adapter.

If you're using a phantom-powered condenser mic, the connections are different. First, turn off any phantom power in your recorder-mixer. Connect your mic to a stand-alone phantom-power supply, and connect the supply output to the impedance-matching adapter.

If your mixer has XLR inputs, they are low-Z balanced. In this case, simply connect the mic to the mixer using a mic cable with a female XLR on the mic end and a male XLR on the mixer end. A low-Z mic input is typically about 1500 ohms, so it provides a bridging load to a mic that is 150 to 300 ohms.

SHOULD I CONSIDER IMPEDANCE WHEN I CONNECT TWO LINE-LEVEL DEVICES?

This is seldom a problem. In most audio devices, the impedance of the line output is low—about 100 to 1000 ohms. The impedance of the line input is high—about 10 kilohms to 1 megohm. So every connection is bridging, and you get maximum voltage transfer. Some audio devices, such as passive equalizers, require a terminating resistor at the input or output for best performance.

CAN I CONNECT ONE SOURCE TO TWO OR MORE LOADS?

Usually yes. You can connect several devices in parallel across one line output. Suppose you connect a mixer output simultaneously to a recorder input, an amplifier input, and another mixer's input in parallel. The combined input impedance of those three loads might be 4000 ohms, which still presents a bridging load to the mixer's 100-ohm output impedance.

Mics are a different story. Suppose you're recording a concert. If you connect one mic to the PA mixer and also to your recording mixer with a Y cable, the combined input impedance will be about 700 ohms or less. This can load down some microphones, reducing the bass in dynamic mics or causing distortion in condenser mics. To prevent the loading, put a 270 ohm 1% resistor in series with Pin 2 and Pin 3 in each leg of the Y cable.

CAN I CONNECT TWO OR MORE SOURCES TO ONE INPUT?

Not recommended. The low output impedance of one source will load down the output of the other source, and vice versa. This can cause level loss and distortion. The same goes for two mics into one mic input. It's better to use a small external mixer.

As another option, put a series resistor in line with each device before combining them. That prevents each device from loading down the other. A minimum resistor value might be 470 ohms per source. If the source is balanced, use one resistor on Pin 2 and one on Pin 3—two 1% resistors per source.

SUMMARY

- Impedance (Z) is the opposition to alternating current, measured in ohms.
- Microphones and line outputs are usually low Z.
- Electric guitars, acoustic-guitar pickups, instrument inputs, and line inputs are usually high Z.
- XLR mic inputs are low impedance; phone jack mic inputs are high impedance.
- Speakers are usually 4 to 8 ohms.
- Equal impedances in parallel result in half the impedance.
- Equal impedances in series result in twice the impedance.
- Connect low-Z sources to low-Z inputs. (A low-Z input is usually 7 to 10 times the source impedance, but it's still called a low-Z input.)
- Connect high-Z sources to high-Z inputs.
- Connect a low-Z source to a high-Z input through a step-up transformer (impedance matching adapter).
- Connect a high-Z source to a low-Z input through a step-down transformer (impedance matching adapter, or direct box).

Where to Learn More

We recommend the following books, magazines, videos, and literature to anyone who wants more education in recording technology.

BOOKS AND VIDEOS

The Library

Go to your local library and search their database for books on recording, home recording, music recording, audio recording, MIDI, or sound production. Read everything you can on the subject.

Pro Audio Books Plus

This company offers an online catalog of all sorts of pro audio and recording books, videos, and courses at www.proaudiobooksplus.com.

Music Books Plus

This is a catalog with a huge number of audio and recording books. The Web site is www.musicbooksplus.com. They also carry tutorial videos on DVD.

howaudio.com

This site offers over 1000 tutorial videos on home recording and related topics.

Focal Press

This is another major audio/recording books catalog and it is found at http://focalpress.com. Focal Press also offers video DVDs about

DAW operation, recording, mic techniques, mixing, and setting up a home studio. If you want to record concerts or gigs, check out the Focal Press book *Recording Music On Location*.

Amazon.com

This online bookstore has a great search engine. In the search field, type in whatever audio subject you are interested in: DVD, MIDI, digital audio, recording, mixing, microphones, etc.

RECORDING MAGAZINES

Recording (home and project studio recording), www.recordingmag.com.

Mix (pro recording and concert sound), www.mixonline.com.

EQ (home, project, and pro recording), www.eqmag.com; *EQ* also offers a recording and sound buyer's guide.

Electronic Musician (home and project studio recording), www.emusician.com; *Electronic Musician* offers a supplement called "Desktop Music Production Guide."

Tape Op (home and project studio creative recording), www.tapeop.com.

Future Music (home, project and pro recording), www.futuremusicmag.com.

Sound On Sound (Britain's premier recording publication), www.soundonsound.com.

In those magazines are ads related to recording products and services.

PRO AUDIO MAGAZINES

Journal of the Audio Engineering Society (*JAES*), pro audio engineering. Scholarly. With its journal, conventions, workshops, and local chapters, the Audio Engineering Society is a tremendous resource, www.aes.org.

Pro Audio Review (reviews of pro audio equipment), www.proaudioreview.com.

Live Sound (concert sound reinforcement), www.livesoundint.com.

Church Production (audio for houses of worship), www.churchproduction.com.

Technologies for Worship (audio for houses of worship), www.tfwm.com.

Church and Worship Technology (audio for houses of worship), http://churchandworship.com.

CONSUMER AUDIO MAGAZINES

Sound and Vision (consumer audio and video), www.soundandvisionmag.com.

Stereophile (high-end audio), www.stereophile.com.

The Absolute Sound (high-end audio), www.avguide.com/the-absolute-sound.

GUIDES, BROCHURES, AND OTHER LITERATURE

Microphone application guides are available from:

Crown International, www.crownaudio.com.
Shure Inc., www.shure.com.
Countryman Associates Inc., www.countryman.com.
AKG Acoustics Inc., www.akg.com.
Sennheiser Electronic Corp., www.sennheiserusa.com.
Audio-Technica U.S. Inc., www.audio-technica.com.
Neumann/USA, www.neumannusa.com.
Schoeps Mikrofone, www.schoeps.de.
Danish Pro Audio, www.dpamicrophones.com (be sure to check out their Microphone University)

You can find valuable information in user manuals and free sales literature provided by manufacturers of recording equipment. Ask your equipment dealer for manufacturers' phone numbers and Web site URLs.

GUIDES TO RECORDING SCHOOLS

The Audio Engineering Society offers a directory of educational programs at www.aes.org/education/directory.cfm.

Each July issue of *Mix* magazine contains a comprehensive directory of recording schools, seminars, and programs. Universities and colleges in most major cities have recording-engineering courses.

An index of recording schools is at www.modrec.com/schools/. Two recent programs are www.recordingconnection.com, and the Bachelor of Science in Music Technology at the University of Saint Francis, School of Creative Arts in Fort Wayne, Indiana. For information call 260-434-3279.

The New Ears/Sound Schools book offers a list of recording schools. The Web site is www.soundschools.com.

THE INTERNET

A great place to ask questions, besides magazines, is on the Internet. In Google, select Groups, then type in rec.audio.pro. You can ask questions and get answers from pro engineers. You may get conflicting answers, because often there are many ways to do the same thing. Also, some who reply are more expert than others. But you'll often find stimulating debates.

Some other valuable Web sites (all starting with www.) are homerecording.com, hometracked.com, prorec.com, kvraudio .com, gearslutz.com, recordbetteraudio.com, recording.org, recordproduction.com, vintagesynth.com, moultonlabs.com, microphones.com, digido.com, prosoundweb.com, digifreq.com, cdrfaq.org, harmonycentral.com, musicplayer.com (forums), and audio-recording-center.com.

Also see pcrecording.com, audiogrid.com, digitalprosound.com, dvdforum.org, recordproduction.com, coutant.org, microphone-data. com, themicrophonevault.com, miclisteningroom.org, audiodirectory. nl/, binaural.com, sweetwater.com (check out publications and forums), cakewalk.com (forums), and www.solorb.com/dat-heads/.

If you lack good multitrack recordings with which to practice mixing, go to www.raw-tracks.com. There you can download individual tracks in WAV or MP3 format, or purchase a CD of raw tracks.

Some online MIDI resources are midi.com, midiworld.com, and harmony-central.com.

The Web sites of audio equipment manufacturers have support sections, online discussion groups, and FAQs with lots of information. Search for the name and model number of equipment you have, or are interested in.

Do a Google search with the keywords audio, recording, recording vocals, audio recording links, MIDI, digital audio, and so on. Also search for FAQs on various audio topics. You'll discover hundreds of audio-related Web sites and links. In Google, you can search Groups as well as Web sites.

RECORDING EQUIPMENT CATALOGS

Here are but a few catalogs from which to order recording gear:

American Musical Supply, www.americanmusical.com
BSW, www.bswusa.com
Guitar Center, www.guitarcenter.com
Musician's Friend, www.musiciansfriend.com
Sam Ash, www.samash.com
Prodigy Pro, www.prodigy-pro.com
B&H Pro Audio, www.bhproaudio.com
Sweetwater Sound, www.sweetwater.com
The Woodwind and the Brasswind, www.wwandbw.com

EXPERIENCE

It's the best teacher. Record all you can with any equipment you have. You can buy a 4-track cassette recorder on eBay for $45, or download free multitrack recording software. Then buy a cheap mic, and practice mic techniques and overdubbing.

Glossary

A weighting See Weighted.

AAC Abbreviation for Advanced Audio Code. A relatively new compressed audio file format, AAC offers better sound quality than MP3, but with less storage space and bandwidth. It's also used with Digital Rights Management technology to help control the copying and distributing of music.

AAF Abbreviation for Advanced Authoring Format; www.aafassociation.org/html/technoinfo/index.html. An industry standard for high-end exchange of video or audio project data.

A-B A listening comparison between two audio programs, or between two components playing the same program, performed by switching immediately from one to the other. The levels of the two signals are matched. See also *Spaced-pair*.

AC-3 Same as Dolby Digital, a perceptual encoding scheme that data-reduces the six surround channels of a 5.1 system to two channels.

Accent microphone See *Spot microphone*.

Access jacks See *Insert jacks*.

Active combining network A combining network with gain. See *Combining network*.

AES Audio Engineering Society.

AES/EBU Also called IEC 988 Type 1. An interface format for digital signals, using a balanced 110-ohm mic cable terminated with XLR-type connectors. See also *S/PDIF*.

AIFF Abbreviation for Audio Interchange File Format. A standard Mac format for uncompressed digital audio files.

Alignment The adjustment of analog tape-head azimuth and of tape-recorder circuitry to achieve optimum performance from the particular type of tape being used.

Alignment tape A prerecorded tape with calibrated tones for alignment of an analog tape recorder.

Ambience Room acoustics, early reflections, and reverberation. Also, the audible sense of a room or environment surrounding a recorded instrument.

Ambience microphone A microphone placed relatively far from its sound source to pick up ambience.

Amplitude, Peak On a graph of a sound wave, the sound pressure of the waveform peak. On a graph of an electrical signal, the voltage of the waveform peak. The amplitude of a sound wave or signal as measured on a meter is 0.707 times the peak amplitude.

Analog-to-digital (A/D) converter A circuit that converts an analog audio signal into a stream of digital data (bitstream).

Anti-alias filter In an A/D converter, a lowpass filter that removes all frequencies above 20 kHz before sampling to prevent audio artifacts called aliasing.

Anti-imaging filter In a D/A converter, a lowpass filter that smooths the voltage steps in the analog signal that was generated by translating digital numbers into analog voltages. The anti-imaging filter recovers the waveform of the original analog signal.

ASIO Abbreviation for Audio Stream Input/Output. Steinberg's computer audio driver spec for Mac and Windows. ASIO has low latency because it interfaces directly between a sound card and the audio application software.

Assign To route or send an audio signal to one or more selected channels.

ATRAC Abbreviation for Adaptive Transform Acoustic Coding. A data compression scheme (used in the MiniDisc) that reduces by 5:1 the storage needed for digital audio. ATRAC is a perceptual coding method that omits data deemed inaudible due to masking.

Attack The beginning of a note. The first portion of a note's envelope in which a note rises from silence to its maximum volume.

Attack time In a compressor, the time it takes for gain reduction to occur in response to a musical attack.

Attenuate To reduce the level of a signal.

Attenuator In a mixer (or mixing console) input module, an adjustable resistive network that reduces the microphone signal level to prevent overloading of the input transformer and mic preamplifier.

Audio interface A device that connects to a computer and converts an audio signal into computer data for storage in memory or on hard disk. The interface also converts computer data into an audio signal. See *Breakout box*, *I/O box*, and *Sound card*.

Autolocate A recorder function that makes the tape or disk go to a program address (counter time) at the press of a button.

Automated mixing A system of mixing in which a computer remembers and updates mixer control settings and moves. With this system, a mix can be performed and refined in several stages and played back at a later date exactly as set up previously.

Auxiliary bus (aux bus) A bus or channel that contains a mix of the aux-send signals of the input modules in a mixer. An aux bus is used to send signals to an effects unit or monitor system. See *Effects bus*.

Auxiliary send (aux-send) A control in a mixer's input module used to send that module's signal to an aux bus. The aux-send level adjusts the amount of effects heard on an instrument, or adjusts the loudness of that instrument in the monitor system.

A/V drive A hard-disk drive meant for audio/video use. It postpones thermal recalibration until the disk is inactive, preventing data errors.

Azimuth In a tape recorder, the angular relationship between the head gap and the tape path.

Azimuth alignment In an analog tape recorder, the mechanical adjustment of the record or playback head to bring it into proper alignment (90 degrees) with the tape path.

Back-timing A technique of cueing up the musical background or a sound effect to a narration track so that the music or effect ends simultaneously with the narration.

Baffled-omni A stereo miking arrangement that uses two ear-spaced omnidirectional microphones separated by a disk, sphere, or foam.

Balance The relative volume levels of various tracks or instruments.

Balanced AC power AC power from a center-tapped power transformer. Instead of one 120 V line and one 0 V line, it has two 60 V lines. They are in phase with each other and sum to 120 V. But they are connected to the center-tap ground out of phase (one is 60 V; the other is −60 V). Any hum and noise on the grounding system cancel out.

Balanced line A cable with two conductors surrounded by a shield in which each conductor is at equal impedance to ground. With respect to ground, the conductors are at equal potential but opposite polarity; the signal flows through both conductors.

Bandpass filter In a crossover, a filter that passes a band or range of frequencies but sharply attenuates or rejects frequencies outside the band.

Basic tracks Recorded tracks of rhythm instruments (bass, guitar, drums, and sometimes keyboard).

Bass management A subwoofer/satellite crossover, usually part of a surround receiver. Bass management routes frequencies above about 100 Hz to the five full-range speakers, and routes frequencies below about 100 Hz from all six channels to the subwoofer. It affects only what you monitor, not what is recorded. Bass management can be done by a surround-receiver circuit, a stand-alone box, a special circuit in a subwoofer, or a software plug-in.

Bass trap An assembly that absorbs low-frequency sound waves in the studio.

Bi-amplification (bi-amping) Driving a woofer and tweeter with separate power amplifiers. An active crossover is connected ahead of these power amplifiers.

Bias In tape-recorder electronics, an ultrasonic signal that drives the erase head. This signal is also mixed with the audio signal applied to the record head to reduce distortion.

Bidirectional microphone A microphone that is most sensitive to sounds arriving from two directions—in front of and behind the microphone. It rejects sounds approaching either side of the microphone. Sometimes called a cosine or figure-eight microphone because of the shape of its polar pattern.

Binaural recording A 2-channel recording made with an omnidirectional microphone mounted near each ear of a human or a dummy head, for playback over headphones. The object is to duplicate the acoustic signal appearing at each ear.

Bit depth The number of bits (ones and zeros) making up a word in a digital signal (such as 16- or 24-bits). Each word is a binary number that is the value of each sample. A sample is a measurement of an analog waveform that is done several thousand times a second during analog-to-digital conversion. High bit depth = long word length = high resolution of the analog signal amplitude = high sound quality. See *Word length*.

Blumlein array A stereo microphone technique in which two coincident bidirectional microphones are angled 90 degrees apart (45 degrees to the left and right of center).

Board See *Mixing console*.

Bouncing tracks A process in which two or more tracks are mixed, and the mixed tracks are recorded on an unused track or tracks. Then the original tracks can be erased, which frees them up for recording more instruments.

Boundary microphone A microphone designed to be used on a boundary (a hard reflective surface). The microphone capsule is mounted very close to the boundary so that direct and reflected sounds arrive at the microphone diaphragm in phase (or nearly so) for all frequencies in the audible band.

Breakout box (I/O box) A group of audio input and output connectors in a chassis, which is wired to a sound card (PCI audio interface) in a computer. Used to interface analog audio signals (and often MIDI and digital signals) with a computer.

Breathing The unwanted audible rise and fall of background noise that may occur with a compressor. Also called pumping.

Bulk tape eraser A large electromagnet used to erase a whole reel of recording tape at once.

Bus A common connection of many different signals. An output of a mixer or submixer. A channel that feeds a tape track, signal processor, or power amplifier.

Bus in An input to a program bus, usually used for effects returns.

Bus master In the output section of a mixing console, a potentiometer (fader or volume control) that controls the output level of a bus. See also *Group fader*.

Bus out The output connector of a bus.

Bus trim A control in the output section of a mixing console that provides variable gain control of a bus, used in addition to the bus master for fine adjustment.

Buzz An unwanted edgy tone that sometimes accompanies audio, containing high harmonics of 60 Hz (50 Hz in Europe).

Calibration See *Alignment*.

Capacitor An electronic component that stores an electric charge. It is formed of two conductive plates separated by an insulator called a dielectric. A capacitor passes AC but blocks DC.

Capacitor microphone See *Condenser microphone*.

Capstan In a tape-recorder transport, a rotating post that contacts the tape (along with the pinch roller) and pulls the tape past the heads at a constant speed during recording and playback.

Cardioid microphone A unidirectional microphone with side attenuation of 6 dB and maximum rejection of sound at the rear of the microphone (180 degrees off-axis). A microphone with a heart-shaped directional pattern.

CD See *Compact disc*.

CD-R Abbreviation for CD-Recordable. A recordable compact disc that can't be rewritten. Once recorded, it can't be erased and reused.

CD-ROM A computer disk drive that plays computer data from CD-ROM disks. The disks are read optically by a laser, as in a compact disc player.

CD-RW Abbreviation for CD-Rewritable. A recordable compact disc that can be rewritten. Once recorded it can be erased and reused.

Channel A single path of an audio signal. Usually, each channel contains a different signal.

Channel assign See *Assign*.

Chorus (1) A special effect in which a signal is delayed by 15 to 35 msec, the delayed signal is combined with the original signal, and the delay is varied randomly or periodically. This creates a wavy, shimmering effect. (2) The main portion of a song that is repeated several times throughout the song with the same lyrics.

Clean Free of noise, distortion, overhang, leakage. Not muddy.

Clear Easy to hear, easy to differentiate. Reproduced with sufficient high frequencies.

Clip To cut off or flatten the tops of a signal waveform, causing distortion. Clipping is caused by setting signal level too high in an audio device or audio software. See also *Region*.

Coincident-pair A stereo microphone, or two separate microphones, placed so that the microphone diaphragms occupy approximately the same point in space. They are angled apart and mounted one directly above the other.

Comb-filter effect The frequency response caused by combining a sound with its delayed replica. The frequency response has a series of peaks and dips caused by phase interference. The peaks and dips resemble the teeth of a comb.

Combining amplifier An amplifier in a mixing console at which the outputs of two or more signal paths are mixed together to feed a single track of a recorder.

Combining network A resistive network in a mixing console at which the outputs of two or more signal paths are mixed together to feed a single track of a recorder.

Compact disc (CD) A read-only optical disc medium for storing digital audio programs up to 80 minutes long. The compact disc stores data in the form of a spiral groove of microscopic pits and is read optically by a laser. CD's digital audio format is 44.1 kHz sampling rate and 16-bit word length.

Comping Recording several musical performances (takes) of a single instrument or vocal on different tracks, and selecting the best parts of each take to be played in order on a composite track during mixdown.

Complex wave A wave with more than one frequency component.

Composite track A track containing the best parts of several takes of a musical-instrument or vocal performance.

Compression (1) The portion of a sound wave in which air molecules are pushed together, forming a region with higher than normal atmospheric pressure. (2) In signal processing, the reduction in dynamic range or gain caused by a compressor. (3) Data compression or data reduction is an encoding scheme to reduce the size of a data file by throwing away audio data deemed inaudible because of masking. ATRAC, MP3, MLP, AAC, RealAudio, OGG, and Microsoft Media are examples of compressed data formats.

Compression ratio (slope) In a compressor, the ratio of the change in input level (in dB) to the change in output level (in dB). For example, a 2:1 ratio means that for every 2 dB change in input level, the output level changes 1 dB.

Compressor A signal processor that reduces dynamic range or gain by means of automatic volume control. An amplifier whose gain decreases as the input signal level increases above a preset point.

Condenser microphone A microphone that works on the principle of variable capacitance to generate an electrical signal. The microphone diaphragm and an adjacent metallic disk (called a backplate) are charged to form two plates of a capacitor. Incoming sound waves vibrate the diaphragm, varying its spacing to the backplate, which varies the capacitance, which in turn varies the voltage between the diaphragm and backplate.

Connector A device that makes electrical contact between a signal-carrying cable and an electronic device, or between two cables.

A device used to connect or hold together a cable and an electronic component so that a signal can flow from one to the other.

Console See *Mixing console.*

Contact pickup A transducer that contacts a musical instrument and converts its mechanical vibrations into a corresponding electrical signal.

Control room The room in which the engineer controls and monitors the recording. It houses most of the recording hardware.

Controller surface A chassis with faders (and sometimes buttons and knobs) that resembles a mixer, used to adjust virtual controls that appear on-screen in computer recording software. Connected to the computer by USB or FireWire, the controller surface might include analog and digital input/output connectors and MIDI connectors.

Convolution reverb (sampling reverb) A reverberation device or plug-in which creates the reverb from impulse-response samples (WAVE files) of real acoustic spaces, rather than from algorithms. The resulting sound quality is very natural.

Crossover An electronic network that divides an incoming signal into two or more frequency bands.

Crossover, active (electronic crossover) A crossover with amplifying components, used ahead of the power amplifiers in a bi- or tri-amped speaker system.

Crossover frequency The single frequency at which both filters of a crossover network are down 3 dB.

Crossover, passive A crossover with passive (non-amplifying) components, used after the power amplifier.

Crosstalk The unwanted transfer of a signal from one channel to another. Crosstalk often occurs between adjacent tracks within a record or playback head in a tape recorder, or between input modules in a console.

Cue, cue send In a mixing-console input module, a control that adjusts the level of the signal feeding the cue mixer that feeds a signal to headphones in the studio.

Cue list See *Edit decision list* (*EDL*).

Cue mixer A submixer in a mixing console that takes signals from cue sends (aux sends) as inputs and mixes them into a composite signal that drives headphones in the studio.

Cuesheet Used during mixdown, a chronological list of mixing-console control adjustments required at various points in the recorded song. These points may be indicated by counter or ABS-time readings.

Cue system A monitor system that allows musicians to hear themselves and previously recorded tracks through headphones.

Damping factor The ability of a power amplifier to control or damp loudspeaker vibrations. The lower the amplifier's output impedance, the higher the damping factor.

DAT (R-DAT) A digital audio tape recorder that uses a rotating head to record digital audio on tape. It is obsolete, having been replaced by Flash memory recorders.

Data compression A data encoding scheme for reducing the amount of data storage on a medium. Same as Data Reduction. See *Compression, ATRAC,* and *MP3*.

DAW Abbreviation for digital audio workstation.

dB Abbreviation for decibel.

Dead Having very little or no reverberation.

Decay The portion of the envelope of a note in which the envelope goes from maximum to some midrange level. Also, the decline in level of reverberation over time.

Decay time See *Reverberation time* (*RT 60*).

Decibel The unit of measurement of audio level. Ten times the logarithm of the ratio of two power levels. Twenty times the logarithm of the ratio of two voltages. dBV is decibels relative to 1 volt, dBu is decibels relative to 0.775 volt, dBv is decibels relative to 0.7746 volt, dBm is decibels relative to 1 milliwatt, dBFS is decibels, full scale (all bits on or off), dBSPL is decibels of sound pressure level, and dBA is decibels, A weighted. See *Weighted.*

Decoded tape A program on analog recording tape that is expanded after being compressed by a noise-reduction system. Such a program has normal dynamic range.

De-esser A signal processor or plug-in that removes excessive sibilance ("s" and "sh" sounds) by compressing high frequencies around 3 to 10 kHz.

Delay The time interval between a signal and its repetition. A digital delay or a delay line is a signal processor that delays a signal for a short time.

Delay compensation Adjusting the timing of a track that is processed by a plug-in, to make it in sync with non-processed tracks. Plug-ins add latency (delay) to a track.

Demagnetizer (degausser) An electromagnet with a probe tip that is touched to elements of an analog recorder tape path (such as tape heads and tape guides) to remove residual magnetism.

Depth The audible sense of nearness and farness of various instruments. Instruments recorded with a high ratio of direct-to-reverberant sound are perceived as being close; instruments recorded with a low ratio of direct-to-reverberant sound are perceived as being distant.

Design center The portion of fader travel (usually shaded), about 10 to 15 dB from the top, in which console gain is distributed for optimum headroom and signal-to-noise ratio. During normal operation, each fader in use should be placed at or near design center.

Designation strip A strip of paper taped near console faders to designate the instrument that each fader controls. Also called a scribble strip.

Desk The British term for mixing console.

Destructive editing In a digital audio workstation, editing that rewrites the data on disk. A destructive edit can't be undone unless a copy of the original data is saved before the edit is done.

DI Abbreviation for direct injection, recording with a direct box.

Diffusion An even distribution of sound in a room.

Digital audio An encoding of an analog audio signal in the form of binary digits (ones and zeroes).

Digital audio workstation (DAW) A computer, audio interface, and recording software that allows you to record, edit, and mix audio programs entirely in digital form. A stand-alone DAW is a digital multitrack recorder-mixer. Stand-alone DAWs include real mixer controls; computer DAWS have virtual controls on-screen.

Digital recording A recording system in which the audio signal is stored as binary digits (ones and zeroes).

Digital-to-analog (D/A) converter A circuit that converts a digital audio signal into an analog audio signal.

Dim To reduce the monitor volume temporarily by a preset amount so that you can carry on a conversation.

Direct box A device used for connecting an amplified instrument directly to a mixer mic input. The direct box converts a high-impedance unbalanced audio signal into a low-impedance balanced audio signal.

Direct injection (DI) Recording with a direct box.

Directional microphone A microphone that has different sensitivity in different directions. A unidirectional or bidirectional microphone.

Direct output, direct out An output connector following a mic preamplifier, fader, and equalizer, used to feed the signal of one instrument to one track of a multitrack recorder.

Direct sound Sound traveling directly from the sound source to the microphone (or to the listener) without reflections. Also, DirectSound is an audio driver for the Windows operating system, intended as an enhancement to MultiMedia Extensions (MME).

DirectX (DX) A package of Windows audio, video, and game-controller drivers, also called multimedia application programming interfaces (APIs).

Distortion An unwanted change in the audio waveform, causing a raspy or gritty sound quality. The appearance of frequencies in a device's output signal that weren't in the input signal. Distortion is caused by recording at too high a level, using improper mixer settings, components failing, or vacuum tubes distorting. (Distortion can be desirable, e.g., for an electric guitar.) In digital recording, distortion called quantization error can also occur at very low signal levels, where there aren't enough bits to record the signal accurately. See also *Clip*.

Dither Low-level noise added to a digital signal to reduce quantization distortion caused by truncating (removing) bits in a digital word. It's a good idea to add dither to a 24-bit program just before converting it to 16 bits for CD release.

Dolby Digital A perceptual coding method using AC-3 data compression, offering 6 discrete channels of digital surround sound. Dolby Digital uses a lossy encoding process to reduce the bit rate needed to transmit the six channels via a 2-channel bitstream. The standard audio format for DVD-Video.

Dolby Pro Logic A surround decoder that decodes the two channels of Dolby-Surround-encoded programs back into four channels (left, center, right, surround).

Dolby Surround A matrix encoding system that combines four channels (left, center, right, surround) into two channels. The Dolby

Pro Logic decoder unfolds the two channels back into four. The surround channel, which is mono and limited bandwidth, is reproduced over left and right surround speakers.

Dolby Tone A reference tone recorded at the beginning of a Dolby-encoded analog tape for alignment purposes.

Doubling A special effect in which a signal is combined with its 15- to 35-msec-delayed replica. This process mimics the sound of two identical voices or instruments playing in unison. In another type of doubling, two identical performances are recorded and played back to thicken the sound.

Drop-frame For color video production, a mode of SMPTE timecode that causes the timecode to match the clock on the wall. Once every minute, frame numbers 00 and 01 are dropped, except every 10th minute.

Drop-out During playback of a tape recording, a momentary loss of signal caused by separation of the tape from the playback head by dust, tape-oxide irregularity, etc. During playback of a hard-disk recording, a momentary loss of signal caused by a buffer memory being emptied. Increasing buffer size usually prevents drop-outs.

Drum machine A device—hardware or software—that plays samples of real drums and includes a sequencer to record rhythm patterns.

Dry Having no echo or reverberation. Referring to a close-sounding signal that hasn't yet been processed by a reverberation or delay device or plug-in.

DSD Abbreviation for Direct Stream Digital. A process that encodes a digital signal in a 1-bit (bitstream) format at a 2.8224 MHz sampling rate. Offers state-of-the-art sound quality; used in the Super Audio CD.

DSP Abbreviation for Digital Signal Processing. Modifying a signal in digital form by doing calculations on the numbers. DSP is used for level changes, EQ, effects, and so on.

DTS Abbreviation for Digital Theater System. A perceptual coding method using data compression, offering 6 discrete channels of digital surround sound. DTS uses a lossy encoding process to reduce the bitrate needed to transmit the 6 channels via a 2-channel bitstream.

DVD Abbreviation for Digital Versatile Disc. A storage medium the size of a compact disc that holds much more data. The DVD stores video, audio, or computer data. DVD-RAM and DVD-R are recordable; DVD-RW is rewritable, and DVD-A is audio only.

DVD-Audio A DVD intended mainly for audio programs. It can use Dolby Digital or DTS encoded programs.

DX See *DirectX*.

DXi Abbreviation for DirectX Instruments. Cakewalk's standard format for plug-in software synthesizers based on Microsoft DirectX technology.

Dynamic microphone A microphone that generates electricity when sound waves cause a conductor to vibrate in a stationary magnetic field. The two types of dynamic microphones are moving coil and ribbon. A moving-coil microphone is usually called a dynamic microphone.

Dynamic range The range of volume levels in a program from softest to loudest.

Earth ground A connection to moist dirt (the ground we walk on). This connection is usually done via a long copper rod.

EASI Abbreviation for Enhanced Audio Streaming Interface. Emagic's sound card driver spec. EASI achieves low latency by interfacing the sound card directly with the audio application software.

Echo A delayed repetition of a signal or sound. A sound delayed 50 msec or more that is combined with the original sound.

Echo chamber A hard-surfaced room, containing a widely separated loudspeaker and microphone, used for creating reverberation.

Edit decision list (EDL) A list of program events in order, plus their starting and ending times.

Editing The cutting and rejoining of magnetic tape to delete unwanted material, to insert silent spaces, or to rearrange recorded material into the desired sequence. Also, the same actions performed on a digital recording with a DAW, hard-disk recorder, or MiniDisc recorder–mixer.

Editing block A metal block that holds magnetic tape during the editing/splicing procedure.

Effects Interesting sound phenomena created by signal processors, such as reverberation, echo, flanging, doubling, compression, or chorus. See *Sound effects*.

Effects bus The bus that feeds effects devices (signal processors).

Effects loop A set of connectors in a mixer for connecting an external effects unit, such as a reverb or delay device. The effects loop includes a send section and a receive section. See *Effects send*, *Effects return*.

Effects return (aux return) In the output section of a mixing console, a control that adjusts the amount of signal received from an effects unit. Also, the connectors in a mixer to which you connect the effects-unit output signal. They might be labeled "bus in" instead. The effects-return signal is mixed with the program bus signal.

Effects send (aux send) In an input module of a mixing console, a control that adjusts the amount of signal sent to an effects device, such as a reverberation or delay unit. Also, the connector in a mixer that you connect to the input of an effects unit. The effects-send or aux-send control adjusts the amount of effects heard on each instrument.

Efficiency In a loudspeaker, the ratio of acoustic power output to electrical power input.

EIA Electrical Industries Association.

Electret-condenser microphone A condenser microphone in which the electrostatic field of the capacitor is generated by an electret—a material that permanently stores an electrostatic charge.

Electrostatic field The force field between two conductors charged with static electricity.

Electrostatic interference The unwanted presence of an electrostatic hum field in signal conductors.

Encoded tape An analog tape containing a signal compressed by a noise-reduction unit.

End-addressed Referring to a microphone whose main axis of pickup is perpendicular to the front of the microphone. You aim the front of the mic at the sound source. See *Side-addressed*.

Envelope The rise and fall in volume of one note. The envelope connects successive peaks of the waves that make up the note. Each harmonic in the note might have a different envelope. A volume envelope in a DAW track is a graph of track volume (fader setting) versus time.

Equalization (EQ) The adjustment of frequency response to alter the tonal balance or to attenuate unwanted frequencies.

Equalizer A circuit (usually in each input module of a mixing console, or in a separate unit) that alters the frequency spectrum of a signal passed through it.

Erase To remove an audio signal from magnetic tape or disk by applying an ultrasonic varying magnetic field to randomize the magnetization of the magnetic particles on the tape or disk.

Erase head A head in a tape recorder or hard drive that erases the signal on tape or disk.

Expander (1) A signal processor that increases the dynamic range of a signal passed through it. (2) An amplifier whose gain decreases as its input level decreases. When used as a noise gate, an expander reduces the gain of low-level signals to reduce noise between notes.

Fade-out To gradually reduce the volume of the last several seconds of a recorded song, from full level down to silence, by slowly pulling down the master fader, or by selecting a fade-out process in a digital editing program.

Fader A linear or sliding potentiometer (volume control), used to adjust signal level.

Feed (1) To send an audio signal to some device or system. (2) An output signal sent to some device or system.

Feedback (1) The return of some portion of an output signal to the system's input. (2) The squealing sound you hear when a PA system microphone picks up its own amplified signal through a loudspeaker.

Feed reel The left-side reel on an analog tape recorder that unwinds during recording or playback.

Filter (1) A circuit that sharply attenuates frequencies above or below a certain frequency. Used to reduce noise and leakage above or below the frequency range of an instrument or voice. (2) A MIDI filter removes selected note parameters.

FireWire A standard protocol for high-speed transfer of data between digital devices. Also called IEEE 1394.

Fixed point In digital signal processing, fixed-point calculations use numbers with a fixed number of digits after the decimal point. See *Floating Point.*

Flanging A special effect in which a signal is combined with its delayed replica, and the delay is varied between 0 and 20 msec.

A hollow, swishing, ethereal effect like a variable-length pipe, or like a jet plane passing overhead. A variable comb filter produces the flanging effect.

Fletcher-Munson effect Named after the two people who discovered it, the psychoacoustical phenomenon in which the subjective frequency response of the ear changes with program level. Because of this effect, a program played at a lower volume than the original level subjectively loses low- and high-frequency response.

Float To disconnect from ground.

Floating point In digital signal processing, calculations using numbers with a fixed number of bits for the mantissa and the remaining bits for the exponent. 32-bit "float" uses a 24-bit mantissa and an 8-bit exponent. It's almost impossible to clip an audio signal in 32-bit floating-point mixing software. When the clip light flashes, the signal isn't clipping in the DAW, but it would be clipped if it were exported to a CD, to an audio interface, or to a 24-bit fixed-point file. See *Fixed point*.

Flutter A rapid periodic variation in tape speed.

Flutter echoes A rapid series of echoes that occurs between two parallel walls.

Flux Magnetic lines of force.

Fluxivity The measure of the flux density of a magnetic recording tape, per unit of track width.

Fly-in (lay-in) To copy part of a recorded track onto another recorder, then re-record that copy back onto the original multitrack tape in a different part of the song, in sync with other recorded tracks. For example, copy the vocal track from the first chorus of the song onto an external DAW or sampler. Re-record (fly-in) that copy onto the multitrack tape at the second chorus. Then the first and second choruses have identical vocal performances.

Foldback (FB) See *Cue system*.

Frequency The number of cycles per second of a sound wave or an audio signal, measured in hertz (Hz). A low frequency (for example, 100 Hz) has a low pitch; a high frequency (for example, 10,000 Hz) has a high pitch.

Frequency response (1) The range of frequencies that an audio device will reproduce at an equal level (within a tolerance, such as ±3 dB). (2) The range of frequencies that a device (mic, human ear, etc.) can detect.

Full-duplex Describing a sound card that can record and play back simultaneously. The card works on two DMA channels.

Full track A single tape track recorded across the full width of an analog tape.

Fundamental The lowest frequency in a complex wave.

FX Abbreviation for effects.

Gain Amplification. The ratio, expressed in decibels, between the output voltage and the input voltage, or between the output power and the input power.

Gain staging Setting all the gain controls in a signal path to get the cleanest signal. At any stage in the signal path, the signal level should be set well above the noise floor but below clipping.

Gap In a tape-recorder head, the thin break in the electromagnet that contacts the tape. In an audio program, the space or silence between songs.

Gate (1) To turn off a signal when its amplitude falls below a preset value. (2) The signal-processing device used for this purpose. See also *Noise gate*.

Gated reverb Reverberation with the reverberant "tail" cut off before it fades out.

General MIDI file (GM file) A MIDI file containing a standard set of musical instrument sounds. A General MIDI file produces the same sounds on any MIDI-capable instrument that supports the GM spec.

Generation A copy of a tape or a bounce of a track. A copy of the original master recording is a first generation tape. A copy made from the first generation tape is a second generation, and so on.

Generation loss The degradation of signal quality (the increase in noise and distortion) that occurs with each successive generation of an analog tape recording.

Gobo A moveable partition used to prevent the sound of an instrument from reaching another instrument's microphone. Short for go-between.

Graphic equalizer An equalizer with a horizontal row of faders; the fader-knob positions indicate graphically the frequency response of the equalizer. Usually used to equalize monitor speakers for the room they are in. Sometimes used for complex EQ of a track.

Ground The zero-signal reference point for a system of audio components.

Ground bus A common connection to which equipment is grounded, usually a heavy copper plate.

Grounding Connecting pieces of electronic equipment to ground. Proper grounding ensures that there is no voltage difference between equipment chassis. An electrostatic shield needs to be grounded to be effective.

Ground loop (1) A loop or circuit formed of ground leads. (2) The loop formed when components are connected together via two ground paths—the connecting-cable shield and the power ground. Ground loops cause hum and should be avoided.

Group (1) To select several faders to make them act in unison. For example, select all the faders for the drum tracks to so that you can adjust the overall level of the drums by pushing one fader. (2) To assign the output of several input modules to a single group or bus, whose level is controlled by a single group fader. For example, assign all the input modules of the drum mics to a single "drums" group. (3) A bus or channel in a mixer that contains the signals from several input modules. For example, a drums group is a group of all the signals of the drum-set mics. See *Submix*.

Group fader (submaster fader) In the output section of a mixing console, a potentiometer (fader or volume control) that controls the output level of a bus or group.

GSIF Abbreviation for GigaSampler Interface. Nemesys' Windows sound card driver that achieves low latency by interfacing directly between the sound card and the audio application software.

Guard band The spacing between tracks on a multitrack analog tape or tape head, used to prevent crosstalk.

Half-track A tape track recorded across approximately half the width of a tape. A half-track recorder usually records two such tracks simultaneously in the same direction to make a stereo recording.

Hard disk A random-access storage medium for computer data. A hard-disk drive contains a stack of magnetically coated hard disks that are read by, and written to by, an electromagnetic head.

Hard-disk recorder (hard-drive recorder) A device dedicated to recording digital audio on a hard-disk drive. A hard-disk recorder–mixer includes a built-in mixer.

Harmonic An overtone whose frequency is a whole-number multiple of the fundamental frequency.

Harmonizer A signal processor that provides a wide variety of pitch-shifting and delay effects.

HD Abbreviation for hard-disk drive.

Head An electromagnet in a tape recorder that either erases the audio signal on tape, records a signal on tape, or plays back a signal that is already on tape. A hard disk drive also has heads with similar functions.

Head gap See *Gap*.

Headphones A head-worn transducer that covers the ears and converts electrical audio signals into sound waves.

Headroom The safety margin, measured in decibels, between the signal level and the maximum undistorted signal level. In a tape recorder, the dB difference between standard operating level (corresponding to a 0 VU reading) and the level causing 3% total harmonic distortion. High-frequency headroom increases with analog tape speed.

Hertz (Hz) Cycles per second, the unit of measurement of frequency.

Highpass filter A filter that passes frequencies above a certain frequency and attenuates frequencies below that same frequency. A low-cut filter.

Hiss A noise signal containing all frequencies, but with greater energy at higher octaves. Hiss sounds like wind blowing through trees. It is usually caused by random signals generated by microphones, electronics, and magnetic tape.

Host A DAW recording program that supports plug-ins. See *Plug-in*.

Hot (1) A high recording level causing slight distortion, may be used for special effect. (2) High average level on a CD making it relatively loud, produced by peak limiting and normalization, or by compression and normalization. (3) A condition in which a chassis or circuit has a potentially dangerous voltage on it. (4) Referring to the conductor in a microphone cable that has a positive voltage on it at the instant that sound pressure moves the diaphragm inward.

Hum An unwanted low-pitched tone (60 Hz or 50 Hz and its harmonics) heard in the monitors. The sound of interference generated in audio circuits and cables by AC power wiring. Hum pickup is caused by such things as placing audio cables near power cables or transformers, faulty grounding, poor shielding, and ground loops.

Hypercardioid microphone A directional microphone with a polar pattern that has 12 dB attenuation at the sides, 6 dB attenuation at the rear, and two nulls of maximum rejection at 110 degrees off-axis.

Image An illusory sound source located somewhere around the listener. An image is generated by two or more loudspeakers. In a typical stereo system, images are located between the two stereo speakers.

Impedance The opposition of a circuit to the flow of alternating current. Impedance is the complex sum of resistance and reactance. Abbreviated as Z.

Input The connection going into an audio device. In a mixer or mixing console, a connector for a microphone, line-level device, or other signal source.

Input attenuator See *Attenuator*.

Input module In a mixing console, the set of controls affecting a single input signal. An input module usually includes an attenuator (trim), fader, equalizer, aux sends, and channel-assign buttons.

Input/output (I/O) console (in-line console) A mixing console arranged so that input and output sections are aligned vertically. Each module (other than the monitor section) contains one input channel and one output channel.

Input section The row of input modules in a mixing console.

Insert jacks Two jacks (send and return) in a console input module or output module that allow access to points in the signal path, usually for connecting a compressor. Plugging into the insert jacks breaks the signal flow and allows you to insert a signal processor or recorder in series with the signal. In many mixers, a single insert jack has both send and return terminals. Also called *Access jacks*.

Intersample peak: An analog signal produced by a digital-to-analog converter that momentarily exceeds the level of the digital samples from which it was converted.

I/O Referring to Input and Output connectors.

I/O box A Breakout Box type of audio interface.

Jack A female or receptacle-type connector for audio signals into which a plug is inserted.

Keyboard workstation Several MIDI components in one chassis—a keyboard, a sample player, a sequencer, and perhaps a synthesizer and disk drive.

Kilo A prefix meaning one thousand. Abbreviated k.

Latency The signal delay through an A/D, D/A converter, through a software program, or through a computer operating system. Monitoring latency is the delay between the time when a musician plays a note and when she hears the monitored signal of that note. Latency can make you play out of sync during overdubs.

Lay-in See *Fly-in*.

Leadering The process of splicing leader tape between program selections on analog tape.

Leader tape Plastic or paper tape without an oxide coating, used for a spacer between takes (for silence between songs) on analog tape.

Leakage The overlap of an instrument's sound into another instrument's microphone. Also called bleed or spill.

LEDE Abbreviation for Live-End/Dead-End, a type of control room acoustic treatment in which the front half of the control room prevents early reflections to the mixing position, while the back half of the control room reflects diffused sound to the mixing position.

LED indicator A recording-level indicator using one or more light-emitting diodes (LEDs).

Level The degree of intensity of an audio signal—the voltage, power, or sound pressure level. The original definition of level is the power in watts.

Level setting In a recording system, the process of adjusting the input-signal level to obtain maximum level on the recording medium without distortion. A VU meter, LED meter, or other indicator shows recording level.

Lightpipe An Alesis connection protocol that transfers 8 digital audio channels at once over a TOSLINK fiber-optic cable.

Limiter A signal processor whose output is constant above a preset input level. A compressor with a compression ratio of 10:1 or greater, with the threshold set just below the point of distortion of the following device. Used to prevent distortion of attack transients or peaks. Also used to decrease the level of signal peaks so that the average level of a program can be raised. See *Look-ahead*.

Line level In balanced professional recording equipment, a signal whose level is approximately 1.23 volts +4 dBm). In unbalanced equipment (most home hi-fi or semi-pro recording equipment), a signal whose level is approximately 0.316 volt (−10 dBV).

Live (1) Having audible reverberation. (2) Occurring in real time, in person.

Live recording A recording made at a concert. Also, a recording made of a musical ensemble playing all at once, rather than overdubbing.

Localization The ability of the human hearing system to tell the direction of a real or illusory sound source.

Locate (Autolocate) A recorder function that makes the tape or disk head go to a specified program address (counter time) at the press of a button.

Look-ahead A limiter plug-in with the "look-ahead" feature checks the upcoming audio for peaks. Audio enters a look-ahead buffer, and the limiter measures that audio signal and reacts quickly to reduce the incoming peaks.

Loop In a sampling program, to play the sustain portion of a sound's envelope repeatedly. Also, a repeated rhythmic or musical pattern.

Loudspeaker A transducer that converts electrical energy (the signal) into acoustical energy (sound waves).

Lowpass filter A filter that passes frequencies below a certain frequency and attenuates frequencies above that same frequency. A high-cut filter.

M Abbreviation for mega, or one million (as in megabytes). Small m (m) is an abbreviation for milli, or one thousandth.

Magnetic recording tape A recording medium made of magnetic particles (usually ferric oxide) suspended in a binder and coated on a long strip of thin plastic (usually Mylar).

Magneto-optical drive (MO drive) A drive that stores data (such as audio) on a 3.5-inch rewritable magneto-optical disk. The drive uses a laser and magnetic head to write data, and a laser to read data.

Mask To hide or cover up one sound with another sound. To make a sound inaudible by playing another sound along with it. Masking is used in many data reduction schemes.

Master A completed tape or CD used to generate tape copies or compact discs.

Master fader A volume control that affects the level of all program busses simultaneously. It's the last stage of gain adjustment before the 2-track recorder.

MD Abbreviation for MiniDisc.

MDM Abbreviation for Modular Digital Multitrack.

Memory A group of integrated circuit chips used to store digital data temporarily or permanently (such as an audio signal in digital format).

Memory recorder A device that records audio on a memory chip, such as Compact Flash. Usually the recorded audio can be uncompressed WAV files or compressed MP3 files.

Memory rewind A tape-recorder function that rewinds the tape to a preset tape-counter position.

Meter A device that indicates voltage, resistance, current, or signal level.

MFX The architecture or protocol for MIDI effects. See *MIDI effects*.

MIC An abbreviation for microphone.

Mic level The level or voltage of a signal produced by a microphone, typically 2 millivolts.

Mic preamp See *Preamplifier*.

Microphone A transducer or device that converts an acoustical signal (sound) into a corresponding electrical signal.

Microphone techniques The selection and placement of microphones to pick up sound sources.

MIDI Abbreviation for Musical Instrument Digital Interface, a specification for a connection between synthesizers, drum machines, and computers that allows them to communicate with and/or control each other.

MIDI channel A route for transmitting and receiving MIDI signals. Each channel controls a separate MIDI musical instrument or synth patch. Up to 16 channels can be sent on a single MIDI cable.

MIDI clock A timing reference that sets a common tempo for all the equipment. The MIDI clock is a series of bytes in the MIDI data stream that conveys timing information. The clock is like a conductor's baton movements, keeping all the performers in sync at the same tempo. The clock bytes are added to the MIDI performance information in the MIDI signal. The clock signal is 24, 48, or 96 pulses per quarter note (ppq). That is, for every quarter note of the performance, 24 or more clock pulses (bytes) are sent in the MIDI data stream.

MIDI controller A musical performance device (keyboard, drum pads, breath controller, etc.) that outputs a MIDI signal designating note numbers, note on, note off, and so on.

MIDI/digital audio software Software that combines a MIDI sequencer with a multitrack digital audio recorder/editor.

MIDI effects (MFX) Non-audio processes applied to MIDI signals, such as an arpeggiator, echo/delay, chord analyzer, quantize, transpose MIDI, event filter, or velocity change. They can be used as real-time, nondestructive plug-ins in MIDI tracks.

MIDI in A connector in a MIDI device that receives MIDI messages.

MIDI interface A circuit that plugs into a computer and converts MIDI data into computer data for storage in memory or on hard disk. The interface also converts computer data into MIDI data.

MIDI out A connector in a MIDI device that transmits MIDI messages.

MIDI thru A connector in a MIDI device that duplicates the MIDI information at the MIDI-IN connector. Used to connect another MIDI device in the series.

MIDI Timecode (MTC) See *MTC*.

Mid-side A coincident-pair stereo microphone technique using a forward-facing unidirectional, omnidirectional, or bidirectional mic and a side-facing bidirectional mic. The microphone signals are summed and differenced to produce right- and left-channel signals.

Mike To pick up with a microphone.

milli A prefix meaning one thousandth, abbreviated m.

MiniDisc (MD) A rewritable, magneto-optical storage medium that is read by a laser. It resembles a compact disc in a 2.5-inch square housing. MD recorders use a data compression scheme called ATRAC.

Mix (1) To combine two or more different signals into a common signal. (2) A control on an effects processor that varies the ratio between the dry (unprocessed) signal and the processed signal.

Mixdown The process of playing recorded tracks through a mixer and mixing them to two stereo channels for recording on a two-track recorder. Also applies to a surround-sound mixdown to 6 or 8 channels.

Mixer A device or software that mixes or combines audio signals and controls the relative levels of the signals.

Mixing console A large mixer with additional functions such as equalization or tone control, pan pots, monitoring controls, solo functions, channel assigns, and control of signals sent to external signal processors.

MLP Abbreviation for Meridian Lossless Packing. Used in DVD-Audio discs, a data-reduction method that compresses six full-range channels of 24-bit, 96 kHz audio without data loss.

MMC Abbreviation for MIDI Machine Control. A set of MIDI commands by which one device can control another. Some commands include Start, Stop, and Locate. MMC doesn't include sync information, but MTC does.

Modeler A device or software that simulates the sound of a microphone, guitar amp, or room.

MO drive See *Magneto-optical drive.*

Modular digital multitrack (MDM) A multitrack tape recorder that records 8 tracks digitally on a videocassette. Several 8-track modules can be linked together to add more tracks in sync. Two examples of MDMs are the Alesis ADAT and TASCAM DA-88. This device is obsolete and has been replaced by multitrack hard-disk recorders.

Monaural Referring to listening with one ear. Often incorrectly used to mean monophonic.

Monitor A loudspeaker in a control room, or headphones, used for judging sound quality. Also, a video display screen used with a computer.

Monitoring Listening to an audio signal with a monitor.

Mono, monophonic (1) Referring to a single channel of audio. A monophonic program can be played over one or more loudspeakers, or one or more headphones. (2). Describing a synthesizer that plays only one note at a time (not chords).

Mono-compatible A characteristic of a stereo program, in which the program channels can be combined to a mono program without altering the frequency response or balance. A mono-compatible stereo program has the same frequency response in stereo or mono because there is no delay or phase shift between channels to cause phase interference.

Moving-coil microphone A dynamic microphone in which the conductor is a coil of wire moving in a fixed magnetic field. The coil is attached to a diaphragm that vibrates when struck with sound waves. Usually called a dynamic microphone.

MP3 (MPEG Level-1 Layer-3) A data compression format for audio. In an MP3 file (.mp3), the data has been compressed or reduced to one-fourth of its original size or less. Compressed files take up less memory, so they download faster. You download MP3 files to your hard drive, then listen to them. MP3 audio quality at a 128 kbps rate is nearly the same as that of CDs (depending on source material).

MP3PRO A data-compression format for audio. MP3Pro is an improvement over MP3. Songs encoded at 64 kbps with MP3Pro are said to sound as good as songs encoded at 128 kbps with MP3. MP3Pro offers faster downloads and nearly double the amount of music you can put on a Flash memory player. MP3 and MP3Pro files are compatible with each other's players, but an MP3Pro player is needed to hear MP3Pro's improvement in sound quality.

M-S RECORDING See *Mid-side*.

MTC Abbreviation for MIDI Timecode. A form of timecode transmitted over a MIDI cable, used for synchronizing MIDI devices, in the form hours:minutes:seconds:frames. MIDI messages that provide SMPTE timecode information, used to sync a sequencer or DAW with other devices that can synchronize to MTC. MTC also syncs those devices to a tape recording that is striped with SMPTE timecode. Unlike SMPTE timecode, MTC isn't sample-accurate. MTC isn't affected by tempo, but MIDI clock is.

Muddy Unclear sounding; having excessive leakage, reverberation, or overhang.

Multi-effects processor See *Multiprocessor*.

Multiple-D Microphone A directional microphone that has multiple sound-path lengths between its front and rear sound entries. This type of microphone has minimal proximity effect.

Multiprocessor A signal processor that can perform several different signal-processing functions.

Multitimbral In a synthesizer, the ability to produce two or more different patches or timbres at the same time.

Multitrack Referring to a recorder with more than two tracks.

Mute To turn off an input signal on a mixer by disconnecting the input-module output from channel assign. During mixdown, the mute function is used to reduce noise and leakage during silent portions of tracks, or to turn off unused performances. During recording, mute is used to turn off tracks that you don't want to hear.

Near coincident A stereo microphone technique in which two directional microphones are angled apart symmetrically on either side of center and spaced a few inches apart horizontally.

Nearfield monitoring A monitor-speaker arrangement in which the speakers are placed very near the listener (usually just behind the mixing console) to reduce the audibility of control-room acoustics.

Noise Unwanted sound, such as hiss from electronics or tape. An audio signal with an irregular, non-periodic waveform.

Noise gate A gate used to reduce or eliminate noise or leakage between notes.

Noise-reduction system A signal processor used to reduce analog tape hiss (and sometimes print-through) caused by the recording process. The Dolby noise-reduction system compresses the high frequencies during recording and expands them in a complementary way during playback.

Noise shaping Filtering the noise added in dithering to make the noise less audible. Usually the filter reduces the level in the upper midrange and increases the level at high frequencies.

Nondestructive editing In a digital audio workstation, editing done by changing pointers (location markers) to information on the hard disk. A nondestructive edit can be undone.

Nonlinear (1) Referring to a storage medium in which any data point can be accessed or read almost instantly in a random fashion, rather than sequentially. Examples are a hard disk, compact disc, and MiniDisc. See *Random access*. (2) Referring to an audio device that is distorting the signal.

Normalize To raise the level of a digital audio signal so that the highest peak in the program is at the highest level allowed by the recording. For example, in a normalized 16-bit recording, the highest peak in the program has all 16 bits on: the highest possible level in a 16-bit recording.

Octave The interval between any two frequencies where the upper frequency is twice the lower frequency.

Off-axis Not exactly in front of a microphone or loudspeaker.

Off-axis coloration In a microphone, the deviation from the on-axis frequency response that sometimes occurs at angles off the axis of the microphone. The coloration of sound (alteration of tone quality) for sounds arriving off-axis to the microphone.

OGG The file extension for Ogg Vorbis, a data-reduction encoding scheme.

OMF A file format for the transfer of digital data between DAW software. You might record a project with Cakewalk Sonar, then export it as an OMF file to be imported into a Pro Tools session.

Omnidirectional microphone A microphone that is equally sensitive to sounds arriving from all directions.

On-location recording A recording made outside the studio, in a room or hall where the music usually is performed or practiced.

Open tracks On a multitrack recorder, tracks that haven't been used yet, or have already been bounced and are available for use.

ORTF Named after the French broadcasting network Office de Radio-diffusion Television Française. A near-coincident stereo mic technique that uses two cardioid mics angled 110 degrees apart and spaced 17 cm horizontally.

Outboard equipment Signal processors that are external to the mixing console.

Output A connector in an audio device from which the signal comes and feeds successive devices.

Outtake A take, or section of a take, that is to be removed or not used.

Overdub To record a new musical part on an unused track in synchronization with previously recorded tracks.

Overhang The continuation of a signal at the output of a device after the input signal has ceased. Sometimes called ringing.

Overload The distortion that occurs when an applied signal exceeds a system's maximum input level.

Oversampling In A/D conversion, oversampling is sampling an audio signal at a higher rate than needed to reproduce the highest frequency in the signal. For example, sampling an audio signal at 8 times 44.1 kHz is called "8 × oversampling." This process is followed by a digital lowpass filter and a gentle-slope analog anti-alias filter. The result is less phase shift and smoother sound compared to a steep, "brick-wall" analog filter used alone. See *Anti-alias filter*. Oversampling also can be applied in D/A conversion. Oversampling D/A converters raise the effective sampling rate by interpolating new samples between the existing samples.

Overtone In a complex wave, a frequency component that is higher than the fundamental frequency.

Pad See *Attenuator*.

Pan pot Shortened term for panoramic potentiometer. In each input module in a mixing console, a control that divides a signal between two channels in an adjustable ratio. By doing so, a pan pot controls the location of a sonic image between a stereo pair of loudspeakers.

Parametric equalizer An equalizer with continuously variable parameters, such as frequency, bandwidth, and amount of boost or cut.

Patch (1) To connect one piece of audio equipment to another with a cable. (2) A setting of synthesizer parameters to achieve a sound with a certain timbre.

Patch bay (Patch panel) An array of connectors, usually in a rack, to which equipment inputs and outputs are wired. A patch bay makes it easy to interconnect various pieces of equipment in a central, accessible location.

Patch cord A short length of cable with a phone plug on each end, used for signal routing in a patch bay.

PCM Abbreviation for Pulse Code Modulation, a method of analog-to-digital conversion in which the instantaneous amplitude of an analog waveform is measured or sampled several thousand times a second, and each measurement is assigned a binary value of a certain number of bits (ones and zeroes).

PDM Abbreviation for Pulse Density Modulation, a method of analog-to-digital conversion in which the instantaneous amplitude of an analog waveform is coded as variations in the average number of fixed-width pulses per unit of time.

Peak On a graph of a sound wave or signal, the highest point in the waveform. The point of greatest voltage or sound pressure in a cycle.

Peak amplitude See *Amplitude, Peak.*

Peaking equalizer An equalizer that provides maximum cut or boost at one frequency, so that the resulting frequency response of a boost resembles a mountain peak.

Peak program meter (PPM) A meter that responds fast enough to closely follow the peak levels in a program.

Period The time between the peak of one wave and the peak of the next. The time between corresponding points on successive waves. Period is the inverse of frequency.

Personal studio A minimal group of recording equipment set up for one's personal use, usually using a 4-track cassette or memory recorder–mixer. Also, a simple 4-track recorder–mixer for one's personal use.

Perspective In the reproduction of a recording, the audible sense of distance to the musical ensemble, the point of view. A close perspective has a high ratio of direct sound to reverberant sound; a distant perspective has a low ratio of direct sound to reverberant sound.

PFL Abbreviation for pre-fader listen. See also *Solo*.

Phantom power A DC voltage (usually 12 to 48 volts) applied to microphone signal conductors to power condenser microphones.

Phase The degree of progression in the cycle of a wave, where one complete cycle is 360 degrees.

Phase cancellation, Phase interference The cancellation of certain frequency components of a signal that occurs when the signal is combined with its delayed replica. At certain frequencies, the direct and delayed signals are of equal level and opposite polarity (180 degrees out of phase), and when combined, they cancel out. The result is a comb-filter frequency response having a periodic series of peaks and dips. Phase interference can occur between the signals of two microphones picking up the same source at different distances, or can occur at a microphone picking up both a direct sound and its reflection from a nearby surface.

Phase shift The difference in degrees of phase angle between corresponding points on two waves. If one wave is delayed with respect to another, there is a phase shift between them of 2piFT, where pi = 3.14, F = frequency in Hz, and T = delay in seconds.

Phasing A special effect in which a signal is combined with its phase-shifted replica to produce a variable comb-filter effect. See also *Flanging*.

Phone plug A cylindrical, coaxial plug (usually 1/4-inch diameter). An unbalanced phone plug has a tip for the hot signal and a sleeve for the shield, which connects to ground. A balanced phone plug has a tip for the signal hot signal, a ring for the return signal, and a sleeve for the shield.

Phono plug A coaxial plug with a central pin for the hot signal and a ring of pressure-fit tabs for the shield or ground. Also called RCA plug. Phono plugs are used on TASCAM modular digital multitrack recorders and on consumer stereo equipment.

Pickup A piezoelectric transducer that converts mechanical vibrations to an electrical signal. Used in acoustic guitars, acoustic basses, and fiddles. Also, a magnetic transducer in an electric guitar that converts string vibration to a corresponding electrical signal.

Pinch roller In a tape-recorder transport, the rubber wheel that pinches or traps the tape between itself and the capstan, so that the capstan can move the tape.

Ping-ponging See *Bouncing tracks*.

Pink noise A noise signal containing all frequencies (unless band-limited), with equal energy per octave. Pink noise is a test signal used for equalizing a sound system to the desired frequency response, and for testing loudspeakers.

Pitch The subjective lowness or highness of a tone. The pitch of a tone usually correlates with the fundamental frequency.

Pitch control A control on a tape recorder that varies the tape speed, thereby varying the pitch of the signal on tape. The pitch control can be used to match the pitch of prerecorded instruments with that of an instrument to be overdubbed. It is also used for special effects, such as "chipmunk voices," and to play prerecorded tracks slowly so that fast musical passages can be overdubbed more easily.

Pitch shifter A signal processor that changes the pitch of an instrument without changing its duration.

Playback equalization In analog tape-recorder electronics, fixed equalization applied to the signal during recording to compensate for certain losses.

Playback head The head in a tape recorder that picks up a prerecorded magnetic signal from the moving tape and converts it to a corresponding electrical signal. The playback head isn't the same as the sel-sync or sync head.

Playlist See *Edit decision list.*

Plug A male connector that inserts into a jack. Also, short for plug-in.

Plug-in Effects software that you load into your DAW recording program (called the host). The plug-in becomes part of the host program and can be called up from within the host. Some manufacturers make plug-in bundles, which are a variety of effects in a single package.

Podcast A digital recording of a radio broadcast or similar program, usually in MP3 format, accessed on the Internet by an RSS feed, for downloading to a computer or personal audio player.

Polar pattern The directional pickup pattern of a microphone. A plot of microphone sensitivity versus angle of sound incidence. Examples of polar patterns are omnidirectional, bidirectional, and unidirectional. Subsets of unidirectional are cardioid, supercardioid, and hypercardioid.

Polarity Referring to the positive or negative direction of an electrical, acoustical, or magnetic force. Two identical signals in opposite polarity are 180 degrees out of phase with each other at all frequencies.

Polyphonic Describing a synthesizer that can play more than one note at a time (chords).

Pop (1) A thump or little explosion sound heard in a vocalist's microphone signal. Pop occurs when the user says words with "p," "t," or "b" so that a turbulent puff of air is forced from the mouth and strikes the microphone diaphragm. (2) A noise heard when a mic is plugged into a monitored channel, or when a switch is flipped.

Pop filter A screen placed on a microphone grille that attenuates or filters out pop disturbances before they strike the microphone diaphragm. Usually made of open-cell plastic foam or silk, a pop filter reduces pop and wind noise.

Portable studio A combination recorder and mixer in one portable chassis.

Post-echo A repetition of a sound, following the original sound, caused by print-through.

Power amplifier An electronic device that amplifies or increases the power level fed into it to a level sufficient to drive a loudspeaker.

Power ground A connection to the power company's earth ground through the U-shaped hole in a power outlet. In the power cable of an electronic component with a 3-prong plug, the U-shaped prong is wired to the component's chassis. This wire conducts electricity to power ground if the chassis becomes electrically hot, preventing shocks. See *Safety ground*.

Preamplifier (Preamp) In an audio system, the first stage of amplification that boosts a mic-level signal to line level. A preamp is a stand-alone device or a circuit in a mixer.

Pre-delay Short for pre-reverberation delay. The delay (about 30 to 150 msec) between the arrival of the direct sound and the onset of reverberation. Usually, the longer the pre-delay, the greater the perceived room size. Using pre-delay on an instrument's or vocal's reverb helps to clarify the sound.

Pre-echo A repetition of a sound that occurs before the sound itself, caused by print-through in analog tape.

Pre-fader/Post-fader switch A switch that selects a signal either ahead of the fader (pre-fader) or following the fader (post-fader). The level of a pre-fader signal is independent of the fader position; the level of a post-fader signal follows the fader position.

Preproduction Planning in advance what you're going to do at a recording session, in terms of track assignments, overdubbing, studio layout, and microphone selection.

Presence The audible sense that a reproduced instrument is present in the listening room. Some synonyms are closeness, definition, and punch. Presence is often created by an equalization boost in the midrange or upper midrange, and by a high direct-to-reverb ratio.

Pressure zone microphone A boundary microphone constructed with the microphone diaphragm parallel with, and facing, a reflective surface.

Preverb A special effect in which the reverberation of a note precedes it, rather than follows it. It turns a snare-drum hit into a whip sound, like ssSSHHK! Chapter 10, Effects and Signal Processors, describes how to create it. Also see *Reverse echo*.

Print To record on tape or disc.

Print-through The transfer of a magnetic signal from one layer of analog tape to the next on a reel, causing an echo preceding or following the program.

Production (1) A recording that is enhanced by effects. (2) The supervision of a recording session to create a satisfactory recording. This involves getting musicians together for the session, making musical suggestions to the musicians to enhance their performance, and making suggestions to the engineer for sound balance and effects.

Program bus A bus or output that feeds an audio program to a recorder track.

Program mixer In a mixing console, a mixer formed of input-module outputs, combining amplifiers, and program busses.

Pro Tools A popular digital audio editing platform for professional use. It offers computer multitrack recording, overdubbing, mixing, editing, and a variety of plug-in effects.

Proximity effect The bass boost that occurs with a single-D directional microphone when it's placed a few inches from a sound source. The closer the microphone, the greater the low-frequency boost due to proximity effect.

Pulse code modulation See *PCM*.

Pulse density modulation See *PDM*.

Punch-In/Out A feature in a multitrack recorder that lets you insert a recording of a corrected musical part into a previously recorded track by going into and out of record mode as the tape or disk is rolling.

Pure waveform A waveform of a single frequency; a sine wave. A pure tone is the perceived sound of such a wave.

Q In a parametric equalizer, Q is the sharpness of the boost or cut. A Q of 1 is a broad boost or cut that covers a wide range of frequencies. A Q to 10 is a narrow boost or cut that covers a small range of frequencies. Q or Quality factor is a filter's center frequency divided by the bandwidth.

Quarter-track A tape track recorded across one-quarter of the width of an analog magnetic tape. A quarter-track recorder usually records two stereo programs (one in each direction).

Rack A 19-inch-wide wooden or metal cabinet used to hold audio equipment.

Radio-frequency interference (RFI) Radio-frequency electromagnetic waves induced in audio cables or equipment, causing various noises in the audio signal.

Random access Referring to a storage medium in which any data point can be accessed or read almost instantly. Examples are a hard disk, compact disc, and MiniDisc.

Rarefaction The portion of a sound wave in which molecules are spread apart, forming a region with lower than normal atmospheric pressure. The opposite of compression.

R-DAT See *DAT*.

RealAudio A highly compressed audio file format used for streaming audio. Generally, RealAudio has lower fidelity (less treble) than MP3, but the fidelity depends on modem speed. RealAudio files (.ra or .rm) are often used as short excerpts or previews of songs.

Real-time recording (1) Recording notes into a sequencer in the correct tempo, for later playback at the same tempo as recorded. (2) A recording made direct to lacquer disc or direct to 2-track without any overdubs or mixdown.

Re-amping Recording a guitar amp that is fed the signal from a direct-recorded electric guitar track. This technique lets you work on the amp's sound during mixdown, rather than during recording.

Recirculation Feeding the output of a delay device back into its input to create multiple echoes. Also, the control on a delay device that affects how much delayed signal is recycled to the input. See *Regeneration*.

Record To store an event in permanent form. Usually, to store an audio signal in magnetic form on magnetic tape or disk, or to store an audio signal in optical form on a CD-R or CD-RW. Recording is also possible on magneto-optical disk, on MiniDisc, in RAM, and on memory cards.

Record equalization In analog tape-recorder electronics, equalization applied to the signal during recording to compensate for certain losses.

Record head The head in a tape recorder or hard drive that puts the audio signal on tape or disk by magnetizing the tape or disk particles in a pattern corresponding to the audio signal.

Recorder–mixer A combination multitrack recorder and mixer in one chassis.

Recording/reproduction chain The series of events and equipment that are involved in sound recording and playback.

Reflected sound Sound waves that reach the listener after being reflected from one or more surfaces.

Regeneration See *Recirculation*.

Region In a digital audio editing program, a defined segment of the audio program. Also called clip or zone.

Release The final portion of a note's envelope in which the note falls from its sustain level back to silence.

Release time In a compressor, the time it takes for the gain to return to normal after the end of a loud passage.

Remix To mix again; to do another mixdown with different console settings or different editing.

Remote recording See *On-location recording*.

Removable hard drive A hard-disk drive that can be removed and replaced with another, used in a DAW or hard-drive recorder to store a program temporarily.

Render To convert one audio format to another in a DAW or hard-drive recorder. Examples: mix a multitrack recording to a 2-channel WAV, AIFF, rm, or MP3 file; convert a track with real-time effects to a track with embedded effects; convert MIDI tracks to audio tracks; combine multiple clips into a single track.

Resistance The opposition of a circuit to a flow of direct current. Resistance is measured in ohms, abbreviated w, and may be calculated by dividing voltage by current.

Resistor An electronic component that opposes current flow.

Return-to-zero See *Memory rewind*.

Reverberation Natural reverberation in a room is a series of multiple sound reflections that make the original sound persist and gradually die away or decay. These reflections tell the ear that you're listening in a large or hard-surfaced room. For example, reverberation is the sound you hear just after you shout in an empty gymnasium. A reverb effect simulates the sound of a room—a club, auditorium, or concert hall—by generating random multiple echoes that are too numerous and rapid for the ear to resolve. The timing of the echoes is random, and the echoes increase in number with time as they decay. An echo is a discrete repetition of a sound; reverberation is a continuous fade-out of sound.

Reverberation time (RT60) The time it takes for reverberation to decay to 60 dB below the original steady-state level.

Reverse echo A multiple echo that precedes the sound that caused it, building up from silence into the original sound. This special effect is created in a manner similar to preverb.

RFI See *Radio frequency interference*.

Rhythm tracks The recorded tracks of the rhythm instruments (guitar, bass, drums, and sometimes keyboards).

Ribbon microphone A dynamic microphone in which the conductor is a long metallic diaphragm (ribbon) suspended in a magnetic field.

Ride gain To turn down the volume of a microphone when the source gets louder, and turn up the volume when the source gets quieter in an attempt to reduce dynamic range.

Ringing See *Overhang*.

RMF Abbreviation for Rich Music Format. A format for a MIDI file with General MIDI sounds plus custom sounds. Designed to be played on a Beatnik player.

Room modes See *Standing wave*.

RT60 See *Reverberation time*.

SACD See *Super Audio CD*.

Safety copy A copy of the master tape or CD, to be used if the master is lost or damaged.

Safety ground See *Power ground*.

Sample (1) To digitally record a short sound event, such as a single note or a musical phrase, into computer memory. (2) A recording of such an event. (3) A measurement of an analog waveform that is done several thousand times a second during a PCM A/D conversion.

Sampling (1) Recording a short sound event into computer memory. The audio signal is converted into digital data representing the signal waveform, and the data is stored in memory chips, tape, or disc for later playback. (2) In PCM digital recording, measuring an analog waveform periodically, several thousand times a second.

Sampling rate In a digital recording, the frequency at which an analog waveform is sampled or measured. The sampling rate of CD-quality audio is 44,100 samples per second. The higher the sampling rate, the higher the high-frequency response of the recording.

Saturation Overload of magnetic tape. The point at which a further increase in magnetizing force doesn't cause an increase in magnetization of the tape oxide particles. Distortion is the result.

Scene automation A form of console automation in which the console settings are stored in memory. A "snapshot" or reading of many of the settings is taken and stored for later recall. In contrast, dynamic automation continuously follows the fader and knob moves, and the automation data is usually stored as a MIDI file.

SCMS Abbreviation for Serial Copy Management System. An anti-copy scheme in consumer digital audio devices (those with S/PDIF connectors). SCMS circuits read flags in the data stream that allow users to make only first-generation copies, not copies of copies.

Scratch vocal A vocal performance that is done simultaneously with the rhythm instruments so that the musicians can keep their place in the song and get a feel for the song. Because it contains leakage, the scratch-vocal recording is usually erased. Then the singer overdubs the vocal part that is to be used in the final recording.

Scrub To manually move an open-reel tape slowly back and forth across a recorder playback head to locate an edit point. Some digital editing software has an equivalent scrubbing function.

SCSI Abbreviation for Small Computer Systems Interface. A standard spec for high-speed computer input and output. Commonly used for hard drives, CD-ROM drives, and CD-R recorders. Supports up to seven devices.

Sensitivity (1) The output of a microphone in volts for a given input in sound pressure level. (2) The sound pressure level a loudspeaker produces at one meter when driven with 1 watt of pink noise. See also *Sound pressure level*.

Sequence A MIDI data file of musical-performance note parameters, recorded by a sequencer.

Sequencer A device or computer program that records a musical performance done on a MIDI controller (in the form of note numbers, note on, note off, etc.) into computer memory or hard disk for later playback. During playback, the sequencer plays synthesizer sound generators or samples.

Session (1) A time period set aside for recording musical instruments, voices, or sound effects. (2) On a CD-R, a lead-in, program area, and lead-out.

Shelving equalizer An equalizer that applies a constant boost or cut above or below a certain frequency so that the shape of the frequency response resembles a shelf.

Shield A conductive enclosure (usually metallic) around one or more signal conductors, used to keep out electrostatic fields that cause hum or buzz. A shield in a mic cable is a cylindrical mesh of fine wires.

Shock mount A suspension system that mechanically isolates a microphone from its stand or boom, preventing the transfer of mechanical vibrations.

Sibilance In a speech recording, excessive frequency components in the 3 to 10 kHz range, which are heard as an overemphasis of "s" and "sh" sounds.

Side-addressed Referring to a microphone whose main axis of pickup is perpendicular to the side of the microphone. You aim the side of the mic at the sound source. See also *End-addressed*.

Signal A varying electrical voltage that represents information, such as a sound.

Signal path The path a signal takes from input to output in a piece of audio equipment.

Signal processor A device that is used to alter a signal in a controlled way.

Signal-to-noise ratio Abbreviated as S/N. The ratio in decibels between signal voltage and noise voltage. An audio component with a high S/N has little background noise accompanying the signal; a component with a low S/N is noisy.

Sine wave A wave following the equation $y = \sin x$, where x is degrees and y is voltage or sound pressure level. The waveform of a single frequency. The waveform of a pure tone without harmonics.

Single-D microphone A directional microphone having a single distance between its front and rear sound entries. Such a microphone has proximity effect.

Single-ended (1) An unbalanced line. (2) A single-ended noise reduction system is one that works only during tape playback (unlike Dolby or dbx, which work both during recording and playback).

Slap, Slapback An echo following the original sound by about 50 to 150 msec, sometimes with multiple repetitions.

Slate At the beginning of a recording, a recorded announcement of the name of the tune and its take number. The term is derived from the slate used in the motion-picture industry to identify the production and take number being filmed.

SMPTE timecode A modulated 1200 Hz square-wave signal used to synchronize two or more recorders or MIDI devices. SMPTE uses the format HH:MM:SS:FF (hours, minutes, seconds, frames) to specify locations in the recorded program. SMPTE is an abbreviation for the Society of Motion Picture and Television Engineers, who developed the timecode.

Snake A multipair or multichannel mic cable. Also, a multipair mic cable attached to a connector junction box.

Snapshot automation See *Scene automation*.

Soft synth A synthesizer in software form.

Solo On an input module in a mixing console, a switch that lets you monitor that particular input signal by itself. The switch routes only that input signal to the monitor system.

Sound Longitudinal vibrations in a medium (such as air) in the frequency range 20 to 20,000 Hz.

Sound card A circuit card that plugs into a computer and converts an audio signal into computer data for storage in memory or on hard disk. The sound card also converts computer data into an audio signal. A type of PCI audio interface (see *Audio interface*).

Sound effects Recordings of non-musical sounds—such as a door slam, gunfire, thunderstorm, car, or telephone—used in dramatic productions, radio spots and commercials. Not to be confused with *Effects*.

Sound module (Sound generator) (1) A synthesizer without a keyboard, containing several different timbres or voices. These sounds are triggered or played by MIDI signals from a sequencer program or by a MIDI controller. A sound module can be a stand-alone device or a circuit on a sound card. (2) An oscillator.

Sound pressure level (SPL) The acoustic pressure of a sound wave, measured in decibels above the threshold of hearing. The higher the SPL of a sound, the louder it is. dB SPL = 20 log(P/Pref), where P = the measured acoustic pressure and Pref = 0.0002 dyne/cm^2.

Sound wave The periodic variations in air pressure radiating from an object vibrating between 20 and 20,000 Hz.

Spaced-pair A stereo microphone technique using two identical microphones spaced several feet apart horizontally, usually aiming straight ahead toward the sound source.

Spatial processor A signal processor that allows images to be placed beyond the limits of a stereo pair of speakers, even behind the listener or toward the sides.

S/PDIF Abbreviation for Sony Philips Digital Interface; IEC 958 Type II. A 2-channel digital signal interface format that uses a 75-ohm coaxial cable with RCA connectors, or an optical cable with TOSLINK connectors. See also *AES/EBU*.

Speaker See *Loudspeaker*.

Spectrum The output level versus frequency of a sound source, including the fundamental frequency and overtones.

SPL See *Sound pressure level*.

Splice To join the ends of two lengths of magnetic tape or leader tape with tape. Also, a splice is the taped joint between two lengths of magnetic tape or leader tape.

Splicing block See *Editing block*.

Split console A console with a separate monitor-mixer section. See also *Input/output (I/O) console*.

Splitter A transformer or circuit used to divide a microphone signal into two or more identical signals to feed different sound systems.

Spot microphone In classical music recording, a close-placed microphone that is mixed with more distant microphones to add presence or to improve the balance.

Standing wave An apparently stationary waveform created by multiple reflections between opposite room surfaces. At certain points along the standing wave, the direct and reflected waves cancel, and at other points the waves add together or reinforce each other.

Stem A submix in Pro Tools lingo—a drums submix, keys submix, left-front submix, etc.

Step-time recording Recording notes into a sequencer one at a time without regard to tempo, for later playback at a normal tempo.

Stereo, Stereophonic An audio recording and reproduction system with correlated information between two channels (usually discrete channels), meant to be heard over two or more loudspeakers to give the illusion of sound-source localization and depth.

Stereo bar, Stereo microphone adapter A microphone stand adapter that mounts two microphones on a single stand for convenient stereo miking.

Stereo imaging The ability of a stereo recording or reproduction system to form clearly defined audio images at various locations between a stereo pair of loudspeakers.

Stereo microphone A microphone containing two mic capsules in a single housing for convenient stereo recording. The capsules usually are coincident.

Streaming audio Audio sent over the Internet in real time. A streaming file plays as soon as you click on its title. A downloaded file doesn't play until you copy the entire file to your hard disk. Streaming audio is heard almost instantly, but usually sounds muffled and can be interrupted by Net congestion. With downloaded audio, you must wait up to several minutes to hear the song, but the sound is high-fidelity and continuous.

Studio A room used or designed for sound recording.

Submaster A master volume control for an output bus or submix.

Submix A small preset mix within a larger mix, such as a drum mix, keyboard mix, vocal mix, etc. Also a cue mix, monitor mix, or effects mix.

Submixer A smaller mixer within a mixing console (or a standalone mixer) that is used to set up a submix, a cue mix, an effects mix, or a monitor mix.

Super Audio CD (SACD) An alternative to DVD-Audio, Super Audio CD was developed by Sony and Philips. It's the size of a CD but offers higher sound quality and 5.1 surround sound. SACD uses the Direct Stream Digital (DSD) process, which encodes a digital signal in a 1-bit (bitstream) format at a 2.8224 MHz sampling rate. This system offers a frequency response from DC to 100 kHz with 120 dB dynamic range.

Supercardioid microphone A unidirectional microphone that attenuates side-arriving sounds by 8.7 dB, attenuates rear-arriving sounds by 11.4 dB, and has two nulls of maximum sound rejection at 125 degrees off-axis.

Supply reel See *Feed reel.*

Surround sound A multichannel recording and reproduction system that plays sound all around the listener. The 5.1 surround system uses the following speakers—front-left, center, front-right, left-surround, right-surround, and subwoofer.

Sustain The portion of the envelope of a note in which the level is constant. Also, the ability of a note to continue without noticeably decaying, often aided by compression.

Sweetening The addition of strings, brass, chorus, etc., to a recording of the basic rhythm tracks.

Sync, Synchronization Aligning two separate audio programs in time, and maintaining that alignment as the programs play.

Sync, Synchronous recording Using a tape record head temporarily as a playback head during an overdub session, to keep the overdubbed parts in synchronization with the recorded tracks.

Sync tone See *Tape sync.*

Sync track A track of a multitrack recorder that is reserved for recording SMPTE timecode. This allows audio tracks to synchronize with virtual tracks recorded with a sequencer. A sync track also can synchronize two audio tape machines or an audio recorder and a video recorder, and can be used for console automation.

Synthesizer A musical instrument (usually with a piano-style keyboard) that creates sounds electronically and allows control of the sound parameters to simulate a variety of conventional or unique instruments.

Tail The end of a reverberation signal where the reverb fades down to silence.

Tail-out Referring to a reel of tape wound with the end of the program toward the outside of the reel. Analog tape stored tail-out is less likely to have audible print-through.

Take A recorded performance of a song. Usually, several takes are done of the same song, and the best one—or the best parts of several—becomes the final product.

Take sheet A list of take numbers for each song, plus comments on each take.

Take-up reel The right-side reel on a tape recorder that winds up the tape as it is playing or recording.

Talkback An intercom in the mixing console for the engineer and producer to talk to the musicians in the studio.

Tape See *Magnetic recording tape*.

Tape loop An endless loop formed from a length of recording tape spliced end-to-end, used for continuous repetition of several seconds of recorded signal.

Tape recorder A device that converts an electrical audio signal into a magnetic audio signal on magnetic tape, and vice versa. A tape recorder includes electronics, heads, and a transport to move the tape.

Tape sync A frequency-modulated signal recorded on a tape track, used to synchronize a tape recorder to a sequencer. Tape sync also permits the synchronized transfer of sequences to tape. See also *Sync track*.

TDIF Abbreviation for TASCAM Digital Interface. An interface protocol for connecting digital audio signals to and from TASCAM DTRS multitrack recorders, such as the DA-78 and DA-88.

Three-pin (3-pin) connector A 3-pin professional audio connector used for balanced signals. Pin 1 is soldered to the cable shield, pin 2 is soldered to the signal hot or in-polarity lead, and pin 3 is soldered to the signal cold or opposite-polarity lead. See also *XLR-type connector*.

Three-to-one rule (3:1 rule) A rule in microphone applications. When multiple mics are mixed to the same channel, the distance between mics should be at least three times the distance from each mic to its sound source. This prevents audible phase interference.

Threshold In a compressor or limiter, the input level above which compression or limiting takes place. In an expander, the input level below which expansion takes place.

Tie To connect electrically, for example, by soldering a wire between two points in a circuit.

Tight (1) Having very little leakage or room reflections in the sound pickup. (2) Referring to well-synchronized playing of musical instruments. (3) Having a well-damped, rapid decay.

Timbre The subjective impression of spectrum and envelope. The quality of a sound that allows us to differentiate it from other sounds. For example, if you hear a trumpet, a piano, and a drum, each has a different timbre or tone quality that identifies it as a particular instrument.

Timecode A signal used to synchronize a tape recorder, hard-disk recorder, MIDI sequencers, or DAW with each other. See also *Tape sync, Sync track, SMPTE timecode, MTC.*

Tonal balance The balance or volume relationships among different regions of the frequency spectrum, such as bass, mid-bass, midrange, upper midrange, and highs.

TOSLINK (Toshiba Link) A fiber-optic cable connection for S/PDIF data transfers.

Track A path on magnetic tape containing a single channel of audio. A group of bytes in a digital signal (on tape, on hard disk, on compact disc, or in a data stream) that represents a single channel of audio or MIDI. Usually one track contains a performance by one musical instrument.

Transducer A device that converts energy from one form to another, such as a microphone, pickup, or loudspeaker.

Transformer An electronic component made of two magnetically coupled coils of wire. The input signal is transferred magnetically to the output, without a direct connection between input and output.

Transient A short signal with a rapid attack and decay, such as a drum stroke, cymbal hit, or acoustic-guitar pluck.

Transient response The ability of an audio component (usually a microphone or loudspeaker) to follow a transient accurately.

Transport The mechanical system in a reproduction device that moves the medium past the read/write heads. In a tape recorder, the transport controls tape motion during recording, playback, fast forward, and rewind.

Trim (1) In a mixing console, a control for fine adjustment of level, as in a Bus trim control. (2) In a mixing console, a control that adjusts the gain of a mic preamp to accommodate various signal levels.

TRS Abbreviation for tip-ring-sleeve. A phone-plug connector used for aux send/return, unbalanced stereo, or balanced mono connections.

Tube A vacuum tube, an amplifying component made of electrodes in an evacuated glass tube. Tube sound is characterized as being "warmer" than solid-state or transistor sound.

Tweeter A high-frequency loudspeaker.

Unbalanced line An audio cable having one conductor surrounded by a shield that carries the return signal. The shield is connected to a circuit's ground.

Unidirectional microphone A microphone that is most sensitive to sounds arriving from one direction—in front of the microphone. Examples are cardioid, supercardioid, and hypercardioid.

Unity gain A condition in which the input and output levels of a device are equal.

USB Abbreviation for Universal Serial Bus. A Mac/PC computer serial interface for connecting external devices such as MIDI interfaces and audio interfaces to a computer. Faster than a standard serial port.

Valve British term for vacuum tube. See *Tube*.

VBR Abbreviation for Variable Bit Rate MP3 encoding. The bit rate varies with the complexity of the audio signal.

Virtual controls Audio-equipment controls that are simulated on a computer monitor screen. You adjust them with a mouse or with a controller surface.

Virtual track A recording of a single take of one instrument or vocal on a random-access medium. The best parts of several virtual tracks can be bounced a single track, forming a complete, perfect performance. This process is called comping tracks.

VST Abbreviation for Virtual Studio Technology. Steinberg's virtual instrument integration standard format for plug-ins.

VU meter A voltmeter with a specified transient response, calibrated in VU or volume units, used to show the relative volume of various audio signals, and to set recording levels in analog tape recorders.

WAV (.wav) A computer audio file format for Windows. It encodes sound without any data reduction, using pulse code modulation. Its audio resolution is 16-bit, 44.1 kHz, or higher.

Waveform A graph of a signal's sound pressure or voltage versus time. The waveform of a pure tone is a sine wave.

Wavelength The physical length between corresponding points of successive waves. Low frequencies have long wavelengths; high frequencies have short wavelengths.

Weber A unit of magnetic flux.

Weighted Referring to a measurement made through a filter with a specified frequency response. An A-weighted measurement is taken through a filter that simulates the frequency response of the human ear.

Windows Media See *WMA*.

Windscreen See *Pop filter*.

WMA Abbreviation for Windows Media Audio. A popular compressed audio file format for streaming audio and for downloads.

Woofer A low-frequency loudspeaker.

Word A binary number that is the value of each sample in an analog-to-digital conversion. A sample is a measurement of an analog waveform that is done several thousand times a second during a PCM A/D conversion. See *Word length*.

Word clock A timing signal for digital audio. Word clock is used to synchronize the digital devices in a large studio for real-time transfer of digital audio.

Word length See *Bit depth*.

Workstation A system of MIDI- or computer-related equipment that works together to help you compose and record music. Usually, this system is small enough to fit on a desktop or equipment stand. See also *Keyboard workstation* and *Digital Audio Workstation*.

Wow A slow periodic variation in tape speed.

Wrapper Software that converts a plug-in from an unsupported format to a supported format. For example, a DAW recording program (host) that can't use VST plug-ins directly might include a wrapper that converts VST plug-ins to Direct-X plug-ins.

XLR-type connector An ITT Cannon part number that has become the popular definition for a 3-pin professional audio connector. See also *Three-pin (3-pin) connector*.

X-Y See *Coincident-pair*.

Y-adapter A cable that divides into two cables in parallel to feed one signal to two destinations.

Z Abbreviation for impedance.

Zone See *Region*.

Index

B

D

F

G

H

X

Y

Z

Practical Recording Techniques Demo Web Site

Welcome to Practical Recording Techniques. This Web site, www.elsevierdirect.com/companions/9780240811444, demonstrates various topics in the book. To access this Web site you will need to enter the passcode BART898DFA9P. All the demos on this Web site are MP3 files encoded at 320 kbps.

All the demos were engineered by Bruce Bartlett, except for the electric-guitar samples (Track 23) recorded by Steve Mills of S.E.M. Labs. All music on this Web site is copyrighted by the original artists and may not be reproduced without permission. Thanks to Joe Probst, Station One, One Word, Ryan Vice, Josh Hilliker, and Grooveside for the use of their recordings.

The recordings on this Web site were produced with Cakewalk® Sonar Producer digital recording software. To provide the best sound quality, the recordings were not processed to maximize their levels.

Demo

1 Introduction
2 Five Ways to Record

Sound, Signals, and Studio Acoustics

3 Amplitude
4 Frequency
5 Harmonics
6 Envelope
7 Frequency Response
8 Noise and Signal-to-Noise Ratio

- **9** Distortion: Digital distortion and analog-cassette-tape distortion
- **10** Echoes and Reverberation
- **11** Leakage
- **12** Hum and Buzz

Microphones

- **13** Transducer Type: Condenser, dynamic, ribbon
- **14** Polar Pattern: Cardioid mic on and off axis; cardioid versus omni at 6 feet
- **15** Microphone Frequency Response: Flat, contoured
- **16** Breath Pops: Effect of pop filter, effect of mic placement

Mic-Technique Basics

- **17** Miking Distance
- **18** Effects of Mic Placement on Tonal Balance
- **19** Stereo Miking the Acoustic Guitar—Close and Distant
- **20** Boundary Mics
- **21** The 3:1 Rule
- **22** Stereo Mic Techniques: Coincident, near-coincident, spaced

Specific Mic Techniques

- **23** Electric Guitar
- **24** Drum Set
- **25** Grand Piano (the PZM recording is a mix of instruments)
- **26** Lead Vocal

Signal Processors and Effects

- **27** EQ
- **28** Compression
- **29** No Effects on Voice
- **30** Echo
- **31** Reverberation
- **32** Preverb
- **33** Chorus
- **34** Flanging
- **35** Pitch Shift

36 Tube Modeling
37 Analog Tape Modeling
38 Guitar-amp Modeling

Mixdown

39 A sample of the finished mix, individual tracks before and after processing, panning, level balancing, the mix before and after processing, and level balancing

Mastering

40 Normalization, peak limiting and normalization, multiband compression (Note: All these examples are MP3 files encoded at 320 kbps for this Web site.)

Data Compression

41 Uncompressed audio file, MP3 data compression, WMA data compression

Decibels

42 0 dB, −10 dB, −6 dB, −3 dB, −1 dB

Outro

43 Thanks for listening!

Total running time 43:58.